Creativity and Technology in Mathematics Education

AF796108

MATHEMATICS EDUCATION IN THE DIGITAL ERA

Volume 10

Series Editors:

Dragana Martinovic, University of Windsor, ON, Canada
Viktor Freiman, Université de Moncton, NB, Canada

Editorial Board:

Marcelo Borba, State University of São Paulo, São Paulo, Brazil
Rosa Maria Bottino, CNR – Istituto Tecnologie Didattiche, Genova, Italy
Paul Drijvers, Utrecht University, Utrecht, The Netherlands
Celia Hoyles, University of London, London, UK
Zekeriya Karadag, Giresun Üniversitesi, Giresun, Turkey
Stephen Lerman, London South Bank University, London, UK
Richard Lesh, Indiana University, Bloomington, USA
Allen Leung, Hong Kong Baptist University, Hong Kong
Tom Lowrie, University of Canberra, Australia
John Mason, Open University, UK
Sergey Pozdnyakov, Saint-Petersburg State Electro Technical University, Saint-Petersburg, Russia
Ornella Robutti, Università di Torino, Torino, Italy
Anna Sfard, Michigan State University, USA & University of Haifa, Haifa, Israel
Bharath Sriraman, University of Montana, Missoula, USA
Anne Watson, University of Oxford, Oxford, UK

More information about this series at http://www.springer.com/series/10170

Viktor Freiman · Janet Lynne Tassell
Editors

Creativity and Technology in Mathematics Education

Springer

Editors
Viktor Freiman
Université de Moncton
Moncton, NB
Canada

Janet Lynne Tassell
Western Kentucky University
Bowling Green, KY
USA

ISSN 2211-8136 ISSN 2211-8144 (electronic)
Mathematics Education in the Digital Era
ISBN 978-3-030-10193-0 ISBN 978-3-319-72381-5 (eBook)
https://doi.org/10.1007/978-3-319-72381-5

© Springer International Publishing AG, part of Springer Nature 2018
Softcover re-print of the Hardcover 1st edition 2018
This work is subject to copyright. All rights are reserved by the Publisher, whether the whole or part of the material is concerned, specifically the rights of translation, reprinting, reuse of illustrations, recitation, broadcasting, reproduction on microfilms or in any other physical way, and transmission or information storage and retrieval, electronic adaptation, computer software, or by similar or dissimilar methodology now known or hereafter developed.
The use of general descriptive names, registered names, trademarks, service marks, etc. in this publication does not imply, even in the absence of a specific statement, that such names are exempt from the relevant protective laws and regulations and therefore free for general use.
The publisher, the authors and the editors are safe to assume that the advice and information in this book are believed to be true and accurate at the date of publication. Neither the publisher nor the authors or the editors give a warranty, express or implied, with respect to the material contained herein or for any errors or omissions that may have been made. The publisher remains neutral with regard to jurisdictional claims in published maps and institutional affiliations.

Printed on acid-free paper

This Springer imprint is published by the registered company Springer International Publishing AG part of Springer Nature
The registered company address is: Gewerbestrasse 11, 6330 Cham, Switzerland

Contents

Part I
Introduction Section: Setting the Stage of Investigation

Chapter 1
Leveraging Mathematics Creativity by Using Technology: Questions, Issues, Solutions, and Innovative Paths

Viktor Freiman and Janet Lynne Tassell

Abstract This introductory chapter aims to introduce the volume providing new insights on creativity while focusing on innovative methodological approaches in research and practice of integrating technological tools and environments in mathematics teaching and learning. This work is being built on the discussions at the mini-symposium on Creativity and Technology at the International Conference on Mathematical Creativity and Giftedness (ICMCG) in Denver, USA (2014), and other contributions to the topic. While presenting a diversity of views, a variety of contexts, angles and cultures of thought, as well as mathematical and educational practices, the authors of each chapter explore the potential of technology to foster creative and divergent mathematical thinking, problem solving and problem posing, creative use of dynamic, multimodal and interactive software by teachers and learners, as well as other digital media and tools while widening and enriching transdisciplinary and interdisciplinary connections in mathematics classroom. Along with ground-breaking innovative approaches, the book aims to provide researchers and practitioners with new paths for diversification of opportunities for all students to become more creative and innovative mathematics learners. A framework for dynamic learning conditions of leveraging mathematical creativity with technology is an outcome of this collective work.

Keywords Mathematical creativity · Technology · Transdisciplinary and interdisciplinary connections · Innovative approaches to teaching and learning

---

V. Freiman (✉)
Université de Moncton, Moncton, NB, Canada
e-mail: viktor.freiman@umoncton.ca

J. L. Tassell
University of Western Kentucky, Bowling Green, KY, USA
e-mail: janet.tassell@wku.edu

© Springer International Publishing AG, part of Springer Nature 2018
V. Freiman and J. L. Tassell (eds.), *Creativity and Technology in Mathematics Education*, Mathematics Education in the Digital Era 10,
https://doi.org/10.1007/978-3-319-72381-5_1

1.1 Interest in Looking into Creativity in Mathematics Education Through Technology Lenses: Questions and Issues

By laying out new directions in mathematics education for the 21st century, Kaput, Hegedus and Lesh (2007) emphasized an infrastructural paradigm shift influenced by new technological tools which would lead to new levels and types of ideas and abilities and to new ways of thinking about traditional concepts and skills. The infrastructural paradigm can, in fact, open doors to more creativity in teaching and learning. For example, Yerushalmy (2009) cites *Math4mobile* mathematical tools as those which enable creative mathematical thinking anywhere anytime for all. Leikin, Levav-Waynberg and Guberman (2011), outline potential benefits in providing technology-based opportunities for inquiry-based learning and therefore, for advancement of students' mathematical creativity.

Inspired by seminal works by French mathematicians Poincaré (1854–1912) and Hadamard (1865–1963) the field of mathematical creativity has been continually expanding and growing throughout the second half of the 20th century and the beginning of the 21st century (Sriraman, 2004; Mann, 2006; Leikin, Levav-Waynberg & Guberman, 2011; Sriraman & Lee, 2011; and Leikin & Sriraman, 2017a).

This constant development and expansion in scholarship on mathematical creativity has led to refining its characteristics, its possible implications in the classroom, and its connection to mathematical giftedness while looking at it from a variety of international and interdisciplinary perspectives. An ongoing work of the International Group for Mathematical Creativity and Giftedness (IGMCG), affiliated with the International Commission for Mathematical Instruction (ICMI) who, in its turn, has established first a Discussion Group on mathematical creativity, then a Topic Study Group (TSG), at its last Congress with a focus, among others, on issues related to communication, information and computer technologies (TSG29, ICME-13, 2016, http://www.icme13.org/files/tsg/TSG_29.pdf).

In fact, entering in the so-called digital era, or perhaps more cautiously, electronic digital era, raises many questions in respect to the teaching and learning mathematics which is the main theme of the book *Series Mathematics Education in the Digital Era (*http://www.springer.com/series/10170*)*. The newest volume in the Series seeks to extend and to deepen a collective understanding for how digital technology shapes and is shaped by multiple views of creativity in mathematics.

In this respect, the following twenty chapters of the book provide reflective and critical overview of research worldwide while focusing on innovative methodological approaches in design and implementation of technological tools and environments in mathematics teaching and learning. It also examines research data on their impact on mathematical creativity. A deeper insight can be gained in terms of the potential of technology to foster creative and divergent mathematical thinking, problem solving and problem posing, creative use of dynamic, multimodal and interactive software by teachers and learners, as well as other digital media and tools while widening and enriching transdisciplinary and interdisciplinary connections in the mathematics classroom. This work originated from discussions at the mini-symposium on Creativity and Technology at the IGMCG-8 conference in Denver, USA (Freiman & Tassell, 2014), which was pursued at the next ICMCG-9 conference in Sinaia, Romania (Singer, Toader, & Voica, 2015). In order to expand the investigation of the topic, we issued an extended call for proposals inviting a larger group of scholars to participate.

Going beyond the design and implementation issues, the authors of the chapters were asked to investigate mathematical creativity in a variety of technology-enhanced contexts, such as online learning, social networking, use of mobile technologies, robotics-based learning, 3D-printing in makerspaces, as well as computer programming. Along with ground-breaking innovative approaches, the book also aims to provide researchers and practitioners with new paths for diversification of opportunities for all students to become more creative and innovative mathematics learners.

As a result, through this collective work we can better grasp opportunities and challenges related to the richness of learning experiences with technology, including online learning. These diverse opportunities can engage not only more students in the mathematics classrooms that offer challenging and enriched programs, but also extend instruction to those students who live in remote areas, often in rural and low-income communities, so they can gain access to a wider range of advanced courses and learn at their own pace according to their interests and abilities (Thompson, 2010; Singer, Sheffield, Freiman, & Brandl, 2016).

When reflecting on possible connections between Mathematics, Technology, and Creativity, one discovers much more diversity and ambiguity related to specific views, understanding and educational contexts and settings. The book takes an opportunity to grapple with varying definitions and understandings, hoping to reveal some poignant differences, and to find some common trends thus attracting a larger international audience of mathematics educators, teacher educators at all levels, and perhaps digital software developers wishing to increase impact of technology on mathematics instruction and learning. But first and foremost, this book is critical for education communities where current issues of mathematics creativity are taken seriously.

1.2 Multiple Views and Approaches: Enlarging and Deepening Our Knowledge

In introducing the book, we strive to spotlight a number of different ways where technology can stimulate and foster creativity in mathematics teaching and learning. As a result, we consider the diversity of contributions coming from educators with different backgrounds, expertise, and experience as a significant advantage of this book. Throughout the chapters, the authors express their particular perspective and understanding of creativity and technology taken from a variety of contexts, angles and cultures of thought, as well as mathematical and educational practices.

Hence, looking at the issues of developing mathematical creativity using technology from the learner's perspective, some of the chapters include details about research and application of problem-solving and problem-posing activities, as well as multiple-solution tasks that have potential to foster the original, flexible, and fluent mathematical thinking while stimulating "invention, innovation, originality, insight, illumination, and imagination" (Leikin & Sriraman, 2017b, p. 1). In such context, digital tools and environments offer opportunities to incorporate, among others, online virtual communities, social media, dynamic geometry software, interactive applets, video-games, robotics programming, and 3D-printing. This could increase opportunities in exploration, modeling, and discussion while eventually affording earlier access to more advanced mathematics thus pushing learning beyond the boundaries of traditional curriculum. We could also find a fostering of interdisciplinary and transdisciplinary connections further stimulating students' interest, motivation, and curiosity, and even making failure 'productive' for deeper learning in "new media technology-pervasive learning environments" (Arnone, Small, Chauncey, & McKenna, 2011, p. 181).

Other chapters provide deeper insights about how technology might enrich mathematics teaching by providing new tools to stimulate creativity in learning. Chapters also include innovative ways of exploring creative potential of these tools thus contributing to new culture of teaching which helps to appreciate a variety of learning paths and build on students' creativity. The TPACK (Technological Pedagogical and Content Knowledge; Biton, Fellus, & Hershkovitz, 2016; Koehler & Mishra, 2015) framework is used to examine the knowledge teachers need, not only to have to teach successfully with technology, but also to emphasize the complexity of transferring of the potential impact of technology and pedagogy of creativity to the mathematics classroom.

Furthermore, Mishra (2012) sees a potential of dealing with this complexity in combining a trans-disciplinary view of creativity with '(in-)disciplinary' and 'indisciplinary' dimensions, the former framing creative work 'in-context' of a particular discipline (in our case, mathematics), and the later allows for "cutting across the boundaries of discipline to emphasize divergent thinking and creativity" (p. 15). According to the authors, this combination re-joins TPACK framework into an inclusive, practical and flexible structure for teaching creatively and effectively with technology (Mishra, 2012, p. 15).

Yet, in many aspects this variety of contexts, theories, and innovative practices does have, in our opinion, a common ground while connecting issues of developing creativity in technology-rich environments to the activities of solving rich problems and aiming at more advanced levels of mathematical thinking. This connection converges to more universal and holistic view of the field in its present state while at the same time offering new developments, along with promising paths for further investigation.

Based on these ideas and the submitted texts, we divided the chapters of the book into five parts, reflecting thematic variety of contributions while grouping them around the issues we find important for outlining the main focus of the book, which is increasing our shared understanding of and deepening our insight into how technology could foster creativity in mathematics.

In the next sub-section, we will examine each of the following parts:

- Teaching Practices and Instructional Strategies to Inspire Authentic Creativity.
- Creativity in Technology-Rich Mathematical Environments.
- New Learning Paths and Creative Teaching Approaches.
- Creativity and Advanced Mathematics.
- Learning from the Theories and Patterns of Students' Creativity.

1.3 Teaching Practices and Instructional Strategies to Inspire Authentic Creativity

The chapters in **Part II** of the book (**Teaching Practices and Instructional Strategies to Inspire Authentic Creativity**) look at how and why teaching and instruction is critical in providing opportunities for students to experience creativity growth, and the complexity of the issues in transferring this awareness into the classroom practice that could influence this growth (Panaoura & Panaoura, 2014). In their contributions, the authors of the chapters consider innovative teaching practices and instructional strategies that might inspire a so-called 'authentic creativity', while providing readers with examples that involve meaningful use of digital tools in mathematics teaching, and showcasing new ways for developing creativity in students.

In Chap. 2, "*Screencasting as a Tool to Capture Moments of Authentic Creativity*," **Dana Cox, Suzanne Harper, and Michael Todd Edwards** use screencasting to capture moments of authentic creativity in an interactive geometry environment. Their case study of two pre-service secondary mathematics teachers, working on The Kaleidoscope Task, exposes four distinct episodes of creativity. At the center of each episode is a moment of insight, categorized by the authors as either representing problem posing or problem solving. This chapter exposes the potential of screencasting to create an auditable trail of problem solving practice. It can be difficult to articulate the activity and thinking that surround moments of insight through verbalized thinking alone. Working in the digital environment affords us new tools to capture and articulate those moments. Screencasting is one technology that can be used to capture not only ambient conversation, but also on-screen action and activity.

The focus of Chap. 3, "*The Create Excellence Framework's Impact on Enhancing Creativity: Examining Elementary Teacher Candidate Mathematics Lesson Planning*," written by **Janet Lynne Tassell, Rebecca Stobaugh, and Marge Maxwell** examines how the Create Excellence Framework helps teacher educators have an impact on the quality of pre-service teachers' lesson plans to enhance creative learning opportunities for children. It builds on four components essential to high-quality lesson plans: Cognitive Complexity, Real-World Learning, Engagement, and Technology Integration. The authors analyze data from two elementary education teacher candidate classes for five semesters. Over the course of the study, for each component, the mean scores increased, and there was a positive statistically significant difference between the scores from the baseline semester to the fifth semester. Increasingly, students were exposed to and utilizing new digital tools to enhance their learning. Using these digital tools along with real-world applications of the content encouraged students to think creatively to solve authentic problems.

In Chap. 4, entitled "*Impacting Mathematical and Technological Creativity with Dynamic Technology Scaffolding*," **Sandra Madden** reflects on studies conducted during the past decade investigating teacher mathematical learning for teaching with technology and its relationship to creativity. Though related to mathematical knowledge for teaching (Ball, Thames, & Phelps, 2008) and technological pedagogical content knowledge (Mishra & Koehler, 2006; Niess, 2005), mathematical learning for teaching with technology has a strong dispositional component coupled with curiosity, creativity, and meaning making (Thompson, 2015). Using design-based research methods, a framework for dynamic technological scaffolding (DTS) has emerged in support of teacher learning. DTS has provided fertile ground for the design and further study of learning trajectories in which learners are exploring and eventually creating cognitively challenging mathematical task sequences in the presence of new (to them) physical and technological tools. By harnessing teachers' motivation to inculcate curiosity, engagement, and learning for their students, these design studies have created conditions where teachers have become curious, creative, and technologically savvy to the point where many have gone on to pursue similar kinds of experiences with their mathematics students. This chapter explores and presents DTS as created and implemented with secondary mathematics teachers, and as creative work pursued by them.

The section is concluded with Chap. 5, "*Three-Act Tasks: Creative Means of Engaging Authentic Mathematical Thinking through Multimedia Storytelling*," in which **Adrienne Redmond-Sanogo, Susan Stansberry, Penny Thompson, and Sheri Vasinda** describe the Three-Act Task approach and the theoretical foundations supporting it. This includes a design process for developing a rubric to evaluate and scaffold the creative multimedia mathematical stories. The rubric draws on four areas of literature for its theoretical grounding: (1) research on selecting and posing high cognitive demand tasks for mathematical problem solving, (2) use of story arc for contextual relevance, (3) research on assessing and measuring creativity, and (4) principles of effective multimedia message design. The rubric developed for assessing the Three-Act Task is designed to serve as a guideline for pre-service and in-service teachers as they select or create Three-Act Tasks to use in their classrooms. The authors conclude their chapter by highlighting several areas of where continuing research is needed, such as professional development of teachers, investigating student outcomes when the tasks are used, and extending the concept beyond the mathematics classroom to fields as diverse as science, English, and social studies.

1.4 Creativity in Technology-Rich Mathematical Environments

The chapters in **Part III** of the book (**Creativity in Technology-Rich Mathematical Environments**) spotlight the chapters surrounding the notion of technology-rich innovative learning environments (Istance & Kools, 2013). While the potential of technology-rich environments to enhance mathematical thinking has been documented in a number of studies (Suh, Johnston, & Douds, 2008), their particular role in fostering creativity in teachers and students remain underexplored. This is why the chapters we grouped in the second part of the book aim to provide innovative ideas for more interactive, explorative, interdisciplinary, and collaborative mathematics lessons. In this respect, our authors, while presenting their unique view of the topic, contribute, collectively to the search for and novel ways of expanding creativity in mathematics education through new types learning spaces enhanced by digital technology tools. It also helps to introduce a new way of thinking about the transformative role of technology by inviting the readers to look beyond the isolated examples for common patterns on how technology could leverage creativity in students' learning of mathematics.

Namely, Chap. 6, "*Interactive Technology to Foster Creativity in Future Mathematics Teachers*," discusses ways in which the use of interactive technology fosters creativity among future secondary mathematics teachers in a problem-based course that integrates mathematics, science, and technology in their first year in college. In their study, **Alfinio Flores, Jungeun Park, and Stephen Bernhardt** found that the use of interactive technology fosters creativity in students naturally when mathematics is taught based on research-based principles to learn mathematics for understanding. Creativity is fostered, promoted and developed when learners use interactive technology to (a) grapple themselves with concepts and make concepts explicit; (b) actively build new understanding on previous knowledge; (c) engage with mathematics as a social process; (d) use multiple representations and connections to enhance their understanding; (e) pose and solve problems; and (f) exercise multiple modes of learning—when they read, talk, write, draw, analyze, apply, present, and reflect. Authors discuss the use of interactive technology and issues related to future teachers' creativity as they solve problems; design experiments and collect, represent, and analyze data; develop mathematical models for phenomena in the physical, biological, and social sciences; and build and program their own robot.

In Chap. 7, "*Creativity and the Design of Music-Mathematics Activities in a Virtual Simulation Learning Environment*," **Trina Davis, Glenn Phillips, and Gerald Kulm** analyzed opportunities for creative and transformative experiences for students of all ages. Mathematics classrooms, in particular, can be fertile places for activities that integrate creativity. While technology and mathematics education are not strangers, there is little work on conceptual frameworks that drive the "math-tech" relationship. This chapter offers a fresh way of considering technology-infused mathematics learning by introducing Koestler (1981) notion of "collision as creation." As two things that are seemingly disconnected collide, the precipitate is a creative moment leading to comedy, discovery, or art. It is in these unique collisions that new knowledge and new ways of knowing come to pass. One such creative moment is explored through student perceptions of a virtual classroom experience. Students engaged within the environment as they connected technology (virtual reality), mathematics, and music. Implications of this work include an expanded idea of what contributes to feelings of efficacy and student success in the mathematics classroom, as well as how music may help students with learning difficult mathematical concepts like fractions and patterns.

Issues regarding preparing teachers to explore possible benefits of interactive tools to support creativity in students are being analyzed in Chap. 8, "*Preparing Teachers to Use Excelets: Developing Creative Modeling Experiences for Secondary Mathematics Students*," thus offering readers an important insight into how such technological experiences can provide learners with an environment that promotes creative thought. Authors **Ginger Watson and Mary Enderson** studied pre-service secondary mathematics teachers' use of excelets (interactive forms of excel) as a tool to promote understanding and creativity in modeling mathematical concepts. By providing pre-service teachers experiences in integrating excelets, how they approached the solving of problems presented a variety of responses and techniques as well as different levels of creativity. It was found that high-tier participants were more creative in their approaches and offered greater insight into their thinking and questioning of mathematics problems. The authors believe that providing pre-service teachers with experiences to use technological tools to explore modeling scenarios is crucial to their development as creative problem solvers and mathematics experts. They will develop creativity in their own mathematical thought as well as have the potential to think about supporting such creativity in their future instruction. Teacher education programs have quite a bit of work to accomplish in developing teachers who understand creativity as it relates to mathematics instruction using technology.

In Chap. 9, "*Creativity in Question and Answer Digital Spaces for Mathematics Education: A Case Study of the Water Triangle for Proportional Reasoning*," which concludes the second part of the book, **Benjamin Dickman** examines the intersection of technology and creativity in mathematics education, using an example of digital spaces that enable creative collaboration. Online Question and Answer (Q&A) sites for the subject of mathematics date back to the late twentieth century, but Q&A spaces organized specifically around mathematics education have emerged as a more recent phenomenon. By bringing together diverse individuals with joint interest in mathematics, these Q&A sites are well-positioned to support productive work; of course, whether the resulting collaboration can be labeled 'creative' is a function of how one interprets this commendation.

Although creativity is often framed with respect to a person, process, product, or environmental press, this chapter advocates for a more inclusive conception in which creativity is viewed through a participatory model. Under this interpretation, creativity is distributed across, for example, various actors, objects, and interactions, and concerns around attribution or eminence are no longer as salient. The chapter on creativity in Q&A digital spaces illustrates some of these notions by describing a routine reference request for the source of a tool for proportional reasoning, posed as a question through the Mathematics Educators Stack Exchange (MESE) platform, and answered by drawing from a confluence of sources that were made possible by recent technologies. In discussing nascent Q&A digital spaces for mathematics education, the chapter opens with three key ideas to be articulated, and closes with three corresponding open avenues for future exploration of these sites, along with modern conceptions of both creativity and technology, continue to evolve.

1.5 New Learning Paths and Creative Teaching Approaches

Recently, Hegedus and Tall (2016) have analyzed how later developments in multimodal learning environments based on haptic and multi-touch technologies can create enhanced learning possibilities for more learners to access core mathematical ideas and think mathematically. The chapters in **Part IV** of the book (**New Learning Paths and Creative Teaching Approaches**) provide important examples of how learning and teaching approaches connect with creative teaching approaches while using new devices and tools, such as 2D- and 3D-manipulatives, 3D-printers, and video-games.

With this section, we turned a focus toward actual strategies that have a creative angle. As we considered the diverse contributions of our authors, we found that teaching, in and of itself, warrants a thoughtful design that may be intentionally employing a creative strategy within the mathematics education classroom, whether it be at the K–12 level or college level. We considered how some of the chapters focus on teaching strategies for elementary through high school teachers, but also include attention toward the preparation of pre-service teachers. The value of these chapters is in emphasizing the importance of remembering that creative instructional practices could eventually lead to the emergence of creative learning approaches to solving rich and authentic mathematics problems. Technology, in this sense, provides affordances for exploration of what students need for better actualization of their learning potential (Martinovic, Freiman, & Karadag, 2013).

For instance, Chap. 10, "*Nurturing Creativity in Future Mathematics Teachers through Embracing Technology and Failure*," by **Marina Milner-Bolotin** discusses how modern educational technologies can open promising opportunities for educating mathematics teachers who are ready and willing to nurture creative and critical thinking in their own students. In particular, the focus is on exploring pedagogies that utilize technologies, such as computer simulations and modeling software, data collection and analysis tools, classrooms response systems, tools for student collaboration on designing multiple-choice questions, and finally tools for collaborative reflecting on video-recorded teacher-candidates' micro-teaching. The chapter discusses how modern technology can help mathematics teachers to turn inevitable student failures in mathematics learning into exciting learning opportunities. The challenges of the implementation of these technologies in mathematics teacher education and the opportunities that they offer for embracing creative mathematical thinking are also discussed.

In the context of early spatial learning, Chap. 11, "*Harnessing Early Spatial Learning Using Technological and Traditional Tools at Home*," by **Joanne Lee, Ariel Ho, and Eileen Wood** reviews extant literature in an effort to identify and describe key ideas and research findings relevant to the use of physical and digital/virtual manipulatives for promoting early spatial development. This context serves as the foundation for exploring the importance of supporting spatial-visual skill development in early childhood years through the complementary use of 2D and 3D spatial play activities. Particular attention is given to the use of 2D spatial-visual iPad® applications to nurture creative thinking afforded by technology and the importance of the design of applications to support positive learning outcomes. They introduce a novel idea that playing with *both* 2D (i.e., iPad® apps) and 3D manipulatives may maximize learning opportunities to foster creative and flexible spatial thinking.

In summary, the author's goal is to provide a foundation for understanding early spatial development in the home and in the context of a technologically rich learning environment. These two contexts provide opportunities for creative expression, discovery and exploration. A key objective is to identify how these aspects of creativity intersect in the current literature and may be important for future study—especially in the critical early years, where home influences are most likely to establish fundamental skills and where touch—screen technologies are increasingly prevalent.

Also aiming at the development of spatial skills, Chap. 12, "*Video Game Play, Mathematics, and Spatial Skills—A Study of the Impact on Teacher Candidates*," by **Janet Lynne Tassell, Elena Novak, and Mengjiao Wu** highlights the importance of spatial abilities in mathematics education, especially among pre-service elementary teachers, and suggests video games as a creative teaching approach for enhancing spatial abilities and mathematics performance. They argue that spatial abilities deserve more attention in mathematics education, as a major predictor of achievements in science, technology, engineering, and mathematical (STEM) fields. To support this notion, the authors introduce recent developments in research on non-educational action video games that promote various cognitive and attentional capabilities that have a potential of improving mathematics achievement. They describe an experimental study that examined the effects of playing such recreational video games on education majors' math problem-solving, math anxiety, working memory, and spatial skills. After 10 hours of playing, both video game intervention groups significantly improved their spatial skills, working memory, and geometry performance from pre- to post-test. In addition, students with low spatial abilities had significantly higher math anxiety. These findings suggest potential impact of video gaming in education and open new horizons for future research that explores how schools and homes working together with strategic gaming plans can help students improve their spatial reasoning and mathematics problem solving. The chapter concludes with future research suggestions on spatial abilities and creativity in mathematics education.

A recent growth of makerspaces in schools created new opportunities for teaching and learning mathematics shown in Chap. 13, "*Prototype Problem Solving Activities Increasing Creative Learning Opportunities Using Computer Modeling and 3D Printing*," written by **Antonia Szymanski**. The purpose of this chapter was, through reviewing literature and presenting a conceptual application, to investigate the relationship of 3D printing with Prototype Problem-Solving Activities (PPSA) to develop creativity in mathematics. Grounded in the makerspaces movement, the potential of 3D printers in PPSA is highlighted in the chapter. PPSA seek to replicate the makers' motivation and curiosity through authentic problems that rely on collaboration and multiple iterations to find optimal solutions. The process of using PPSA allows the students to develop higher order thinking skills of analysis and synthesis in their mathematical understanding.

1.6 Creativity and Advanced Mathematics

The chapters in **Part V** of the book (**Creativity and Advanced Mathematics**) incorporate actual problems and focus on the learning of mathematics. This part, while providing several examples of creative problems in the mathematics classroom, initiates conversations about possibility of learning more advanced mathematics. Technology enters in here in varying ways and levels, providing a context to consider that situations are unique to schools, classrooms, and individual children. A number of promising opportunities provided by technological tools also require a critical examination of possible pitfalls teachers and students should be aware of. We particularly found these chapters intriguing as they focus on the problems and scenarios, with samples provided for how the conversations happened in the authors' classrooms, thus shedding light on advanced mathematics topics from a creativity perspective.

Hence, in Chap. 14, "*Can a Kite Be a Triangle? Aesthetics and Creative Discourse in an Interactive Geometric Environment*," **Hope Gerson and Paul Woo Dong Yu** claim that creativity is essentially aesthetic rather than cognitive. In this chapter, the authors set out to study creativity within a dynamic geometry environment using Sinclair (2006) aesthetic framework to identify creativity within a class discussion. Careful analysis of the case stories shows various elements of aesthetic sensibilities in the students' responses: student interest, genuine curiosity, brainstorming, and motivation. They see creativity in the three aesthetic roles, generative, motivational, and evaluative, as the students generate ideas, the teacher makes instructional choices, and then students and teacher together resolve the mathematical discussion. In this episode, the inquiry began with the generative aesthetic with the motivational aesthetic at its peak during the climax and then continuing through to the resolution with the evaluative aesthetic playing the major role.

The authors also found the generative nature of the dynamic geometry environment to be particularly rich in allowing students to activate the generative aesthetic to change their focus, look for patterns, and reorganize their thinking in different ways. The generative aesthetic was driving the creativity forward allowing students to view the triangle-kite from many different perspectives. The students generated ideas which activated motivational responses and lead to resolution. The creative and aesthetic qualities of open inquiry, the Geometers' Sketchpad, and teacher moves, created a setting where students and the teacher made aesthetically motivational, generative, and evaluative choices to build understanding of geometric properties of kites and triangles as well as the limitations of sets of geometric properties in classifying geometric shapes.

In a context of teacher education, **Sergei Abramovich**, in Chap. 15, "*Technology and the Development of Mathematical Creativity in Advanced School Mathematics*," reflects on his work with pre-service teachers enrolled in a secondary mathematics education course which is enhanced by powerful computing technology. In this context, the development of creativity is considered through the lenses of the theory of affordances frequently used by mathematics education researchers when talking about the affordances of digital tools. A concern has been raised that because of the availability of mathematical software capable of solving rather advanced problems almost at the push of a button, technology provides a negative affordance by facilitating problem solving to an extent that mathematics appears giving up its creative flavor. With this in mind, the chapter focuses on a new type of problems which stem from the modification of traditional ones in a sense that they are both technology immune (TI) and technology enabled (TE). In other words, when a computer does not provide the final answer, its use in solving a problem is critical.

In Chap. 16, "*'Integrating' Creativity and Technology through Interpolation*," the authors **Bharath Sriraman and Daniel Lande** attempt to show the paradox of the digital age where mathematical information is readily available, as well as tools to compute classical results through symbolic mathematical software, such as Mathematica, devoid of the motivation or origins of these results. Simply accessing information or invoking a function or a routine does not reveal the process via which these results were obtained. Using examples which high school students might encounter in a Calculus textbook, particularly when learning to integrate basic circular functions, they uncover how original hand calculations using first principles can result in deep insights that present students with the opportunities of learning and understanding the origins and necessity of these functions. To further understand what these results mean, they employ 21st century tools to visually represent functions that were obtained via mathematical interpolation without the aid of modern technology. Original techniques are contrasted with modern graphing techniques for the same functions.

In a particular context of Mathematics Village, which has existed in a natural environment in Turkey for ten years, in Chap. 17, "*Ancient School without Walls: Collective Creativity in Mathematics Village*," **Elçin Emre-Akdoğan and Gönül Yazgan-Sağ** have dealt with the role environment plays in the development of mathematical creativity. The Mathematics Village, a unique institution in Turkey, offers short but intense courses, and at which high school and college students are taught by professors and engage in seminar discussions. The authors have examined how the Mathematics Village could promote mathematical creativity, as well as the transformation of the culture of mathematical creativity that emerged from the Mathematics Village (non-virtual environment) into Social Media (virtual environment).

The context of this study, as well as its findings, reveal that the Mathematics Village promotes mathematical creativity of students and enables mathematicians to activate their own creativity. From that perspective, having an educational setting that provides freedom can positively affect students' state of mind and creativity. The results of this study contribute to the literature due to the fact that this study examines how providing a context, in which people with different levels of education and interest in mathematics could activate mathematical creativity, and how the creative characteristics of the Village could be potentially enhanced by a virtual environment using social media through: (1) Information-gathering, (2) Interaction and sharing, and (3) Social networking.

1.7 Learning from the Theories and Patterns of Students' Creativity

Despite important advances on understanding of learning and teaching potential of the use of technology in mathematics, little is known about the particular ways students' creativity could emerge in such environments. In this respect, the chapters in **Part VI** (**Learning from the Theories and Patterns of Students' Creativity**), the last part of the book, give an angle for us to consider as we focus on the student authentic learning. It also shares empirical evidence and theoretical thoughts regarding what mathematics educators could learn from these authentic learning experiences. The authors within this section provide theoretical insight on approaches that are grounded in multiple data sources collected over the years, but also being stretched by the concept of the interaction of the three areas, namely technology, creativity and mathematics.

In this respect, Chap. 18, "*APOS Theory: Use of Computer Programs to Foster Mental Constructions and Student's Creativity*" by **Draga Vidakovic, Ed Dubinsky, and Kirk Weller** refers to Piaget (1981, p. 227) saying that "the body of mathematics is a model of creativity, and it also rests on a process of reflective abstraction." From Piaget's perspective, mathematical thought, which he believed to be inherently creative, is driven by an individual's ability to contemplate ideas and to make abstractions based on those ideas. This perspective is the main premise behind APOS, a constructivist theory of mathematical learning. The letters that make up the acronym APOS, represent the four basic mental structures—Action, Process, Object, and Schema. The general theory serves as a framework to describe individual cognition for specific mathematical concepts. On this basis, an instructor can design activities that align with student learning of a concept. As students construct new mental structures, they can achieve new insights, engage in higher level reasoning, and apply techniques or approaches in new, or novel, ways. Typically, this involves having students write, test, and discuss the effects of running computer programs or, in some cases, of making adjustments to code provided by the system. Getting the solution to a problem by pushing buttons and having the computer cough up the solution does not really go beyond just listening to an instructor present the solution, or reading about it in a text, neither of which fosters creativity or learning. The authors argue that these types of activities constitute the essence of creative thinking.

From Vygotskian and Piagetian perspective, in Chap. 19, "*The Nature of Knowledge and Creativity in a Technological Context in Music and Mathematics: Implications in Combining Vygotsky and Piaget's Models*," **Yves de Champlain, Lucie DeBlois, Xavier Robichaud, and Viktor Freiman** take a cross-disciplinary look at constructivist and cognitive approaches to creativity in order to understand its implications in the learning process. Indeed, creativity necessarily implies at some point to see the world in a way that is different than what we previously knew. And the cognitive act of changing our view about what we know implies changing our relationship with what we know and even with knowledge itself. So, hypothesizing that technology has an influence with our relationship with knowledge in a very deep way, the questions these authors asked are: How can technology contribute to change the learner's relationship with knowledge in a learning setting? How can technology contribute to change our relationship with creativity itself?

To answer these questions, the authors went back to Piaget's and Vygotsky's founding models of our relationship with knowledge. They considered that if we understand the nature of the learning process, then we may also understand the nature of creativity in the learner's perspective and the impact of technology in this relationship. These authors studied creativity form the learner's perspective in two apparently opposite curricula: mathematics and music. Apparently opposite because mathematics is often associated with right and wrong answers while music is usually associated with freedom and self-expression. But beyond appearances, the cognitive processes of creativity share a large common ground in both fields.

Over the past decades, educational systems have continually worked on integrating technology into mathematics education. Creativity, on the other hand, was —more often than not—less attended to. In Chap. 20, titled, "Putting the horses before the cart: Technology, creativity, and authorship harnessed three abreast," **Osnat Fellus and Yaniv Biton** frame technology, creativity, and authorship as analogous to horses pulling the cart of mathematics education. Fellus and Biton build on Latour's conception of technology, Vygotsky's perception of creativity, and Bakhtin's notion of authorship to suggest a departure from traditional ways of viewing creativity in mathematics education as arrogated to giftedness, and a shift to a unification of the notions of technology, apprenticeship, and authorship to allow for expressions of creativity for all learners. Using the teachings of the three classic scholars, the authors, widen our perception and conception of technology, creativity, and learning; show how the very intertwining of the notions allows for a far wider understanding of the idea of creativity in mathematics education; and frame the learning of mathematics as learning connections in a network of mathematical ideas. In this chapter, the authors first denote the terms technology, creativity, and authorship and discuss how the interconnectivity among the three is paramount in the context of mathematics, they then showcase how this interconnectivity is co-constructed in an episode drawn from the work of the second author as a mathematics instructor in the virtual high school in Israel. The chapter concludes with a discussion of some possibilities this meeting point among Latour's precepts of technology, Vygotsky's concept of creativity, and Bakhtin's construct of authorship may hold for the teaching and learning of mathematics.

Rounding out the last section is Chap. 21 from **Dominic Manuel**. Here he explores "*Virtual Learning Communities of Problem Solvers: A Potential for Developing Creativity in Mathematics?*" Researchers in mathematics education argue that mathematics is a school subject that can support the development of creativity in students. Yet this opportunity is still under-used in mathematics classrooms. In this chapter, the author explores the virtual community CAMI (Communauté d'apprentissages multidisciplinaires interactifs) as a potentially rich environment form of developing mathematical creativity in the context of problem solving.

The author developed a conceptual framework to analyze the richness of mathematical problems in CAMI including its open-ended-ness, complexity, contextuality, ill-defined-ness, and possibility for multiple interpretations. In addition, using Leikin's (2007) notion of collective solution space as well as criteria for mathematical creativity such as fluency (number of correct possible answers to a problem), flexibility (number or appropriate strategies to solve a problem), and originality (correct answers and strategies less frequently used in solutions to a problem) the author has assessed the mathematical creativity of the students' solutions to these problems while looking whether a link exists between richness of the problem and creativity of solutions. Results show that, in general, richer problems seem to bring different correct answers and more original solutions in the CAMI virtual community. They also imply potentially promising practice teachers could develop by analyzing this diversity of solutions with their students thus providing further support for the development of mathematical creativity.

1.7.1 In What Way Technology Can Foster Creativity in Mathematics: Searching a Common Ground

Based on our own reflection of each contribution to this book, we present in this section a model which attempts to illustrate an interaction between different contexts, theories, and practices discussed by our authors, in a hope to make some common sense of the field and draw preliminary conclusions about what has been done thus far and what are some promising paths for future investigations. Since early 1980s, the technology (or ICT) is considered as driver of change and innovation in schools, in general, and in teaching and learning mathematics, in particular (Bottino, Artigue, & Noss, 2009). According to the authors, it might potentially contribute not only to improving existing curriculum goals and classroom practices but also to changing a nature of knowledge itself, shifting our questioning from 'How' to 'What' in a variety of contexts (Papert, 2006; cited by Bottino et al., 2009). The call for change is also coming from the world outside the school where "novel kinds of mathematical knowledge, techno-mathematical literacies have become of critical importance" (idem, p. 75).

Kaput and Thompson (1994) have identified three aspects of electronic technology for '**enabling**' "deep change in the experience of doing and learning mathematics" (p. 678), one of which is *interactivity*, when computer is providing students with the reaction to her or his action allowing, in turn, further interpretation, reflection and further action from the students. Here we find the sources for '**engaging**' students in more active learning. A second source of power of change mentioned by the authors is related to the *design* which provided aid for thinking or problem solving, along with intelligent feedback or context-sensitive advice, actively linking representation systems, control physical processes from the computer, all this could influence students' mathematical experiences more deeply than ever before (Thompson, 1991; cited by Kaput & Thompson, 1994). We call this '**empowering**' the students' learning experiences.

Regarding the 3rd aspect, it is surprising that already in 1994, the visionaries could identify *connectivity* as yet another source of change when linking "teachers to teachers, students to students, students to teachers, and perhaps most important, that link the world of education to the wider worlds of home and work" (idem, p. 679) which has potential of '**enriching**' mathematical experiences while promoting discourse among students and among students and teacher and eventually '**encouraging**' them to develop higher-order thinking (Niess et al., 2009).

These potentially transformative aspects of pedagogy based on digital technology, according to Kaput (1998), bring a number of challenges to mathematics education which he grouped in four categories, namely, (1) representations and modeling, (2) curriculum structures and prerequisites, (3) content shifts due to the computational medium, and (4) the move to ubiquitous, heterogeneous, connected technologies. Projected by Kaput on the period of 5–10 years, these challenges seem still be in center of educational debates nowadays, almost two decades later. For instance, in a similarly titled paper, Wade, Rasmussen, and Fox-Turnbull

(2013) emphasize technology integration as 'cultural transformation' from traditional teacher-directed to innovative student-centered learning where students are engaged in learning through teamwork, critical thinking, and problem solving.

We could also document the potential of technology to provoke changes in teaching and learning based on our own work on the design and the development of the CAMI virtual learning community (Freiman & Lirette-Pitre, 2009), then analyzing students' work in solving problem, elaborating authentic solutions, and inventing strategies within informal online learning environments (Freiman, 2009; LeBlanc & Freiman, 2011; Pelczer & Freiman, 2015) and in-class when exploring robotics-based learning tasks (Savard & Freiman, 2016). We also noticed the potential of these technology-rich environments and technology-enhanced experiences to be used by caring teachers who seek to contribute to fostering students' mathematical creativity (Freiman & DeBlois, 2014). Yet, there are also possibilities to use technology to support lecturing-type presentations and exercise-type activities within learning management systems, portals, video and intelligence in tutoring systems which reproduce rather traditional way of teaching and learning (Kynigos & Daskolia, 2014). In order to become real enablers of creative mathematical thinking, digital technologies have to be combined with other processes, mechanisms and tools of school education, among which the design and use of appropriate educational resources (Kynigos & Daskolia, 2014), as well as changing approaches to teacher education, assessment and educational policy (Henrikson, Mishra, & Fisser, 2016).

In a similar way to Flavin's approach to analyze disruptive character of technology-enhanced learning, Martinovic et al. (2013), in the opening Volume of the Series, used activity theory, along with affordances theory in order to grasp innovative practices in visual mathematics and cyberlearning. In continuation of this work, refereeing to the issue of fostering creativity in mathematics, we consider the emergence of Dynamic Learning Conditions, as a result of paradigmatic shift within a Technology-Rich Environment which is being created by novel (and sometimes new) forms of teaching and learning activities, approaches, as well as assessment practices and has potential to increase opportunities for creativity and self-directed learning (Mishra, Fahnoe, & Henriksen, 2013).

We represent this complex phenomenon by the schema shown in (Fig. 1.1). By "approach" we mean to include models of the following: Informal, Open, Virtual, Visual, Dynamic, Collaborative, Constructive, Connected, Experiential. By "activity" we include tasks the learners will investigate through Dynamic Explorations, Simulations, Modeling, Problem Solving and Problem Posing, Inquiry, Discussion, Collaboration, Questioning, Hands-on. By "assessment" we mean the emergence of new ways of capturing and scaffolding moments of learning insights: Formative, Alternative, Immediate Feedback, Feeding Forward, Qualitative Rubrics.

Moreover, the chapters provide multiple examples of how an interaction of these three components could potentially transform teaching and disrupt learning be means of enabling, empowering, encouraging, engaging, and enriching factors, often mentioned in the literature on successful technology integration. In the

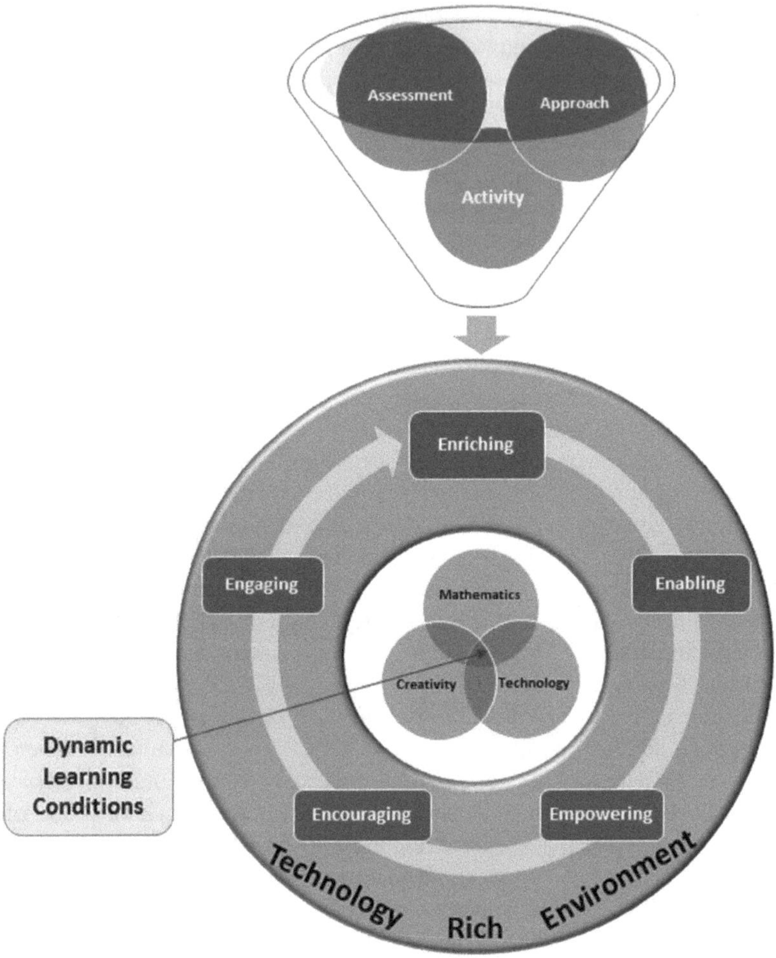

Fig. 1.1 Dynamic learning conditions of leveraging mathematical creativity with technology

following paragraphs, we will describe how these factors, across a variety of topics brought by our authors, converge toward increased synergy between mathematics and technology that could potentially contribute to increase opportunity for fostering creativity in learners.

1.7.1.1 Enabling

We learn from the chapters how technology can enable deeper understanding, authentic context, transformation in knowledge (discovering new knowledge), communicating reasoning about math problems, reflection and critical thinking, alternative ways to calculate properties, generate ideas thus directly contributing to mathematical creativity in both, students and teachers. In fact, technology might provide students with new paths in their journey to mathematical knowledge while enabling them to use and experience powerful cognitive tools: authentic context; authentic activity, access to expert thinking and performance, multiple roles and perspectives, reflection, collaboration, articulation, coaching and scaffolding, integrated authentic assessment (Herrington & Herrington, 2007). Through many chapters, these aspects are illustrated in connection to the opportunity in gaining a deeper understanding of geometry by using the Dynamic Geometry Software (Bokosmaty, Mavylidi, & Paas, 2017) and the Interactive Geometry Software (IGS) (Chap. 2). Exploring mathematics creativity through inquiry, in an environment of interactive geometry, can also enable mathematical discussion by the students (Chap. 14).

In a context or pre-service teachers' education (Chap. 8), solving modeling tasks using Excelets, an interactive form of an Excel spreadsheet allows for the manipulation of data and the visualization of changes in numeric, graphic, and symbolic for thus enabling performing mathematical experiments to the student and help him/her discover new mathematical knowledge (Dijanic & Trupcevic, 2017). Many digital spaces, and Question and Answer (Q&A) sites such as Mathematics Educators Stack Exchange (MESE) (Chap. 9) provide a platform through which those interested in the teaching and learning of mathematics can harness new technologies to address novel queries thus enabling exchange of ideas. Play and manipulatives (Chap. 11) can further foster creative educational experiences by providing a learning environment that encourages engagement and enables understanding (Cockett & Kilgour, 2015).

Videogames as a creative teaching approach could potentially enhance spatial abilities and mathematics performance (Chap. 12) while providing multifaceted and multimodal access to information (Lowrie, 2015). In a context of makerspaces, 3D printing tools can enable specific connections to technological literacy that moves students beyond being mere consumers of information to generating ideas and reflecting on thinking which requires higher levels of thinking, innovation and creativity (Huleihil, 2017).

A potential of technology to enable mathematical creativity for more advanced mathematics is conditioned by the development of students' ability to go beyond simply entering correctly all data into a computer when solving problems thus experiencing fundamental definitions of mathematical concepts (Chap. 15). The value of hand calculations is therefore highlighted since this enables one not only to derive mere results, but also grasping the process while contrasting the 'original' techniques with the 'modern graphing' techniques (Chap. 16). The learner's own mathematical creativity (Chap. 17) is impacted through access to rich mathematical problems allowing for communication of student reasoning (Chap. 21), apprenticeship and authorship (Chap. 20), and reflection and collaboration (Chap. 18).

Overall, by different examples of problems and tools that enable their solution, learners can "generate and multi-directionally link different representations in order to explicitly and dynamically reveal the different facets of the complex ideas embedded in the solutions of a mathematical problem" (Hirashima, Hoppe, & Young, 2007, p. 409).

1.7.1.2 Enriching

An online resource for mathematical enrichment NRICH sets up the goals to "promote an interest in mathematics, to raise the standards of achievement in school mathematics, to assist the mathematical development of children who have the potential to go on to study mathematical subjects at university, and to support the special educational needs of exceptionally able children" (Jones & Simons, 1999, p. 3). In a context of mathematical problem-solving, this type of technology may provide students with an access to rich and challenging problems and eventually spark mathematical creativity which can be considered as flexible, fluent, and original approach to the solution (Chap. 21), or lead to the increased role of imagination, and consequentially to the authenticity and complexity of students' mathematical reasoning (Chap. 19). In such contexts, students' creativity is inspired either by the development of new solutions or strategies (Problem Solving Insights) or spurred new questions (Problem Posing Insights) (Chap. 2), a phenomenon, Freiman (2009) calls Enrichment.

In a more interdisciplinary perspective, the use of authentic problems and interdisciplinary approaches to problem solving can simulate real-life connections in the STEM fields (Chap. 13) eventually leading to more holistic and interdisciplinary view of mathematical creativity (Sriraman & Dahl, 2009). Through enrichment, students can also go beyond the typical curriculum, as it is the case of the three-act tasks (Chap. 5) where rich mathematical tasks can be presented by means of different types of multimedia while engaging students in an authentic process of finding solutions, and then comparing them with the 'real' story. In a similar way, by modifying traditional problems from advanced mathematics, designing a curriculum which is both technology-immune and technology-enabled in the sense that whereas software can facilitate problem solving, its direct application is not sufficient for finding an answer (Chap. 15). This can eventually contribute to transformation that occurs in knowledge when using technologies as a part of creative process for all learners (Chap. 19).

Richer learning could also happen in a more informal context, like activities elicited during play by parents and early childhood educators both in the context of traditional 3-dimensional play (e.g., blocks and puzzles) environments and virtual 2-dimensional digital formats (e.g., iPads® and computers) (Chap. 11). Lee and Ferrucci (2012) argue that virtual manipulatives in a context of mathematics teaching and learning could "enrich and transform learning environments of the students' and further enrich their learning experiences due to dynamic nature of thinking, which may enhance students' thinking and creativity" (p. 127). Chapter

12 gives an example of creativity fostered when a partnership is formed between school and home in a context of video gaming helping students enrich their spatial reasoning and problem-solving abilities (Lowrie, 2015).

1.7.1.3 Encouraging

The role of *encouraging* is to provide a caring learning and teaching environment for students in turn allowing for more creativity. Technology tools could be incorporated through a safe and user-friendly setting (Kazakoff, Orkin, Bundschuh, & Schechter, 2017). If students are provided with such kind of environment, they become enabled to cut to the authentic learning and intuitive exploration of richer and real-world connected mathematics (Baker & Galanti, 2017). The case of Mathematics Village describes, in its turn (Chap. 17), an educational setting that provides freedom that can positively affect students' state of mind and creativity which can be further expanded by means of social media. Collaboration is another factor illustrating the role of technology in the encouragement of creative mathematical work, as shows the example of APOS (Action, Process, Object, and Schema) where collaborative activity is facilitated within a computer environment (Chap. 18).

For example, a supportive environment where pre-service teachers could receive feedback and coaching on their lesson, deliver the lesson to students, then debrief the lesson design can create conditions where teachers have become "curious, creative, and technologically savvy", willing to pursue similar kinds of experiences with their future students (Chap. 3, p. 89). On the other side, a classroom simulation example involving teaching experiences in a virtual setting incorporating technology and music can be incorporated into pre-service teacher education in a more creative way (Chap. 7). Overall, a component of a technology-rich environment, like Geometers' Sketchpad, can engage students into an open inquiry where students and the teacher make "aesthetically motivational, generative, and evaluative choices" (Chap. 14, p. 347). Another example of 3-D printing in makerspaces (Chap. 13) shows how the teacher can "encourage students who may not perceive themselves as exceptional in mathematics by providing new ways in which to demonstrate mathematical thinking" (p. 323). Technology can also help to scaffold teacher-candidates and consequently mathematics learners in experiencing and overcoming 'productive failures' and thus encourage deeper insight into mathematics (Chap. 10).

1.7.1.4 Empowering

The outcome of the learning experience is the empowerment while being enabled through the instruction that makes learners feel stronger as result of productive struggle and perseverance (Sengupta-Irving & Agarwal, 2017). For example, using Dynamic Technology Scaffolding (DTS) (Chap. 4, p. 89) increases teachers' capacity to explore and create "cognitively challenging mathematical task sequences in the presence of new physical and technological tools." In a similar way, Three-Act Task rubric can empower pre-service and in-service teachers as they select or create these types of tasks to use in their classrooms (Chap. 5). Always referring to the context of teacher education, our authors show how future teachers become confident in solving problems; designing experiments and collecting, representing, and analyzing data; developing mathematical models for phenomena in the physical, biological, and social sciences; and building and programming their own robot (Chap. 6). At the same time empowerment is fraught with inevitable challenges and productive failures, at other times it can be filled with exhilarating discoveries and new insights (Chap. 10).

Self-efficacy growth is another outcome of mathematical creativity empowered by technology connecting music with challenging mathematical concepts like fractions and patterns while contributing to the emergence of feelings of efficacy and success in students in the mathematics classroom (Chap. 7), an observation which corroborates with findings from Chap. 12 where higher spatial abilities impacted self-efficacy for individuals having significantly higher confidence in learning mathematics. Also, the 3-D printing context seems to empower students' ability to "self-assess and reflect on their understanding" and "develop higher order thinking skills of analysis and synthesis in their mathematical understanding" (Chap. 13, p. 323).

1.7.1.5 Engaging

Related to motivation and interest, *engaging* gets at how the instruction is designed to elicit student involvement in the learning. Does the design of instruction provide a way for students to approach learning with verve and excitement? For engagement to occur, we look for captivating students, whether by real-world and authentic learning (Lombardi, 2007), novel connections, or creative use of tools (Bray & Tangney, 2017). Often, an engaging task in mathematics involved being "hands-on, minds-on." For example, through the instructional design, the pre-service teachers focus on a key component of the Create Excellence design, as they "learned to design instruction around authentic tasks, where cognitive levels and engagement are also increased" (Chap. 3, p. 59).

Another example (Chap. 7) shows how "…environmental (technology) and conceptual (music) frameworks can be juxtaposed to mathematics teaching to create more engaged and productive learning" (p. 181). Use of technology can also engage mathematics teacher-candidates in exploring how it can facilitate productive mathematical thinking (Chap. 10). Collaboration again comes upfront, in connection to engagement in a context of "Question and Answer (Q&A) sites such as Mathematics Educators Stack Exchange (MESE) where members can engage collaboratively with others who share their interests" (Chap. 9, p. 233).

1.8 Conclusion

The goal of the book was to further explore connections between mathematics, creativity, and technology in the digital era. It is not surprising that investigation of novel fields of research and practice, like in our case, connections of technology to creativity in mathematics brings a variety of contexts, issues, practical examples, theoretical approaches, and research findings. Yet, with the above presented general schema which reflects our reading of the whole book some common trends seem to emerge. First, we observe increasing opportunities provided by technology-enhanced environments and tools to access to richer, sometimes more advanced, and real-life connected mathematics for more students.

Along with an increasing variety of mathematical problems and tasks, as well as multiplicity and authenticity of the approaches and methods of their solutions in such new types of digital learning ecosystems might thus emerge, where mathematical investigations, enabled by technology, become multimodal, interactive, dynamic, and process-oriented which could, in its turn, lead to important mathematical discovery while empowering deeper understanding through reflection and critical thinking. Such environments become more attractive, motivating, and overall engaging for students who are encouraged to explore, tinker, produce conjectures, and looking for explanations while being able to share their discoveries with their peers and teachers, and increase opportunities to collaborate.

All this brings additional challenges to the traditional forms of teaching and learning while prompting paradigmatic shifts and transformations in pedagogical approaches and assessment practices. This is why it is not surprising to find many chapters devoted to teacher education and professional development. While it is not yet very clear, in what sense, living multiple experiences of enriched mathematics infused by technology thus developing more creative teaching approaches, teachers would become better prepared to foster creativity in their students, several examples provided in the book could lead to more focused discussions, as well as initiate new projects and research paths.

Moreover, new digital spaces, being analyzed as parts of the digital learning ecosystems seem to have potential to disrupt traditional learning routine engaging students (and teachers!) into an open-ended process of posing and solving problems where the search for solution is not well-structured but rather ill-defined which may lead to so-called 'productive failures' but also make learning more authentic, meaningful, goal-oriented, interdisciplinary and self-directed, and overall more creative.

When considering the contributions from the holistic perspective, we can notice continuously evolving conceptualization of creativity in mathematics empowered by technology. By bridging creative teaching approaches and students' inquiry and problem-based learning, creativity itself is becoming more diverse, multifaceted, collaborative, and even distributive, a phenomenon, not yet very well understood in research and practice, which needs more investigation.

In conclusion, the book demonstrates the need for continuation in examining the relationships and possibilities for closer connections between mathematics, creativity, and technology their potential in teaching and learning in the digital era. Along with the connectedness, it also shows that within the mathematics education community, there exist varying views about what constitutes technology-enhanced learning and authentic creativity experiences in the mathematics classroom and beyond which need to be critically reflected in terms of extended research and further impact on practice. In this respect, the variety of the perspectives provided by our authors does contribute to a healthy debate and pushes the reader for deeper investigation of the topic.

References

Arnone, M. P., Small, R. V., Chauncey, S. A., & McKenna, H. P. (2011). Curiosity, interest and engagement in technology-pervasive learning environments: A new research agenda. *Educational Technology Research and Development, 59*(2), 181–198.

Baker, C. K., & Galanti, T. M. (2017). Integrating STEM in elementary classrooms using model-eliciting activities: Responsive professional development for mathematics coaches and teachers. *International Journal of STEM Education, 4*(10). Retrieved from https://doi.org/10.1186/s40594-017-0066-3.

Ball, D. L., Thames, M. H., & Phelps, G. (2008). Content knowledge for teaching: What makes it special? *Journal of Teacher Education, 59,* 389–407.

Biton, Y., Fellus, O., & Hershkovitz, S. (2016). Border crossing: Bringing together pre-service teachers' technological, pedagogical, and content knowledge through the use of digital textbooks in mathematics. In C. Csikos, A. Rausch, & J. Szitanyi (Eds.), *Proceedings of the 40th Conference of the International Group of the Psychology of Mathematics Education* (Vol. 2, pp. 107–114). August 2016. Szeged, Hungary: PME.

Bokosmaty, S., Mavylidi, M-F., & Paas, F. (2017). Making versus observing manipulations of geometric properties of triangles to learn geometry using dynamic geometry software. *Computers & Education, 113*, 313–326.

Bottino, R. M., Artigue, M., & Noss, R. (2009). Building European collaboration in technology-enhanced learning in mathematics. In N. Balacheff, S. Ludvigsen, T. de Jong, A. Lazonder, & S. Barnes (Eds.), *Technology-enhanced learning—Principles and products* (pp. 73–87). Milton Keynes, UK: Springer.

Bray, A., & Tangney, B. (2017). Technology usage in mathematics education research—A systematic review of recent trends. *Computers & Education, 114*, 255–273.

Cockett, A., & Kilgour, P. (2015). Mathematical manipulatives: Creating an environment for understanding, efficiency, engagement, and enjoyment. *TEACH Collection of Christian Education, 1*(1). Retrieved from http://research.avondale.edu.au/cgi/viewcontent.cgi?article=1004&context=teachcollection.

Dijanić, Ž., & Trupčević, G. (2017). The impact of using GeoGebra interactive applets on conceptual and procedural knowledge. In *The 6th International Scientific Colloquium Mathematics and Children*, Osijek, Croatia. Retrieved from https://bib.irb.hr/datoteka/878910.Dijanic_Trupcevic_Impact_GeoGebra_knowledge_finish.pdf.

Freiman, V. (2009). Mathematical enrichment: Problem-of-the-week model. In R. Leikin, A. Berman, & B. Koichu (Eds.), *Creativity in mathematics and the education of gifted students* (pp. 367–382). Rotterdam, The Netherlands: Sense Publishing.

Freiman, V., & DeBlois, L. (2014). Le site Internet CAMI: Une ressource virtuelle pour soutenir la résolution de problèmes chez les élèves francophones du Nouveau-Brunswick. *Cahiers des sciences de l'éducation. Université de Liège, 36,* 115–142.
Freiman, V., & Lirette-Pitre, N. (2009). Building a virtual learning community of problem solvers: Example of CASMI community. *ZDM, 42*(1–2), 245–256.
Freiman, V., & Tassell, J. L. (2014). Using technology to develop mathematical creativity and giftedness. In *Proceedings of the 8th MCG Conference* (pp. 175–186). https://drive.google.com/file/d/0B1W-oFlzfvpobU5Ma0g5RjZPRWc/view.
Hirashima, T., Hoppe, U., & Young, S. (2007). *Supporting learning flow through integrative technologies*. Oxford, UK: IOS Press.
Hegedus, S., & Tall, D. (2016). Foundations for the future: The potential of multimodal technologies for learning mathematics. In L. D. English & D. Kirshner (Eds.), *Handbook of international research in mathematics education* (3rd ed., pp. 543–562). New York, NY: Routledge.
Henriksen, D., Mishra, P., & Fisser, P. (2016). Infusing creativity and technology in 21st century education: A systemic view for change. *Educational Technology & Society, 19*(3), 27–37.
Herrington, A., & Herrington, J. (2007). What is an authentic learning environment? In L. A. Tomei (Ed.), *Online and distance learning: Concepts, methodologies, tools, and applications* (pp. 68–76). Hershey, PA: Information Science Reference.
Huleihil, M. (2017). 3D printing technology as innovative tool for math and geometry teaching applications. *IOP Conference Series: Materials Science and Engineering, 164.* Retrieved from http://iopscience.iop.org/article/10.1088/1757-899X/164/1/012023/pdf.
Istance, D., & Kools, M. (2013). OECD Work on Technology and Education: Innovative learning environments as an integrating framework. *European Journal of Education, 48,* 43–57.
Jones, K., & Simons, H. (1999). *Online mathematics enrichment: An independent external evaluation of the NRICH project 1998–99*. Southampton, England: University of Southampton.
Kaput, J. J. (1998). *Technology as a transformative force in education: What else is needed to make it work?* Paper for the NCTM 2000 Technology Working Group.
Kapu,T. J., Hegedus, S., & Lesh, R. (2007). Technology becoming infrastructural in mathematics education. In R. Lesh, E. Hamilton, & J. Kaput (Eds.), *Foundations for the future in mathematics education* (pp. 173–192). Mahwah, NJ: Lawrence Erlbaum.
Kaput, J. J., & Thompson, P. W. (1994). Technology in mathematics education research: The first 25 years in JRME. *Journal for Research in Mathematics Education, 2,* 676–684.
Kazakoff, E. R., Orkin, M., Bundschuh, K., & Schechter, R. L. (2017). Fostering engagement in educational technologies through developmental theory and program data. In R. D. Roscoe, S. D. Craig, & I. Douglas (Eds.), *End-user considerations in educational technology design* (pp. 99–122). Hershey, PA: IGI Publishing.
Koehler, M. J., & Mishra, P. (2015). TPACK (technological pedagogical content knowledge). In J. Spector (Ed.), *The SAGE encyclopedia of educational technology* (pp. 783–786). Thousand Oaks, CA: SAGE Publications, Inc. https://doi.org/10.4135/9781483346397.n318.
Koestler, A. (1981). The three domains of creativity. In D. Dutton & M. Krausz (Eds.), *The concept of creativity in science and art* (pp. 1–17). Dordrecht, The Netherland: Springer.
Kynigos, C., & Daskolia, M. (2014). Supporting creative design processes for the support of creative mathematical thinking—Capitalising on cultivating synergies between math education and environmental education. In *Proceedings of the 6th International Conference on Computer Supported Education* (CSEDU 2014) held in conjunction with SMARTGREENS 2014, CLOSER 2014 and WEBIST 2014, April 1–3, 2014, Barcelona, Spain. Paper #256. https://doi.org/10.5220/0004965603420347.
LeBlanc, M., & Freiman, V. (2011). Mathematical and didactical enrichment for pre-service teachers: Mentoring online problem solving in the CASMI project. *The Montana Mathematics Enthusiast, 8*(1–2), 291–318.
Lee, N. H., & Ferrucci, B. (2012). Enhancing learning of fraction through the use of virtual manipulatives. *The Electronic Journal of Mathematics & Technology, 6*(2), 126–140.

Leikin, R. (2007). Habits of mind associated with advanced mathematical thinking and solution spaces of mathematical tasks. In D. Pitta-Pantazi & G. Philippou (Eds.), *Proceedings of the Fifth Congress of the European Society for Research in Mathematics Education* (pp. 2330–2339). Larnaca, Cyprus: The authors.

Leikin, R., Levav-Waynberg, A., & Guberman, R. (2011). Employing multiple-solution-tasks for the development of mathematical creativity: Two comparative studies. In M. Pytlak, T. Rowland, & E. Swoboda (Eds.). *Proceedings of the Seventh Congress for European Research in Mathematics Education*, Rzeszow, Poland.

Leikin, R., & Sriraman, B. (Eds.). (2017a). *Creativity and giftedness, interdisciplinary perspectives from mathematics and beyond.* Cham, Switzerland: Springer International Publishing.

Leikin, R., & Sriraman, B. (2017b). Introduction to intersdisciplinary perspectives to mathematical creativity and giftedness. In R. Leikin & B. Sriraman (Eds.), *Creativity and giftedness: Interdisciplinary perspectives from mathematics and beyond* (pp. 1–6). Cham, Switzerland: Springer International Publishing.

Lombardi, M. (2007). Authentic learning for the 21st century: An overview. *EduCause Learning Initiative.* ELI Paper 1.

Lowrie, T. (2015). Digital games, mathematics, and visuospatial reasoning. In T. Lowrie & R. Jorgensen (Eds.), *Digital games and mathematics learning: Potential, promises and pitfalls* (Vol. 4, pp. 71–92). *Mathematics Education in the Digital Era Book Series, MEDE.* Dordrecht, The Netherland: Springer.

Mann, E. (2006). Creativity: The essence of mathematics. *Journal for the Education of the Gifted, 30*(2), 236–260.

Martinovic, D., Freiman, V., & Karadag, Z. (2013). Visual mathematics and cyberlearning in view of affordance and activity theories. In D. Martinovic, V. Freiman, & Z. Karadag (Eds.), *Visual mathematics and cyberlearning* (Vol. 1, pp. 209–238). *Mathematics Education in the Digital Era (MEDE) Book Series.* Dordrecht, The Netherland: Springer.

Mishra, P. (2012). Rethinking Technology & Creativity in the 21st Century: Crayons are the Future. *TechTrends, 56*(5), 13–16.

Mishra, P., Fahnoe, C., & Henriksen, D. (2013). Creativity, self-directed learning and the architecture of technology rich environments. *TechTrends, 57*(1), 10–13.

Mishra, P., & Koehler, M. J. (2006). Technological pedagogical content knowledge: A framework for teacher knowledge. *Teachers College Record, 180*(6), 1017–1054.

Niess, M. L. (2005). Preparing teachers to teach science and mathematics with technology: Developing a technology pedagogical content knowledge. *Teaching and Teacher Education, 21*(5), 509–523.

Niess, M. L., Ronau, R. N., Shafer, K. G., Driskell, S. O., Harper, S. R., Johnston, C., et al. (2009). Mathematics teacher TPACK standards and development model. *Contemporary Issues in Technology and Teacher Education, 9*(1), 4–24.

Panaoura, A., & Panaoura, G. (2014). Teachers' awareness of creativity in mathematical teaching and their practice. *Issues in the Undergraduate Mathematics Preparation of School Teachers, 4.* Retrieved from https://files.eric.ed.gov/fulltext/EJ1043048.pdf.

Papert, S. (2006). Afterward: After how comes what. In K. R. Sawyer (Ed.), *Cambridge handbook of the learning sciences* (pp. 581–586). Cambridge, UK: Cambridge University Press.

Pelczer, I., & Freiman, V. (2015). Assessing difficulty level in mathematical problem solving in informal online environments: Case of the CASMI community. In K. S. Sullenger & S. Turner (Eds.), *New ground. Pushing the boundaries of studying informal learning in science, mathematics, and technology* (pp. 129–152). Rotterdam, The Netherlands: Sense Publishers.

Piaget, J. (1981). Creativity: Moving force of society [Talk presented at 1972 Eisenhower Symposium, Johns Hopkins University, Baltimore, MD]. Appendix to J. M. Gallagher and D. K. Reid. *The learning theory of Piaget and Inhelder.* Monterrey, CA: Brooks/Cole.

Savard, A., & Freiman, V. (2016). Investigating complexity to assess student learning from a robotics-based task. *Digital experiences in mathematics education*, 2(2), 93–114.

Sengupta-Irving, T., & Agarwal, P. (2017). Conceptualizing perseverance in problem solving as collective enterprise. *Mathematical Thinking and Learning, 19*(2). https://doi.org/10.1080/10986065.2017.1295417.

Sinclair, N. (2006). *Mathematics and beauty: Aesthetic approaches to teaching children*. New York, NY: Teachers College Press.

Singer, F. M., Toader, F., & Voica, C. (Eds.). (2015). *The Proceedings of the 9th Mathematical Creativity and Giftedness International Conference*, Sinaia, Romania. http://mcg-9.net/pdfuri/MCG-9-Conference-proceedings.pdf

Singer, F. M., Sheffield, L. J., Freiman, V., & Brandl, M. (2016). *Research on and activities for mathematically gifted students*. ICME-13 Topical Survey. Dordrecht, The Netherland: Springer.

Sriraman, B. (2004). The characteristics of mathematical creativity. *The Mathematics Educator, 14* (1), 19–34.

Sriraman, B., & Dahl, B. (2009). On bringing interdisciplinary ideas to gifted education. In L. V. Shavinina (Ed.), *The international handbook of giftedness* (pp. 1235–1256). New York, NY: Springer.

Sriraman, B., & Lee, K. (Eds.). (2011). *The elements of creativity and giftedness in mathematics*. Rotterdam, The Netherlands: Sense Publishers.

Suh, J. M., Johnston, C. J., & Douds, J. (2008). Enhancing mathematical learning in a technology-rich environment. *Teaching Children Mathematics, 15*(4), 235–241.

Thompson, D. L. (2010). Beyond the classroom walls: Teachers' and students' perspectives on the key instructional practices that facilitate a successful online learning experience for gifted students. *Journal of Advanced Academics, 21*(4), 662–712.

Thompson, P. W. (1991). To experience is to conceptualize: Discussions of epistemology and experience. In L. P. Steffe (Ed.), *Epistemological foundations of mathematical experience* (pp. 260–281). New York, NY: Springer.

Thompson, P. W. (2015). Researching mathematical meanings for teaching. In L. D. English & D. Kirshner (Eds.), *Third handbook of international research in mathematics education* (pp. 435–461). London, UK: Taylor and Francis.

Wade, W. Y., Rasmussen, K. L., & Fox-Turnbull, W. (2013). Can technology be a transformative force in education? *Preventing School Failure, 57*(3), 162–170. Retrieved from https://doi.org/10.1080/1045988X.2013.795790.

Yerushalmy, M. (2009). Educational technology and curricular design: Promoting mathematical creativity for all students. In R. Leikin, A. Berman, & B. Koichu (Eds.), *Creativity in mathematics and the education of gifted students* (pp. 101–113). Rotterdam, The Netherlands: Sense Publishers.

Part II
Teaching Practices and Instructional Strategies to Inspire Authentic Creativity

Chapter 2
Screencasting as a Tool to Capture Moments of Authentic Creativity

Dana C. Cox, Suzanne R. Harper and Michael Todd Edwards

Abstract In the context of working with preservice secondary mathematics teachers (PSMTs) in a course on mathematical problem solving with technology, we tested the potential of technology to both inspire and capture moments of authentic creativity in the mathematics classroom. In a case study of two PSMTs working in partnership to solve a task using Interactive Geometry Software (IGS), we documented a rich narrative based on four episodes of creativity. These four episodes can be characterized as moments of creative insight because they represent moments that inspired either the development of new solutions or strategies (Problem Solving Insights) or spurred new questions (Problem Posing Insights). At the heart of the case is a task that requires constant negotiation and discussion in a digital workspace. Capturing an authentic narrative can be challenging with verbalized thinking alone, as the articulation of insight is not always possible. Screencasts are a tool that captures verbalized thinking as well as on-screen activity. This case study illustrates the power that this tool has in preserving the authenticity of those moments, but also in creating a record of practice to which both students and teachers might refer when making learning processes explicit.

Keywords Preservice teacher education · Problem solving · Interactive geometry software · Screencasting · Modeling

D. C. Cox (✉) · S. R. Harper · M. T. Edwards
Oxford, USA
e-mail: dana.cox@miamioh.edu

S. R. Harper
e-mail: harpersr@miamioh.edu

M. T. Edwards
e-mail: Edwardm2@MiamiOH.edu

© Springer International Publishing AG, part of Springer Nature 2018

V. Freiman and J. L. Tassell (eds.), *Creativity and Technology in Mathematics Education*, Mathematics Education in the Digital Era 10,
https://doi.org/10.1007/978-3-319-72381-5_2

2.1 Introduction

The themes of creativity, insight, struggle, risk, and ambition emerge continuously in a course we teach for preservice secondary mathematics teachers (PSMTs) on mathematical problem solving with technology. Bolden, Harries and Newton (2010) recommend explicit opportunities for preservice mathematics teachers to develop a sense of what creativity in mathematics means and what it looks like in the classroom as they may not have been given an opportunity to do so in their own education. This, in part, is the goal of our course as students encounter mathematics through the lens of technology such as GeoGebra, Geometer's Sketchpad, Desmos, Fathom 2L, and Gapminder in the process of solving novel mathematics problems and engaging in mathematical modeling. Classroom activity is intended to necessitate insight, provoke struggle, allow for creativity, and inspire PSMTs to pursue personal mathematical inquiry as they work.

In the act of problem solving, we have found that student ambition and creativity are often hampered by feelings of risk, as many are conditioned to value a produced solution over the actual process of building one. This works against the essence of mathematics (Dreyfus & Eisenberg, 1996; Ginsburg, 1996). In our experiences with enacting problem solving tasks with preservice teachers, the fear that a solution will not be found (and thus an assignment not completed) is so great at times, that it seems to impair PSMTs' willingness to ask and pursue ambitious mathematical questions, to create ambitious models for data or natural phenomena, or to move beyond the boundaries of what they already know into an area where creativity is imperative and where insight may be profound. This is not far removed from other research findings about the interactions between self-efficacy, anxiety, and problem solving performance (Hoffman, 2010).

At the core of this dilemma is the need to target beliefs about the nature of mathematics and what it means to do mathematics. In a previous study, we found that PSMTs draw heavily from their personal experiences to illustrate beliefs about the role technology plays in mathematics instruction (Cox & Harper, 2016). We concluded that in order to change beliefs, it was not enough to provide PSMTs with opportunities to experience mathematics from a new perspective, we must also help them become aware of their achievement and explicit about the meaning of classroom activity and learning; these experiences must be a part of a relatable story. We can project, then, that in order for PSMTs to understand the meaning of creativity in mathematics, we would have to create opportunities for them to experience it, but also reflect upon that experience.

In this study, we set out to create an environment where our PSMTs would experience moments of creative insight and test the capability of technology to both support creative insight and capture it for study. Capturing insight would serve two purposes. First, it would help us study those moments with the intent to better understand the genesis of creativity in the mathematics classroom. Second, as an artifact of the act of problem solving, such recordings could give PSMTs a personal and explicit sense of the meaning behind creativity in mathematics.

Capturing insight, we hypothesized, would require capturing mathematical activity as well as the intention behind that activity. This became a two-fold problem of study: (1) What does it look like when PSMTs engage creatively with a mathematical task, and (2) What advantages does technology give us when documenting PSMTs mathematical processes and intentions during the act of problem solving? In this chapter, we will give an account of what happened when we asked our PSMTs to create a model using Interactive Geometry Software, and then used screencasting as a medium through which to capture their insights.

2.2 Designing for Creativity in the Mathematics Classroom

To inspire creative action among PSMTs, teacher educators must build lessons around tasks that inspires such behavior. In this section we will define creativity in such a way that differentiates problem solving and problem posing as distinct, creative acts. Then we will describe the framework on which we built our creative task. We are using the term *creative task* to refer to tasks that foster creativity in mathematics.

2.2.1 *Defining Creativity*

Liljedahl and Sriraman (2006) define creativity in mathematics as both the process by which original or novel solutions are found for given problems and the generation of new problems or perspectives on existing ones. In this sense, it is in the act of solving problems and in posing new ones that students of mathematics find outlets for creativity (Silver, 1997). Wagner (1993), in writing about the purposes of educational research, puts forward two metaphors that further distinguish these acts. The act of problem solving is like filling in a *blank spot*, finding an answer to a problem that has already been posed or a question already asked. Creative insight is responsible for the miraculous turn from impossibility to solvability. On the other hand, we as scholars have *blind spots* which are "areas in which existing theories, methods, and perceptions actually keep us from seeing phenomena as clearly as we might" (Wagner, 1993, p.16). The act of problem posing is the result of asking a question that, prior to creative insight, we had not thought to ask. By posing the question, we become aware of our blind spots. In a complementary way, Torrance (1966) frames creativity as a process of becoming more aware of something we do not know (exposing a blind spot) and then searching for ways to fill that deficiency and communicating the results to others (filling in a blank spot). Thus, creativity in the mathematics classroom is directly observable through acts of problem solving and problem posing. In our study, we wanted to create opportunities for and capture both types of creative action.

2.2.2 Developing a Creative Task

Before we could capture moments of creative insight, we needed to develop a creative task. In the search for ways to inspire creativity in the mathematics classroom, Leikin and Pitta-Pantazi (2013) reviewed three different strategies from the literature. First, creativity can be developed through open-ended problems and by encouraging divergent thinking (Kwon, Park, & Park, 2006). Second, model-eliciting activity has some potential for developing creativity (Chamberlin & Moon, 2005). Third, there is potential in offering students non-routine, novel, or ill-defined problems (Chiu, 2009).

While considering these recommended strategies, we developed a framework for the design of creative tasks (Harper & Cox, 2017). A creative task should:

1. be framed as an open-ended problem that permits multiple strategies and solutions;
2. be ill-defined so that students are given opportunities to make decisions about where to put their focus, creating an environment that fosters diversity rather than conformity of thinking; and
3. enable the act of problem posing. By reformulating the task as a specific problem to be solved, solvers are able to imprint their own perceptions of a real-life phenomenon into the task (Silver, 1997).

The *Kaleidoscope Task*, shown in Fig. 2.1 is a problem solving scenario designed to encourage creativity in the mathematics classroom based on this framework. First, it is open-ended. This means that the task is not focused on a specific answer and has no expected strategy. Here, there is no idealized product or exemplar for which all are aiming and there are multiple productive paths to take. Second, the task is ill-defined since the direction to take depends on the facets of a kaleidoscope students wish to represent. There is room for decision making and there is potential for diverse thinking. Third, problem posing enters the process as students pose questions of both mathematical and technological possibility. These often sound like, "I wonder if we can get it to…" or "How might we use rotation here instead of reflection?" These perceptions provoke problem statements which motivate and require problem solving to answer.

This problem solving task was also developed around the context of a kaleidoscope because it is rich in terms of the mathematics that is accessible to PSMTs as they create their models. Within a Kaleidoscope, we encounter mathematical

The Kaleidoscope Task: With a partner, create one interactive geometry sketch that, for the two of you, embodies a "Kaleidoscope". You will have 20 minutes to create your sketch before demonstrating it to the group.

Fig. 2.1 The kaleidoscope task

topics such as polygons, symmetry, rigid transformations, dilations, congruence, random behavior, and even algebraic relationships (as found in the animation techniques available to make models dynamic). We realize that modeling a kaleidoscope is not a novel task, and in fact many mathematics educators have used a similar task with their mathematics students in varying degrees. Some concentrate on building a fully-functioning, physical kaleidoscope (e.g., Kaplan, Gross, & McComas, 2015); while others create static models and focus on patterns that are present in a single image created by a kaleidoscope (e.g., Graf & Hodgson, 1990). Other models are a *cookbook lesson* (Harper & Edwards, 2011) describing how to create a kaleidoscope using interactive geometry software with the mathematical focus on simulating random behavior, rotating about a center, and/or coordinating complex systems of animation (Wert, 2011). The task we have developed differs from each of these as the desired end-product is not a physical or static model; furthermore, the lesson does not include step-by-step technology directions for the PSMTs to create their kaleidoscope.

Interactive Geometry Software (IGS) is a versatile tool for PSMTs to use when modeling the behavior of a kaleidoscope. We perceive three specific roles that IGS plays in this task. First, the creation of geometric objects as well as access to geometric tools enables PSMTs to realize a mathematical model of a kaleidoscope more quickly and reliably than other methods (Graf & HoIGSon, 1990). More so, the ability to animate geometric objects enables a fuller range of models to be created, including some that capture the dynamic and randomly shifting qualities of kaleidoscopic images (Moreno-Aremella, Hegedus, & Kaput, 2008). Second, IGS provides feedback to the PSMTs and encourages them to design models iteratively, tweaking and adjusting features based on comparisons between their intended result and the results given by their model (Laborde, Kynigos, Hollebrands, & Strässer, 2006). Third, IGS acts as a collaborative work tool. As PSMTs present projected sketches to classmates, they share their work in an environment that enables in-the-moment tweaks and edits. As classmates make suggestions or offer conjectures about the way the sketch works, these can be considered and tested giving everyone immediate access to the results (Beatty & Geiger, 2010).

Having shared the task to optimize the potential to create opportunities for creative insight, we now turn our attention to delineating how we enacted the task in our classroom and how we used screencasting to document their work.

2.3 Methodology

This study was designed to better understand the genesis of creativity in the mathematics classroom by capturing moments of creative insight and our students' reactions to them. We utilized technology in both the design of the task and in data collection. This became a two-fold problem of study: (1) What does it look like when PSMTs engage creatively with a mathematical task, and (2) What advantages does technology give us when documenting PSMTs mathematical processes and intentions during the act of problem solving?

2.3.1 Setting and Participants

PSMTs pursuing licensure to teach secondary mathematics at our institution take a course, typically in their sophomore year, that focuses on mathematical problem solving with technology. PSMTs have taken Calculus I and Calculus II prior to enrolling in this course. Some may have taken, or are taking, other mathematics courses such as linear algebra or discrete mathematics, but these are not prerequisites. As our PSMTs enroll in the majority of their education courses in the junior year, those enrolled in our course have not taken any methods courses or completed any field placements in a clinical or school setting, nor are they expected to take such courses concurrently with our course.

We presented the *Kaleidoscope Task* to 15 PSMTs enrolled in the mathematical problem solving with technology course described above. The PSMTs were in the seventh week of exploration at the end of a unit on mathematical problem solving with IGS. PSMTs have had prior experiences constructing geometric objects such as points, lines and polygons; transforming geometric objects with reflections, rotations, translations and dilations; and using sliders as algebraic parameters as well as tools to animate mathematical objects. These are described as 'prior experiences' to evoke a sense of exploration and activity rather than direct instruction. Very rarely are either mathematical or technological demonstrations conducted in this course. Demonstrations are generally limited to those conducted by students presenting solutions to posed problems along with the techniques by which they were obtained.

2.3.2 Data Collection

We asked the PSMTs to work in partners at one computer. The intent was for students to have the environment to "engage in discussion mediated through a mathematical object that reflects changes in theories, which then allows for a back and forth of problem solving, theorizing, testing, and checking" (Beatty & Geiger, 2010, p. 272).

After reading the task and establishing partnerships, PSMTs were left largely on their own to work with instructors available for consultation. Following a 15-min construction period, each partnership introduced the class to their sketch. While projecting their work on a screen at the front of the room, they were invited to verbally address (a) how their sketch embodies a kaleidoscope, (b) mathematical or technological problems that emerged as they worked, (c) mathematical or technological insights they reached while working, and (d) limitations or struggles that prevented them from realizing their vision. Each partnership then entertained questions or comments from the class, with several facilitating whole-class problem solving.

Three sources of data were collected during the 80-min class period. For each partnership, we collected completed IGS sketches which were submitted electronically. Partnerships electronically submitted *creation screencasts* documenting all creative and on-screen construction activity—including all verbal communication—in the IGS environment. In this study, we define screencast as is a digital recording of computer screen output, also known as a *video screen capture*, often containing audio narration (Udell, 2005). Specifically, students used Screencast-O-Matic (http://screencast-o-matic.com/home), a freely available, web-based platform, to create their screencasts. The platform works passively in the background and does not interfere with IGS activity. Traditionally, the audio narration of a screencast is scripted. Such was not the case in this study as the audio narration consisted solely of spontaneous conversation between partners as they worked. As each partnership shared their sketches to the whole class, we video recorded group discussions.

An additional source of data was collected after the class period. Individual PSMTs created *reflective screencasts* (screen output with voiceover) where they addressed, from an individual standpoint, questions similar to those posed in the whole class discussion, but including additional prompts regarding the construction process: (1) the process by which they created their model, (2) the facets of a kaleidoscope they had hoped to model, (3) what they had hoped to accomplish but could not and the limitations they perceived, and (4) insights they had along the way and the impact of those insights on their model.

2.3.3 Data Analysis

For the purposes of this chapter, we focus our attention on developing a rich and detailed case study of one partnership: Abby and Olivia. Relevant to this case, we include the sketch submitted by the PSMTs, the creation screencast, a video of the presentation and subsequent discussion of their model to the class, and two reflection screencasts.

To develop the case study, we initially focused on the creation screencast. We transcribed the creation screencast, however it was inadequate for full data analysis. Abby and Olivia often moved into a shorthand method of speaking that relied on spoken words, gesture (such as pointing with the mouse cursor), technological activity (such as constructing a mathematical object), and technological events (such as the moment Olivia animates the kaleidoscope for the first time). Thus, all analysis was completed using the creation screencast that contained additional data useful for interpreting the form of gesture and technological activity and events.

Data from the IGS sketch, the recorded class discussion, and the reflective screencasts were used to interpret the creation video. The sketch was powerful because it enabled us to document different facets of a kaleidoscope that Abby and Olivia attended to, as well as the relationship of these facets to mathematical constructions within the created models. Because the sketch was data with which we could interact, it also enabled us to form and test our own conjectures about how the sketch worked and how closely it resembled the physical phenomenon of a kaleidoscope.

The reflective data collected during whole class discussion and reflective screencasts enabled us to listen to PSMTs describe first-hand the intentions and emotion behind their creative work. Furthermore, the data were useful in our analysis as we compared and contrasted the way problems were posed by the partnerships during the creation screencast with the way individual PSMTs articulated those problems after-the-fact.

2.4 The Case of Abby and Olivia

We present here the case of Abby and Olivia. Abby and Olivia are second-year PSMTs who worked together on this task by choice rather than assignment. They had previous experience working together in this course. Their case was selected over six others because it demonstrates the potential of technology to both facilitate and document moments of creative insight, but also for the variety of such moments that it provides for analysis. In this sense, their case is remarkable in its clarity.

We present the case chronologically in four parts, identifying and providing evidence of four distinct moments of creative insight: *The Symmetry Decision*, *The Turning Dilemma*, *A Tool For Turning*, and *A Return to Symmetry*. Data to support each section comes from creation screencasts, submitted IGS sketches, reflection screencasts, and a video of the presentation of the kaleidoscope to classmates. Additionally, Olivia made herself available for a brief interview after the course had concluded. In this interview she provided additional insights into the construction of their model. Olivia was then given an opportunity to engage in member check (Lincoln & Guba, 1985) to strengthen the validity of our interpretations.

2.4.1 The Symmetry Decision

The *Kaleidoscope Task* provides a quick entry for the PSMTs into the process of modeling. Abby and Olivia agreed almost immediately that symmetry and circles would play important roles in their model. After one minute of work, they negotiated a rough sketch that included a circle divided into six congruent sectors. A transcript of this work accompanies Fig. 2.2, a screenshot illustrating the construction process.

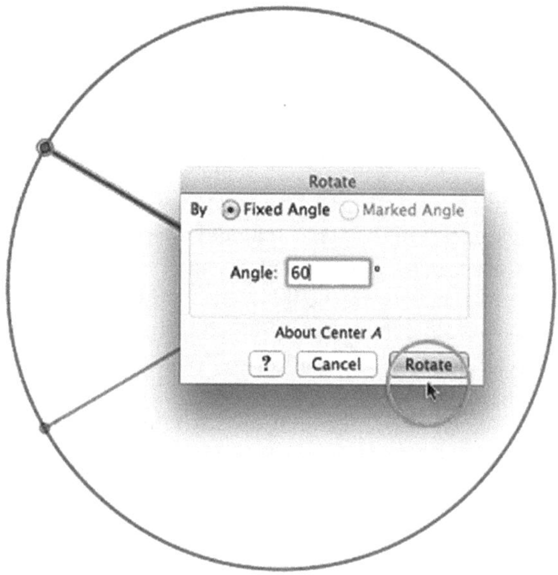

Fig. 2.2 Rotating radius segment 60 degrees about the center of the circle

(00:02) A: So we know that…it has symmetry. So do we just wanna like…
(00:05) O: It's usually circular, so do you want to make a circle?
(00:08) A: Yep. Ok.
(00:12) O: …and how do we want symmetry to work?
(00:19) A: Do we want to start with like, making it kind of like a pie?
(00:55) O: Ok.
(00:57) A: So now

Once the PSMTs had divided their circle into six congruent sectors by rotating a radius using 60 degree intervals (see Fig. 2.2), Olivia questioned their initial decision to rotate the segments (radii). Here, she recognized dissonance between the act of rotating an object within the model and her expectations about its eventual overall symmetry. In the middle of expressing her concern (01:06), she suspended her argument and acknowledged that, at this point, the segments, constructed initially through rotation, do not limit the model to rotational symmetry and can be repurposed in the future as lines of reflection. This moment of creative insight into the duality of these segments created a situation wherein the PSMTs could continue to design the model while suspending *The Symmetry Decision*.

(01:00) O: Should we have rotated or reflected so it's the same in all of them?
(01:05) A: What do you mean?
(01:06) O: Should we have it reflected around… like… instead of rotated cause otherwise we can make the reflection lines. We can make these reflection lines.
(01:16) A: Oh. I know what you mean. Ok.
(01:17) O: So, I guess we draw…
(01:19) A: Add a shape?
(01:22) O: That I guess stays on these lines?

(01:24) A: Yeah. That's so cool. Yeah. Do that.
(01:30) O: Here.
(01:31) A: Ok.
(01:32) O: So we can make some different shapes I guess and see what happens in a second

Following this excerpt, the PSMTs work on designing one sector of the circle. The image shown in Fig. 2.3 depicts their emerging kaleidoscopic model after some time has passed (02:16). The image includes three shapes created by Olivia with the polygon tool. In the presentation to the whole class, Abby describes on video, "And then from there we started with one of those triangles and we put like a polygon in there and, like, we made it however we wanted and then we did that a couple times but, like, different colors. We had three, purple, pink, and green." When she concludes, the instructor questions, "What do you mean by "we made it however we wanted?", and Abby clarifies, "So like, instead of just making it like a regular polygon, we like…manipulated it so that…went like ooooh…something like that" (gestures on the whiteboard, makes four imaginary points in a figure eight).

Thus, we know from the sketch, as well as Abby's summary of the process, that the shapes were each created in a haphazard way using points on the radii and circumference of the circle and that one shape is particularly indicative of a back-and-forth selection process where dots were placed on radii in an alternating pattern before the figure was closed. The triangle is formed using one point on each radii and a third point on the circumference.

It is at this moment that Olivia is ready to return to *The Symmetry Decision*. Instead of debating the mathematics in theory, she suggests a trial and error "let's see what happens" approach to save time. "Ok. So let's see this reflection thing

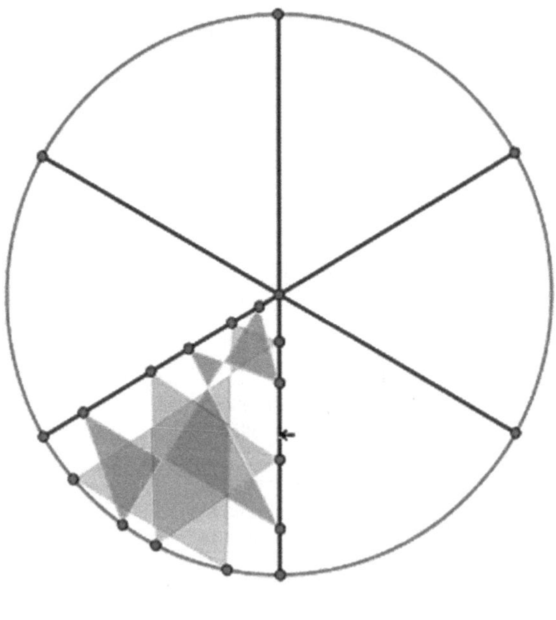

Fig. 2.3 Creating the initial sector of the kaleidoscope

before we get farther. So if I …let's just see what happens if I reflect this whole thing. It might not work, but it'd be quicker" (02:17).

The inarticulate transcript of the 38-second-long discussion that ran parallel to this design experiment does not do justice to the mathematical processes at work as the PSMTs systematically reflected the three polygons over each of five constructed radii. Even after the fact, Abby is inarticulate and imprecise when she describes the process to the class. For instance, she notes that, "We like highlighted that whole triangle and reflected it into the next… like the adjacent triangle." Later, in her reflection screencast, Olivia says, "And then what I did was I selected all the shapes within that circle. Those three. And I reflected them over each of these radiuses (sic). And that created so to have a mirror image" (Fig. 2.4).

Intermittent laughter, timed to specific events on screen, indicate that both PSMTs were aware of and engaged in the process of moving from one designed sector to a full model. The eventual solution, though quickly produced, did not take a linear path and required the PSMTs to attend to the following: precision, as they selected the points to be reflected; replication, as they identified and repeated a reflection procedure, and problem solving, as they encountered unexpected results and identified procedures that had not been adequately followed. The only full sentence comes at the completion of the static image occurs when Abby interjects, "Ta da, holy cow that's so cool" (02:55).

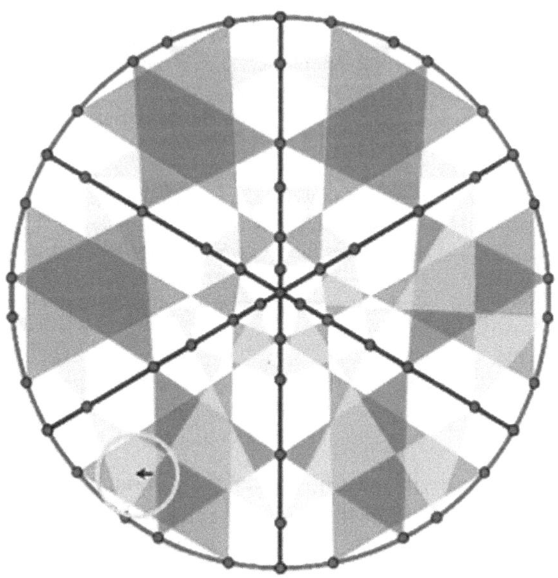

Fig. 2.4 Initial sector reflected over the radii of the circle to create a static kaleidoscope

2.4.2 The Turning Dilemma

Once Abby and Olivia created a first complete draft of their kaleidoscope sketch, they analyzed their model with a critical eye. With animation features of the sketch engaged, both PSMTs were drawn to the motion of two sets of points: those placed directly on the radii of the circle, which we will call radius points, and those placed on the circumference, which we call circumference points. The questions Abby and Olivia ask about those points and their motion are distinct and stem from the connections they make between objects in the dynamic sketch and physical facets of a kaleidoscope.

Abby questioned whether the radius points should be allowed to move. She based this question on previous experiences with kaleidoscopes, associating the circumference of the circle with the "tube" of a kaleidoscope. Since it is the tube that twists, she wondered if the motion should be limited to circumference points. Olivia also focused on circumference points, but instead questioned whether they should be allowed to move beyond the original sector. As such, Olivia focused on an entirely different facet of the construction: the mirrors. This conversation is captured in the creation screencast transcript beginning as the PSMTs contemplate their model.

(03:39) A: Wait. Is that technically considered a kaleidoscope by doing that…?
(03:45) O: Possibly….
(03:50) O: I would say yes.
(03:51) A: Think of a kaleidoscope when you turn it…
(03:56) O: It has to stay within the back lines.
(03:58) A: So wait. Should? … should only? …
(04:02) O: They are crossing over.
(04:03) A: Should only the points on the outside circle be moving? Do you know what I mean?
(04:10) O: Yeah. I'm also wondering if they are allowed to move between the segments. If they are allowed to move beyond just…let's stop where it looks like.
(04:22) [O stops the animation at a strategic, intentional point and gestures to a point as she asks a question. A screenshot of this moment is shown in Fig. 2.5]
(04:23) O: I'm wondering if they are allowed to move beyond, go around the circle past the sector. It's really a mirror so they have to stay within it. Um. It looks cool.
(04:33) A: It does.
(04:34) O: Um.
(04:36) A: Here, maybe let's move some of these points down so we can see better.
(04:51) O: This is, like, mind boggling. I wonder if, like…I wonder if I only want these points to move on this triangle. Like, within the segment.

After Olivia and Abby asked their initial questions, both acknowledged that they were pleased with the way the animated model looked. There was a tacit understanding that the questions they asked should not be taken as reasons to reject the model, but rather a means for *refine* it. Abby's move (04:36) to isolate the radius points from those on the circumference indicated that she, like Olivia, was willing to examine this subset of the model critically.

Here, the transcript becomes inadequate as a means for isolating a moment of Abby's creative insight.

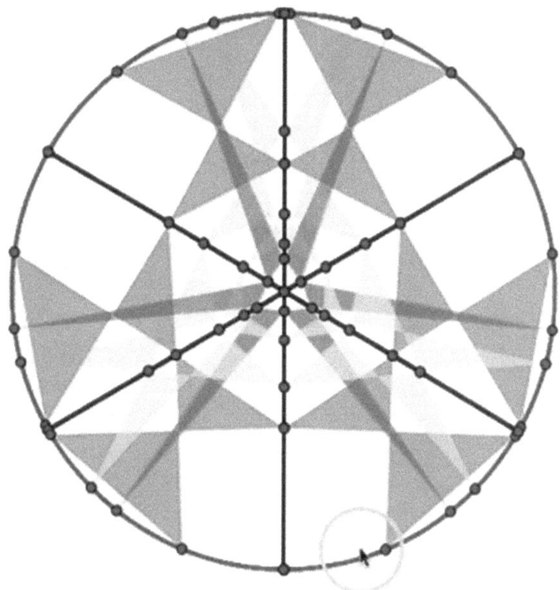

Fig. 2.5 Olivia stops the animation to focus on the behavior of the circumference points

(05:04) A: I see what you mean. I don't know… Can we move that point out?
(05:14) O: Which one? That worked! It did what you wanted it to.
(05:21) A: Wait, can you move that again? I just want to see if that's what I'm talking about. Yeah!! That's what we want. We want it to rotate like that.
(05:31) O: Is that how a kaleidoscope works?
(05:35) A: Yeah, because when you look in a kaleidoscope, what you do is you turn it…
(05:44) A: (continues) I wonder if there's a way to, like, watch a kaleidoscope in action

Abby expressed a new perspective on the model when she noted that, "When you look in a kaleidoscope, what you do is turn it" (05:35). We use the screen output to help interpret why this moment was important and why we consider it a moment of creative insight. In our review of the screencast, Abby asks Olivia to move a point. When Olivia responds, "which one?", rather than try to explain which of many points she wants moved, it seems as if Abby takes over the mouse herself and targets a point on the original circle shown in Fig. 2.6. This point is (1) is the unique point that was constructed to determine the original size of the circle and (2) is the endpoint of a radius used as a line of reflection, the radius that Olivia associates with a mirror (04:23). It is a very specific point and when it is dragged, the circle changes size and the objects within the circle move, creating an illusion of rotation about the center of the circle. This visual effect provokes Abby to think about the model from a new perspective and prompts within her a desire to convey to Olivia a new problem: How can we "turn" the model? We refer to this moment as *The Turning Dilemma*.

Olivia was still unable to rectify images within the animated sketch with Abby's statement "*what you do is turn it.*" Hence, Abby suggested they do a Google search to find a better way to convey this insight to Olivia. While search results were

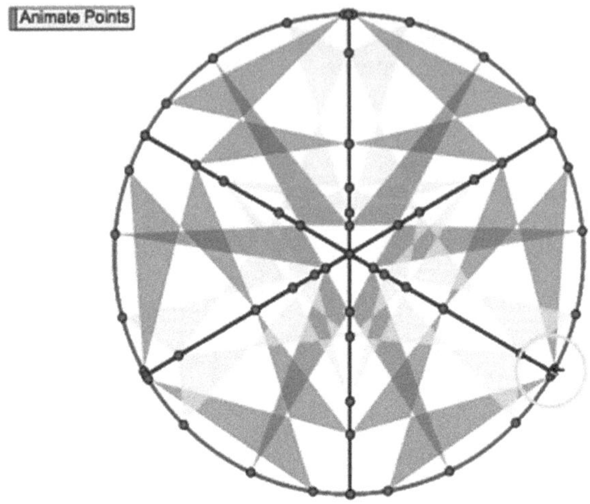

Fig. 2.6 Abby selects a specific point

produced on screen, the results of the searches weren't pursued. However, images associated with those results led to a desirable outcome for Abby. As Olivia viewed the images in Google, she noted, "I see what you mean, you rotate the entire thing," (06:48). At this instant, the PSMTs implicitly posed a problem to be solved: namely, *what technological processes are desired and available to "turn" a kaleidoscopic model?* Abby and Olivia each reflected on *The Turning Dilemma* in their reflection screencasts:

> **Olivia:** One of the things we were trying to accomplish is maybe to have the entire kaleidoscope rotate in a circle not just have the points rotating around the kaleidoscope, but we couldn't figure out a way to do that without having the shape…with the shapes because we couldn't get them to stay fixed.
>
> **Abby:** I think our goal was that we knew we had to have symmetry and another thing was that we wanted to make sure that the points on…er…we wanted to make sure there was rotation. Olivia and I both remember that when we were younger we would have little kaleidoscopes made out of paper towel tubes. When you would rotate them you'd see new shapes. We thought that was so cool. We wanted to make sure that was included.

2.4.3 A Tool for Turning

The third moment of creative insight we captured occurred while the PSMTs were solving *The Turning Dilemma*. Olivia had a hidden moment of insight (06:58) that we refer to as *A Tool for Turning*. The actual moment at which Olivia realized that a key to solving *The Turning Dilemma* was the creation of an external tool apart from their model is mysterious and undocumented. She did not fully express her insight nor did she articulate her plan to Abby. Olivia conveyed her insight in an almost entirely visual way. Whereas the *The Turning Dilemma* was initiated by Abby's direct request for on-screen activity, *A Tool for Turning* is produced by Olivia's intentional and yet unarticulated on-screen actions.

The tool created by Olivia to model a turning motion was a larger circle constructed using a new point as the center and an existing point on the circumference and radius of the original circle (see Fig. 2.7).

The original circle seemed to be on the interior of the larger circle. Olivia articulated her intentions after-the-fact while reflecting on the model:

> The circle was just a mechanism. I wanted the kaleidoscope to rotate around the inside of the larger circle. I wanted the smaller circle to always remain tangent to the larger circle at that point where they were connected, and it would always be on the interior and go around and around. The point would just travel along the circumference of the larger circle.

Figure 2.7 and the transcript that follows depict events leading to the construction. Abby expresses "Oooh" in such a way that invokes (to us, the observer) delight, surprise, and happiness. It is at this "Oooh" moment that Abby had her own creative insight and seemed to make the same realization as Olivia—that an external tool might be helpful in resolving the *Turning Dilemma.*

(06:58) O: Could we put this circle…wait a second. So. If I …I hate finding that point…
(07:07) A: I wonder if you can animate just like one of the points on the …
(07:11) O: I want to make a tinier circle. How do I do that? [O shrinks the kaleidoscopic model.] There. So what if I…What would happen…
(07:25) A: If you just animate … [O connects the larger circle to the indicated point in Fig. 2.7.]
(07:27) A: Ooooh!

Animating the tangent point created erratic behavior for the circles. The animated point was an independent object on which the circles have been built. As it was independent of the circles, it moved freely in the plane rather than along either

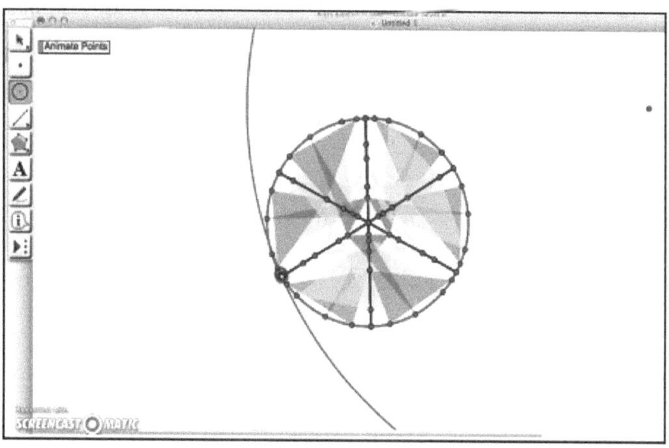

Fig. 2.7 Turning the kaleidoscope tube

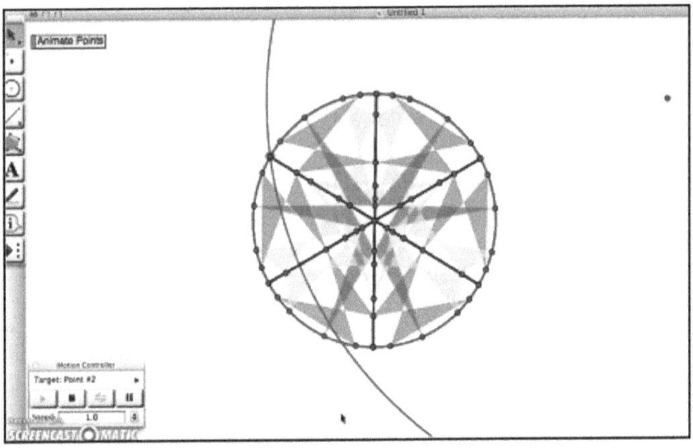

Fig. 2.8 First attempt to "turn the kaleidoscope"

circumference and brings the circles along with it. That caused the circles to change position as well as size, and also destroyed the illusion of tangency between the circles (see Fig. 2.8).

(07:37) A: I'm confused.
(07:37) O: Oooooooh. It's changing the length this thing in the circle, too.
(07:41) O: Do you see what I was trying to do though?
(07:42) A: I see what you mean, yeah

In her reflection screencast, Abby summarized, "And so, that was really frustrating…we were like awww! It's not what we want. We were confused as to what we needed to do to fix it." Even though the tool did not work as intended, its creation had the consequence of conveying insight and establishing a trajectory for the partnership for subsequent problem solving. The remaining time was spent in careful experimentation and investigation of alternative rotating mechanisms.

2.4.4 A Return to Symmetry

In the creation screencast around the 3:30 min mark, Abby and Olivia observe their kaleidoscope model after they have animated the radius and circumference points for the first time. While watching all of the points and shapes move in the interior of the circle, Olivia poses a question about the model. Specifically, she asks,

(04:10) O: I'm also wondering if they are allowed to move between the segments. If they are allowed to move beyond just…let's stop where it looks like.
[O stops the animation at a strategic, intentional point and gestures to a point as she asks a question. A screenshot of this moment is shown in Fig. 2.5]
(04:23) O: I'm wondering if they are allowed to move beyond, go around the circle past the sector. It's really a mirror so they have to stay within it.

Olivia's question goes unanswered as the PSMTs begin to focus on *The Turning Dilemma*.

After watching the creation screencast, as researchers, we had a moment of creative insight and became curious about Olivia's question. In fact, the creative insight was captured in our researcher notes while transcribing the creation screencast (see Fig. 2.9).

This creative insight produced two subsequent questions: (1) what was Olivia "seeing" in the motion of the model for her to question whether or not the model was valid? Specifically, what points or regions prompted her to ask if "they are allowed to move between the segments;" and, (2) could we use their original IGS sketch to determine how the points were moving and test the mathematical validity of their kaleidoscope model?

Since we had access to Abby and Olivia's original IGS file, we were able to conduct a "thought experiment" to find out whether the circumference points were really "moving beyond" the 60-degree arc. This helped us to better understand what Olivia was seeing and describing in the creation screencast. We also wanted to determine whether their kaleidoscope construction was mathematically valid and to know how the motion was limited in ways by the IGS that matched (or failed to match) our physical expectations of a kaleidoscope.

We gained insight by labeling points that were not originally labeled in the students' sketch and observing the movement of the animated points. We were able to identify one complete set of corresponding points (namely C, C', … C''''') on the circle (see Fig. 2.10).

Furthermore, we noticed that points C, C'' and C'''' traveled around the circle in a counterclockwise manner; whereas points C', C''', and C''''' traveled around the circle in a clockwise manner. When the points are not labeled, this movement is consistent with a kaleidoscope with a point along the circle contained along a 60-degree arc and image points reflected over the radii of the circle. However, when the points are labeled, it is clear that the points along the circle have not been reflected; as such, this is not a valid mathematical model for a kaleidoscope. The final (unlabeled) animation seems to be visually correct, however, it does not use a valid mathematical construction to model a "real" kaleidoscope.

In a brief interview after the course had concluded and during our data analysis phase, Olivia provided additional insights into the construction of their model. She proudly mentioned that she had recreated a new kaleidoscope, one where the "shapes do not go outside the pie." Considering this comment carefully, we are now

> The part where Olivia questions "moving beyond" is really interesting. It seems as if the girls are watching the animated points and imagining that they move from one sector into another. In reality, it's an optical illusion and the animated points, if labeled, would clearly just move within the sector ...Wait? Do the points on the outside of the circle limit themselves to one sector, or do they move out? Now I need to check it! … This is really worthy of some more conversation in Author 2's office. (Author 1 Research Notes)

Fig. 2.9 Researcher's notes while transcribing Abby and Olivia's creation screencast

Fig. 2.10 Corresponding points labeled in Abby and Olivia's kaleidoscope

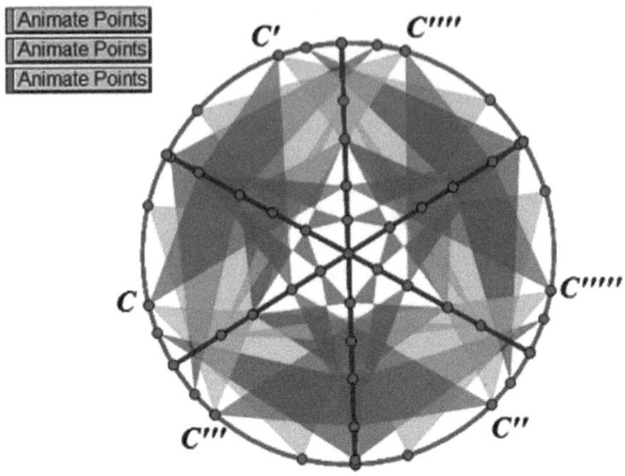

confident that Olivia was concentrating on the polygons that spanned multiple sectors of the circle when she asked "if they are allowed to move beyond, go around the circle past the sector." We went back to Abby and Olivia's original creation screencast and captured an image of the initial polygons spanning multiple sectors (see Fig. 2.11). Since there were no constraints for the circumference points, these points could move "beyond" the 60-degree arc of the circle, yielding polygonal images in the dynamic model that are never seen in a real kaleidoscope.

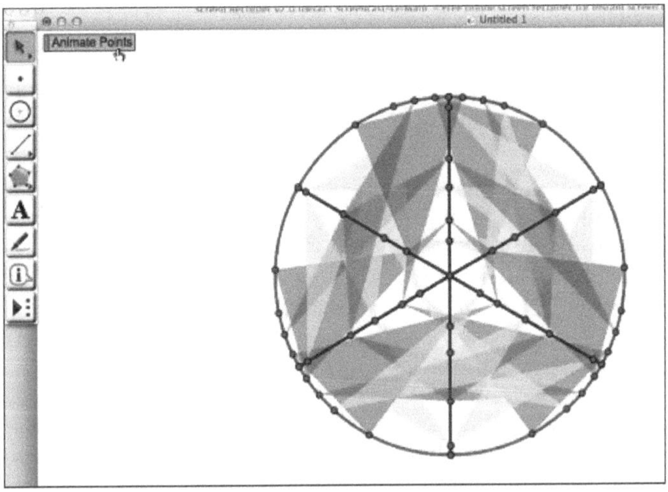

Fig. 2.11 Notice in the kaleidoscope that polygons span multiple sectors of the circle

2.5 Discussion

In the previous section, we presented four different moments of creative insight. First, *The Symmetry Decision* was presented as a moment in which PSMTs were poised to make a decision between creating a design based on rotational or reflectional symmetry. Second, *The Turning Dilemma* is a moment where a new perspective provokes the PSMTs to reconsider their model and ask how they might model the turning action required to operate a kaleidoscope. Third, *A Tool for Turning* is a moment where Olivia first incorporates an external tool that exists outside of the model and whose purpose is to take action upon the model rather than to exist within. Fourth, *A Return to Symmetry*, is a moment wherein we, the researchers, are inspired to engage in a post hoc examination of the model and inquire after its mathematical validity, and after Olivia's unacknowledged question about the sanctity of sectors.

These four moments are examples of creative insight because they represent moments that inspired either the elimination of blank spots through the development of new solutions or strategies (*The Symmetry Decision, A Tool for Turning*) or the exposure of blind spots which spurred new questions (*The Turning Dilemma, A Return to Symmetry*).

In the following sections, we return to our original questions: (1) What does it look like when PSMTs engage creatively with a mathematical task, and (2) What advantages does technology give us when documenting PSMTs mathematical processes and intentions during the act of problem solving? We found that the Kaleidoscope Task did provoke moments of creative insight leading up to episodes of problem solving and problem posing. The difference between these two types of insights are important and the role(s) that technology played in those moments varied with respect to this difference. Here, we parse our discussion to reflect this important distinction.

2.5.1 Problem Solving Insights

We characterize those insights that lead to the elimination of blank spots as *Problem Solving Insights*, new perspectives on existing problems that either enable a solution or illuminate a new approach or strategy. In this paper, we have described two such moments: *The Symmetry Decision* and *A Tool for Turning*.

In *The Symmetry Decision*, Olivia recognizes the flexibility of their model based on a circle divided into an even number (6) of sectors. The insight is first conveyed to Abby as a move to temporarily suspend the decision. Later, the insight is transformed into an experimental strategy where the PSMTs decide to use reflectional symmetry tools in the IGS and examine their results in the sketch. In *A Tool for Turning*, Olivia creates an external tool to act upon their model. By using IGS to create one possible tool, she is able to convey that insight to Abby without verbal articulation. While the tool Olivia initially creates is flawed, the strategy persists and the partners go on to build and test additional animation tools and processes.

In both of these episodes, technology plays a key role in both the inception of creativity and in its capture. First, suspending *The Symmetry Decision* and moving forward as if the radii could operate under either rotational or reflectional symmetry was a decision that could have been far more complex to make without the aid of the IGS environment. The technology enabled the students to take a design approach and make the decision to use reflectional symmetry without making a formal commitment to the mathematics. What we mean by this is that they understood that once they had conducted an experiment, if the result did not match their perceptions/concept images (Tall & Vinner, 1981) of a kaleidoscope, they could revise their work (Laborde, Kynigos, Hollebrands, & Strässer, 2006). Second, in *A Tool for Turning*, a moment of creative insight is provoked by unarticulated action taken within the IGS environment. The lack of articulation renders the screencast essential in documenting the nature of this insight and its provocation. Furthermore, it is only through watching the screencast that a full analysis of either moment is possible, the transcript leaving far too much to be interpreted.

2.5.2 Problem Posing Insights

In contrast to Problem Solving Insights, we characterize *Problem Posing Insights* as those that expose the existence and nature of blind spots and yield new questions that had previously gone unasked. In this paper we have described two such moments: *The Turning Dilemma* and *A Return to Symmetry*.

In *The Turning Dilemma*, a creative insight provoked the PSMTs to reconsider their model from a new perspective as they attempted to incorporate an additional facet of a physical kaleidoscope—the turning of the tube. The resulting question, *how to turn the model*, is only articulated after the creative insight and the exposure of a blind spot. In *A Return to Symmetry*, the insight occurs for us, the research team, as we reconsider our assumptions about the model as well as our initial interpretations of Olivia's concerns. Once we became aware of our assumptions, we were able to articulate new questions about the motion of the circumference points and also about the geometric objects to which Olivia was attending.

Technology continues to play a central role in the inception and creation of these Problem Posing Insights. In *The Turning Dilemma*, the initial moment of insight occurred as a result of on-screen action, causing Abby to see additional potential in the model. Olivia's haphazard dragging of one key point elicited motion that Abby felt would make their model behave in a more realistic way. Through a Google search, Abby finds a way to convey her insight to Olivia. Although no links were followed, Olivia is able to ascertain Abby's meaning by glimpsing the provided visual images that featured cardboard tubes and advertised DIY instructions for school-aged children.

In *A Return to Symmetry*, it is only because of the IGS artifact, the sketch, that a post hoc analysis of the mathematical model was possible. Furthermore, the moment of insight was provoked by the necessity of interpreting the screencast, a digital recording of activity within the IGS environment that was, in turn, necessary to interpret the recorded dialogue. It is a new perspective on that screencast and the underlying construction that provokes this moment of insight, and a return to the original digital environment to investigate the newly-posed questions.

2.6 Concluding Thoughts

In the following sections, we will conclude by discussing implications for teacher preparation, limitations of the study, and directions for further research.

2.6.1 Implications for Teacher Preparation

In the United States, there are two standards documents that influence the opportunities students have to problem solve, specifically, *Principles and Standards for School Mathematics* (National Council of Teachers of Mathematics, 2000) and the *Common Core State Standards for Mathematics* (National Governors Association Center for Best Practices, Council of Chief State School Officers, 2010). These documents have framed a professional conversation around the role of problem solving and problem posing in mathematics education, but more needs to be done to define what experiences PSMTs need with respect to each. Given the increased attention placed on modeling as a mathematical practice (National Governors Association Center for Best Practices, Council of Chief State School Officers, 2010), it is valuable to make it explicit that these are creative pursuits. More can be done to capture episodes of creativity in classroom observations to use as examples for the field and cases for use in the preparation of future teachers.

Beyond examples and research studies, we think that screencasting and IGS create an environment where PSMTs can not only engage in creative mathematics, but can create artifacts and records of practice such as creation screencasts and IGS files for use in presenting their work, but also for individual or group reflection focused on identifying and describing moments of creative insight. Identifying moments of creative insight is not intuitive or natural work and helping PSMTs reconstruct their thinking process is fraught. In this case, both PSMTs utilized the sketch created during class as evidence in their reflection, however neither Abby nor Olivia identified any insights that they had when solving the Kaleidoscope Task in their reflection casts. While Abby ignored the prompt entirely, Olivia went so far as to deny that insight occurred, which stands in contradiction to the data presented here from the creation screencast. What power we give to our PSMTs when we enable them to go back in time and watch the insight unfold.

By making students aware of their own creative insights, perhaps we could push their expertise further into the realm Lewis (2006) describes as aesthetic.

> Pre-service teacher education programs in technology education ordinarily do not include coursework on creativity. Thus, most teachers do not have preparation that is sufficient enough to allow them to inject creativity into their teaching. Teachers may introduce design/problem solving activities into their teaching, but the competence they bring to the classroom is more in the realm of the technical than the aesthetic. There is a clear need here for professional development activities aimed at helping teachers see possibilities for introducing creative elements into the curriculum, and into instruction (p. 47).

We would go even further, however, and say that making creativity explicit is the only way to change deeply-held beliefs (Philipp, 2007) about the nature of mathematics and instruction. Tharp, Fitzsimmons and Ayers (1997) found that that philosophical beliefs about the nature of mathematics and what it means to learn mathematics influenced teachers recognition of the value of technology for instruction. We feel that it would be more powerful to utilizing technology to examine creative insight, thus having the potential to impact not only views about the role of technology in the classroom, but also the potential of technology to spur and support creative insight for students.

2.6.2 Limitations

Although the research has documented creative insights as PSMTs engage in authentic problem solving, there were some unavoidable limitations to this study. First, due to the research design and questions, the results of the study are not indicative of the work of all PSMTs. It is difficult, if not impossible, to draw conclusions beyond Abby and Olivia's case. Second, because of the specific screencast software used, the PSMTs were only given 15 min to create their kaleidoscope model, which may have impeded the PSMTs creative insights or problem solving abilities. The existence of a time limit may have impacted some of the PSMTs' abilities to construct their kaleidoscope models. An argument can be made in the opposite direction, however, that the shortened time period may have also helped to alleviate some anxiety about the final product. If PSMTs interpret the shortened time frame as preventing them from achieving a polished, "finished" sketch, then they may work more freely, uninhibited by assumptions about assessment. Finally, the data collected in the individual PSMTs screencasts were limited. Not all students addressed each of the reflection questions; hence, some interpretation remains centered around a single data source.

2.6.3 Directions for Further Research

The creation screencast is as valuable in research as it is in the mathematics classroom and in the preparation of PSMTs. The screencast, operating in a completely transparent and backgrounded way, is a tool used to capture authentic moments of problem solving and problem posing. This is particularly true when used in the environment described here and by Beatty and Geiger (2010). The collaborative context worked in conjunction with a task that required constant negotiation and discussion to create a rich narrative captured in verbalized thinking as well as on-screen activity. Without access to the creation screencast, we would have to rely on Abby and Olivia's memory and interpretation of problem solving events after-the-fact, and creative insights may have gone uncaptured simply because they were so subtle as to fail even to register as important.

We see a great deal of possibility in pursuing research projects that utilize the screencast to document episodes of collaborative problem solving and posing in a IGS environment. In this chapter, we have presented one case study that illustrates the power that this tool has in preserving the authenticity of those moments, but also in creating a record of practice to which both students and teachers might refer when making learning processes explicit.

We are currently analyzing the ways PSMTs articulate mathematics differently in *creation screencasts*, *presentation videos,* and *reflection screencasts.* It is possible that these shifts in the precision and validity of mathematical language might indicate the possibility of cognitive gains from the act of personal reflection about mathematical problem solving. More data will be required to explore this conjecture.

It is exciting to imagine the possibility of using yet another layer of screencasting to capture the reflective process, particularly if it were to be shared between teachers and students. What would happen if teachers and students sat down together to verbally annotate a creation screencast? For us, this process is reminiscent of the *director's cut* videos that are often published alongside full length feature films on DVD wherein movie directors watch the final version of the film and provide a stream-of-consciousness discussion of the film and its creation. What sort of metacognitive insights might we be able to capture in these annotations? The impact of those metacognitive insights on content knowledge and pedagogical beliefs may be powerful and welcome in the field of mathematics education.

References

Beatty, R., & Geiger, V. (2010). Technology, communication, and collaboration: Re-thinking communities of inquiry, learning and practice. In C. Hoyles & J.-B. Lagrange (Eds.), *Mathematics education and technology-rethinking the Terrain: The 17th ICMI study* (pp. 251–284). New York, NY: Springer.

Bolden, D. S., Harries, T., & Newton, D. P. (2010). Pre-service primary teachers' conceptions of creativity in mathematics. *Educational Studies in Mathematics, 73*(2), 143–157.

Chamberlin, S. A., & Moon, S. M. (2005). Model-eliciting activities as a tool to develop and identify creatively gifted mathematicians. *Journal of Advanced Academics, 17*(1), 37–47.

Chiu, M.-S. (2009). Approaches to the teaching of creative and non-creative mathematical problems. *International Journal of Science and Mathematics Education, 7*(1), 55–79.

Cox, D. C., & Harper, S. R. (2016). Documenting a developing vision of teaching mathematics with technology. In M. L. Niess, S. Driskell, & K. Hollebrands (Eds.), *Handbook of research on transforming mathematics teacher education in the digital age* (pp. 166–189). Hershey, PA: IGI Global.

Dreyfus, T., & Eisenberg, T. (1996). On different facets of mathematical thinking. In R. J. Sternberg & T. Ben-Zeev (Eds.), *The nature of mathematical thinking* (pp. 253–284). Hillsdale, NJ: Lawrence Erlbaum Associates.

Ginsburg, H. P. (1996). Toby's math. In R. J. Sternberg & T. Ben-Zeev (Eds.), *The nature of mathematical thinking* (pp. 175–282). Hillsdale, NJ: Lawrence Erlbaum Associates.

Graf, K.-D., & Hodgson, B. R. (1990). Popularizing geometrical concepts: The case of the kaleidoscope. *For the Learning of Mathematics, 10*(3), 42–49.

Harper, S. R., & Cox, D. C. (2017). Quickfire challenges to inspire problem solving. *Mathematics Teacher, 110*(9), 686–692.

Harper, S. R., & Edwards, M. T. (2011). A new recipe: No more cookbook lessons. *Mathematics Teacher, 105*(3), 180–188.

Hoffman, B. (2010). "I think I can, but I'm afraid to try": The role of self-efficacy beliefs and mathematics anxiety in mathematics problem-solving efficiency. *Learning and Individual Differences, 20*(3), 276–283.

Kaplan, G., Gross, R., & McComas, K. K. (1996). Mathematics through the lens of a kaleidoscope: A student centered approach to building bridges between mathematics and art. In K. Delp, C. S. Kaplan, D. McKenna, & R. Sarhangi (Eds.), *Proceedings of Bridges 2015: Mathematics, Music, Art, Architecture, Culture* (pp. 573–580). Phoenix, AZ: Tessellations Publishing.

Kwon, O. N., Park, J. S., & Park, J. H. (2006). Cultivating divergent thinking in mathematics through an open-ended approach. *Asia Pacific Education Review, 7*(1), 51–61.

Laborde, C., Kynigos, C., Hollebrands, K., & Strässer, R. (2006). Teaching and learning geometry with technology. In A. Gutierrez & P. Boero (Eds.), *Handbook of research on the psychology of mathematics education: Past, present and future* (pp. 275–304). Rotterdam, The Netherlands: Sense Publishers.

Leikin, R., & Pitta-Pantazi, D. (2013). Creativity and mathematics education: The state of the art. *ZDM, 45*(2), 159–166.

Lewis, T. (2006). Creativity: A framework for the design/problem solving discourse in technology education. *Journal of Technology Education, 17*(1), 36–53.

Liljedahl, P., & Sriraman, B. (2006). Musings on mathematical creativity. *For the Learning of Mathematics, 26*(1), 17–19.

Lincoln, Y. S., & Guba, E. G. (1985). *Naturalistic inquiry*. Newbury Park, CA: Sage Publications.

Moreno-Armella, L., Hegedus, S. J., & Kaput, J. J. (2008). From static to dynamic mathematics: Historical and representational perspectives. *Educational Studies in Mathematics, 68*(2), 99–111.

National Council of Teachers of Mathematics. (2000). *Principles and standards for school mathematics*. Reston, VA: Author.

National Governors Association Center for Best Practices, and Council of Chief State School Officers. (2010). *Common core state standards for mathematics*. Washington, DC: Authors.

Philipp, R. A. (2007). Mathematics teachers' beliefs and affect. In F. K. Lester (Ed.), *Second handbook of research on mathematics teaching and learning* (pp. 257–315). Reston, VA: National Council of Teachers of Mathematics.

Silver, E. A. (1997). Fostering creativity through instruction rich in mathematical problem solving and problem posing. *ZDM Mathematics Education, 29*(3), 75–80.

Tall, D., & Vinner, S. (1981). Concept image and concept definition in mathematics with particular reference to limits and continuity. *Educational Studies in Mathematics, 12*(2), 151–169.
Tharp, M. L., Fitzsimmons, J. A., & Ayers, R. L. B. (1997). Negotiating a technological shift: Teacher perceptions of the implementation of graphing calculators. *Journal of Computers in Mathematics and Science Teaching, 16*(4), 551–575.
Torrance, E. P. (1966). *Torrance test on creative thinking: Norms-technical manual, Research Edition*. Lexington, MA: Personal Press.
Udell, J. (2005, November 16). *What is screencasting? O'Reilly Media archive*. Retrieved from http://archive.oreilly.com/pub/a/oreilly/digitalmedia/2005/11/16/what-is-screencasting.html.
Wagner, J. (1993). Ignorance in educational research or, how can you not know that? *Educational Researcher, 22*(5), 15–23. https://doi.org/10.3102/0013189X022005015.
Wert, T. (2011). *Creating a kaleidoscope with geometer's sketchpad*. [Video file]. Retrieved from https://www.youtube.com/watch?v=Wgxyd0Iveu4.

Chapter 3
The Create Excellence Framework's Impact on Enhancing Creativity: Examining Elementary Teacher Candidate Mathematics Lesson Planning

Janet Lynne Tassell, Rebecca Stobaugh and Marge Maxwell

Abstract The focus of this research is to examine the impact of an instructional instrument to improve the quality of pre-service teachers' lesson plans to enhance creative learning opportunities for children. The Create Excellence Framework focuses on four components essential to high-quality lesson plans: Cognitive Complexity, Real-World Learning, Engagement, and Technology Integration. The research study examined data from two elementary education teacher candidate classes for five semesters to measure the impact of the instrument on instructional planning for mathematics or mathematics and science integration. Over the course of the five semesters, for each component, the mean scores increased, and there was a positive statistically significant difference between the scores from the baseline semester to the fifth semester. In the fifth semester, the component of Technology Integration had the largest increase and Real-World Learning has the highest mean score. As students learned to design instruction around authentic tasks, cognitive levels and engagement also increased. Students were exposed to and utilizing new digital tools to enhance their learning. Using these digital tools along with real-world applications of the content encouraged students to think creatively to solve authentic problems.

Keywords Lesson planning · Real-world learning · Creativity
Technology integration · Math instruction

J. L. Tassell (✉) · R. Stobaugh · M. Maxwell
Western Kentucky University, Bowling Green, KY, USA
e-mail: janet.tassell@wku.edu

R. Stobaugh
e-mail: rebecca.stobaugh@wku.edu

M. Maxwell
e-mail: marge.maxwell@wku.edu

© Springer International Publishing AG, part of Springer Nature 2018
V. Freiman and J. L. Tassell (eds.), *Creativity and Technology in Mathematics Education*, Mathematics Education in the Digital Era 10,
https://doi.org/10.1007/978-3-319-72381-5_3

3.1 Introduction

Consider this scenario in a typical United States intermediate classroom:

> Allison walks into her 5th grade classroom at the beginning of the year, excited to be progressing to a grade level with more intensive mathematics. However, she finds that mathematics is excessive worksheets and old textbooks. Allison has not ever had an opportunity to do a creative project or use hands-on manipulative to solve real-world problems other than solving word problems from the textbook. She sees no connection between mathematics and technology which can lead to student inquiry and engagement. Allison sees mathematics as memorization, and she thinks that technology is used only for taking quizzes, locating information, and word processing papers. She sees the teacher use the ActivBoard to show a PowerPoint presentation or to have students come to the board one at a time to circle an answer. The teacher typically presents the lesson as a lecture everyday while the students sit in rows taking notes. Allison always works alone, never partnering with another student or working in a group on any type of assignments or discussions. When Allison or another student asks a question, the teacher is always the one to answer the question with no discussion. She had hoped to experience more real-world problem solving and interactive classroom discussions and explorations like her cousin at another school talks about.

Children are naturally curious and desire to learn through meaningful experiences (Division of Elementary, Secondary, and Informal Education, 2000). When given the opportunity to gather and use data from authentic scenarios, the students more readily experience passion for and higher degrees of learning in mathematics. However, mathematics classrooms are still experiencing a trend of sterile worksheet curricula environments which do not allow for creativity nor use of technology, both of which can allow for sense making as advocated by Wood, Merkel, and Uerkwitz (1996). With this worksheet curriculum in mind, there may be a long-awaited solution for teachers, consequently appealing to parents and students alike. Through this solution, students will experience challenging questions centered around authentic projects. In this chapter, an instructional framework is provided, supported with research, and discussed so that teachers can use it with children to help facilitate potential for more meaningful learning and mathematical understanding via a real-world, creative angle, while integrating technology (Tassell, Maxwell, & Stobaugh, 2013).

Technology integration is now more of an expectation rather than an option. Many United States teaching standards require effective technology integration (Tennessee Department of Education, 2007; Texas Education Agency, 2014). Schools are spending large portions of their budgets to purchase various technology capabilities, all in hopes that students engage in deeper learning that connects with the real world. Unfortunately, the primary use of technology is oftentimes for teacher presentations to garner student attention rather than for "student use" of technology to advance student learning to higher cognitive levels. For students to succeed in the formative up through pre-college years, teachers need to be considering how to embrace the new challenges they are facing in the mathematics classroom. Much of this can be tackled through a lens of a creative instructional disposition. Students filling classrooms are part of a "'creative, multimedia' generation" (Rosen, 2010, p. 218). The iGeneration is craving even more from education than ever before with technology and creativity (Oblinger, 2003; Prensky, 2010), yet many our mathematics teachers have not kept up with the awareness *and* learning curve (Shriki, 2010, 2013).

Teachers need to accept and embrace that students *love* to create (Rosen, 2010). Some students are wanting to channel this creativity in their coursework through technology in forms of movies, podcasts, webpages, and other digital products, and not the outdated technology formats of the past (Prensky, 2010). Students also want choice in their assignments and projects. When students have the responsibility of making choices, it increases engagement levels (Wood, 2010). Freedom to work at their own pace with support and partnership of the teacher is appealing to the students. Students enjoy space and time to creatively explore the content (Rosen, 2010).

To guide the integration of technology in the classroom, the International Society for Technology Education established standards for teachers (ISTE, 2008) and for students (ISTE, 2007). Both of these sets of standards promote students using technology to be creative, communicate, collaborate, and think critically. Another framework of skills, the Partnership for 21st Century Skills (2009), promotes students working collaboratively to create media products while engaging in critical thinking. For the teaching angle, the Partnership for 21st Century Skills (2009) stated that a learning experience should be one that "Enables innovative learning methods that integrate the use of supportive technologies, inquiry- and problem-based approaches and higher-order thinking skills" (p. 8). Therefore, when teachers are designing tasks, they need to consider these new expectations that indicate higher student competence when using technology to collaborate with students on cognitively demanding learning tasks about real-world topics. All of this leads to a broader and more inclusive view of technology—where technology integration is connected to higher-order thinking, real-world learning experiences, and engaged learning. However, the reality is that there is a gap between curriculum standards and instructional practices. The disconnect forms and urgency for the foundation of the Create Excellence Framework.

3.2 Review of Research on How Teachers Teach Creativity Through Real-World Lessons, Collaboration, and Intellectual Risks

Creativity as defined by Pink (2005) is a necessity in thinking through complexities of our interconnected world. Sternberg (2006) stated that educational researchers and psychologists profess the benefits of creative thinking on emotional, cognitive, and professional areas of life. However, even though there is an elevated focus on creativity, *teaching* in a way that supports creativity is still an anomaly (Henrickson & Mishra, 2013). With a focus on high-stakes stakes testing and published/scripted curricula, creativity is not the focus in most classrooms in the United States (Giroux & Schmidt, 2004).

At Michigan State University, a study was conducted for how to integrate creativity into classroom and the role of teachers in enhancing children's creativity (Mishra, Koehler, & Henriksen, 2011; Mishra, Henricksen, & The Deep-Play Research Group, 2012). Their focus is on embedding creativity into the context of the content area, and not just in a general sense of creativity instruction (Mishra et al., 2012). The goal is to help teachers learn how to teach their students to be the kind of creative people that can look beyond the boundaries of their content area of expertise and make connections back to that field to create new ideas (Henrickson & Mishra, 2013).

In a study conducted from 2000 to 2010 of eight United States award-recognized teachers by Henrickson (2011), research revealed that 90% of the teachers noted creativity as their main teaching mantra and gave examples of how creativity was taught through instruction in their classroom. Davidovitch and Milgram (2006) go on to emphasize that for instruction to be "effective", it must be "creative".

From the study of the eight teachers, ten key creative teaching approaches emerged (Henrickson & Mishra, 2013). One of these practices is: "link lessons to real-world learning." For this to happen, authentic experiences must be incorporated so that creativity is woven in relevant learning. The teachers in the study all stated that "real-world" learning is creative, offering novel opportunities for learning. Another approach to teaching that emerged is "valuing collaboration." The rationale was that successful design teams do their best work through collaborative efforts. These teachers also brought up concerns of working in isolation, emphasizing the importance of discussing and sharing ideas with others as a creative catalyst in learning. A third approach connected to our study is "taking intellectual risks." The teachers emphasized the idea of modeling new ideas and approaches in their classroom, showing that they were open to failure.

In this chapter, we will share the impact of the instructional planning support, Create Excellence Framework, on teacher candidates in designing their mathematics and integrated mathematics/science lessons. We begin with giving an overview of the Create Excellence Framework with details and research for the four components supported by research: Cognitive Complexity, Real-World Learning, Engagement, and Technology Integration. The next phase of the chapter shares the research of five semesters of working with teacher candidates in their lesson planning with this model. The overarching goal is to consider how these components connect to enhancing student creative thinking opportunities through real-world lesson plans.

3.3 The Create Excellence Framework

The Create Excellence Framework includes four components: Cognitive Complexity, Real-World Learning, Technology Integration, and Engagement. All the components important for adding depth to learning and planning comprehensive lessons are addressed in this framework. This instrument draws ideas from Moersch (2002) who originally developed the HEAT instrument (Maxwell, Constant, Stobaugh, & Tassell, 2011). The Create Excellence instrument measures five levels of integration of each component (see Fig. 3.1 for Create Excellence Framework). Each component covers the same five levels of increasing complexity to help the teacher target growth in his or her instructional development of tasks and projects: (1) Knowing, (2) Practicing, (3) Investigating, (4) Integrating, and (5) Specializing.

A target level of 3 or higher on the Create Excellence Framework was established because students are using higher-level thinking (Analyze or higher), engaging in learning where students experience choice and differentiation, simulating real-world experiences, and creating technology products even if they are an add-on to the lesson. At higher levels on the Create Excellence Framework students are more responsible for their own learning, beginning to think like experts, planning their own learning experiences while learning is embedded in the real world, and technology is seamlessly integrated and a necessary part of the learning experience. In the Technology Integration component *student* use of technology is emphasized, instead of teacher use of technology. The Cognitive-Complexity component also incorporates higher-level thinking skills (Maxwell, Stobaugh, & Tassell, 2015).

Tasks are small classroom activities while *projects* are more complex and use several instructional strategies, have open-ended solutions, involve more student choice and decision making, and take longer to complete. The lower levels of the framework are teacher directed (levels 1–3), whereas higher levels are more student directed (levels 4–5) with the teacher partnering with students to design projects and assignments (Tassell et al., 2013). The target levels for consistent student learning are levels 3 and 4, which are shaded in tables depicting the framework levels throughout the book. While level 3 is still teacher-directed, students are engaging in higher cognitively complex tasks and projects. Students are beginning to take more responsibility for their learning in level 4. Level 5 is attained after consistent learning at levels 3 and 4 and could be accomplished a few times a year (Maxwell et al., 2015).

3.3.1 Cognitive Complexity

The student's level of thinking with the content is vital to comprising a quality task. When objectives, activities, and assessments are properly aligned at higher levels of cognitive thinking, not only does instruction improve but student learning has a better chance of improving as well (Raths, 2002). The Cognitive Complexity component within the Create Excellence Framework is based on the revised Bloom's Cognitive Taxonomy (Anderson & Krathwohl, 2001). The revised Bloom's taxonomy includes six levels (Remember, Understand, Apply, Analyze, Evaluate, and Create along with nineteen cognitive processes classified within its six levels) (Anderson & Krathwohl, 2001).

CREand Te Excellence

	CReaTE Levels	Cognitive Complexity (Bloom's Cognitive Processes in parentheses)	Real World	Engagement	Technology Integration	
Lower Cognitive Complexity	**Level 1 Knowing**	• Teacher directs student interaction with content/standard at REMEMBER level (Recognizing; Recalling)	• Non-relevant problems using textbook/ worksheets	• Teacher lecture or questioning and students take notes OR • One correct answer expected	• Code 0 for no technology • Teacher uses technology for demonstration or lecture OR • Student technology use at Remember level OR • Technology is a student option but not required or used for keyboarding	Teacher-Directed
Lower Cognitive Complexity	**Level 2 Practicing**	• Teacher directs student interaction with content/standard at UNDERSTAND level (Interpreting; Exemplifying; Classifying; Summarizing; Inferring; Comparing; Explaining) and APPLY (Executing; Implementing)	• Provide some application to real world using real objects or topics	• Students engaged in a task directed by the teacher AND • Multiple solutions for one task are accepted	• Students use technology for Understand or Apply thinking tasks OR • Students use technology for gathering information	Teacher-Directed
Higher Cognitive Complexity	**Level 3 Investigating**	• Teacher directs student interaction with content/standard at an ANALYZE level (Differentiation; Organizing; Attributing), EVALUATE level (Checking; Critiquing), or CREATE level (Generating; Planning; Producing)	• Learning simulates the real-world (such as a class store or assuming the role of a political commentator)	• Student choice for task AND • Task differentiated by content, process, and/or product (such as addressing learning preferences, interests, or ability levels).	• Technology use appears to be an add-on or alternative—not essential for task completion AND • Students use technology for Analyze, Evaluate, or Create thinking tasks.	Teacher-Directed
Higher Cognitive Complexity	**Level 4 Integrating**	• Student-generated questions/ projects with content/standard at ANALYZE level (Differentiation; Organizing; Attributing), EVALUATE (Checking; Critiquing), or CREATE level (Generating; Planning; Producing)	• Learning emphasizes and impacts the classroom, school, or community AND • Learning is integrated across subject areas	• Students partner with the teacher to define the content, process, and/or product AND • Student inquiry-based approach AND • Students collaborate with other students	• Student technology use: o is embedded in content and essential to project completion AND o promotes collaboration among students and partnership with teacher AND o helps them solve authentic problems at the Analyze, Evaluate, or Create levels.	Student-Directed
Higher Cognitive Complexity	**Level 5 Specializing**	• Student-generated questions/ projects with content/standard at CREATE level (Generating; Planning; Producing) AND • Complex thinking like a content expert OR open-ended, global learning emphasis	• Learning has a positive impact on a national or global issue or problem AND • Collaborates with experts in a field or discipline	• Students initiate their own inquiry- based learning projects; thorough immersion; full implementation from topic to solution AND • Students initiate appropriate collaborations pertaining to their project	• Student-directed technology use: o is seamlessly integrated in content at the o Create level AND o has several technologies AND o include ______ expert o and/or global organizations to find solutions to an in-depth "real" problem	Student-Directed

© Maxwell, Stobaugh, Tassell 2014

Fig. 3.1 Create Excellence Framework

Create is the highest level on Bloom's revised taxonomy. It involves organizing information in a new way to design a product or novel solution, hence creative thinking. There are three Create-level cognitive processes within Bloom's taxonomy, and they occur sequentially: (1) generating, (2) planning, and (3) producing. When students engage in the generating cognitive process, they explore various ideas or solutions to solve an ill-defined problem through hypothesizing and exploring various relevant options. To begin this process, the topic must be researched and thoroughly understood so the ideas generated logically connect to the identified topic. The ideas should also be varied, unique, and detailed (Swartz & Parks, 1994). Planning is the second step in the creation process. Students will take the best idea they generated and decide on a plan to carry out the project. Often there is more than one way to solve the problem. Also, during the planning process, students often realize they must revise their idea or consider a new idea. The final step is to follow through with the plan and produce the product.

At levels 1 and 2 of the Create Excellence Framework in the Cognitive Complexity component, learners are engaged in teacher-directed learning experiences and Bloom's Taxonomy levels of Remember, Understand, and Apply level. While level 3 of the Create Excellence Framework is teacher-directed, students are engaging in the higher levels of Bloom's Taxonomy—Analyze, Evaluate, and Create. At the student-directed levels of the Create Framework (levels 4 and 5), students employ the top three cognitive levels (Analyze, Evaluate, and Create). At these two highest levels the *students*, instead of the teachers, are identifying the questions, tasks, or projects. On level 4 and 5, students generate projects on the Create level while thinking like an expert focused on an open-ended, global learning emphasis.

In the mathematics field specifically, Bloom's revised taxonomy helps teachers with instruction by providing steps and ideas for math questions worth asking, to know the difference between open and closed questions (Petti, 2017). As teachers work on their "good questions" that are worth asking, these questions lend themselves to exploration and more questions that students can reflect on and grow as inquirers. The outcome may then be the students are better mathematical thinkers and engaged, lifelong learners.

Table 3.1 provides an example for Cognitive Complexity in the Create Excellence Framework at the level 5. As students simulate and perform tasks and projects like professionals in the field, they often naturally engage higher-order thinking skills as they analyze, evaluate, and solve problems just like skilled workers.

3.3.2 Real-World Learning

Real-World Learning is where the student learns from, interacts with, and has an impact on the real world (Maxwell, Stobaugh, & Tassell, 2017). The goal of real world learning is for the student to interact with the real world to solve real

Table 3.1 Example of Cognitive Complexity with mathematics in Create Excellence Framework

Create level 5 description for Cognitive Complexity	Sample task/project
• Students generate questions or projects with content at Bloom's Create level (Generating, Planning, Producing) • Students engage in complex thinking like a content expert or with content that has an open-ended, global-learning emphasis	Have you ever wondered how polls are done? How did they calculate that 70% of Americans like a certain food or type of car? Do they ask every single person in America? NO, they use a polling percentage or a "sample" (or part of the population). Groups of students in your class will create their own poll, ask students around your school about their opinion on a specific school issue, and then predict the percentage of student opinions about that issue at your school (Maxwell et al., 2017)

problems and experience authentic learning. For example, students may learn letter-writing skills when they want to write a letter to their senator urging him/her to support water conservation near their town. This experience teaches the students that real-world solutions are complex—they may not always work, may not always please everyone, and may have consequences that impact other areas (Maxwell et al., 2017). Elements of real-world learning incorporate learning integrated across subject areas, learning as close to the real world as possible, and collaborating with experts in the field or discipline being studied.

3.3.2.1 Integrated Learning

Educators Barton and Smith (2000) state that interdisciplinary learning "provide[s] authentic experiences in more than one content area, offer[s] a range of learning experiences for students, and give[s] students choices in the projects they pursue and the ways they demonstrate their learning" (p. 54). Interdisciplinary units enable teachers to use classroom time more efficiently and address content in depth while giving students the opportunity to see the relationship between content areas and engage in authentic tasks and projects (Maxwell et al., 2017).

Students immersed in authentic-learning activities cultivate the kind of portable skills that are applicable in new and different situations, settings, or connections. These skills include judgment to distinguish reliable from unreliable information, patience to follow longer arguments and assignments, ability to recognize relevant patterns in unfamiliar contexts, and flexibility to work across disciplinary and cultural boundaries to generate innovative solutions (Jenkins, 2009).

In problem-based learning, students work for an extended period of time to investigate and respond to a complex questions, problem, or challenge. Problem-based learning is the center of medical students' training as they develop work skills—collaborating, chairing a group, listening, recording, cooperating, respecting colleagues' views, critically evaluating literature, self-directing learning and use of resources, and presenting on and engaging in real medical tasks and projects (Wood, 2003).

Students involved in authentic learning are motivated to persevere despite initial frustration, as long as the project embodies what really counts to them—a social structure they enjoy, topics and activities of personal interest, and a feeling that what they are doing is important and valued (Herrington, Oliver, & Reeves, 2003; Prensky, 2010). By confronting students with uncertainty, ambiguity, and conflicting perspectives, instructors help them develop more mature mental models that coincide with the problem-solving approaches experts use. Be aware that the balance of challenge and uncertainty must be just right so that students are sufficiently engaged but not overwhelmed. Authentic-learning exercises expose the messiness of real-life decision making, where there may not be a right or a wrong answer per se, although one solution may be better or worse than others depending on the particular context or consequences. Such a nuanced understanding involves considerable reflective judgment, a valuable lifelong skill that goes well beyond content memorization (Keyek-Franssen, 2010).

3.3.2.2 Learning in the Real World

When a student learns from, interacts with, and has an impact on the real world, higher retention of learning will occur. Real-world learning is organized around complex activities built on multiple themes and academic disciplines and requires multiple steps over an extended duration of time. Students have a real audience for their work. They use real data and learn content through working on projects and real problems that interest them (Schools We Need Project, n.d.). Take, for example, the fourth-grade class featured in the opening vignette of this chapter that decided to design landmarks for local heroes. This would be a level 4 real-world learning project in which learning impacts the school and community. Learning is integrated across subject areas—language arts, mathematics, science, economics, and social studies (Maxwell et al., 2017).

As another example, students may investigate and create projects to solve community issues such as developing a local walking trail, promoting girls' inclusion in community athletics, or endorsing stricter policies on littering in the community. This would also be a level 4 real-world learning project since it is student directed and the students are having an impact on their community (Maxwell et al., 2017).

Students prefer real, not just relevant, learning. Relevant means that students can relate, connect, or apply the content you are teaching to something they know about (for example, sports, music, social networking, movies, or games). The problem with relevance is it does not go far enough to make learning meaningful and engaging. As education innovator Prensky (2010) says, "Real means that there is a continuous perceived connection by the students between what they are learning and their ability to use that learning to do something useful or impact the real world" (p. 72). For students to actively attend to and retain information, it must be

Table 3.2 Example of flow in mathematics classroom from traditional to real learning

Traditional learning	Relevant learning	Real learning
Teacher assigns: Assigns problems about geometry from the textbook	Teacher scenario: We have been studying about how a city involves geometry in architecture. How could you help design blueprints for our city? Assume the role of an architect who is designing a new neighborhood for the city. Create a Voki to give your pitch to the decision panel	Teacher scenario: After studying how cities are planned and the geometry involved, students brainstorm building and neighborhood designs, and ways to be "green". One team decides to investigate how the city can be more efficient in using building materials. They work with a house planner to help troubleshoot issues in the city. They design posters with Glogster.com or Kerpoof.com to encourage citizens to conserve materials and go green. The mayor and "Go Green" director judge the posters and select one to duplicate and display around the city

relevant to their interests or foreseeable future needs (Sousa, 2006). In fact, traditional learning will usually fall under level 1 or 2, relevant learning under level 2 or 3, and real learning under level 4 or 5, depending on the level of impact. Table 3.2 provides a sample topic to illustrate the differences among traditional, relevant, and real learning.

3.3.2.3 Collaborating with Field or Discipline Experts

Real-world problems comprise complex tasks that students investigate over a sustained period of time. Students locate their own resources and are not given a finite list of resources. Collaboration is integral to authentic learning, where teamwork is critical to making decisions, solving problems, creating products, and maneuvering the social aspects of learning with a team. Collaboration between the teacher and students is essential to select the content, design the tasks or projects, and construct the assessment. Finally, authentic learning usually culminates in the creation of a whole product; however, the process is just as valuable to student learning as the product. For example, in a conservation unit, each student may document how much water his or her family uses each week, study personal water use habits, and make recommendations to his or her family about water conservation at home. The process of studying one conservation method at home could lead to other conservation efforts at home. It shows students that they can learn about topics that affect them and make informed decisions about many aspects of their lives (Maxwell et al., 2017).

Table 3.3 Example of Real World learning with mathematics in Create Excellence Framework

Create level 4 description for Real World learning	Sample task/project
• Learning emphasizes and impacts the classroom, school, or community AND • Learning is integrated across subject areas	Elementary students created an organic garden at their school in collaboration with a local organic farmer. Students implement their design (including geometric patterns and measurements), grow the vegetables, and sell their products at the local farmers' market. The organic farmer helped the students by reviewing their designs and giving feedback, advising about pricing and keeping accurate records of sales, and how to use the data to plan for next year's garden (Maxwell et al., 2017)

True collaboration with experts in the field is invaluable in student acquisition of the knowledge, skills, and dispositions necessary to develop discipline, work ethic, and collaboration proficiencies. Collaboration with these experts could occur in person at the school, through a field trip to the expert's work location, or via video conferencing with Skype. Teachers of a specific discipline may find themselves collaborating with other teachers and experts from other disciplines (Maxwell et al., 2017). Table 3.3 provides an example of collaboration with an expert.

3.3.3 Engagement

The Engagement component of the Create Excellence Framework is concerned with the degree to which learners take responsibility for their own learning; partner or collaborate with the teacher, other students, or outside experts; and use/manage resources such as teachers, experts in the discipline, and tools/technology. Teachers can help the student differentiate their interests and make choices in how they approach the task. They can also support the student by helping them identify resources and collaboration opportunities (Maxwell et al., 2017).

Student engagement has become an important quality in creating effective schools and advancing student achievement. Educators know now that students simply staring at the teacher or completing worksheets does not equal engaged learning, and just because students are quiet and busy, that does *not* mean they are engaged in their learning. Activities that focus on procedures and rudimentary tasks as opposed to cognitively demanding learning opportunities have been found to actually impede student engagement (Blumenfeld & Meece, 1988). Engaged learning involves students solving problems or creating solutions to ill-structured, multidisciplinary, real-world problems. There are several facets of engaged learning, including inquiry-based learning, student-directed learning, collaboration within and beyond the classroom–students collaborate or partner with other students, teachers, or outside experts, and differentiated learning (Maxwell et al., 2017).

3.3.3.1 Inquiry-Based Learning

Student engagement is connected to a movement in education toward inquiry-based learning. With inquiry-based learning, students are engaging with real-world issues while solving problems or creating solutions to develop deep understandings. According to biology instructor Schamel and research associate Ayres (1992), students learn in a more effective manner when they generate their own questions based on their observations rather than developing a solution to a situation or problem with a predetermined answer. The National Science Education Standards (1996) state, "Inquiry is something that students do, not something that is done to them" (p. 21). Since inquiry-based learning is student directed, it would be placed at the Integrating level (4) of the Create Excellence Framework if students are collaborating with the teacher and other students. It would be considered level 5 (Specializing) if students are collaborating beyond the classroom (Maxwell et al., 2017).

The basis of inquiry-based learning is that students are key planners and designers in the learning process. Table 3.4 shows the comparisons between traditional and inquiry-based learning with students directing the learning, the teacher facilitating the learning, and students having input in the assessment (Maxwell et al., 2017).

3.3.3.2 Student-Directed Learning

Student-directed learning is another key component of student engagement. Student-directed learning places the learning focus directly on the students and less heavily on the teacher's actions. As incorporated in all elements of inquiry-based learning, students are active learners, take responsibility for their own learning, and constantly formulate new ideas and refine them through their collaboration with others (Hung, Tan, & Koh, 2006). In project-based learning, students have voice and choice. Students help teachers set clear expectations so that they know what success looks like. Students articulate the targets or goals and examine targets in their own work (Antonetti & Garver, 2015).

Table 3.4 Comparison of traditional and inquiry-based learning

Traditional	Inquiry based
Teacher directed	Student directed
Teacher as giver of knowledge	Teacher as facilitator of learning
Content mastery	Content mastery and beyond
Vertical and linear learning path	Learning is more web-like; concept development ranges from linear to spiral
Teacher-created assessment	Assessment requires student input

Finding the spark—a real-world subject, idea, or project that makes a student light up—is the key to customizing learning experiences and engaging individual students. In order to tailor learning to meet students' educational needs and aspirations, teachers seek and develop knowledge of each student's unique tendencies, circumstances, and interests through both formal processes (such as surveys or advisories) and informal processes (casual conversations and insight from partner or cooperating organizations, community members, or other teachers) (Martinez, 2014). For example, on a level 4 project, students might partner with the teacher to decide which tasks they need to complete or determine what type of products they might produce.

Student-directed learning in comparison to teacher-directed approaches increases students' depth of understanding, increase critical-thinking skills, improve long-term retention, and increase students' positive feelings toward the subject studied (Crie, 2005). At the highest levels of student-directed learning, students establish the learning goals based on their interests or questions they pose. At this level of self-directed learning, students may also co-construct knowledge, assume varied roles and tasks, and participate in self-monitoring and assessment (Maxwell et al., 2017).

The inquiry process identifies several levels based on the level of student input. Open inquiry involves the top level of student engagement in the learning process with *no predetermined questions since students propose and pursue their own questions*. This level could correlate with Create framework levels 4 or 5 in the student-engagement component, depending on the amount of student initiation of inquiry and collaboration. In the second level, guided inquiry, the teacher decides on the topic, but the students can decide how they will approach the topic and investigate the problem. This level could connect with Create framework level 3 or 4, depending on the amount of teacher input or student collaboration. At the third level, structured inquiry, the teacher determines the topic and method for investigation and students explore various solutions. This level could correlate with Create framework level 2 or 3, depending on task choices and differentiation. In the lowest level, limited inquiry, students follow the directions and make sure their results match those given in the text. This level would be Create framework level 2 since students are engaged in a teacher-directed task (Maxwell et al., 2017).

3.3.3.3 Collaborating Within and Beyond the Classroom

Collaboration is the third key component to student engagement. In engaging tasks, students should collaborate within the classroom with other students and teachers or beyond the classroom with outside experts. Teachers and experts provide real-world tools, techniques, and support that allow for open communicating and sharing (Hung et al., 2006).

Extending learning beyond the traditional classroom provides students with real-world learning experiences that allow them to communicate with experts, take ownership of their learning, and extend their support networks. Educators, including principals, act as consummate networkers throughout the process—searching for meaningful resources that meet school's learning goals and student interests in places like museums, colleges, and community organizations. For many educators, tapping these resources has been done to arrange internships or mentorships, but the Create Excellence Framework encourages teachers and principals to use their networking skills for deeper learning (Martinez, 2014).

3.3.3.4 Differentiated Learning

Opportunities for choice combined with a broad variety of instructional strategies result in the highest levels of engagement (Raphael, Pressley, & Mohan, 2008). When students are given choices, they have a sense of ownership of their personal learning. A diverse collection of instructional strategies should be paired with students' prior knowledge and readiness to learn in order to promote student engagement. However, the level and complexity of the varied instructional strategies and activities must also be challenging (Gregory & Chapman, 2007).

Differentiation begins at level 3 with the teacher differentiating content, process, or product. At level 4, students partner with the teacher to define their own content, process, or product. At level 5, students design and implement their own inquiry-based projects from topic to full implementation to solution. Students initiate their own outside collaborations with field experts. (See Table 3.5 for an example of Level 5.) With both of these top levels, instruction is differentiated as students choose what content to examine, what processes they will use to find the solution, and how they will demonstrate their learning (product) (Maxwell et al., 2017).

3.3.4 Technology Integration

With advances in technology doubling every eighteen months (McGinnis, 2006), there is a plethora of technologies available to schools. Internationally there is quite a variance of integration of technology based on factors including access to technology, government prioritizing and investing in technologies, and varying comfort levels and beliefs in the importance of utilizing digital tools for K-12 learning. According to a report by the European Commission (2013) in the European Union 63% of nine-year-olds do not study at a "highly digitally equipped school." Among the European countries, there is a large variance in the average ratio of computers available for educational purposes. The average for the European Union is 5:1, but in Greece it's 21:1.

Traditionally, technology in classrooms has been a gadget to obtain students' attention or inserted as an add-on to instruction to meet curriculum or teaching

Table 3.5 Example of Engagement with mathematics in Create Excellence Framework

Create level 5 description for Engagement	Sample task/project
Students initiate their own inquiry-based learning projects with thorough immersion and full implementation from topic to solution, and students initiate appropriate collaborations pertaining to their project	Students were disturbed after watching a documentary about students in a Kenyan school who did not have chairs for their classroom. The documentary deeply moved these fourth graders. The students wanted to raise funds for chairs for the African students. The teacher and students used Coggle (https://coggle.it), an online mind-mapping tool, to brainstorm ways to raise the funds. One student's idea was to sponsor a math day at school where students paid fifty cents for solving a math problem. Another student contacted his uncle, a member of a civic club, to help them. The students also participated in an event at the county fair to raise funds. The students kept careful records on a spread-sheet, set up formulas to calculate the total and amount still needed. The teacher con-tacted an international humanitarian group for the students to work with to purchase and ship the chairs. The humanitarian group delivered the chairs (with desktops) and made a video of the excited African students to share with the fourth graders

standards, but it fails to meaningfully impact instruction when teachers use it in that capacity. Technology used to deliver teacher-directed content (as a glorified blackboard) and digital worksheets has not delivered the rate of return expected for the millions of dollars spent on technology (Schwartzbeck & Wolf, 2012). Without sound application of technology integration, money spent on technology is wasted. Authors Greaves, Hayes, Wilson, Gielniak, and Peterson (2010) state, "Although educational technology best practices have a significant positive impact, they are not widely and consistently practiced" (p. 12). Technology is a *tool* to reach an educational goal; technology is not the goal itself. Author, educator, and technology administrator Richardson (2013) comments, "It's not about the tools. It's not about layering expensive technology on top of the traditional curriculum. Instead, it's about addressing the new needs of modern learners in entirely new ways" (p. 12).

Our research shows high correlation of technology integration with the other three components of the framework (Maxwell et al., 2011). Technology should be used not simply as an add-on but to meaningfully support the work to more efficiently and effectively accomplish the task, just as it is in the professional world. Authors Jukes, McCain, and Crockett (2010) state that the revised Bloom's taxonomy reflects the "new era of creativity that has been facilitated by the emergence of the online digital world" (p. 69). Technology paired with critical thinking, student engagement, and real-world learning provides opportunities for students to produce novel products to address authentic problems (Maxwell et al., 2017).

Schools must have a planned approach in order to maximize the impact of these technologies to enhance student learning (Pence & McIntosh, 2010). Educators, however, struggle to integrate technology in meaningful ways that involve higher-order thinking, collaborative tasks, and authentic problem solving (The United Nations Educational, Scientific and Cultural Organization [UNESCO], 2004). Optimally, technology integration is a seamless component of instruction to engage students in authentic, creative-thinking tasks (Maxwell et al., 2017).

The Create Excellence Framework's technology-integration component advocates for this new approach, incorporating real-world tasks that are naturally infused with critical thinking and student engagement. Effective technology integration seamlessly embeds technology tools as part of the instructional design in order to engage students with significant content at high levels of thinking, whereby students use varied technologies to collaborate with others, explore solutions to real-life problems, and share their results in an authentic manner. While some may view technology as helpful in building basic foundations of knowledge through online games that reinforce basic applications of content, students more effectively use technology to design solutions and create new products, which are high-level thinking activities. Technology tools have the potential to enhance student learning, but they must be implemented in a research-based framework to ensure sound implementation.

Jukes et al. (2010) developed a list of 21st century competencies that include students thinking creatively to address real-world issues, critically assessing the quality of digital content, and creating their own digital projects. The U.S. 21st Century Workforce Commission's (2000) National Alliance of Business maintains that "the current and future health of America's 21st century economy depends directly on how broadly and deeply Americans reach a new level of literacy—21st century literacy" (p. 5). Their alliance identifies 21st century literacy as including digital literacy, inventive thinking, and results-based thinking.

At the highest level on the Create Excellence Framework, students design projects where (a) technology is seamlessly integrated into content at the Create level of Bloom's taxonomy, (b) several technologies are used, and (c) students collaborate with field experts and/or global organizations to find solutions to an in-depth "real" problem. Teachers can partner with students to design open-ended assignments that have no single right answer, require students to design solutions to problems that require higher-level thinking, and naturally embed technology. Table 3.6 provides a description of the Create Level 4 along with a sample task/project.

Table 3.6 Example of Technology Integration with mathematics in Create Excellence Framework

Create level 4 description for Technology Integration	Sample task/project
• Student technology use… – Is embedded in content and essential to project completion AND – Promotes collaboration among students and partnerships with teacher AND – Helps them solve authentic problems at the Analyze, Evaluate or Create levels	You will find and investigate five different apps or websites that you think could help you, your classmates, and other third-graders practice and understand the concept of fractions. You will then review the apps or websites you researched and rank the top five programs. To share your thoughts, you will publish a review of the five best apps for learning fractions in our classroom newsletter and on our class website. After the newsletter is published, the class will choose the top five apps/websites out of all of those collected and critiqued to use for the next month to practice fractions (Maxwell et al., 2017)

3.4 Research Study

3.4.1 Purpose of the Research

As teacher candidates utilize the Create Excellence Framework, they can design higher quality lessons sparking student creativity. Use of the Framework can enhance the teacher's knowledge of intentional lesson plan design and can positively impact teacher candidate instructional planning performance, in turn providing opportunities for real-world learning opportunities that provide authentic learning opportunities for creative thinking. The authentic learning experiences can then inspire an environment for students to develop and tap into their creativity as applied to real-world learning and meaningful cognitive challenge. Creativity inspires discovery learning, an inquiry-based learning method where students discover facts and relationships for themselves (Bruner, 1961). For over ten years Robinson and Aronica (2015) has been saying that we are preparing students for careers that don't yet exist. Learning how to be more creative (and thus adaptable)—now that's what prepares students for life beyond the classroom. Business executives say that creativity is valued as the most important business skill in the modern world (Robinson & Aronica, 2015).

The researchers analyzed lesson plans developed by pre-service teacher education students at a southeastern university based on their level of the Create Excellence implementation over five semesters. Through utilizing the Create Excellence Framework in a pre-service Elementary Mathematics Methods course and an Elementary Education Senior Project course, the intention is that the participants should possess greater abilities to design higher-level thinking lessons around authentic topics that integrate student design with technology while employing creative thinking skills. The students, or pre-service teachers, were different students each semester. Therefore, the research is not following the students through the five semesters, but rather following the effect of instruction with the use of the Create Excellence Framework for impact on pre-service teachers' lesson plans.

3.4.2 Research Questions

The research questions for this study were as follows:

1. Is there a significant difference in pre-service teachers' mathematics/science lesson plan scores over the five semesters for the Cognitive Complexity component to enhance opportunities for creative thinking?
2. Is there a significant difference in pre-service teachers' mathematics/science lesson plan scores over the five semesters for the Real-World Learning component to enhance opportunities for creative learning in authentic situations?
3. Is there a significant difference in pre-service teachers' mathematics/science lesson plan scores over the five semesters for the Engagement component to enhance creativity in working with others?
4. Is there a significant difference in pre-service teachers' mathematics/science lesson plan scores over the five semesters for the Technology Integration component to enhance creative opportunities in learning?

3.4.3 Research Method

Over the course of five semesters in two different pre-service teacher preparation courses (Elementary Mathematics Methods and Elementary Education Senior Project), pre-service teachers were instructed on components of the Create Excellence Framework. These two undergraduate elementary program area courses required pre-service teacher education students to develop lesson plans as part of the typical course requirements. The pre-service teachers in the Elementary Education Senior Project course were required to design one mathematics or mathematics and science integrated lesson plan that embedded the Create Excellence Framework components at a level 3 or higher. In the Elementary Mathematics Methods course, the pre-service teachers were required to design a problem-solving lesson with the Create Excellence Framework components. The researchers then began using the Create Excellence Framework for instruction with the preservice teachers and continued four more semesters of data collection beyond the baseline semester: Spring 2010, Fall 2010, Spring 2011, and Fall 2011. The Fall 2009 semester data established a baseline before any instruction occurred on the Create Excellence Framework. In the study, a total of 253 pre-service teachers' lesson plans were collected from five semesters from the two courses. Researchers analyzed plans to identify the level for each component in the Create Excellence Framework.

Preservice teacher names were removed from the lesson plans, numbered, and randomly divided. Next, blind scoring was conducted by the researchers and scorers. The 253 samples were scored after the Fall 2011 semester. In total, eleven evaluators rated the lessons—three were the researchers, two were Assistant Superintendents, and the others were P-12 teachers. The researchers trained the other scorers on the use of the Create Excellence Framework for the scoring of the lesson plans. A main focus of the training was on calibration of the scoring of the evaluators. To establish the calibration, the researchers chose four anchor lessons with agreed upon ratings, and trained and discussed these in detail for scoring calibration of the application of the framework. The new members of the scoring team each scored the "training" lesson plans, shared and discussed their ratings for each of the four Create Excellence Framework components. At this point in the study, the calibration goal was to score two consecutive lessons with Create component ratings no more than one level apart on each component from the score set by the researchers. After each training lesson, the discussion provided opportunities to refine the understanding of the Create Excellence Framework. After three training lessons, the calibration goal was met.

After the calibration was established, three teams of scorers were paired together with one researcher in each of the pairs. The lesson plans were randomly distributed among the three scoring teams. A scoring team evaluated the same set of lessons—giving every lesson in the study two sets of scores. The ratings were recorded on spreadsheets. The scores were averaged when the scorers did not agree upon a score. (see results in Table 3.7.)

Table 3.7 Mean of each Create Excellence Framework component across five semesters

	Cognitive Complexity m	Real-world learning m	Technology integration m	Engagement m
Fall 2009 N = 43	2.00	1.674	.791	1.465
Spring 2010 N = 44	2.068	1.977	1.273	1.727
Fall 2010 N = 47	2.191	2.042	1.702	1.894
Spring 2011 N = 46	2.283	2.174	1.957	1.891
Fall 2011 N = 73	2.425	2.726	2.247	2.110
Increase in m from Fall 2009 to Fall 2011	.425	1.052	1.456	.645

m Mean—rounded to third decimal place

The scores were analyzed in SAS for statistical difference with an Analysis of Variance (ANOVA), followed by a Tukey Studentized Range (HSD) Test to determine where differences occurred. The researchers were primarily investigating if there was a difference between the first, or baseline, semester and the last semester in the five-semester sequence.

3.5 Results

The forthcoming results share the analysis of the four research questions, one question at a time. The data are analyzed in a progression of mean and standard deviation, ANOVA, and *Tukey's Studentized Range* (*HSD*). Following the Results section, the Discussion and Conclusions sections share more details and thoughts for interpretation.

3.5.1 *Research Question One*

Is there a significant difference in pre-service teachers' mathematics/science lesson plan scores over the five semesters for the Cognitive Complexity component to enhance opportunities for creative thinking?

The means of the scores of pre-service teachers' lesson plans of the Cognitive Complexity component of the framework increased from 2.00 in the Fall 2009 semester to 2.425 in the fifth semester in Fall 2011 (see Table 3.8). Although this component had the least growth over the five semesters, Table 3.9 reveals a significant difference among the means of the semesters. The significant difference did occur between the Fall 2009 and Fall 2011 semesters (see Table 3.10).

Table 3.8 Descriptive data for Cognitive Complexity component

Semester	N	Mean *m*	Standard deviation *SD*
Fall 2009	43	2.00000000	0.37796447
Spring 2010	44	2.06818182	0.66113811
Fall 2010	47	2.19148936	0.44907140
Spring 2011	46	2.28260870	0.58359208
Fall 2011	73	2.42465753	0.52487586

Table 3.9 ANOVA for the Cognitive Complexity component

Source	*df*	*SS*	*MS*	*F*	*p*
Model	4	6.37098940	1.59274735	5.71	<0.0002
Error	248	69.23375369	0.27916836		
Total	252	75.60474308			

Table 3.10 Tukey's studentized range (HSD) for Cognitive Complexity component

Group comparison	Difference between means	Simultaneous 95%	
		Confidence	Limits
Fall 2011–Fall 2009	0.14555	0.42466	0.70377

Only significant results reported for difference between Fall 2009 and Fall 2011

3.5.2 Research Question Two

Is there a significant difference in pre-service teachers' mathematics/science lesson plan scores over the five semesters for the Real-World Learning component to enhance opportunities for creative learning in authentic situations?

Compared to other components, Cognitive Complexity has the highest mean score at the beginning of the study while the Real-World Learning means were the highest at the end of the study. The Real-World Learning component also showed steady increase with a baseline score of 1.674 (Fall 2009) and moving to 2.726 (Fall 2011) four semesters later (Table 3.11). This was an increase of 1.052 over the five semesters. The ANOVA results (Table 3.12) indicated a significant difference. The Tukey results confirmed a significant difference between the Fall 2009 and the Fall 2011 semester means (Table 3.13). Therefore, the pre-service teacher lesson plan scores did significantly increase from the baseline to the fifth semester of the study.

Table 3.11 Descriptive data for Real-World Learning component

Semester	N	Mean	Standard deviation
Fall 2009	43	1.67441860	0.56572458
Spring 2010	44	1.97727273	0.45691770
Fall 2010	47	2.04255319	0.41480466
Spring 2011	46	2.17391304	0.60752130
Fall 2011	73	2.72602740	0.55927220

Table 3.12 NOVA for the Real-World Learning component

Source	*Df*	*SS*	*MS*	*F*	*P*
Model	4	35.43000102	8.8575026	31.62	<0.0001
Error	248	69.4632704	0.2800938		
Total	252	104.8932806			

Table 3.13 Tukey's studentized range (HSD) for Real-World Learning component

Group comparison	Difference between means	Simultaneous 95%	
		Confidence	Limits
Fall 2011–Fall 2009	1.05161	0.77204	1.33118

Only significant results reported for difference between Fall 2009 and Fall 2011

3.5.3 Research Question Three

Is there a significant difference in pre-service teachers' mathematics/science lesson plan scores over the five semesters for the Technology Integration component to enhance creative opportunities in learning?

Technology Integration had the lowest baseline score with a .79 average in Fall 2009 and the highest mean increase (1.456) of all four components (Table 3.14). Each semester the mean scores on the Technology Integration component progressively increased with the highest mean gain in the Spring 2010 semester when the Create Framework was first introduced. A significant difference was revealed by the ANOVA (Table 3.15) and confirmed by the Tukey test between the Fall 2009 and Fall 2011 semesters (Table 3.16).

3.5.4 Research Question Four

Is there a significant difference in pre-service teachers' mathematics/science lesson plan scores over the five semesters for the Engagement component to enhance creativity in working with others?

Table 3.14 Descriptive data for Technology Integration component

Semester	N	Mean	Standard deviation
Fall 2009	43	0.79069767	0.67464769
Spring 2010	44	1.27272727	0.81735923
Fall 2010	47	1.70212766	0.85757225
Spring 2011	46	1.95652174	0.75884479
Fall 2011	73	2.2467534	0.79548765

Table 3.15 ANOVA for the Technology Integration component

Source	*df*	*SS*	*MS*	*F*	*P*
Model	4	68.2748986	17.0687246	27.64	<0.0001
Error	248	153.1480263	0.6175324		
Total	252	221.4229249			

Table 3.16 Tukey's studentized range (HSD) for Technology Integration component

Group comparison	Difference between means	Simultaneous 95%	
		Confidence	Limits
Fall 2011–Fall 2009	1.4559	1.0408	1.8710

Only significant results reported for difference between Fall 2009 and Fall 2011

Table 3.17 Descriptive data for Engagement component

Semester	N	Mean	Standard deviation
Fall 2009	43	1.46511628	0.50468459
Spring 2010	44	1.72727273	0.54404328
Fall 2010	47	1.89361702	0.63362458
Spring 2011	46	1.89130435	0.60473174
Fall 2011	73	2.10958904	0.39305229

Table 3.18 ANOVA for the Engagement component

Source	*df*	*SS*	*MS*	*F*	*P*
Model	4	12.11609114	3.02902279	10.81	<0.0001
Error	248	69.47284166	0.28013243		
Total	252	81.58893281			

Table 3.19 Tukey's studentized range (HSD) for Engagement component

Group comparison	Difference between means	Simultaneous 95%	
		Confidence	Limits
Fall 2011–Fall 2009	0.64447	0.36488	0.92406

Only significant results reported for difference between Fall 2009 and Fall 2011

In Fall 2009, the Engagement component had a 1.465 average and increased to 2.110 in Fall 2011, a .645 increase (Table 3.17). Engagement scores continued to increase each semester, excluding a minimal decrease in Spring 2011. The ANOVA (Table 3.18) and Tukey (Table 3.19) again revealed a significant increase from the Fall 2009 to the Fall 2011 semesters.

3.6 Discussion and Conclusions

The means of pre-service teachers' lesson plans using the Create Excellence Framework demonstrated significant increase on all four components from the first to the last semester of the research period. This finding suggests that pre-service teachers can learn to increase these components in their lesson planning.

The Cognitive Complexity dimension had the least increase of all dimensions over the course of five semesters, but had steady increases each semester. To deepen students' understanding of Bloom's taxonomy (Bloom, 1956; Bloom, Englehart, Furst, Hill, & Krathwohl, 1956), the professors engaged teacher candidates in determining the Bloom's level of sample tasks. In addition, in other teacher candidate lessons for the classes, they were expected to demonstrate their ability to design instruction above the Remember and Understand level of Bloom's taxonomy; hence, they had multiple times to practice designing lesson with rigorous learning outcomes. It is not surprising that the teacher candidates had difficulty designing lesson plans beyond a Create Framework level 2 on average. A Create Framework level 3 lesson plan requires pre-service teachers to design instruction that challenges students to think within the top three levels of the Revised Bloom's Taxonomy (Anderson & Krathwohl, 2001): Analyze, Evaluate, and Create. Most of the pre-service teachers were able to design instruction on the Understand and Apply level of Bloom's taxonomy, but were not able to reach the top three levels on Bloom's taxonomy or develop student-generated tasks on the higher levels. As mentioned earlier by Henrickson and Mishra (2013) in the study of how teachers teach creativity, a related area to the Create Framework's Cognitive Complexity is a teacher's willingness to "Take Intellectual Risks" and model new ideas and approaches in their classroom, showing that they were open to failure. Perhaps for the teacher candidate, the risk-taking is too much of a leap.

The Engagement component also showed significant increase over the course of the five semesters, with just a slight dip one semester. The professors incorporated differentiation and grouping techniques for the teacher candidates to use within the lessons, along with encouraging different forms of collaboration. The teacher candidates were also encouraged to work toward a student-directed lesson versus a teacher-directed lesson. As mentioned earlier with the Henrickson and Mishra (2013) study of teacher awardees, a related area to Engagement that emerged as an indicator of excellent teaching of creativity was their area of "Collaboration." The master teachers emphasized that their students needed to learn about the benefits of working together to hear the ideas of others and solve problems with integrated strengths.

The Real-World Learning component had the highest mean scores in the final year of the study. Professors challenged students to determine a real-world situation as the context of their lesson and then design their plan. The Create Framework's Real-World Learning component is connected to the Henrickson and Mishra (2013) study, "Link Lessons to Real-World Learning," as it asserts that authentic experiences must be incorporated so that creativity is woven in relevant learning with creative and novel learning opportunities.

The Technology Integration component had the highest mean increase of the all the components. As professors modeled new technologies in class and adopted higher expectations for integration, teacher candidates quickly began utilizing digital tools to enhance instruction. As teacher candidates moved to higher levels of integration of each component, the instruction became more engaging via the use of technology and provided more opportunity for students to express their creativity and learning in a variety of ways.

With each subsequent semester the researchers gained expertise in various technologies, critical-thinking strategies, and new ways to engage students in authentic tasks, which were then incorporated into the class. Professors also started to have students select unique technologies for in-class presentations to challenge students to investigate the uses of various technologies. The researchers began asking students to create a sample student product of the Create Excellence lesson they designed, which then challenged the pre-service teachers to carefully analyze their task directions.

To increase the quality, professors displayed and discussed more examples of quality Create Excellence Framework lesson planning assignments. Students also critiqued each other's work suggesting ways to improve the assignments. This formative feedback increased the quality of the final product.

3.7 Implications for Future Research

With new technologies providing more efficient and effective methods for learning, students are able to utilize digital tools to creatively solve authentic problems. However, oftentimes, P-12 instruction is teacher-directed with little freedom to produce diverse solutions. In mathematics classrooms teachers are frequently pressured to meet the content standards by covering content instead of uncovering deep learning while also lacking instructional tools for developing students' creativity (Shriki, 2010, 2013). However, when instruction is designed focused on high levels of Cognitive Complexity, Real-Learning Learning, Engagement, and Technology Integration, students are empowered to direct their own learning, while solving real-world problem using relevant digital tools. These instructional experiences can invigorate even reluctant learners as learning becomes more than memorizing but bursting with opportunities for creative expressions (Mann, 2006).

Teacher candidates and practicing teachers in many states are now faced with changing evaluation systems for teacher quality (Danielson, 2007). Through this process, the hope is that the Create Excellence Framework will provide a tool to help both pre-service teachers and current teachers be prepared to perform in the upper echelon of these more rigorous teaching standards. This framework has been designed to be a tool to support raising P-12 student achievement. However, a lingering question is: By use of the framework, do teachers actually improve in their instruction? A possible next step for future research is to study the impact of the Create Excellence Framework implemented by teachers to thoughtfully and intentionally plan for the four components of instruction to ensure a well-rounded lesson and deeper student learning. As teachers and administrators look for another tool to plan and improve instruction, the Create Excellence Framework may be the answer!

3.8 Final Thoughts

So, let us check back in with Allison in her 5th grade classroom, as introduced in the opening scenario of this chapter. After learning of the Create Excellence Framework, her teacher was able to implement an authentic STEM project where students were able to have opportunities to expand in creative learning and choices through the four components of Cognitive Complexity, Real-World Learning, Engagement, and Technology Integration.

The teacher informs the class that the classroom student desks are going to be replaced due to being old/ineffective. Groups are assigned to develop the optimal student desk that would meet the needs of 5th grade students. Allison's group identifies what qualities a desk should have to meet the needs of students in their classroom. Students brainstorm various conceptual designs. Students evaluate which concept is most likely to meet their needs and is cost effective. Using their engineering skills, students make calculations of size of the desk, cost of materials, and build a prototype. Each group tests and evaluates their prototype; then they restructure and improve the original design. While Allison's group is formulating their conceptual design, they use a free online program, Google Sketch-Up, to develop their design. The class is told that the desk-constructing groups are in competition for a "school choice" award. Their persuasive presentation for the principal is created in Animoto (online presentation program). Allison's team collaborates with the teacher constantly to ensure the group is progressing on their solution and Skype with a furniture designer to pose questions about their prototype and get feedback from the designer.

In Allison's classroom instruction make-over, the Real-World Learning component is found in the real-world mathematics lesson of desk redesign scenario and the competition for student choice award along with the persuasive presentation. Cognitive Complexity is seen in the students generating their own questions and design; students evaluating which concept will meet needs; students at the creating level of thinking; students test and evaluate the prototype, then restructure and improve their plan. For Engagement, the students work as a team and collaborate with the teacher; they consult with a furniture designer. For Technology Integration, students use Google sketch-up, Skype, Animoto for presentation, and build a prototype of the desk.

The example of Allison's classroom experience provides a picture of how instruction can be enhanced in the mathematics classroom—infusing opportunities for children to creatively engage in authentic learning experiences.

References

Abrami, P. C., Lou, Y., Chambers, B., Poulsen, C., & Spence, J. C. (2000). Why should we group students within-class for learning? *Educational Research and Evaluation, 6*(2), 158–179.

Anderson, L. W., & Krathwohl, D. R. (2001). *A taxonomy for teaching, learning and assessing: A revision of Bloom's taxonomy of educational objectives*. New York, NY: Addison Wesley Longman.

Antonetti, J. V., & Garver, J. R. (2015). *17,000 classroom visits can't be wrong: Strategies that engage students, promote active learning, and boost achievement*. Alexandria, VA: Association for Supervision and Curriculum Development.

Barton, K. C., & Smith, L. A. (2000). Themes or motifs? Aiming for coherence through interdisciplinary outlines. *Reading Teacher, 54*(1), 54–63.

Bloom, B. S. (1956). *Taxonomy of educational objectives*. Philadelphia, PA: David McKay Publishing Company.

Bloom, B. S., Englehart, M. D., Furst, E. J., Hill, W. H., & Krathwohl, D. R. (1956). *Taxonomy of educational objectives: The classification of educational goals. Handbook 1: Cognitive domain*. New York, NY: David McKay.

Blumenfeld, P. C., & Meece, J. L. (1988). Task factors, teacher behavior, and students' involvement and use of learning strategies in science. *Elementary School Journal, 88*(3), 235–250.

Bruner, J. S. (1961). The act of discovery. *Harvard Educational Review, 31,* 21–32.

Crie, M. (2005). *Inquiry-based approaches to learning*. New York, NY: Glencoe. Retrieved from http://www.glencoe.com/sec/teachingtoday/subject/inquiry_based.phtml.

Danielson, C. (2007). *Enhancing professional practice: A framework for teaching*. Alexandria, VA: ASCD.

Davidovitch, N., & Milgram, R. M. (2006). Creative thinking as a predictor of teacher effectiveness in higher education. *Creativity Research Journal, 18*(3), 385–390.

Division of Elementary, Secondary, and Informal Education. (2000). *Foundations: A monograph for professionals in science, mathematics, and technology education*. Arlington, VA: Author. Retrieved from http://www.nsf.gov/pubs/2000/nsf99148/htmstart.htm.

European Commission. (2013). Survey of schools: ICT in education. *European Schoolnet*. Retrieved from https://ec.europa.eu/digital-single-market/news/survey-schools-ict-education.

Giroux, H. A., & Schmidt, M. (2004). Closing the achievement gap: A metaphor for children left behind. *Journal of Educational Change, 5,* 213–228.

Greaves, T. W., Hayes, J., Wilson, L., Gielniak, M., & Peterson, R. (2010). *The technology factor: Nine keys to student achievement and cost-effectiveness*. Chicago, IL: Market Data Retrieval.

Gregory, G. H., & Chapman, C. (2007). *Differentiated instructional strategies: One size doesn't fit all* (2nd ed.). Thousand Oaks, CA: Corwin Press.

Henrickson, D. (2011). *We teach who we are: Creativity trans disciplinary thinking in the practices of accomplished teachers* (Doctoral dissertation, Michigan State University). Retrieved from ProQuest Dissertations and Theses (3489807).

Henricksen, D., & Mishra, P. (2013). Learning from creative teachers. *Educational Leadership, 70* (5). Retrieved from http://www.ascd.org/publications/educational-leadership/feb13/vol70/num05/Learning-from-Creative-Teachers.aspx.

Herrington, J., Oliver, R., & Reeves, T. C. (2003). Patterns of engagement in authentic online learning environments. *Australian Journal of Educational Technology, 19*(1), 59–71.

Hung, D., Tan, S. C., & Koh, T. S. (2006). Engaged learning: Making learning an authentic experience. In D. Hung & M. S. Khine (Eds.), *Engaged learning with emerging technologies* (pp. 29–48). Dordrecht, The Netherlands: Springer.

International Society for Technology in Education. (2007). *The national educational technology standards for students*. Eugene, OR: International Society for Technology in Education. Retrieved from http://www.iste.org/standards/nets-for-students.aspx.

International Society for Technology in Education. (2008). *The national educational technology standards for teachers*. Eugene, OR: International Society for Technology in Education. Retrieved from http://www.iste.org/standards/nets-for-teachers.aspx.

Jenkins, H. (2009). *Confronting the challenges of participatory culture: Media education for the 21st century*. Cambridge, MA: MIT Press.

Jukes, I., McCain, T., & Crockett, L. (2010). *Understanding the digital generation: Teaching and learning in the new digital landscape*. Thousand Oaks, CA: Corwin Press.

Keyek-Franssen, D. (2010, December 15). Clickers and CATs: Using learner response systems for formative assessments in the classroom. *Educause Review*. Retrieved from http://www.educause.edu/ero/article/clickers-and-cats-using-learner-response-systems-formative-assessments-classroom.

Mann, E. L. (2006). Creativity: The essence of mathematics. *Journal for the Education of the Gifted, 30*(2), 236–260.

Martinez, M. (2014). *Deeper learning: The new normal*. Retrieved from http://www.advanc-ed.org/source/deeper-learning-new-normal.

Maxwell, M., Constant, M., Stobaugh, R., & Tassell, J. L. (2011). Developing a HEAT framework for assessing and improving instruction. In C. Maddux (Ed.), *Research highlights in technology and teacher education 2011* (pp. 13–20). Chesapeake, VA: SITE.

Maxwell, M., Stobaugh, R., & Tassell, J. L. (2017). *Real-world learning framework for elementary schools: Digital tools and practical strategies for successful implementation*. Bloomington, IN: Solution Tree.

Maxwell, M., Stobaugh, R., & Tassell, J. L. (2015). *Real-world learning framework for secondary schools: Digital tools and practical strategies for successful implementation*. Bloomington, IN: Solution Tree.

McGinnis, J. O. (2006). Age of the empirical. *Policy Review, 137,* 47–58.

Mishra, P., Henriksen, D., & Deep-Play Research Group. (2012). Rethinking technology and creativity in the 21st century: On being (in)disciplined. *Tech Trends, 56*(6), 18–21.

Moersch, C. (2002). Measures of success: Six instruments to assess teachers' use of technology. *Learning & Leading with Technology, 30*(3), 10–18.

Oblinger, D. (2003). Boomers, gen-xers, and millennials: Understanding the "new students". *Educause Review, 38*(4), 37–47.

Partnership for 21st Century Skills. (2009). *P-12 framework definitions*. Retrieved from http://www.p21.org/storage/documents/P21_Framework_Definitions.pdf.

Pence, H. E., & McIntosh, S. (2010). Refocusing the vision: The future of instructional technology. *Journal of Educational Technology Systems, 39*(2), 173–179.

Petti, W. (2017). Math questions worth asking. *Education World*. Retrieved from http://www.educationworld.com/a_curr/mathchat/mathchat026.shtml.

Raphael, L. M., Pressley, M., & Mohan, L. (2008). Engaging instruction in middle school classrooms: An observational study of nine teachers. *Elementary School Journal, 109*(1), 61–81.

Pink, D. H. (2005). *A whole new mind*. New York, NY: Riverhead Books.

Prensky, M. (2010). *Teaching digital natives: Partnering for real learning*. Thousand Oaks, CA: Corwin.

Raths, J. (2002). Improving instruction. *Theory into Practice, 41*(4), 233–237.

Richardson, W. (2013). Students first, not stuff. *Educational Leadership, 70*(6), 10–14.

Robinson, K., & Aronica, L. (2015). *Creative schools: The grassroots revolution that's transforming education*. New York, NY: Penguin.

Rosen, L. (2010). *Rewired: Understanding the iGeneration and the way they learn*. New York, NY: Palgrave Macmillan.

Schamel, D., & Ayres, M. P. (1992). The minds-on approach: Student creativity and personal involvement in the undergraduate science laboratory. *Journal of College Science Teaching, 21* (4), 226–229.

Schools We Need Project. (n.d.). *An engaging program of real world learning*. Retrieved from http://schoolsweneed.wikispaces.com/Real+World+Learning.

Shriki, A. (2010). Working like real mathematicians: Developing prospective teachers' awareness of mathematical creativity through generating new concepts. *Educational Studies in Mathematics, 73,* 159–179.

Shriki, A. (2013). A model for assessing the development of students' creativity in the context of problem posing. *Creative Education, 4*(7), 430–439.

Schwartzbeck, T. D., & Wolf, M. A. (2012). *The digital learning imperative: How technology and teaching meet today's education challenges*. Washington, DC: Alliance for Excellent Education.

Sousa, D. A. (2006). *How the brain learns* (3rd ed.). Thousand Oaks, CA: Corwin Press.

Sternberg, R. (2006). The nature of creativity. *Creativity Research Journal, 18*(1), 87–98.

Swartz, R. J., & Parks, S. (1994). *Infusing critical and creative thinking into content instruction: A lesson design handbook for the elementary grades*. Pacific Grove, CA: Critical Thinking Press.

Tassell, J. L., Maxwell, M., & Stobaugh, R. (2013). CReaTE excellence: Using a teacher framework to maximize STEM learning with your child. *Parenting for High Potential, 3*(2), 10–13.

Tennessee Department of Education. (2007). *Educator licensure standards*. Retrieved from https://www.tn.gov/assets/entities/education/.../lic_educator_licensure_standards.pdf.

Texas Education Agency. (2014). *Texas teaching standards*. Retrieved from https://www.ritter.tea.state.tx.us/rules/tac/chapter149/ch149aa.html.

The United Nations Educational, Scientific and Cultural Organization. (2004). *Integrating ICTs into education: Lessons learned—A collective case study of six Asian countries*. Bangkok, Thailand: Author. Retrieved from http://www.unescobkk.org/fileadmin/user_upload/ict/e-books/ICTLessonsLearned/ICT_integrating_education.pdf.

U.S. 21st Century Workforce Commission. (2000). *A nation of opportunity: Building America's 21st century workforce*. Washington, DC: Author. Retrieved from http://digitalcommons.ilr.cornell.edu/cgi/viewcontent.cgi?article=1003&context=key_workplace.

Wood, D. F. (2003). ABC of learning and teaching in medicine: Problem based learning. *British Medical Journal, 326*(7384), 328–330.

Wood, J. (2010, April 11). *Rewiring education and connecting with the iGeneration*. Retrieved from JoeWoodOnline Blog: http://www.joewoodonline.com/rewiring-education-connecting-with-the-igeneration/.

Wood, T., Merkel, G., & Uerkwitz, J. (1996). Creating a context for talking about mathematical thinking. *Educacao e matematica, 4*, 39–43.

Chapter 4
Impacting Mathematical and Technological Creativity with Dynamic Technology Scaffolding

Sandra R. Madden

Abstract This chapter reports on studies conducted during the past decade that investigated mathematical learning for teaching with technology (MLTT) and its relationship to creativity. Though related to mathematical knowledge for teaching (MKT) (Ball et al. in J Teach Educ 59, 389–407, 2008) and technological pedagogical content knowledge (TPCK) (Mishra and Koehler in Teachers Coll Rec 180 (6):1017–1054, 2006; Niess in Teach Teach Educ 21(5):509–523, 2005), mathematical learning for teaching with technology has a strong dispositional component coupled with curiosity, creativity, and meaning making (Thompson in Third handbook of international research in mathematics education. Taylor and Francis, London, pp. 435–461, 2015). Using design-based research methods (Cobb et al. in Educ Researcher 32(1):9–13, 2003) a framework for dynamic technological scaffolding (DTS) has emerged in support of teacher learning. DTS has provided fertile ground for the design and further study of learning trajectories in which learners are exploring and eventually creating cognitively challenging mathematical task sequences in the presence of new (to them) physical and technological tools. By harnessing teachers' motivation to inculcate curiosity, engagement, and learning for their students, these design studies have created conditions where teachers have become curious, creative, and technologically savvy to the point where many have gone on to pursue similar kinds of experiences with their mathematics students. This chapter will explore and present DTS as created and implemented with secondary mathematics teachers and DTS as creative work pursued by teachers.

Keywords Curiosity · Creativity · Dynamic technology scaffolding
Provocative tasks · Design research

S. R. Madden (✉)
University of Massachusetts, Amherst, USA
e-mail: smadden@educ.umass.edu

© Springer International Publishing AG, part of Springer Nature 2018
V. Freiman and J. L. Tassell (eds.), *Creativity and Technology in Mathematics Education*, Mathematics Education in the Digital Era 10,
https://doi.org/10.1007/978-3-319-72381-5_4

4.1 Introduction and Literature Connecting Curiosity, Creativity, Mathematical Learning for Teaching with Technology, and Dynamic Technology Scaffolding

The research in this chapter addresses this books' focus on mathematical creativity when using dynamic cognitive tools (DCTs), where the participants are secondary level mathematics teachers. Cognitive tools are technologies, that enhance the cognitive powers of human beings during thinking, problem solving, and learning and help learners organize and represent what they know (National Governors Association Center for Best Practices, Council of Chief State School Officers, 2010; Jonassen, 1994; Jonassen & Reeves, 1996). Dynamic cognitive tools include digital (e.g., *GeoGebra, Core-Math Tools, Cabri Geometry, Geometers' Sketchpad, TinkerPlots, Fathom, Desmos, NLVM,* etc.) and physical tools (e.g., compasses, hula-hoops, ribbon, pipe cleaners). The class of DCTs that are computer-based technological tools are considered computational technologies (Santos-Trigo & Machin, 2013). The chapter explores secondary teachers, DCTs, and creativity in two primary ways: (1) **teachers as mathematical learners** in novel computational technology environments navigating unfamiliar mathematical and technological terrain, and (2) **teachers as designers** of lesson sequences for students in novel computational technology environments navigating unfamiliar mathematical and technological terrain. Creativity as examined in this chapter includes learners' self-discoveries that give rise to new and meaningful insights (Beghetto & Kaufman, 2009).

DCTs allow mathematical and statistical objects to be represented, constructed, linked, and manipulated in order to explore relationships and transcend the limitations of the mind (Pea, 1985). Some technological DCTs are pre-built, expert-constructed models for users to explore, such as Java applets, microworlds, or simulations. These are considered exploratory modeling environments, with what are sometimes called route-based tools (Bakker, 2002; Doerr & Pratt, 2008). Other technological DCTs are environments for users to construct their own models. These are considered expressive modeling environments, with what some refer to as landscape tools (Bakker, 2002; Doerr & Pratt, 2008). It is the expressive modeling environments where learners are free to construct representations and relationships that illuminate their mathematical understanding and creative thinking and reasoning. It is, however, a nontrivial matter to convince teachers to invest in technologically intensive mathematical learning for the purpose of innovation in their classrooms. This chapter addresses research aimed at supporting teachers as learners toward this end.

Through retrospective analyses of implemented learning trajectories based upon hypothetical learning trajectories (Simon & Tzur, 2004) that have been refined carefully over the past decade, this chapter will illuminate particularly promising practices with secondary mathematics teachers as they begin to explore mathematical and statistical terrain with technological tools. In service to the focus of this book on creativity, this work highlights creative mathematical and technological thinking, reasoning, and meaning as mathematical tasks in technologically rich environments are designed and implemented.

Research participants were 64 pre- or in-service mathematics teachers from the northeast U.S. who participated in design-based research studies conducted during a graduate level course intended to prepare teachers for the demands of utilizing mathematically oriented technology in their teaching for the support of their students' mathematical learning. Pre-course surveys indicated that very few learners from this environment had prior experiences with DCT as learners (Madden, 2013). Consequently, DCTs were not used with students in their own classrooms prior to the course. The course introduced students to a variety of DCTs, tasks, and learning environment conducive to supporting their facility with tools for reasoning mathematically in preparation for doing similar work with their students. Approximately half of the semester addressed learning with dynamic geometry technologies and the other half explored dynamic statistical technologies. All names are pseudonyms. Throughout the chapter, "students," "teachers," and "learners" refer to participants in the course, unless specifically noted. "Instructor" refers to the teacher of the course, also the author.

In this learning environment, mathematical tasks and innovative curriculum materials [e.g., Core-Plus Mathematics Project (CPMP) (Hirsch, Fey, Hart, Schoen, & Watkins, 2015)] challenge teachers' notions of what it means to know and do mathematics and consequently, what it means to teach mathematics. Teachers are introduced to mathematical habits of mind as they engage in mathematical activity (Cuoco, Goldenberg, & Mark, 1996). Some important mathematical habits of mind in this environment include: pattern sniffing, looking at algebraic objects geometrically, looking at geometric objects algebraically, and modeling with mathematics and statistics. Connections within and among big mathematical ideas are explored (Steen, 1990). There is a recurring aspect of computational thinking (Wing, 2006) that permeates the teacher learning environment explored in this chapter and promotes reconsideration of the nature of what counts as mathematical thinking and meaning making.

The following sections provide a brief review of literature on curiosity and creativity, mathematical learning for teaching with technology, and dynamic technology scaffolding in preparation for sections exploring the design of task sequences to support teacher learning and creativity, evidence of teachers as creative learners of mathematics, and teachers as designers of technologically rich learning environments.

4.1.1 Curiosity and Creativity

Due to the nature of creativity explored in this chapter, the connection between curiosity and creativity is examined. Loewenstein (1994) introduced the notion of "information gap" as antecedent to curiosity. He maintained that curiosity "arises when one's informational reference point in a particular domain becomes elevated above one's current level of knowledge" (p. 87). He calls the information gaps produced by the feeling of deprivation "curiosity" and suggests that curious individuals are motivated to eliminate the feeling of deprivation by obtaining more information. "Curiosity involves an indissoluble mixture of cognition and motivation" (pp. 94–95). The information gap perspective is particularly relevant in the study of mathematics teacher learning and curriculum development because of teachers' professional desire to feel competent and "curiosity is particularly strong when it comes to knowledge pertaining to one's own competence" (p. 93). Loewenstein discusses the importance of "priming the pump," asserting the need for experiences for learners to bump into ideas they know something about but exposing gaps in their knowledge. This perspective is especially relevant with respect to teacher learning and the use of dynamic cognitive tools to support mathematical learning.

Kashdan and Fincham (2002) suggest that curiosity may be characterized "as a self-regulatory mechanism that facilitates intrinsic goal effort, perseverance, personal growth, and, under the right conditions, creativity" (p. 373). It is those "right conditions" that are under consideration in this chapter. Self-determination (Ryan & Deci, 2000) as connected to autonomy, competence, and relatedness is within the mix of interest, motivation, and curiosity as teachers navigate new learning. As described by Madden (2013), tasks that teachers find contextually, mathematically, or technologically provocative have proven useful in supporting learning because they excite epistemic curiosity which may lead learners in the direction of creativity.

Much of the mathematical creativity literature explores advanced mathematical thinking or that of mathematically gifted learners. However, like Silver (1997), "mathematics educators can view creativity not as a domain of only a few exceptional individuals but rather as an orientation or disposition toward mathematical activity that can be fostered broadly in the general school population" (Silver, 1997, p. 79). There are many models of which a few relevant works are described below.

Some have theorized that mathematicians' creative processes follow a four-stage Gestalt model of *preparation-incubation-illumination-verification* (Sriraman, 2009). Though not a definition of mathematical creativity, the four-stage model highlights the importance of sustained thought and engagement as one contemplates and works toward a creative insight. "… creativity is closely related to deep, flexible knowledge in content domains; is often associated with long periods of work and reflection rather than rapid, exceptional insight; and is susceptible to instructional and experiential influences (Holyoak & Thagard, 1995; Sternberg, 1988)" (Silver, 1997, p. 75). Periods of incubation may be an essential aspect of creativity requiring inquiry-oriented, creativity-enriched mathematics curriculum and instruction (Silver, 1997).

Ervynck (1991) suggested that mathematical creativity develops across three stages:

- Stage 0, the *preliminary technical stage*, where practical or technical application of mathematical rules or procedures precedes genuine mathematical activity. For example, in Ancient Mesopotamia and Egypt staking out a right angle was accomplished using triangle side lengths of 3, 4, and 5 units prior to more formal mathematical learning.

- Stage 1, *algorithmic activity*, where procedures and techniques are carried out routinely. "As with the tool-object dialectic, it is essential that the tool become familiar in action before it becomes the focus of reflective activity" (Ervynck, 1991, p. 43). Though the tools referred to by Ervynck tend to be mathematical procedures, it is conceivable that other kinds of tools (e.g., DCTs) may serve a similar orienting purpose.
- Stage 2, the *creative (conceptual, constructive) activity* stage, the stage where mathematical creativity is conjectured to occur. This is the stage at which the underlying conceptual structure of a problem is utilized and non-algorithmic steps and decisions are made to advance a theory. This complex activity is a hallmark of mathematical creativity.

Ervynck (1991) describes mathematical creativity as "the ability to create mathematical objects, together with the discovery of their mutual relationships" (p. 46). Whereas Ervynck's focus is on advancing mathematics, his perspective may be taken as relative to the learner, where the advanced mathematical perspective is in relation to the current mathematical perspective of the learner. Researchers de Freitas and Sinclair (2014), suggest "A creative act:

1. *introduces or catalyses the new* – quite literally, it brings forth or makes visible what was not present before;
2. *is unusual*, in the sense that is must not align with current habits and norms of behavior;
3. *is unexpected or unscripted* – in other words, without prior determination or direct cause;
4. *is without given content*, in that its meaning cannot be exhausted by existent meanings" (p. 89).

The perspective of creative work as discussed by Csikszentmihalyi (2000) of that which is both novel and valuable is a useful frame for this chapter (Sriraman, 2004). This definition is exploited through the design, implementation, and study of learning environments with teachers. In the context of mathematical problem solving, problems are places where information gaps become evident. When learners experience such a gap, in the presence of tools to augment their investigation, the potential exists for the expression of their creative ideas in novel forms to solve problems. This is mathematically creative action using technology. For teachers, creative action is also related to design as they design units comprised of lessons and task sequences. The design process of solving a problem through the creation of something for someone is inherently creative (Sullivan, 2017). In agreement with Vygotsky (2004),

> Any human act that gives rise to something new is referred to as a creative act, regardless of whether what is created is a physical object or some mental or emotional construct that lives within the person who created it and is known only to him (p. 7).

The literature suggests that creative individuals tend to be attracted to complexity, which most school math curricula has very little to offer.

> Classroom practices and math curricula rarely use problems with an underlying mathematical structure and all students a prolonged period of engagement and independence to work on such problems … for mathematical creativity to manifest in schools, students should be given the opportunity to tackle non-routine problems with complexity and structure, which require not only motivation and persistence but also considerable reflection (Sririman, 2004, p. 26).

For these kinds of classroom practices to become reality, it is imperative that teachers have similar kinds of experiences in their learning (Sowder, 2007).

4.1.2 Mathematical Learning for Teaching with Technology (MLTT)

It is well established that mathematics teacher preparation should attend to developing teachers' mathematical knowledge for teaching (MKT) (Ball et al., 2008) and technological pedagogical content knowledge (TPCK) (Mishra & Koehler, 2006; Niess, 2005). Collectively, these frameworks identify types of knowledge required of teachers and help practitioners and researchers understand that knowledge of one's discipline is a necessary but insufficient condition for teaching. Desirable mathematics teacher knowledge includes knowing mathematics deeply and flexibly, ability to anticipate and help shape students' mathematical thinking and reasoning, and the capacity to design tasks, select and use tools for mathematical investigating, reasoning and problem solving.

In addition to knowing mathematics or having mathematical knowledge for teaching, it is productive to consider *how* one knows mathematics. For example, learners will say they *know* or *understand* what is meant by, "perpendicular bisector." Some can construct one, others find it relevant for making various geometric arguments, yet many are unable to invoke the property that all points on the perpendicular bisector of a segment are equidistant from its endpoints, during a situation that may call for it (e.g., when constructing a parabola given a focus point and directrix). That one may recall a definition or remember a property does not suggest robust knowledge or understanding suggested by MKT. How does a kind of deep, flexible and robust knowledge, especially related to the incorporation of technological tools for teaching and learning, develop?

Thompson (2015) encourages a shift from thinking about what teachers *know* to what teachers *mean*. "This shift is essentially from a philosophically mainstream view of knowledge as justified, true belief and about things external to the knower to a Piagetian perspective in which meaning and knowledge are largely synonymous, and both are grounded in the knower's schemes" (p. 436). Furthermore, "With meaning defined appropriately, a focus on meanings positions us to help teachers focus on creating instruction that helps students develop productive

Table 4.1 Thompson and Harel's definitions of understanding, meaning, and way of thinking

Construct	Definition
Understanding (in the moment)	Cognitive state resulting from an assimilation
Meaning (in the moment)	The space of implications existing at the moment of understanding
Understanding (stable)	Cognitive state resulting from an assimilation to a scheme
Meaning (stable)	The space of implications that results from having assimilated to a scheme. The scheme is the meaning. What Harel previously called Way of Understanding
Way of thinking	Habitual anticipation of specific meanings or ways of thinking in reasoning

Thompson et al. (2014), Thompson (2015)

meanings" (p. 438). Thompson draws heavily on Piaget's notion of assimilation to a scheme to understand mathematical meanings as see in Table 4.1. The movement from *understanding in the moment* to *stable meaning* suggests time and experiences are needed for one to develop stable meaning and also that meanings grow iteratively.

It is useful to consider learning as an evolving process of *coming to know*, continuing to *grow meaning*. As Thompson (2015) reminds us, "The mathematical knowledge that matters most for teachers resides in the mathematical meanings they hold. Teachers' mathematical meanings constitute their images of the mathematics they teach and intend that students have" (p. 437).

The rationale for introducing mathematical meanings into the discussion of creativity and technology is that many of the mathematics teachers in the studies described herein have initially demonstrated mathematical meanings through their activity that appear relatively fragile, inflexible, and insufficient when considering the demands as teachers. As learners, their mathematical meanings are functions of the opportunities they have had to learn, so this finding is not surprising (Presmeg, 2007; Zbiek & Hollebrands, 2008). Their ways of thinking about mathematics tend to be disassociated with the use of DCT to support understanding and meaning (Madden, 2013). Based on pre-course survey data, computer software or even web-based tools have been nearly non-existent in classrooms of teachers in this study. Classrooms with SmartBoards are often used to show static PowerPoint slides and the occasional YouTube video or teacher-directed demonstration (Martinovic & Zhang, 2012). Even though professional organizations have recommended school access to technology to support the learning of mathematics for decades (NCTM, 1989, 2000), many people desiring to be mathematics teachers as well as practicing mathematics teachers do not yet provide access to these tools for their students. There is a dominance of pencil and paper work, hence mathematical meanings are often constrained to those resulting from a relatively static medium (Ekmekci, Corkin, & Papakonstantinou, 2015; Goldin, 2003; Niess, 2006).

There are a host of reasons why teachers do not incorporate technological tools in their classrooms (Zbiek & Hollebrands, 2008). Research has indicated that many teachers simply have not had the opportunity to explore mathematical or statistical concepts and relationships with technology (Madden, 2013). Even those who have used graphing calculators, the only tool research participants regularly report familiarity with, either as students or with students, have demonstrated a very basic sense for what a tool like this might be helpful (e.g., calculating, graphing, tables, some statistical plots and computations). Hence, any mathematical meanings they enjoy are limited to the non-DCT experiences that helped to generate them.

When learners are introduced to technological DCTs, there is a space, frequently significant, between getting started and "becoming one with the tool." This could be considered a significant information gap (Loewenstein, 1994). Elsewhere (Madden, 2013), conditions found to support learners' instrumental genesis (Guin & Trouche, 1998) in new technological environments, where "The core of instrumental genesis in mathematics education is understanding the mathematics of the technology and being able to use it for one's own purposes" have been elaborated (Zbiek, Heid, Blume, & Dick, 2007, p. 1179). In particular, the use of *provocative tasks* (Madden, 2011), *dynamic technology scaffolding* (Madden, 2008), and *sustained intellectual press* (Madden, 2013) in a technologically resourced environment contribute to co-creating curious and intellectually stimulating learning experiences for teachers.

Provocative tasks are those that may elicit surprise, curiosity, controversy, or cognitive conflict for learners (Madden, 2011). Tasks may be mathematically or statistically provocative, contextually provocative, technologically provocative, or in the intersection of two or more categories simultaneously. When tasks are provocative, they promote engagement, discussion, collaboration, and often creativity as solution strategies are considered and pursued. *Sustained intellectual press* is a pedagogical commitment in which a classroom community is co-constructed such that its members come to embrace (1) educative discomfort (Frykholm, 2004), (2) cognitive conflict (Johnson & Johnson, 2009), (3) cognitive overload (Sweller, 1988), (4) undefined endpoints, (5) shared authority for knowledge (Wilson & Lloyd, 2000), (6) individual and group potential (Cohen, 1994), (7) multiple solution paths and ways of knowing (Boaler & Greeno, 2000), (8) collective intelligence, and (9) sharing not comparing (Madden, 2013). Together with *dynamic technology scaffolding*, which is described in detail in the next section, these characteristics combine to contribute to a classroom ecology where learners grapple with non-routine mathematics in unfamiliar ways. They share a safe intellectual space where taking risks and false starts are celebrated (NCTM, 2014). With growing capacity to reason with DCT, learners harness their creativity and express their understanding in new and novel ways in dynamic technological environments. Mathematical and technological discoveries are generated as members co-construct an environment that serendipitously serves as a model for teachers' classrooms—a very different model (Beghetto & Kaufman, 2009). They are **learning mathematics for teaching with technology**. As the following section will illustrate, dynamic technology scaffolding was (1) a framework for the creation of provocative tasks, (2) a sequence of tasks designed to engage novice technological learners in the creation of mathematical meaning, and (3) a tool for teachers to express their creative mathematical and pedagogical insights.

4.1.3 Dynamic Technology Scaffolding: What Is It and How Does It Impact Creativity?

Dynamic technology scaffolding (DTS) (Madden, 2008, 2010; Madden & Gonzales, 2017) is a design principle that supports learners' expanding mathematical meanings while expanding their facility with dynamic technological tools for learning and teaching mathematics and statistics. DTS assumes a modeling stance toward mathematics by prioritizing physical and technological environments for representing and manipulating mathematical relationships (Fig. 4.1). The left-to-right thickening of arrows pointing to the Central Mathematical Big Idea indicate that with each new layer, the mathematical big idea is further developed. DTS task sequences first introduce learners to a big mathematical or statistical idea through exploration with a physical model. The investigation then extends to the dynamic technological environment with an already constructed exploratory model (e.g., expert built) that is isomorphic or nearly isomorphic to the physical modeling experience. Finally, learners in a dynamic expressive (e.g., user constructed) modeling environment establish and demonstrate mathematical meanings and understanding through their

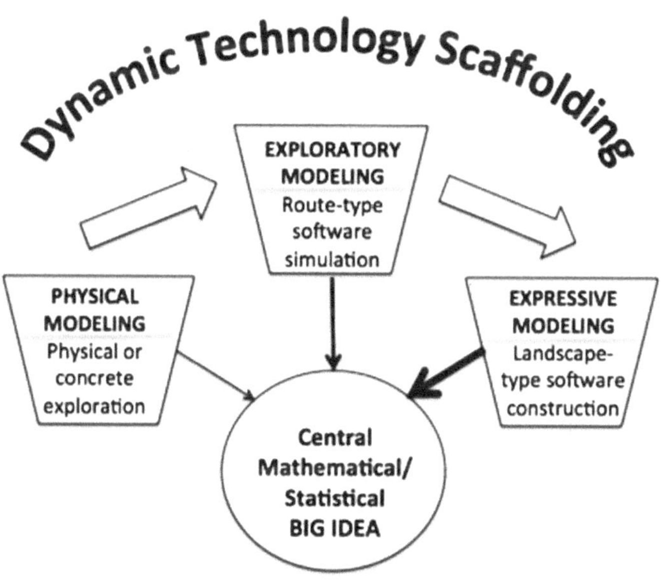

Fig. 4.1 Dynamic technology scaffolding model (Madden, 2008, 2013)

constructions and linking of mathematical or statistical objects as they create their own functioning model in furtherance of their investigation.

DTS has been utilized to provide learners access to complex mathematical or statistical ideas and to support technological facility of novice users of DCT as they navigate new and challenging content and technological terrain (Madden, 2008, 2013). Mathematical creativity emerges during the expressive part of the sequence as learners often struggle to expand their mathematical meanings en route to engineering a working digital model.

DTS has been used both as a pedagogical design device for the construction of mathematical learning trajectories for pre- and in-service teachers and also a framework for those teachers to create DTS trajectories for their students (Madden, 2008, 2013). As a framework, it places the development of big mathematical or statistical ideas centrally with ways of modeling aspects of the content and supporting the development of using cognitive tools in the service of mathematical meaning making as seen in Figs. 4.1 and 4.2. As learners' facility with dynamic cognitive tools (e.g., *GeoGebra*) grows, so does the potential to creatively pursue mathematical solutions to complex problems (Madden 2013). As Sect. 4.3 will demonstrate, DTS has contributed to significantly increasing learners' technological facility and thus their capacity to express technological mathematical creativity.

In additional to supporting mathematical and technological learning, **designing** DTS task sequences is a mathematically creative endeavor. Especially because so much of what is called "mathematics" in schools (and oftentimes college) is still paper and pencil work with largely procedural goals, access to task sequences involving physical, exploratory, and expressive modeling **requires someone to create them**. A number of DTS task sequences have been described elsewhere (Madden, 2010, 2011, 2013), thus a fresh example is presented here (additional examples are provided in Sect. 4.2). Consider the problem based on one from Polya (1945), as an example of a task explored by teachers:

> **Inscribe a square in a given triangle**. Two vertices of the square should be on the base of the triangle, the two other vertices of the square on the two other sides of the triangle, one on each. Your solution should apply for any triangle.
>
> Physical Model: Learners are asked to draw any triangle on a sheet of paper and to try to approximate the location of a square inscribed within it as specified, using straight edges,

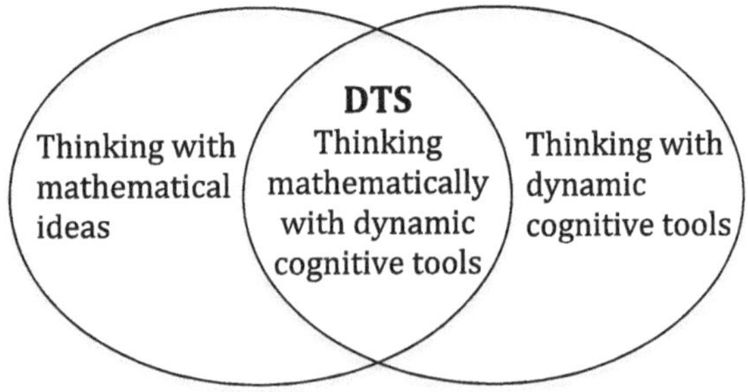

Fig. 4.2 DTS as the intersection of thinking with mathematical ideas and thinking with dynamic cognitive tools

compasses, patty paper, and hand gestures. They compare sketches and typically conclude that they have a good idea about the problem and feel they can work on it.

Exploratory Model: Learners are then provided an exploratory model of Polya's triangle that could be manipulated by dragging the vertices of the triangle to explore variants of a representation differently from the paper versions and to provide evidence that such a construction in a dynamic geometry software environment is possible (Fig. 4.3).

Expressive Model: Finally, learners are asked to use dynamic geometry software (e.g., *Cabri II* or *GeoGebra*) to construct a working model in order to further explore solutions.

In every implementation of this task-to-date with teachers, never has a teacher communicated a strategy for a general solution prior to working on their own dynamic construction. The problem is accessible to a large range of learners, but deceptively challenging. Importantly, the choice of this problem for teachers is not only for them to experience a problem to solve, but to simultaneously orient them to a technological environment where their current facility with a tool is both sufficient to engage, but will require mathematical insights to yield a solution (Ervynck, 1991; Sriraman, 2009). The tool does not yield a solution; the tool in the hands of a creative mathematical thinker yields a solution.

Teachers have been able to create a range of reasonable constructions as they work toward solution, but a solution has never emerged until someone constructs something similar to the objects in Fig. 4.4.

Here a moveable point P is placed on the base of a triangle and a line through P perpendicular to the base of the triangle is constructed to determine the necessary length of the side of a square with P as a vertex. The compass tool and perpendicular line tool in the dynamic geometry software are used to create the additional sides of the square with one side on the base and a vertex on one side of the triangle. Then by moving P along the base, different squares emerge with one of them as the solution. Oftentimes students move P back and forth along the base looking for clues until someone eventually notices that the resultant square does not vary randomly (Santos-Trigo & Machin, 2013). They notice the unattached vertex of the

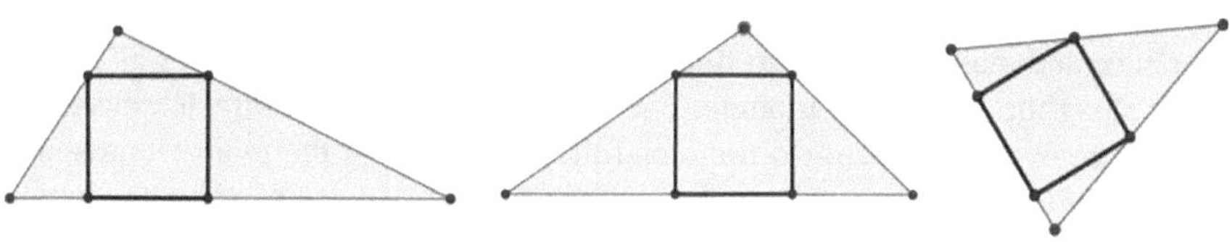

Fig. 4.3 Three images from an exploratory model of Polya's triangle problem

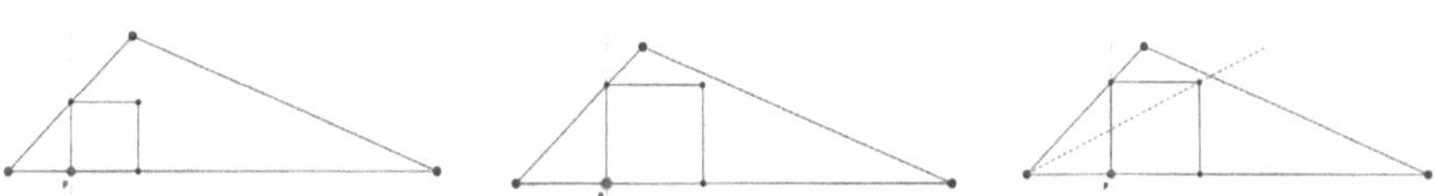

Fig. 4.4 Images representing a dynamically constructed square leading to a locus-related solution

square moves smoothly along a line and often use the locus tool in the software to confirm this. From there, the solution is still not obvious, but someone will inevitably say something like, "Oh, okay, that means that to get THE square I want, I can start with any square with a side on the base and a vertex on one side, draw a line from the vertex of the triangle through the hanging vertex and locate THE point that will define the solution."

It is this Gestaltist moment of insight, where the creative solution emerges (Sriraman, 2009). The four-stage Gestalt model of *preparation-incubation-illumination-verification* (Sriraman, 2009) can be seen as the physical and exploratory models support *preparation*. Creative insights often require periods of work and *incubation* (Silver, 1997). The use of the locus tool makes visible to the learner what was not present before—the regularity of the behavior of the system—*illumination* (de Freitas & Sinclair, 2014). It is also during the struggle of the construction and search for structure where some discomfort can result, but it is recognized that,

> Problems should sometimes be easy and straightforward, so that students come to feel powerful and confident. But sometimes problems should lead to impasse, evoking puzzlement, bewilderment, and frustration, yet offer the possibility of proceeding with renewed determination and achieving the elation of sudden insight or the satisfaction of performing a difficult feat (Goldin, 2003, p. 282).

Collectively unpacking the solution leads to exploration of more formal mathematical relationships related to scaling, slope, proportional reasoning, and dilations in addition to general problem solving. Thus, mathematical *verification* helps to complete the creative solution process.

As learners work on the Polya triangle task sequence, three stages of mathematical creativity can also be witnessed (Ervynck, 1991). Stage 0, the *preliminary technical stage* occurs as learners estimate empirical solutions with paper and tools and make further observations with an exploratory model. Stage 1, *algorithmic activity* occurs as they began to apply principles of construction in a DCT, eventually arriving at a working model. Stage 2, the *creative activity* stage can be seen as a student manipulates his/her construction, looking for patterns, structure, or invariance and makes a choice to find the locus of the unattached vertex as point P moves along the base of the triangle. The technology has allowed the learner to see structure below the surface—to transcend the limitations of the mind (Jonassen & Reeves, 1996; Pea, 1985). Through the construction of hot-linked mathematical objects according to the rules imposed by the builder, an external representational system has been engineered to explore (Goldin, 2003). Some aspects of the system are obvious while others are subtle, requiring an inquiring eye to seek pattern and invariance (Santos-Trigo & Machin, 2013). In a DCT environment, the builder is in control and has flexibility to add emergent details to a design to further investigation. The disposition for pattern sniffing (Cuoco et al., 1996) is fostered and the tool offers possibilities to extend an investigation.

Experience with tasks like Polya's triangle task serve to nurture the idea of locus as a set of points in a dynamic situation as a function of another object moving along a path—a particularly powerful mathematical representation and idea, one which has not been well-developed in the learners in the studies discussed here. However, once learners have the opportunity to experience the power of the locus tool as they explore various problem contexts (e.g., locus definition of parabola or ellipse, Grashof's principle), they begin to gravitate to it when they suspect some regularity within a system. Polya's triangle problem as a DTS sequence contributed to learners' evolution of mathematical thought and prepared them with more tools for future creative problem solving. The tool did not solve the problem, rather it afforded the learner a plethora of opportunities to interrogate and amend their mathematical representations and meanings en route to more stable ways of thinking (Thompson, 2015).

The Polya triangle task supports recognition that a solution to the problem may be an extension of stretching and shrinking through the lens of proportional reasoning. The application of a familiar mathematical idea (e.g., proportional reasoning) in a novel context (e.g., Polya triangle task) provokes learners to re-evaluate and extend their own mathematical meanings related to proportional reasoning. Polya's triangle task is just one example of a DTS sequence. The following sections examine DTS as a creative lever for expanding teachers' ideas about learning mathematics for teaching with technology and then teachers as designers of DTS tasks as acts of mathematical and technological creativity.

4.2 Dynamic Technology Scaffolding as Creative Lever for Expanding Teachers' Ideas of Learning Mathematics for Teaching with Technology

Table 4.2 contains a partial list of DTS sequences that have been enacted in research studies (e.g., Madden, 2008, 2010, 2011, 2013). The set illustrates a wide variety of mathematical ideas being explored and a range of physical and technological tools and is provided to stimulate thinking about the design of DTS tasks. Each of the DTS sequences arose from a thought experiment during which the author, as designer, considered the potential for learners' mathematical development, technological development, and pedagogical development. Seeds of inspiration for DTS tasks may be found in many contexts (Madden & Gonzales, in press). Many digital exploratory models are now freely available. Some particularly useful websites include http://nlvm.usu.edu/en/nav/vlibrary.html, http://www.nctm.org/coremathtools/, https://illuminations.nctm.org, http://www.cut-the-knot.org, http://www.shodor.org/interactivate/, and https://phet.colorado.edu/en/simulations/category/math.

As teachers begin to experience and then imagine mathematical tasks like those from Table 4.2 in the presence of DCT, the world of mathematics meaning takes on additional dimensions (Thompson, 2015). Optimization problems are not restricted to techniques of calculus, p-values become meaningful through physical

Table 4.2 Sample dynamic technology scaffolding task sequences

Problem context	Mathematical ideas	Physical model	Exploratory model	Expressive model
Soda can rack	Geometric constructions, proof	Steely balls and cardboard box	*Illuminations* website	*Cabri II Plus Geometry* (or *GeoGebra*)
Parabola as a locus of points	Locus definition	Patty paper	*Cabri II Plus* demo or applet	*Cabri Plus Geometry* (or *GeoGebra*)
Varignon's parallelogram	Properties of quadrilaterals, proof	Ribbon, elastic, and humans	*Cabri II Plus* demonstration (often not needed)	*Cabri Plus Geometry* (or *GeoGebra*)
Quadrilateral linkages	Grahof's theorem	Geostrips	*CPMP-Tools* Design a linkage	*Cabri Plus Geometry* (or *GeoGebra*)
Steroid testing	Probability, empirical sampling distributions	Dice, coins, signs, humans	Graphing calculator simulation or *Fathom 2* simulation	*Fathom 2 TinkerPlots*
Orbital express	Randomized controlled experiments, randomization testing	Paper towels, measuring device index cards	*CPMP-Tools* Randomization test	*Fathom 2*
Moving ladder, path of cat on ladder	Rate of change	Meter stick, wall	Graphing calculator, applet	*GeoGebra*
Slicing a cube	3D modeling, composing, decomposing shapes	Plastic shapes, colored water	*Cabri 3D* video	*GeoGebra 3D*
Triangle orthogonal projection	Optimization	Rulers	*GeoGebra* virtual worksheet	*GeoGebra*
Polya's triangle	Problem solving, proportional reasoning	Paper, triangular cutouts	*Cabri II* demo	*GeoGebra*
Ellipse	Conic sections, locus definition	Ribbon, humans	*Cabri II* demo	*Cabri II* or *GeoGebra*
Oil well problem	Minimization, reflections	Miras	Graphing calculators	*Cabri II* or *GeoGebra*

(continued)

Table 4.2 (continued)

Problem context	Mathematical ideas	Physical model	Exploratory model	Expressive model
Rotations as product of reflections over intersecting lines	Transformations	Miras	*National Library of Virtual Manipulatives*	*Cabri II* or *GeoGebra*
Three point semicircle problem (see Sect. 4.3)	Probability	Pipe cleaner, cereal pieces	*Cabri II* or *Cut-the knot applet*	*Fathom, Cabri II* or *GeoGebra*

randomization approaches, transformational geometry contributes to elegant solutions in place of (or in addition to) extensive algebraic manipulation. The word, "imagine" is fitting because it is through imagination and thought experiments that creative task sequences are born (Vygotsky, 2004). In Sect. 4.3, an extended example illustrating teachers' exploration of the semicircle problem will expand the discussion of mathematical meanings, DCTs, and creative mathematical and technological approaches to non-routine mathematical tasks.

4.3 Technology, Creativity, and Mathematical Meaning Collide

During the second half of the course, after developing some initial facility with the dynamic statistical software, *Fathom*, students read, *Can You Fathom This?* (Edwards & Phelps, 2008). Questions were posed at the end of the article and students were asked to select at least one task to work on. Serendipitously, many students gravitated to the following task:

> **Semicircle problem**: Three points are chosen at random on the unit circle. Find the probability that all three points lie on some semicircle (pp. 215–216).

Ponder your own response to this task. Figure 4.5 depicts a physical model used to build some intuition to the problem. As students worked on this problem, no exploratory model was utilized because the reading served as a scaffold. Students were encouraged to use tools of their choice and prepare to share solution approaches.

Figure 4.6 contains one student's *Fathom*-facilitated creative solution to the problem. This student used algebraic relationships to generate sets of three random points to be tested against the semicircle criteria. As exemplified by the attribute definitions below, the student created a way to generate and test sets of points. She exhibited a need to verify her setup through a visual geometric representation in addition to the numerical data. As illustrated in the figure, she created a way to

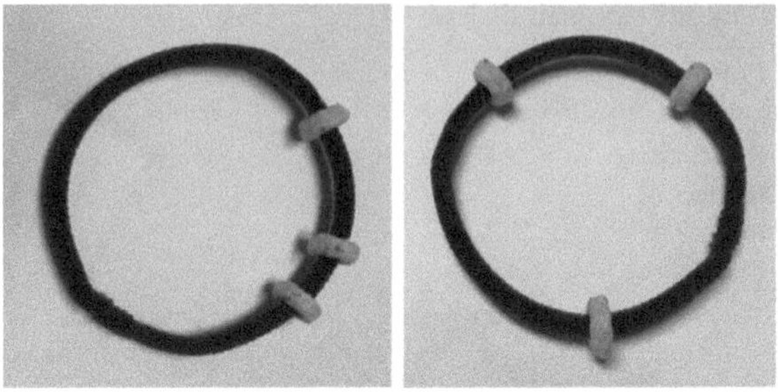

Fig. 4.5 Physical models of semicircle task made from pipe cleaners and cereal pieces. This model can be modified to explore a line segment approach to the problem

leverage the numerical capabilities within *Fathom* to geometrically represent three unit circles in order to determine whether her simulation worked correctly, a move highlighting representational competence and flexibility. Once convinced, she constructed an empirical sampling distribution (generated samples, created measures, and collected measures) to correctly estimate the theoretical solution. Her solution was highly creative, highly mathematical, highly visual, fascinating, correct, and impossible without technology.

Another student excitedly explained a different approach to the class as peers asked questions about her method.

Ella So, I found two cases where there's 100% probability, after choosing the first two points, that there's 100% probability that you'll have a semicircle by choosing that third point. So those two cases are if the two points are one on top of each other, that no matter where I put that third point, that would be true, so that's a 100% probability. And then the other case was if we have a semicircle they're on exactly opposite sides, so perfectly opposite, then if the point fell either on the top or the bottom, that would also be 100% probability. And I also looked at the extreme case for when the probability was as small as possible. So that's if the two points are just slightly less than exactly opposite each other. So that, like maybe that would be a semicircle (referring to a drawing) so the point would have to lie on here (referring to a drawing), which is slightly more than 50%, this is about 50% chance. And so because I knew that these were uniformly chosen, we should have, it should be uniform somewhere between 100, somewhere between 50 and 100%, uniformly distributed, so I knew that the probability would have to be 75%.

Ella had logically arrived at an elegant solution to the problem and she expressed confidence in her solution. However, as a novice Fathom user, she was challenged to find a method of solution using the tool. Here technology was mediating her mathematical reasoning in ways that were in the moment unfamiliar, but about to become powerful.

Ella **But I wanted to figure out how to run a *Fathom* program to do that**. So I had the Fathom program run, to choose a random number between 50 and 100% and I did that by having it be one minus random, which is a

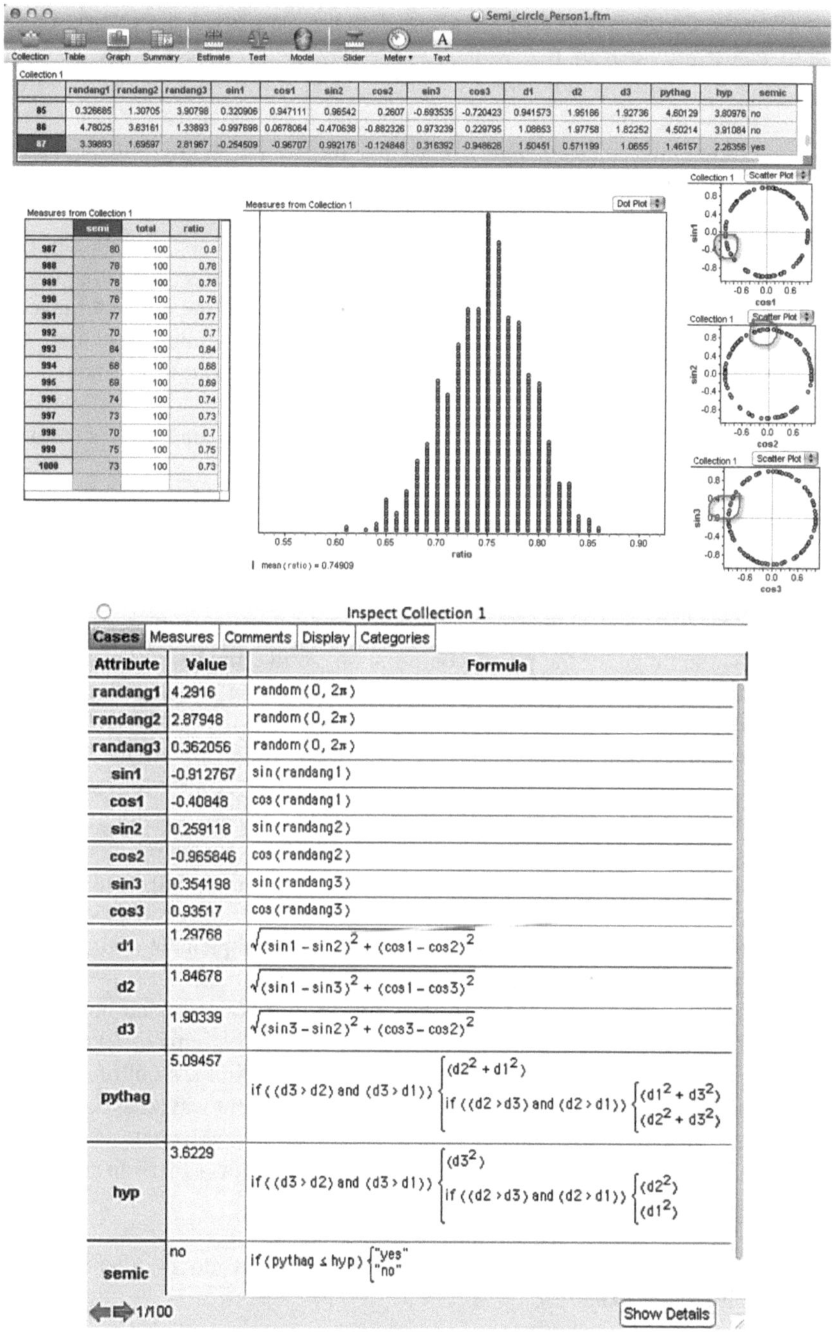

Fig. 4.6 *Fathom* simulation for estimating the probability for the semicircle task using brute force along with geometric representation. Attribute definitions illustrate mathematically creative meanings and goals in the technological environment

randomly chosen number between 0 and 1, and divided that by two. So random divided by two would give us anywhere between 0 and 50% and the one minus would make it between 50 and 100%. And then I had it run a second measure, so this was my measure 1 [referring to *Fathom* syntax, [(1-random())/2], and then my second measure was choosing a random number between 0 and 1, and then, so that was essentially where my third point would fall. And so I said that if this number (random()) was equal to or smaller than this number (referring to measure 1), then it's falling on a semicircle, so that's an allowed case. And if it was larger than this number (again referring to measure 1), then it was outside of the allowed range for possible semicircles. So then I created a measure, semicircle or not. And then I had it collect those measures. And that came out to be about 7, 7½ out of 10 times.

Nick So just to kind of go through measures. Your first measure kind of sets your first two points.

Ella Yes.

Nick And then your second [measure] places your third [point] and it asks whether or not it's a semicircle, which your third measure kind of gives you that answer.

Ella Yeah. Well so my first measure what basically was setting up what the possible range for the points could fall on, so like [makes a sketch] if I had a circle and these were my two points [referring to sketch]. Then this [referring to one of the points] could either be the start of the circle [meant semicircle] and go to the other side, here [referring to the end of a semicircle]. Or this could be the start of a circle [again meant semicircle] and it would end over here [referring to sketch], so my total range for possible locations for that third point would be anywhere in that range [between endpoints of the two semicircles on the diagram]. And the range that would be outside of the possible, so my circle formation would be, if the point fell anywhere in here [referring to the complement of the previous set on the sketch]. Yeah [responding to peer question].

Kaley How did you display the data once you had all your yes and no's?

Ella Well, so I collected these [measures 1, 2, and 3] in groups of 10 and then I created a measure to count the number of yeses. And it's out of 10, which isn't the best setup, I would like to do it in a different way, maybe later, but then I just had it tally up the number of each number of counts of yes out of 10, and I got, so that ranged from 4 to 9, I think or 4 to 10 in my couple hundred samples, I think it was a 100 sample.

Wes So you did like a 100 samples of (long pause).

Ella 10, yeah. Which wasn't ideal, I would probably do it differently later (Classroom applause).

Ella forcefully explained and defended her *Fathom* construction and illustrated her growing understanding of the use of empirical sampling distributions to explore probabilistic phenomena. Her explanation also included a drawing that was later

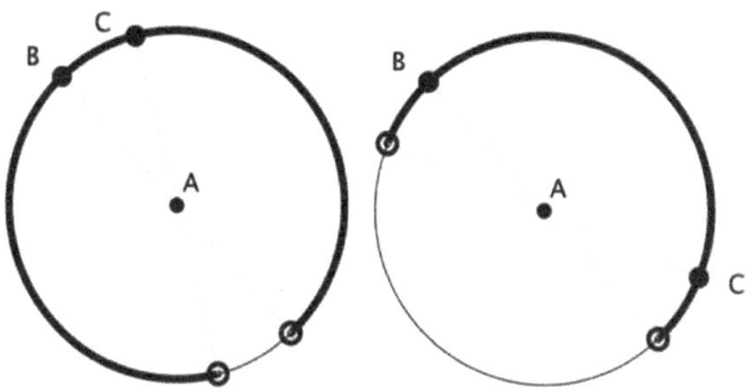

Fig. 4.7 Dynamic geometric model for the semicircle problem. The highlighted arc indicates the space of possibilities for a third point to lie on a semicircle

turned into a dynamic geometric construction to illustrate her thinking. In Fig. 4.7, dragging points B or C generates the range of locations that a third point could take to complete a semicircle. This move represents a technological turn toward a geometric representation and a creative construction using *GeoGebra* to augment a previous numerical approach. Creating this working simulation required a way of thinking about the problem that took into account the original pipe cleaner representation as well as Ella's thinking that led to her numerical solution. Through the process of engineering a system that worked, the learner(s) continued to interrogate and explore additional mathematical relationships (Santos-Trigo & Machin, 2013).

Finally, another student approached the problem by cutting the circle at the location of the first random point, straightening it to a segment and imagining it with length 1 unit. Then two other random points were generated and tested against the semicircle condition by invoking measurement conditions. This student also built an empirical sampling distribution to estimate the theoretical probability of a semicircle (Fig. 4.8). This particular strategy leads nicely to a non-technological mathematical argument.

For these students, *Fathom* and *GeoGebra* became very powerful tools for expressing their creations and provided a landscape for their mathematical meanings to emerge and thrive. These kinds of technological approaches to problems became commonplace toward the end of the course. As students shared approaches, others took them up or modified them to suit their thinking (Vygotsky, 2004). Students' creative technological and mathematical solutions to problems became shared, thus contributing to a growing repertoire of strategies and insights that simultaneously enhanced learners' mathematical understandings AND technological facility.

In addition to exploring new and challenging content and technological terrain as learners, these teachers were also tasked with creating DTS trajectories for supporting learning in their own classrooms, an instructional move that has both challenged and inspired mathematics teachers. The cognitive demand (Stein, Smith, Henningsen, & Silver, 2000) of creating these trajectories is exceptionally high. As discussed in following section, teacher-constructed DTS task sequences have been analyzed and characterized as mathematically creative, technologically creative, or mathematically and technologically creative. When teachers design mathematical task sequences incorporating DCT and DTS, they open a window to their mathematical meanings and simultaneously extend their mathematical learning for teaching with technology.

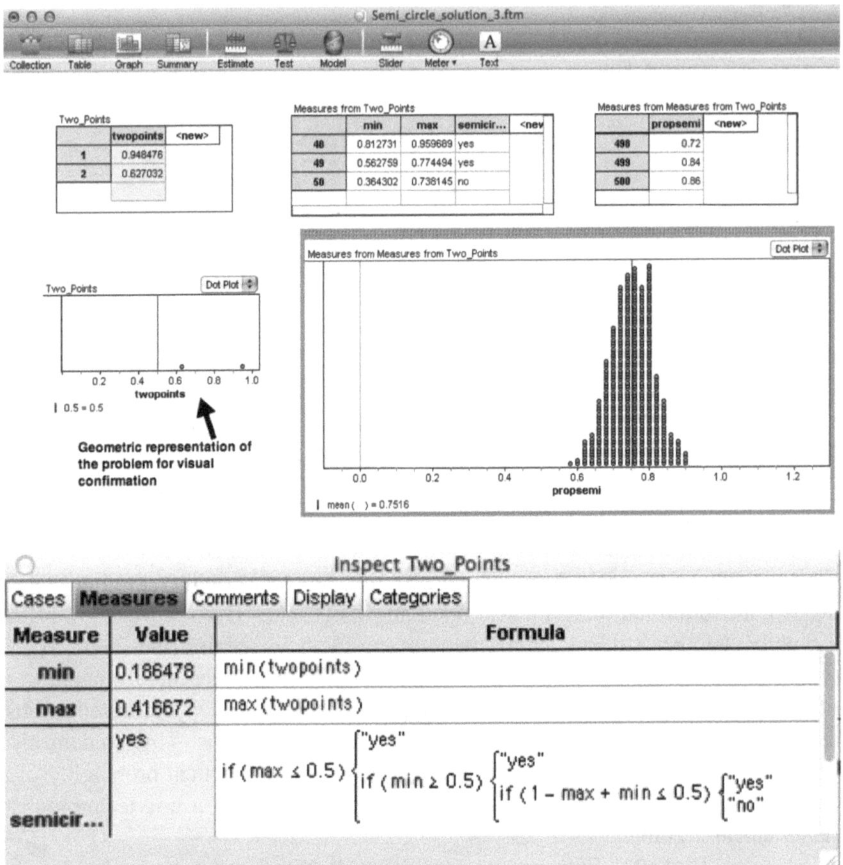

Fig. 4.8 Fathom simulation for estimating the probability for the semicircle task using the segment representation. The measure definitions illustrate the learners' thinking about the mathematical relationships in the context of the digital tool as well as creative and computationally appropriate insight into a mathematical solution

4.4 Teachers Designing Dynamic Technology Scaffolding Sequences: Acts of Mathematical and Technological Creativity

Teachers were assigned to develop two DTS sequences midway through the course. Having experienced 8–10 DTS prototypes as learners (e.g. Table 4.2), teachers were asked to assume a *Doers to Designers* (Kadijevich & Madden, 2015) perspective in order to design DTS task sequences. Though teachers' mathematical meanings and instrumental genesis with tools had increased since beginning the course, the creative demands of designing DTS sequences did not go unnoticed by teachers. For example,

> In selecting my developmental tasks for Project 1, I kept reflecting on the Dynamic Scaffolding Model. I wanted to build a physical model, followed by an exploratory activity and pull it together with an expressive activity. In all honesty, I was stressed. Not because I didn't understand the model, but because I felt constrained for the following reasons: (1) I'm not creative and (2) I'm limited when it comes to technology; the tool still remains an artifact. I have not yet obtained instrumental genesis. I am not one with the tool … The final phase of the dynamic technology scaffolding model is designated as the landscape type software construction phase. Although I feel I addressed the physical exploration and software simulation phases of the model, I believe that my expressive portions of my tasks seemed weak. (Shannon).

Teachers were asked to envision task sequences similar to those they had experienced during the course, but to support the exploration of a different mathematical idea. They had complete autonomy over content and context selected, tools selected, and the development of ideas, but design is highly creative work (Dorst & Cross, 2001). Because teachers have to imagine the development of a mathematical task sequence that purposely deepens the level of mathematical concept development as the inquiry transitions from physical environment to exploratory modeling environment to expressive modeling environment, they are designing task sequences unlike those typical in their own prior mathematical learning or teaching experiences. Shannon, like many others, poignantly expressed her personal uncertainty with the project as she announced her aversion to creativity. Imagination is a highly complex process and to see a decline in creativity in adulthood is common, so her response, similar to many others, may be anticipated (Vygotsky, 2004).

Since 2010, 64 teachers in this environment have designed 128 DTS task sequences. Creativity in this context was viewed as a novel or inventive use of tools, materials, or trajectory to develop a mathematical or statistical idea (Csikszentmihalyi, 2000; de Freitas & Sinclair, 2014; Silver, 1997). Each task was analyzed and categorized according to the nature of creativity exhibited using the following criteria:

- **Routine exploration of mathematical or statistical ideas**—direct application of typical models to routine or widely familiar textbook tasks. Limited creativity.
- **Technologically creative**—illustrates novel technological approach; physical model may be creative; a combination of different exploratory and expressive tools may be coordinated; may significantly extend facility with a tool. Mathematical context is routine.
- **Mathematically creative**—involves non-routine context, is cognitively demanding, has multiple solution paths. Technological components are borrowed directly from other sources or approaches are trivial with the tool.
- **Mathematically and technologically creative**—involves non-routine context, is cognitively demanding, has multiple solution paths AND—illustrates novel technological approach; physical model may be creative; a combination of different exploratory and expressive tools may be coordinated; may significantly extend facility with a tool.

Of 128 DTS sequences analyzed, 51 (40%) were categorized as routine exploration, 30 (23%) as technologically creative, 15 (12%) as technologically creative, and 32 (25%) as mathematically and technologically creative. A significant proportion of the tasks created by teachers indicated a tentative approach, much like Shannon's, to the project. That the highest proportion of tasks were coded as routine exploration tasks indicating teachers' propensity to reach for very familiar mathematical ideas (e.g., triangle midpoint theorem, sum of angles of triangles, etc.) and construct task sequences that illustrate DTS, but in a way that does not appear to push their mathematical understanding or creativity a great deal. Though less creative than tasks in other categories, routine exploration tasks still require coordination of physical and virtual tools for mathematical activity and for teachers this *is* creative activity (Beghetto & Kaufman, 2009). For them, it is novel and useful (Csikszentmihalyi, 2000) and contributes to solving a problem of practice related to providing students' physical and digital access to mathematical ideas.

Approximately 23% (30/128) of tasks were categorized as technologically creative. Figure 4.9 contains one teacher's technologically creative DTS task sequence. Here, the physical model is quite creative for the context of the exploration, but the extension to the virtual environment is direct and not particularly challenging, hence not mathematically creative. This categorization recognizes creative approaches using virtual or physical technologies, but, as in this case, the mathematical demand of the task is limited or routine.

Together, 45/128 (35%) of DTS sequences were either mathematically creative or technologically creative, but not both. Task sequences in these categories tended to address mathematical content that was highly familiar to teachers (e.g., Pythagorean Theorem). Task sequences in these categories explored the use of combinations of less familiar physical materials, applets, and virtual construction environments, or addressed less familiar mathematical content (e.g., from innovative curriculum materials or more obscure mathematical problems) using technological approaches that are not seen as novel (e.g., physical materials are paper and pencil, exploratory model is teacher constructed and students simply replicate in the expressive phase). Frequently in these categories of creation, teachers have identified interesting content or relevant tools, but the scaffolding approach either simply replicated the solution to a mathematical problem across the three different modeling phases, thus not advancing the mathematical learning of the learner in any significant manner or involved some confusion of the DTS principle. However, teachers demonstrated a higher degree of creativity here than the routine exploration group through their choice of problem contexts, physical and exploratory tools.

The remaining 25% (32/128) of DTS tasks represent highly creative work that illuminates learners' grasp of DTS as a design principle through the selection of cognitively demanding mathematical ideas scaffolded with thoughtful physical materials, appropriate exploratory models, all leading to learners constructing in an expressive environment in ways that will extend their mathematical and technological facility. The example in Fig. 4.10 illustrates a creative use of physical materials to explore conic sections and an appropriate exploratory model to further the exploration. The expressive model provides a potentially useful site for

Activity: Students will discover the relationship between central and inscribed angles in a circle. This is an important concept that is in many word problems in Geometry and can easily be overlooked/forgotten. The idea that the vertex can move without changing the angle measure can also be discussed/proved using the models.

Physical Model: I have constructed a model of this using a bicycle wheel, a rubber band, two broken pencils, and duct tape. It allows the ability to make any angle within the circle, The vertex can attach to the center of the wheel to replicate the central angle. The rubber band allows a student to "stretch" the vertex down to intersect with the circle and become an inscribed angle.

Interactive model: This also is a Cabri demonstration that students can watch get manipulated on a projector. There is a circle constructed from a "driver" with an angle and its measure given. The vertex of the angle is movable and can be placed anywhere (including the center and edge of the circle). The two endpoints are fixed on the circle but can be adjusted.

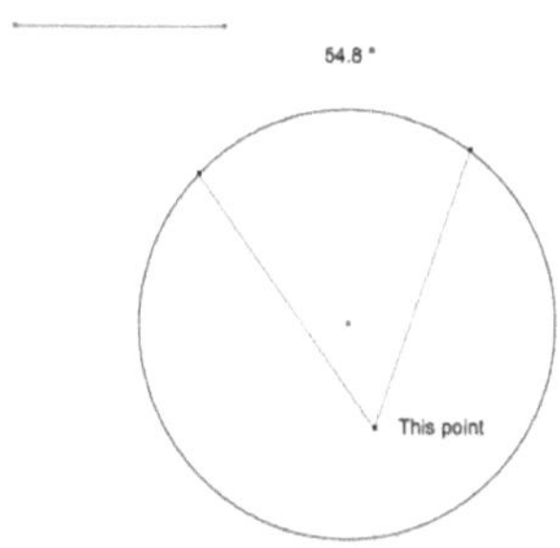

This behaves just like the physical model. There would be merit to having students construct this themselves in the "third step" on Cabri, as it will deepen their understanding of the concept and get them more comfortable with the software. Using a driver, the circle itself can be manipulated to be different sizes and students can see that it still preserves the property.

Fig. 4.9 Teacher's DTS task illustrating a creative physical model with only basic construction demands in the expressive modeling stage

exploration; however does not maintain fidelity to the model of the mathematics under consideration, from a DTS perspective. Still it illustrates a teacher's potential goals for students in a technological environment and a creative approach to the development of tasks for students using DCT. Even though DTS was not fully realized in this application, the teacher demonstrated creative use of DCTs for the exploration of conic sections.

The example in Fig. 4.11 represents an ambitious task for a beginning designer and another mathematically and technologically creative task. Though the student

Task 1: Exploring conic sections

1. Physical Exploration: Materials needed: Party hats, scissors, paper, markers

Objective 5 minutes:
Pass out the materials to all students in the class. The big idea or objective of the lesson is to explore conic sections and manipulate the plane cutting through them to see properties and bring forward thinking about important features of the circle, ellipse, and parabola. This is an introductory lesson.

Opener 10 minutes: Students will be given three party hats so they can see each conic section. Have each student cut his or her party hat in a random way. When they are done, have them place their paper on top of the cut end and use their marker to outline the shape created on the cone. Have students share with their groups (preferably in groups of 3 or 4) their shapes and write down what they see the same or different about their shapes and their group members.

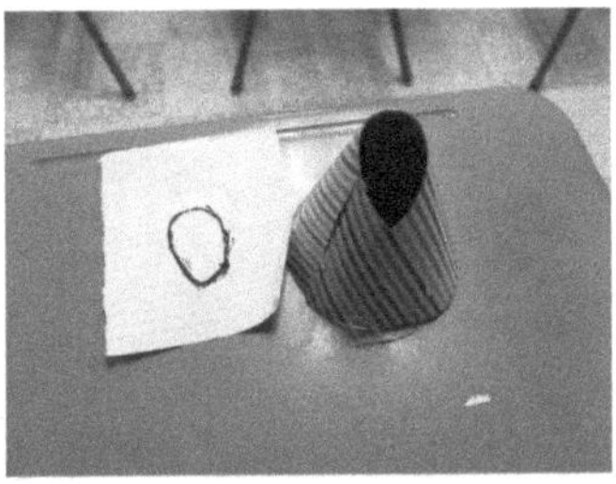

Mini Lesson 20 minutes:
Introduce to students the three conic sections and their associated definitions. Ask students if their paper or "planes" represent any of these conic sections. Have students try to cut their party hats to obtain the other two conic sections on the board and ask them to write down how cutting it plays a role in the development of the different sections. Does angle play a role? Moving the scissors up or down make it different? What about the ends of the hats or "cones"? How does the "plane" or paper play a role? This gets students thinking about moving the plane or "slicing" the cone in ways to develop the other conic sections.

2. Route type software simulation: Materials needed: Ipad/laptops, internet, graphing paper

Mini Lesson 20 minutes:
Have students go to http://illuminations.nctm.org/Activity.aspx?id=3506 and introduce the model to the students. Ask students how this model is like their own. Ask student what they think height and slant mean under the cone parameters. Ask students what m and b stand for in relation to the plane (may need to refresh $y = mx + b$ here). Play with the parameters and have students watch what happens in the plane next to it. It is key to notice whether or not students are making the connection between the plane and "paper" that is cutting the cone and relate it to the physical models that began with. After introducing the model have students play and explore the model on their own. Ask them to write down any relationships that they observe between the cone view and the graph view.

Does angle play a role? Is there a connection between m and angle? What about b and the shift of the paper up and down? These would be essential questions for students who seem to be struggling with concept and/or have students dig deep for more connections. Have students use the graphing paper to have the images and further the connections that need to be made before continuing to the next part.

> Explore the different conic sections and their graphs. Use the **Cone View** to manipulate the cone and the plane creating the cross section, and then observe how the **Graph View** changes. (**http://illuminations.nctm.org/Activity.aspx?id=3506)**

Fig. 4.10 Teacher's DTS task illustrating development of mathematical ideas, multiple modeling environments; limited development of technological development with the expressive model

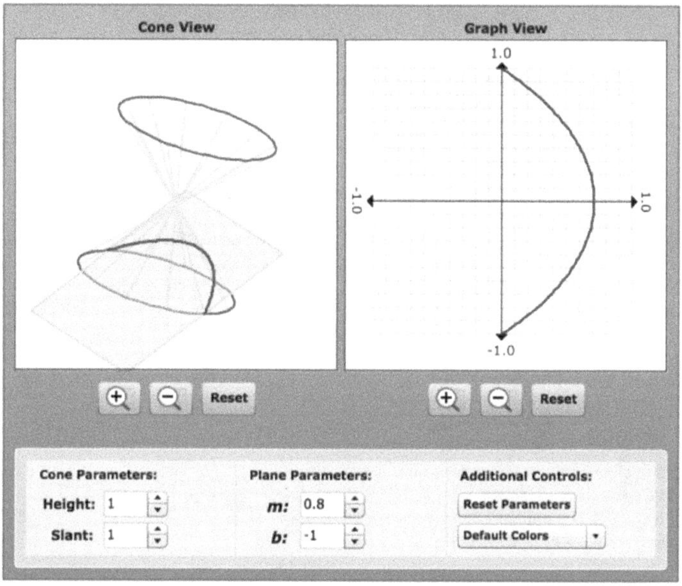

3. Landscape type software construction: Materials needed: Ipad/laptops, internet, access to Geogebra

Mini Lesson 20 minutes:
Have students access Geogebra through the Internet and ask them to construct the conic through 5 points. Have students play with the model and ask them how this relates to the first two models. Why is it constructed through 5 points? How did students construct it? How does constructing it this way connect to the ideas you developed in the first two tasks? Have students explore and develop connections by manipulating points and asking what do these points mean in terms of the models you looked at previously?

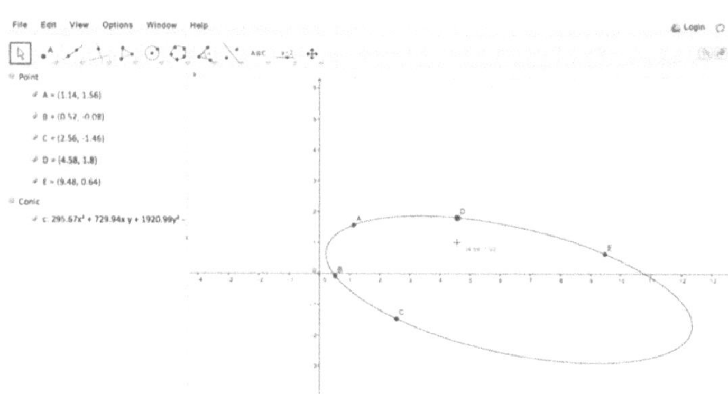

Fig. 4.10 (continued)

Task 1: Locomotive Walschaerts Valve Gear

Tool:
Designed using Cabri; adaptable to Geometer's Sketchpad, used by my classroom.

Objectives Addressed
G-CO 12. Make formal geometric constructions with a variety of tools and methods (compass and straightedge, string, reflective devices, paper folding, dynamic geometric software, etc.).
G-MG 1. Use geometric shapes, their measures, and their properties to describe objects

The primary goal of this task is to improve students' use of Geometer's SketchPad in an expressive way. Thus far, my students have primarily used dynamic geometry software in an exploratory way. Students are generally given step-by-step directions. In this activity, students are challenged to use DGS tools to express phenomena they observe or experience. This will help them gain expertise in DGS that helps them achieve instrumental genesis.

Physical or concrete exploration:
Students will examine moving diagram here:
http://en.wikipedia.org/wiki/File:Walschaerts_motion.gif

I was unable to create a physical model so far. I think that I would be able to with some bicycle wheels mounted in a line through their centers. I would have a link made from piece of wood connected to a spoke of each wheel that would force the wheels to move together. I would also have a driver piece that could move along the wheels' mount (as shown in the screen-cap of the Cabri model) that is connected to the link.

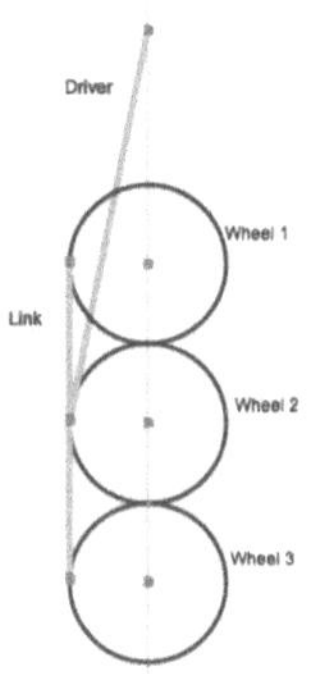

Route-type software simulation: Cabri demonstration attached

Instructions for students
Observe the online animation, and explore the physical model.
Pick out the essential elements of the model. Your own model should include the wheels, the driver, and the link between the three wheels.
You may look at and move my model, but may not show any hidden lines.
Students will be given instructions about the "Compass" which they are not yet familiar with.
Use any tools you think are appropriate to create the model. Ask me if you have any questions.

If you complete the basic construction shown here, add other components of the valve gear.

Fig. 4.11 Teacher's DTS task illustrating development of mathematical ideas, multiple modeling environments; ambitious context, potential to extend technological facility through modeling

had not yet successfully created a physical model, she made strides in that direction and navigated a challenging construction in *Cabri II* that she included with her written work, demonstrating her productive modeling evolution in *Cabri II.*

Across these three examples, despite the varying degrees of attending to the DTS framework as intended, teachers demonstrated growth and development in their creative use of physical tools for modeling, appropriate uses of exploratory models to extend limitations to the physical world, and increasing capacity to imagine students constructing and creating mathematical models in expressive environments (Madden, 2013). Teachers' reflections in the following section will further corroborate the impact of DTS 'doing and designing' on teachers' views, beliefs, and practice.

4.5 Mathematical Meaning, Creativity, Technology, and Teaching

Excerpts from a small but representative sample of students' final reflection papers for the course are shared. These artifacts illustrate some of the ways students' mathematical and technological creativity was evoked and experienced and how their mathematical meanings expanded to impact their teaching, thus increasing their TPCK. Creativity abounds in their words as they communicate profound transformations in their beliefs and actions related to teaching mathematics with various technologies. As designers of learning spaces, their introduction of new technologies to their students represents a creative act (Jonassen, 1994). It is novel and useful (Csikszentmihalyi, 2000) and is unusual compared to past practices (de Freitas & Sinclair, 2014). Participants' words provide evidence for many of the claims made throughout this chapter regarding mathematics prior knowledge and dispositions, experience with tools as learners and designers, characteristics of mathematics learning environments, and connections to some of the DTS tasks mentioned in Sect. 4.2. Signs of vulnerability, apprehension, excitement, curiosity, reflectiveness, awareness, and growth permeate their writing. The excerpts are direct quotes and all names are pseudonymns.

4.5.1 Shelly

> At first I just saw the computer software as fun, but through this course have realized how much meaning it can bring to the classroom. When I say that my views on teaching and learning mathematics have changed, I am inclined to ask how could they not have. I was never exposed to the dynamic technology of Cabri II, Fathom, or Tinkerplots in my K-12 years, so I was sure (discontentedly so) that I would be teaching mathematics the same way most of my teachers did lecture-style, straight from the book. I now believe I have the resources (through the in-class activities, assignments, and our texts) to implement technology other than calculators into my classroom and can see the copious amounts of benefits they will have for the students in my lessons.

Shelly recognizes a shift in her thinking about the work of mathematics teaching to include new technologies and indicates a newly found appreciation for dynamic technologies for teaching mathematics. For her, DCTs and innovative resources moved her to consider new and different ways of teaching, that is, they provided the inspiration for her to **create** a new model of herself as a teacher using tools and resources that she now believes will be useful in support of student learning (Csikszentmihalyi, 2000).

4.5.2 Caitlyn

> My attitude toward the use of dynamic technology, since I began using Fathom in my mathematics classes, has been courageously optimistic. I have observed firsthand the potential of these technological tools to enhance student learning. In my experience, the use of dynamic technology supports students' reasoning and sense-making of the mathematical concepts. My students are generally more engaged, more self-directed, and more interested in the content. They are more motivated learners in this technological arena. My classroom is more student-centered as the students are allowed to take ownership of their learning. Zbiek and Hollebrands (2008) highlight the need for teachers to create learning environments in which students' thinking is valued...This semester they [students] reasoned about the quadratic models from the steroid study in lesson two of Mathematical Modeling Our World that was introduced during the course. The students used Fathom to create simulations and algebraic models that represent their findings.

Caitlyn, an eight-year veteran teacher, recognizes her teaching has been transformed by the intentional use of *Fathom* to teach mathematics. She has intentionally worked to create conditions through the use of new curriculum materials and computational technology to encourage a more student-centered learning environment allowing for student creation of digital mathematical representations and solutions.

4.5.3 Jason

> This course has been pretty eye opening for me. Coming in, I had no idea what any of the programs we used were. By being introduced to them I not only have other sources for the students to be able to connect with but something that opened my mind to other possibilities, connecting physical manipulations to the computer programs, setting up problems for students to explore, etc... One of the biggest moments I had with Cabri was when we were given the three problems to work on using Cabri in class time. They were the oil problem, investigating rotations, and the feed and water problem. For each problem I didn't think I was going to use Cabri. I really didn't want to use Cabri, I was confident in my own abilities to figure it out alone. For each problem, though, Cabri ended up being pivotal in finding the answers and completely switched my beliefs around. It wasn't all Cabri either, just something that clued me in and helped me think about the problem in a way that I couldn't visualize.

Jason conveyed awareness of the way in which *Cabri* impacted his thinking en route to solving several optimization problems. Inherent in his response is a reluctance to use tools, perhaps because he may hold a somewhat purist view of mathematics. Nonetheless, his use of *Cabri* sparked a realization that a dynamic geometry tool could help him "see" a solution in a new and powerful manner (Kashdan & Fincham, 2002). His creation of dynamic visual representations extended his realm of reasoning and provoked him to reconsider his position about the use of technology (Madden, 2013).

4.5.4 Connor

> This past semester has provided me with a great insight on building a true understanding of a concept. Three-level scaffolding using dynamic technology provides such a clear connection from ideas to physical models to a dynamic, virtual model. Talking about an idea can give some insight to a topic with minimal understanding. Building physical models can further force knowledge out of the brain, but it's the dynamic, virtual model that takes it to that incredible level… Graphing calculators, computer software, and website applications are in the forefront of the dynamic technology available to mathematics learners. These technologies take the concepts into a realm that is well beyond most imaginations – allowing learners to observe examples well beyond the limitations of a classroom.

Connor recognizes the power of the technology to transcend the limitation of the mind (Pea, 1985). He acknowledges the importance of imagination and the idea that technologies extend one's imagination to new and valuable territory.

4.5.5 Ellen

> The only technology that I had ever used in the classroom was a TI-84 and My Math Lab software. I didn't realize the world of technology stretched far beyond these two "artifacts. … The most effective application of technology that I learned from this course was the dynamic technology scaffolding projects that we did in class. We did paper folding to learn about parabolas. We used string to work with quadrilaterals and parallelograms. We did a complete lesson on Steroid testing. All of these are examples of lessons that I would love to use in my own classroom… This class showed me that my understanding of mathematics doesn't run as deeply as I had thought. For example, we discussed the definition of a parabola as being a shape such that the distance of the focal point to any point on the parabola is the same as the distance from that point on the parabola to a pre-determined straight line used to create the parabola… Not only is this a definition that I had never heard of, but this is a characteristic of a parabola that I didn't even recognize. Once this definition was established, it was an absolute battle to construct this shape in Cabri. Cabri showed me the difference between drawing and constructing shapes. Of course, I could draw the parabola, or even the soda can object, but constructing these was a completely different story. I needed to understand relationships and properties of these objects in order for them to behave properly.

Ellen elaborated on how her mathematical meaning of parabola evolved from an understanding in a moment to that of a more stable, connected and robust perspective (Thompson, 2015) as well as the ways in which dynamic technology scaffolding impacted her thinking and reasoning. She alludes to her creative insight while constructing versus drawing after intense struggle with the soda can and parabola investigations and acknowledges the need for deep, flexible knowledge and time to make sense of her own understanding (Jonassen, 1994; Silver, 1997; Sriraman, 2009).

4.5.6 Karen

> I knew I was capable of solving the task even though at times I doubted myself. This class has taught me that this doubt is an important step in problem solving for students. So many students doubt themselves and their work in a math classroom. I think this could help them reach a deeper understanding. Technology may add the needed challenge for students to make important connections. Working on this project in class helped me think about how students would react to a problem like that. I could see exactly how each step helped students make deeper connections and formed a greater understand than the previous step. Having manipulatives is a very helpful beginning for most students. I have seen in my own practice that these visuals help students make necessary connections to form deeper understanding. For example in my own lesson I could see how algebra tiles really helped students visualize quadratics and how to solve them. After engaging students with the manipulatives moving to technology now seems like a natural next step. They have started to form conjectures and are ready to put them to the test. I lived through these experiences in class and realize how important it is for students to do the same.

Karen pointed to the importance of physical dynamic cognitive tools (manipulatives) for supporting student learning of mathematics. She acknowledged the necessity of productive struggle in problem solving environments (NCTM, 2014) and though she seems hesitant to move toward using technology with her students, she is on the brink.

4.5.7 Christa

> The activities in [the course] helped to further solidify the idea that dynamic software can help students to build reasoning skills; mainly because these activities help me to see mathematical concepts from different perspectives and to relearn reasoning skills. For example, the "Paper Folding and Technology Investigation" Dynamic Technology Scaffolding (DTS) task opened my eyes to a property of parabolas that I did not know. … Working on the physical model by folding the paper and creating a parabola planted the seed for me to see this property. Then, by analyzing an exploratory model I made further progress towards understanding the concept. When it came to constructing the expressive model, however, it became necessary to fully understand this reflective property because otherwise the construction did not work. This is an aspect that I believe makes working in a dynamic geometry software environment so useful for reasoning; for the student to construct a properly working model of the object being proposed, he/she must establish the necessary mathematical properties of that object. Additionally, the software allows for students to explore multiple representations in a short period of time, which can facilitate the discovery of the mathematical properties intended for the student to learn.

Christa describes the way DTS with the parabola task affected her. She recognized the limitation of her prior mathematical meaning (Thompson, 2015) associated with parabola and appears to recognize both the novelty and usefulness of having to express one's meaning in an expressive technological environment in order to expand such mathematical meaning. Her own creation seemed to convince her (de Freitas & Sinclair, 2014).

4.5.8 Summary of Teachers' Final Reflections

Perhaps the most profound shift is expressed in the recognition of what mathematics learning *can be* as opposed to *what it has been*. Teachers leaving this learning environment were more than inspired; they were prepared. Surely there is more to learn, but they have started the journey towards thinking differently and more flexibly about mathematics through using technological and non-technological tools strategically (National Governors Association Center for Best Practices, Council of Chief State School Officers, 2010). They have **experienced** and **created** learning sequences for their students and have demonstrated capacity to approach the design of learning sequences creatively and with the desire to expand the mathematical landscape for their students (Beghetto & Kaufman, 2009). They have been carefully thinking about creating conditions for students to have access to important mathematical ideas and resources to enhance that access. From the perspective of Csikszentmihalyi (2000), teachers have grappled with mathematical problems and tools to generate solutions that are novel and useful, that is, creative.

4.6 Discussion

In the spirit of Pasteur, from a lecture given at the University of Lille in 1854, "*Chance favors the prepared mind.*" In the 21st century, it is imperative that mathematics teachers are prepared to use and support their students' use of computational technologies for mathematical investigation (NCTM, 2014). If our students are to become competent and creative solvers of problems in a data- and computationally-intensive world, they must have opportunities to develop their facility for doing so through engagement with tasks, tools, and collaborative environments in which to thrive.

Formal introduction to mathematics tends to happen in schools and is shaped heavily by teachers. The research presented here represents a series of mathematically focused design-based studies with pre- and in-service teachers for nearly 10 years during which hypothetical learning trajectories were created, implemented, studied, refined, and continue to evolve. The mathematical and statistical landscape in the research has included big ideas of pattern, shape, change, and uncertainty (Steen, 1990) with connections among these big ideas. Dynamic modeling with physical artifacts and technological tools of both exploratory and expressive modeling types (Doerr & Pratt, 2008) have been integral to the research and learning environment and the impacts on teachers' facility with tools and mathematical development have been positive and strong. The learning environment has positively impacted teachers' *learning mathematics for teaching with technology* and improved their TPCK, more generally.

Through experiences such as those described in this chapter, teachers encountered mathematical ideas in new contexts and with multiple tool affordances that allowed them to translate these experiences to create more robust mathematical meanings (Thompson, 2015). They further prepared their minds—to notice, connect, and create (Sriraman, 2009). Dynamic technology scaffolding was an important ingredient in supporting mathematical learning for teaching with technology because investigating mathematics with dynamic cognitive tools has served as a space for "information gap" (Loewenstein, 1994), which has been harnessed to inculcate curiosity and creativity with teachers. Teachers' words and actions suggest their experiences in this environment have begun to profoundly impact their own practices.

DTS allowed learners to wrestle with their own mathematical conceptions, exposing a need to know more, and to persist when using multiple physical and technological tools to formulate and justify a mathematical solution for which he or she is confident. Moments of *incubation* were followed by *illumination* as the result of exploration using multiple physical and virtual tools. The mathematical habits of mind of looking at geometric objects algebraically and vice versa and modeling geometrically, algebraically, and even statistically through the use of stochastic devices for simulating random behavior were all evident in teachers' explorations (Cuoco et al., 1996). They have been engaged in mathematical creativity (Beghetto & Kaufman, 2009) mediated by the use of tools. They close their own "information gap" through mathematically creative action.

Through the design of DTS tasks for their classrooms, teachers' demonstrated a developing disposition toward actively engaging students in mathematical activity, much like that experienced during their own mathematical investigation. DTS became a design tool to utilize in lesson planning and teachers have demonstrated *mathematical learning for teaching with technology* in a manner that highlights their own mathematical curiosity, creativity, and tenacity. However, with approximately a third of the DTS task sequences designed by teachers showcasing limited creativity with respect to mathematics or technology, it is clear that teachers may need additional support to pursue the design of high quality, technologically rich, worthwhile mathematical tasks for their classrooms.

Nearly all teachers had undergraduate or graduate degrees in mathematics prior to participating in these studies, but that training provided teachers little or no prior opportunity to think or reason with dynamic cognitive tools. This finding is alarming, especially because of the alternative pathways to licensure that have become common for teachers. It underscores the importance for substantive opportunities for teachers to be introduced to mathematical thinking with tools as well as the challenges associated with teachers transitioning to become designers of these types of learning environments for students.

4.7 Conclusion

Dynamic technology scaffolding has been a productive framework for supporting teachers' creativity as learners and task designers. It has provided teachers with opportunities to grow many new mathematical meanings. Because providing all learners in one's care with access to mathematical ideas is essential to facilitate growth in mathematical meanings, supporting teachers' capacity to design with DTS in mind may be especially beneficial. Access to ideas and representations is essential for creative thought, yet many students fail to engage in mathematics classrooms for the simple reason that they do not have access to the ideas required of them to engage (Cohen, 1994). With DTS task sequences, nearly everyone can gain access to relevant mathematical ideas through physical explorations. Nearly everyone can extend his or her thinking through use of an appropriate exploratory model. With the use of physical and exploratory models, a wide range of students may be invited into mathematical investigation and innovative thought. Students may then fruitfully access mathematical relationships and representations to consider, discuss, and extend to their own creative constructions using computational technologies. Moving to expressive modeling often raises the cognitive demand of the task by leaps and bounds; however, as it has been with learners in these studies, oftentimes a cognitive lock gets opened through the physical and exploratory modeling activity, building scaffolds for further investigation in an expressive modeling space. Using DTS, learners are invited into the world of mathematical meaning making with tools. Their creative ideas are elicited as they solve complex and interesting problems. By asking teachers to become DTS task designers, their mathematical and technological creativity are invoked and their mathematical learning for teaching with technology simultaneously extended.

References

Bakker, A. (2002). *Route-type and landscape-type software for learning statistical data analysis.* Paper presented at the International Conference on Teaching Statistics, Cape Town, South Africa.

Ball, D. L., Thames, M. H., & Phelps, G. (2008). Content knowledge for teaching: What makes it special? *Journal of Teacher Education, 59,* 389–407.

Beghetto, R. A., & Kaufman, J. C. (2009). Do we all have multicreative potential? *ZDM Mathematics Education, 41,* 39–44.

Boaler, J., & Greeno, J. (2000). Identity, agency, and knowing in mathematical worlds. In J. Boaler (Ed.), *Multiple perspectives on mathematics teaching and learning: International perspectives on mathematics education* (pp. 171–200). Westport, CT: Ablex Publishing.

Cobb, P., Confrey, J., DiSessa, A., Lehrer, R., & Schauble, L. (2003). Design experiments in educational research. *Educational Researcher, 32*(1), 9–13.

Cohen, E. G. (1994). *Designing groupwork: Strategies for the heterogeneous classroom.* New York, NY: Teachers College Press.

Csikszentmihalyi, M. (2000). Implications of a systems perspective for the study of creativity. In R. J. Sternberg (Ed.), *Handbook of creativity* (pp. 313–338). Cambridge, UK: Cambridge University Press.

Cuoco, A., Goldenberg, E. P., & Mark, J. (1996). Habits of mind: An organizing principle for mathematics curricula. *Journal of Mathematical Behavior, 15,* 375–402.

de Freitas, E., & Sinclair, N. (2014). *Mathematics and the body: Material entanglements in the classroom.* New York, NY: Cambridge University Press.

Doerr, H. M., & Pratt, D. (2008). The learning of mathematics and mathematical modeling. In M. K. Heid & G. W. Blume (Eds.), *Research on technology and the teaching and learning of mathematics: Volume 1. Research syntheses* (Vol. 1, pp. 259–286). Charlotte, NC: Information Age Publishing Inc.

Dorst, K., & Cross, N. (2001). Creativity in the design process: Co-evolution of problem-solution. *Design Studies, 22*(5), 425–437.

Edwards, M. T., & Phelps, S. (2008). Can you fathom this? Connecting data analysis algebra, and probability simulation. *Mathematics Teacher, 102*(3), 210–217.

Ekmekci, A., Corkin, D., & Papakonstantinou, A. (2015). *Technology use of mathematics teachers at urban schools.* Paper presented at the 2015 Annual Meeting of American Educational Research Association, Chicago, IL.

Ervynck, G. (1991). Mathematical creativity. In D. Tall (Ed.), *Advanced mathematical thinking* (pp. 42–53). Dordrecht, The Netherlands: Kluwer Academic Publishers.

Frykholm, J. (2004). Teachers' tolerance for discomfort: Implications for curricular reform in mathematics. *Journal of Curriculum and Supervision, 19*(2), 125–149.

Goldin, G. A. (2003). Representation in school mathematics: A unifying research perspective. In J. Kilpatrick, W. G. Martin, & D. Schifter (Eds.), *A research companion to principles and standards for school mathematics* (pp. 275–285). Reston, VA: NCTM.

Guin, D., & Trouche, L. (1998). The complex process of converting tools into mathematical instruments: The case of calculators. *International Journal of Computers for Mathematical Learning, 3*(3), 195–227.

Hirsch, C. R., Fey, J. T., Hart, E. W., Schoen, H. L., & Watkins, A. E. (2015). *Core-plus mathematics: Contemporary mathematics in context, Course 1.* Columbus, OH: McGraw-Hill Education.

Holyoak, K. J., & Thagard, P. (1995). *Mental leaps: Analogy in creative thought.* Cambridge, MA: MIT Press.

Johnson, D. W., & Johnson, R. T. (2009). Energizing learning: The instructional power of conflict. *Educational Researcher, 38*(1), 37–51.

Jonassen, D. H. (1994). Technology as cognitive tools: Learners as designers. *ITForum Paper, 1,* 67–80.

Jonassen, D. H., & Reeves, T. C. (1996). Learning with technology: Using computers as cognitive tools. In D. H. Jonassen (Ed.), *Handbook of research for educational communications and technology* (1st ed., pp. 693–719). New York, NY: Macmillan.

Kadijevich, D. M., & Madden, S. R. (2015). Comparing approaches for developing TPCK. In *New directions in technological and pedagogical content knowledge research: Multiple perspectives* (pp. 125–146). Charlotte, NC: Information Age Publishing.

Kashdan, T. B., & Fincham, F. D. (2002). 'Facilitating creativity by regulating curiosity': Comment. *American Psychologist, 57*(5), 373–374.

Loewenstein, G. (1994). The psychology of curiosity: A review and reinterpretation. *Psychological Bulletin, 116*(1), 75–98.

Madden, S. R. (2008). *Dynamic technology scaffolding: A design principle with potential to support statistical conceptual understanding.* Paper presented at the 11th International Congress on Mathematics Education, Monterrey, Mexico.

Madden, S. R. (2010). Designing mathematical learning environments for teachers. *Mathematics Teacher, 104*(4), 274–282.

Madden, S. R. (2011). Statistically, technologically, and contextually provocative tasks: Supporting teachers' informal inferential reasoning. *Mathematical Thinking and Learning, 13*(1–2), 109–131.
Madden, S. R. (2013). Supporting teachers' instrumental genesis with dynamic mathematical software. In D. Polly (Ed.), *Common core mathematics standards and implementing digital technologies* (pp. 295–318). Hershey, PA: IGI Global.
Madden, S. R., & Gonzales, A. (2017). Nurturing persistent problem solvers: Heart work. *Mathematics Teacher, 111*(3).
Martinovic, D., & Zhang, Z. (2012). Situating ICT in the teacher education program: Overcoming challenges, fulfilling expectations. *Teaching and Teacher Education, 28*(3), 461–469.
Mishra, P., & Koehler, M. J. (2006). Technological pedagogical content knowledge: A framework for teacher knowledge. *Teachers College Record, 180*(6), 1017–1054.
National Governors Association Center for Best Practices, and Council of Chief State School Officers. (2010). *Common core state standards (Mathematics)*. Washington, D.C.: National Governors Association Center for Best Practices, Council of Chief State School Officers. Retrieved from http://www.corestandards.org/Math/.
NCTM. (1989). *Curriculum and evaluation standards for school mathematics*. Reston, VA: The National Council of Teachers of Mathematics.
NCTM. (2000). *Principles and standards for school mathematics*. Reston, VA: The National Council of Teachers of Mathematics.
NCTM. (2014). *Principles to actions: Ensuring mathematical success for all*. Reston, VA: National Council of Teachers of Mathematics.
Niess, M. L. (2005). Preparing teachers to teach science and mathematics with technology: Developing a technology pedagogical content knowledge. *Teaching and Teacher Education, 21*(5), 509–523.
Niess, M. L. (2006). Guest editorial: Preparing teachers to teach mathematics with technology. *Contemporary Issues in Technology and Teacher Education, 6*(2), 195–203.
Pea, R. (1985). Beyond amplification: Using the computer to reorganize mental functioning. *Educational Psychologist, 20*(4), 167–182.
Polya, G. (1945). *How to solve it: A new aspect of mathematical method*. Princeton, NJ: Princeton University Press.
Presmeg, N. (2007). The role of culture in teaching and learning mathematics. In F. K. Lester (Ed.), *Second handbook of research on mathematics teaching and learning* (Vol. 1, pp. 435–460). Charlotte, NC: Information Age Publishing.
Ryan, R. M., & Deci, E. L. (2000). Self-determination theory and the facilitation of intrinsic motivation, social development, and well-being. *American Psychologist, 55*(1), 68–78.
Santos-Trigo, M., & Machin, M. C. (2013). Framing the use of computational technology in problem solving approaches. *Mathematics Enthusiast, 10*(1), 279–302.
Silver, E. A. (1997). Fostering creativity through instruction rich in mathematical problem solving and problem posing. *Zentralblatt für Didaktik der Mathematik, 29*(3), 75–80. https://doi.org/10.1007/s11858-997-0003-x.
Simon, M. A., & Tzur, R. (2004). Explicating the role of mathematical tasks in conceptual learning: An elaboration of the hypothetical learning trajectory. *Mathematical Thinking and Learning, 6*(2), 91–104.
Sowder, J. (2007). The mathematical education and development of teachers. In F. K. Lester (Ed.), *Second handbook of research on mathematics teaching and learning* (pp. 157–224). Charlotte, NC: Information Age Publishing.
Sriraman, B. (2004). The characteristics of mathematical creativity. *The Mathematics Educator, 14* (1), 19–34.
Sriraman, B. (2009). The characteristics of mathematical creativity. *ZDM Mathematics Education, 41*, 13–27. https://doi.org/10.1007/s11858-008-0114-z.
Steen, L. (Ed.). (1990). *On the shoulders of giants: New approaches to numeracy*. Washington, D. C.: National Academy Press.

Stein, M. K., Smith, M. S., Henningsen, M., & Silver, E. A. (2000). *Implementing standards-based mathematics instruction: A casebook for professional development*. New York: Teachers College Press.

Sternberg, R. J. (Ed.) (1988). *The nature of creativity: Contemporary psychological perspectives*. New York, NY: Cambridge University Press.

Sullivan, F. R. (2017). *Creativity, technology, and learning*. New York, NY: Routledge.

Sweller, J. (1988). Cognitive load during problem solving: Effects on learning. *Cognitive Science, 12*, 257–285.

Thompson, P. W. (2015). Researching mathematical meanings for teaching. In L. D. English & D. Kirshner (Eds.), *Third handbook of international research in mathematics education* (pp. 435–461). London, UK: Taylor and Francis.

Thompson, P. W., Carlson, M. P., Byerley, C., & Hatfield, N. (2014). Schemes for thinking with magnitudes: A hypothesis about foundational reasoning abilities in algebra. In K. C. Moore, L. P. Steffe, & L. L. Hatfield (Eds.), *Epistemic algebra students: Emerging models of students' algebraic knowing. WISDOMe Monographs* (Vol. 4, pp. 1–24). Laramie, WY: University of Wyoming.

Vygotsky, L. S. (2004). Imagination and creativity in childhood. *Journal of Russian and East European Psychology, 42*(1), 7–97.

Wilson, M., & Lloyd, G. M. (2000). Sharing mathematical authority with students: The challenge for high school teachers. *Journal for Research in Mathematics Education, 15*(2), 146–169.

Wing, J. (2006). Computational thinking. *Communications of the ACM, 49*(3), 33–35.

Zbiek, R. M., Heid, M. K., Blume, G. W., & Dick, T. P. (2007). Research on technology in mathematics education: A perspective of constructs. In F. K. Lester (Ed.), *Second handbook of research on mathematics teaching and learning* (Vol. 2, pp. 1169–1208). Charlotte, NC: Information Age Publishing.

Zbiek, R. M., & Hollebrands, K. (2008). A research-informed view of the process of incorporating mathematics technology into classroom practice by in-service and prospective teachers. In M. K. Heid & G. W. Blume (Eds.), *Research on technology and the teaching and learning of mathematics: Volume I. Research Syntheses*. Charlotte, NC: Information Age Publishing Inc.

Chapter 5
Three-Act Tasks: Creative Means of Engaging Authentic Mathematical Thinking Through Multimedia Storytelling

Adrienne Redmond-Sanogo, Susan Stansberry, Penny Thompson and Sheri Vasinda

Abstract Three-Act mathematics tasks provide opportunities for P–12 learners to engage in creative problem posing, exploration, and problem solving through video storytelling. Because they are innovative and relatively new, preservice and inservice teachers may not be familiar with evaluating, creating, and implementing Three-Act Tasks. In this chapter, we describe our design process for developing a rubric to evaluate and scaffold these creative multimedia mathematical stories. The rubric draws on four broad areas of literature for its theoretical grounding: (1) research on selecting and posing high cognitive demand tasks for mathematical problem solving, (2) use of story arc for contextual relevance, (3) research on assessing and measuring creativity, and (4) principles of effective multimedia message design and use of story arc. The rubric developed insures a Three-Act Task attends to mathematical concepts, effective use of digital technologies, and creative thinking. It is designed to serve as a guideline for preservice and inservice teachers as they select or create Three-Act Tasks to use in their classrooms.

Keywords Three act mathematics tasks · Multimedia message design
Problem-posing · Creativity · Digital technologies

A. Redmond-Sanogo (✉) · S. Stansberry · P. Thompson · S. Vasinda
Oklahoma State University, Stillwater, USA
e-mail: adrienne.redmond@okstate.edu

S. Stansberry
e-mail: susan.stansberry@okstate.edu

P. Thompson
e-mail: penny.thompson@okstate.edu

S. Vasinda
e-mail: sheri.vasinda@okstate.edu

© Springer International Publishing AG, part of Springer Nature 2018

V. Freiman and J. L. Tassell (eds.), *Creativity and Technology in Mathematics Education*, Mathematics Education in the Digital Era 10,
https://doi.org/10.1007/978-3-319-72381-5_5

5.1 Introduction

How does the multimedia story of a creature sneaking into a kitchen to eat cookies (http://gfletchy.com/the-cookie-monster/) engage second graders in learning to creatively pose, model, solve, and discuss challenging mathematical problems? Schiro (2004) suggests that stories provide opportunities for children to imagine themselves in the context of a narrative, where they solve problems alongside the characters, reflect upon, and retell the problem-solving endeavors. In this way, they make meaning of the vicariously-lived stories and bring meaning to their own lives. Wells (1987) argues that storytelling is a *primal* act of mind that we engage in both consciously and unconsciously; it is the way that the mind works. Smith (1992) reinforces this primal nature of storytelling as meaning making, positing that "thoughts flow in terms of stories" (p. 62) and that "the brain is a story-seeking and story-creating instrument" (p. 63).

The Three-Act Math Task (Meyer, 2013a) is a recent pedagogical technique that echoes Schiro's (2004) belief that good storytelling and good math instruction are related. Three-Act Tasks harness both the power of well-told stories and appropriate technology integration. These tasks engage students of all ages in mathematical creative question posing, authentic and engaging problem solving, and mathematical modeling through the use of teacher-created video and digital images that tell a story in three acts. In Act One (Fig. 5.1), an intriguing short video clip introduces the problem and characters in a math drama. The purpose is to set the stage for students to wonder, pose questions, and estimate both probable and improbable solutions.

Act Two (Fig. 5.2) adds a photo or video clip that provides an additional event, or rising action, revealing enough information to solve a number of problems that students have posed. Students use the new information to solve the problems they

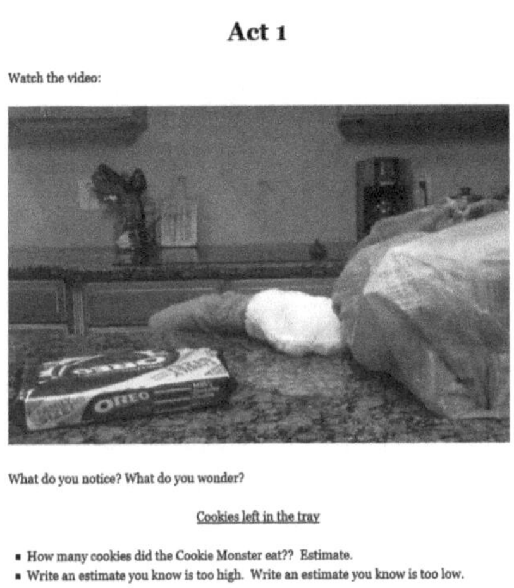

- Students watch a video of the cookie monster eating cookies.
- Problem posing: The teacher asks the students what they notice and wonder?
- The teacher records students thoughts and chooses a question to explore. In this case, the teacher chooses the question *"How many cookies did the Cookie Monster eat?"*
- Estimation: Teacher asks students to write an estimate that is too high and one that is too low so that students can have a range or appropriate solutions.

Fig. 5.1 Act one of the Cookie Monster, permission granted from Graham Fletcher

- Students decide what resources and information that they will need to solve the problem.
- Students work with a partner to solve the problem.
- The teacher monitors the students' solutions and strategically decides the order in which to have students share their solutions and strategies.
- The teacher has students share solutions and makes connections between students' strategies.

Fig. 5.2 Act two of the Cookie Monster, permission granted from Graham Fletcher

- Act three is the great reveal. The teacher shares the solution to the problem and sets up a sequel.
- Students use this as an opportunity to verify their solutions.

Fig. 5.3 Act three of the Cookie Monster, permission granted from Graham Fletcher

have posed, compare the solution to their estimations, and discuss and compare their approach to that of their peers.

Finally, Act Three (Fig. 5.3) provides a concluding video, photo, or graphic that resolves the story and provides a solution to the mathematics problem. Students compare their processes and solutions with the third act information with the option of setting up an extension, or sequel.

Three-Act Tasks are growing in popularity because of their engaging and creative nature revealing mathematics in authentic contexts; therefore, we now support preservice teachers in our program on how to create and use them. Because our experience with preparing preservice teachers gave us insight into the guidance needed on how to create an engaging and effective Three-Act Task, we developed a rubric for evaluating quality in these types of lessons. Our goal was to develop a rubric that could be used both by those who support teacher development and by the preservice teachers themselves when they create their own Three-Act Tasks or select from among the many available on the Internet. In this chapter, we explain Three-Act Tasks from a theoretical perspective of storytelling as a teaching and learning tool. We focus on how mathematics pedagogy (including high-cognitive demand tasks), creativity, and multimedia design contribute to the creation of an effective Three-Act Task. We then describe the development of our rubric, and present the rubric in its final form. We conclude by discussing implications for research and practice.

5.2 History of the Integration of Mathematics and Storytelling

Schiro (2004), tracing the use of stories in mathematics education in the 20th and early 21st centuries, found that stories took second place to decontextualized numeric exercises considered to be the essence of important mathematics teaching. Sets of numeric exercises were often followed by unrelated story problems with disjointed storylines and underdeveloped characters and plots. Students were either exhausted by the numeric problem sets or uninterested in these mathematical "stories" (Schiro, 2004).

At the end of the 1980s, The National Council of Teachers of Mathematics (NCTM) issued a statement influenced by constructivist learning theory that inspired changes in the teaching of mathematics. With the understanding that learners construct meaning and that teachers create the environment for this, teachers found that many children's picture books included mathematical concepts upon which they could capitalize. Unlike the "story problems" described above, these stories captured children's interest and provided a contextual springboard for a mathematics unit of study. Unfortunately, in most classrooms, once the unit was launched and some context was established, traditional practices were resumed and there was no return to the stories in the children's literature. "There is nothing wrong with using children's literature in this way—as a springboard into mathematics—but doing so limits the power that mathematical stories can have in children's lives" (Schiro, 2004, p. 48). Schiro's (2004) own work advocates for the use of oral storytelling in teaching and learning mathematics, citing several advantages to storytelling from Sarah Cone Bryant's 1905 work. These advantages include:

- The storyteller is free; the reader is bound to the book.
- The storyteller can interact with the listeners in a responsive way.
- The storyteller can include the audience in the telling and make changes based on their reactions/responses.
- The storyteller can craft the story to his/her own needs—especially if she cannot find a book that meets her needs.

Using the responsiveness of oral story telling in terms of crafting stories that highlight math concepts and relate to a particular group of students, Three-Act Tasks provide a visually engaging provocation, where the visuals elicit curiosity that can lead to inquiry. Provocations, a concept inspired by the preprimary schools of Reggio Emilia, Italy, involve some sort of stimulus that sparks interest, wonder, discussion, investigation, theory building, and critical thinking (Malaguzzi, 1998). Using gaps between each act of the story arc, Three-Act Tasks provide opportunities for problem-solver interaction that involves the audience in the telling. These tasks also allow teachers to engage in their own creative thinking within the content and context of their classroom. "The best teachers are the best storytellers" (Smith, 1992, p. 62). As Istenic, Starcic, Cotic, and Volk (2016) noted, "In solving mathematical problems, situational storytelling provides a semantic structure for the principles that are to be practiced in solving the problem" (p. 32).

5.3 High Cognitive Demand Tasks and Student Mathematical Learning

The *Principles to Actions: Ensuring Mathematical Success for All* (NCTM, 2014, p. 3) describes many unproductive teaching and learning mathematics situations that are pervasive in today's classrooms. These situations include: "too much focus on learning procedures without any connection to meaning, understanding or the applications that require these procedures" (NCTM, 2014, p. 3); narrow curricula and low expectations; lack of access to materials, tools and technology; too much focus on high stakes testing with a lack of emphasis on problem solving and reasoning; and limited professional development and coaching opportunities. *Principles to Actions* also looked at beliefs teachers hold about the teaching and learning of mathematics and categorized them as productive and unproductive, positioning unproductive beliefs as an obstacle to effective teaching practices. Ineffective practices, such as teaching through a review, demonstration, and practice paradigm, can impede the mathematical learning of all students (Boaler & Staples, 2008; Stein, Smith, Henningsen, & Silver, 2009) and do not help students develop the skills they need to be successful in the workforce of tomorrow. The P21 Partnership for 21st Century Learning (2015) published a summary of the skills that are critical for every student to learn in order to be successful in future jobs. Those 21st Century skills include problem solving, creativity, analytical thinking, collaboration, and communication. Mathematics classrooms that focus on developing these skills better prepare students to be effective and productive global citizens. However, when teachers teach through unproductive practices, students often leave their P-12 experiences believing that mathematics is a static, segmented discipline for those who are naturally inclined to be successful at solving mathematical problems and memorizing steps, formulas, and procedures (Allen, 2011; NCTM, 2014). Recent literature (Kisa & Stein, 2015; NCTM, 2014; Henningsen & Stein, 1997) highlights the importance of posing challenging tasks that encourage students to think and reason mathematically, discuss their thinking, consider the thinking of others, and seek out the connections between and within mathematics concepts. Kapur (2014) found that students who engaged in problem-based learning prior to receiving direct instruction demonstrated stronger conceptual understanding and ability to transfer learning to new situations than their direct instruction only peers. Padmavathy and Mareesh (2013) found similar results when examining the influence of problem-based learning on middle school students' understanding of mathematics. Problem-solving activities provide students with opportunities to think critically, communicate their mathematical thinking, and think creatively

(Krulik & Rudnik, 1999; Boaler, 2016); these are skills which may be lacking in a classroom that focuses solely on direct instruction.

Fishman, Marx, Best, and Tal (2003) found that teachers tend to provide surface-level experiences to students even when the teachers believe they are developing student-centered activities. For example, teachers will initially engage students in an exploration with manipulatives, but then tell students exactly how to use them to solve the problem. Research (Boaler & Staples, 2008; Hiebert & Wearne, 1993; Stein et al., 2009) suggests that tasks with a high cognitive demand on student thinking are more effective at encouraging higher-level thinking and reasoning. "Mathematical tasks are viewed as placing higher-level cognitive demand on students when they allow students to engage in active inquiry and exploration or encourage students to use procedures in ways that are meaningfully connected with concepts or understanding" (NCTM, 2014, p. 19).

Using cognitive and constructivist theories of learning as a foundation, the American Psychological Association Work Group (APAWG) (1997) developed fourteen Learner-Centered Principles in an effort to bring research supporting this pedagogical stance to classrooms. Six of these principles reflect research on the importance and power of providing rich and engaging tasks to students (Polly & Hannafin, 2010):

1. Nature of the Learning Process: When students are provided with rich learning experiences, they are able to learn more effectively.
2. Construction of Knowledge: Students make connections between existing knowledge and new learning when allowed to do so in meaningful ways.
3. Context of Learning: Learning is contextual. Culture, technology, and instructional practices influence what is learned.
4. Motivational and Emotional Influences on Learning: Students who are motivated to learn will learn more effectively.
5. Intrinsic Motivation to Learn: Novel and challenging tasks with an element of choice and capitalizing on student interest spark natural curiosity leading to intrinsic motivation and student's use of creativity and higher-order thinking.
6. Effects of motivation on effort: Engaged and motivated students are more likely to persist in solving difficult tasks.

Three-Act Math Tasks provide students with contextually rich and connected understanding of mathematics (England, 2015). The tasks prompt students to "pose their own mathematical questions and reason about the world around them, wade through the messiness of real life for needed information, model situations using the mathematics they have learned, predict and estimate, and to be driven by curiosity" (England, 2015). This scenario of posing questions, seeking information, and finding solutions mirrors the problem solving process (Ortiz, 2016). Three-Act Math Tasks provide students a problem solving opportunity. However, when creating a new Three-Act Task or selecting an existing one available on the internet, teachers need to clearly identify the mathematical goals and determine if the Three-Act Task chosen will lead to a rich problem-posing and solving experience for the learners (Hiebert, Morris, Berk, & Jansen, 2007). The teacher should understand what mathematics students need to learn, why it is important, how it fits into the structure of knowledge that students already have, and what learning will take place next (NCTM, 2014). At the same time, they also need to focus on tasks that will be motivating and challenging to their students so the students will be able to persist when solving them. Mathematics educators recognize the power of stories to engage students in mathematics problem solving (Schiro, 2004).

5.4 Integration of Mathematics and Technology: New Opportunities and New Challenges

The word *technology* can have a variety of meanings depending on the context in which it is used. The Merriam Webster online dictionary defines the term as (1) "the practical application of knowledge especially in a particular area," (2) "a manner of accomplishing a task especially using technical processes, methods, or knowledge," and (3) "the specialized aspects of a particular field of endeavor," with educational technology listed as an example of the third definition. Spector (2012) defines technology broadly as "the practical application of knowledge for a purpose" (p. 5). Spector (2012) also emphasized change as "a basic aspect of technology, since knowledge is generally progressing and the goals and intentions of people are dynamic" (p. 5). This element of change might explain why the common, colloquial understanding of *technology* often centers on the newest tools, or, in the well-known quote by Alan Kay, "anything that wasn't around when you were born" (Greelish, 2013). Currently, the word technology in general, and educational technology in particular, often brings to mind the application of digital tools such as the internet, digital video, or image editing software for the purpose of improving learning.

For the purpose of this chapter, we take the broad view of technology that emphasizes the application of knowledge for a specific purpose, while recognizing that the standards for technology integration in the schools (as described below) focus more narrowly on the use of digital tools. The Three-Act Math Task described here represents a technology in the broader sense, as it applies knowledge of mathematics, learning theory, and literacy in a systematic way to facilitate deep understanding of mathematical concepts and mathematical thinking. At the same time, the purposeful, supportive use of digital technology, such as digital video, is central to the implementation of the Three-Act Math Task.

Across the globe, more than 20 years of research devoted to the study of educational technology indicates that expected transformations for student learning have yet to be realized (Clark-Wilson, Robutti, & Sinclair, 2014). International studies (Mullis, Martin, & Foy, 2008; Ofsted, 2008) show that teachers are not using technology in the teaching and learning of mathematics despite the fact that their standards explicitly call for the integration of technology in the classroom.

To change this trend, it is important to help preservice teachers become comfortable teaching and learning in environments where students have access to digital tools, like the camera and Internet access needed to produce and post a Three-Act lesson. Students benefit when their teachers integrate technology into the classroom, as they gain "a greater sense of ownership of the mathematics that they are learning, since the applications promote a sense of shared enterprise in the learning of mathematics" (NCTM, 2014, p. 79). Teachers also benefit from this integration of technology since it allows them to share and locate resources on the web and through social media. The National Centre for Excellence in the Teaching of Mathematics (NCETM) (2011) keeps a posted list of technology resources to support various aspects of mathematics teaching and learning. However, NCTM (2014) cautions that using video lectures, simulations, and video presentations are no more effective at teaching mathematics than having a teacher lecture if sense-making and problem solving are not the focus. Thus, it is important that teachers critically analyze the resources they have readily available.

The International Society for Technology in Education Standards for Teachers (ISTE-T) also address the need to develop skills in appropriate technology integration:

- ISTE-T Standard 2. Design and develop digital age learning experiences and assessments.
  - 2a. Design or adapt relevant learning experiences that incorporate digital tools and resources to promote student learning and creativity.
- ISTE-T Standard 3. Model digital age work and learning.
  - 3c. Communicate relevant information ideas effectively to students, parents, and peers using a variety of digital age media and formats.
- ISTE-T Standard 4. Promote and model digital age citizenship and responsibility.
  - 4a. Advocate, model, and teach safe, legal, and ethical use of digital information and technology, including respect for copyright, intellectual property, and the appropriate documentation of sources (International Society for Technology in Education, 2008).

ISTE-T defines the skills and knowledge all digital-age teachers must master to be effective, and the teacher's ability to create a quality Three-Act Task provides evidence of meeting these standards. For example, a Three-Act Task is designed using digital tools to convey a mathematical story that provokes students to wonder, think critically, develop theories and create solutions to a problem addressing the development of digital age learning experiences (part of ISTE-T 2a). Additionally, a Three-Act Task provides an opportunity to convey relevant information and ideas (ITSE-T 3c) through a contextual digital video. Teachers have the opportunity to "advocate, model, and teach safe, legal, and ethical use of digital information and technology" (ISTE-T Standard 4a) as they select resources for the video that meet

Fair Use Guidelines and/or are appropriately cited. Digital media technologies have become a primary mode of sharing stories (Jewitt, 2008), as evidenced by the popularity of tools such as Snapchat, YouTube, Flickr, Pinterest, and Facebook (Brabazon, 2016).

Three-Act Tasks have the potential to encourage a culture of curiosity in the mathematics classroom (Meyer, 2013b). While the technical aspect of taking a photo or shooting a video with any available camera is simple, the design and development of these multimedia stories requires creativity and critical thinking from the designers. Student creativity then emerges as they are working through the Three-Act Task. Act One is where the problem posing occurs, but in Act Two students need to participate in mathematical modeling and problem solving. "If students aren't grappling with the question, 'What's important here and how would I get it?' they may be doing lots of valuable mathematics but they aren't modeling" (Meyer, 2013c). Thus, the teacher developing the Three-Act Task must anticipate students' quests for information and have a way to help them get their hands on it through images, videos, or searching the Web. Act One and Act Two are where students do their work, but Act Three brings it all together for the learner and sets up a sequel. The teacher needs to make sure to revisit students' estimates and questions posed in Act One. The mathematics and mathematical academic language must be formalized and ideas need to be connected in Act Three. Finally, the teacher must set up a sequel that will "entice and activate the imagination" (Meyer, 2013d) and has the potential for further inquiry. Creating a Three-Act Task is therefore a deceptively complex endeavor, and preservice teachers need guidance when learning to create them, or even to select and implement examples created by others.

5.5 Creating a Three-Act Math Task

An effective Three-act Math Task includes the creative use of high-quality multimedia, specifically video, to tell an engaging story that highlights and presents opportunities to uncover mathematical concepts. In addition to strong mathematical content knowledge, this requires an understanding of creativity and its importance in teaching and learning, as well as an understanding of how to implement creative ideas effectively through multimedia story telling. The rubric presented in this chapter, therefore, features an explicit evaluation of the exposition of the story to reveal mathematical problem-posing opportunities, overall quality of the video narratives, the strength of the mathematical content knowledge, and creative elements used to create the task. This section provides a summary of the theoretical concepts that informed the development of these items in the rubric.

5.5.1 Measuring Creativity in a Three-Act Math Task

Designing and producing an effective, motivating Three-Act Task requires creativity on the part of the teacher/creator, but the task itself should elicit curiosity and creative problem posing and solving on the part of the students. The Three-Act Task is inherently a creative product due to its emphasis on finding and solving problems and evoking curiosity. A focus on improving creativity can be seen across business, industry, and education. Yet, defining (Plucker, Beghetto, & Dow, 2004; Friedel & Rudd, 2005) and assessing (Makel, 2009; Turner, 2013; Rubenstein, McCoauch, & Siegle, 2013; Koehler & Mishra, 2008) creativity, as well as building teachers' creative self-efficacy (Stansberry, Thompson, & Kymes, 2015), remain barriers to effectively instilling creative habits in P–12 students. Creativity is defined as "the interaction among aptitude, process and environment by which an individual or group produces a perceptible product that is both *novel* and *useful* as defined *within a social context*" (Plucker, Beghetto, & Dow, 2004, p. 90). Because Three-Act Tasks begin by evoking curiosity in students and then leading them through problem posing and problem solving, an assessment rubric designed to evaluate Three-Act Tasks should include means for measuring the effectiveness of these tasks. The teacher must embed the math problem *within a particular context* that is relevant to students and captures their interest. The visual information provided in Acts One and Two must be provocative or *novel* enough to capture students' attention and spur curiosity toward potential problems and resolutions. The visuals used in the Three-Act Tasks must be *useful* in initially providing just enough information to spark curiosity and problem posing followed by additional information to support students in solving the problem.

Henriksen, Mishra, and Mehta (2015) propose a framework for evaluating creativity in lesson plans based on the degree to which they are *novel*, *effective*, and *whole*, or NEW. Their framework encapsulates the widely accepted belief that creativity involves something that did not exist before (Amabile, 1988; Oldham & Cummings, 1996; Zhou & George, 2001), is useful (Fox & Fox, 2010; Cropley, 2001), and is valued aesthetically in a specific context (Koehler and Mishra, 2008). According to the NEW framework, a lesson plan high in *novelty* demonstrates uniqueness, excitement, and interest. A lesson high in *effectiveness* "makes subject matter clear and comprehensible to most learners and presents it in interesting ways that make the subject come alive" (p. 477). A product high in *wholeness* has "excellent or exceptional aesthetic qualities [and] flawless or near-perfect production values" and "provide[s] aesthetically cohesive, or 'whole' learning that is exciting, thoughtful and stimulating to [students]" (p. 478). This framework operationalizes elements of creativity that can be measured or evaluated, and was therefore foundational to our thinking as we developed the rubric described in this chapter.

5.5.2 *Multimedia Message Design*

Appropriate integration of multimedia content is another fundamental feature of the Three-Act Task (Meyer, 2013a). The videos and images need to elicit observation of and curiosity about mathematical concepts. This requires the creator to develop technology skills (e.g., digital video production) in the context of a coherent multimedia mathematics lesson. The elements of multimedia message design described below enhance the effectiveness of a Three-Act Task.

5.5.2.1 Video Quality and Film Grammar

Research has shown that multimedia can either support or detract from learning, depending on how well it integrates with the purpose of a lesson (Mayer, 2002, 2005), so attention must be given to the quality of the video used. In the context of the Three-Act Task, this means the video needs to be of adequate quality to avoid posing a distraction, and should conform to the basic principles of film grammar (Birth of Image, 2010; Chandler, 2015). For example, the camera distance and height should be appropriate for showing the desired level of detail, the camera direction should be consistent enough to give the scene coherence, and camera movements should be smooth. These guidelines are considered necessary but not sufficient characteristics of engaging video. They help avoid drawing viewers' attention to extraneous aspects of video production, but the content of the video must still be designed to direct viewers' attention to the most important elements.

5.5.2.2 Directing Viewers' Attention

The Limited Capacity Model of Motivated Mediated Message Processing (LC4MP) (Lang, 2000, 2006) explains how a recipient's response to a media message is influenced by human information processing. Due to the limits of short-term memory, viewers do not have the capacity to process all the information present in the environment, so in order for a mediated message to be processed, it must be selected, or given attention. Messages that contain "novelty, change, and intensity" (Lang, 2000, p. 49), for example, are more likely to gain attention than messages that do not have these features.

The LC4MP also posits there are two motivational systems—appetitive (positive and appealing) and aversive—that affect the way the message is processed. Appetitive motivation orients viewers to "information intake" and to processing "as much information as possible about the stimulus… and the surrounding environment" (Lang, 2006, p. 562). Thus, the ideal video for a Three-Act Task would contain a positive, appealing form of novelty or surprise that would both gain viewers' attention and promote careful observation of the information presented.

5.5.2.3 Presenting a Coherent Narrative

Three-Act storytelling is inherent in the design of this task and multimedia production provides an opportunity to enhance it visually with a focus on a narrative. Narrative has been proposed as a way to transform learning into an aesthetic experience (Hobbs & Davis, 2012; Parrish, 2009), so an engaging narrative is important in a Three-Act Task, and video provides the medium for storytelling. This does not mean the video must have an extensive elaborate plot, but it should have a discernible beginning, middle, and end (Parrish, 2009). Cohn (2013) breaks the structure of a visual narrative unit into the following five steps (p. 8):

- An *establisher* sets the scene for the action (e.g., a bag of cookies sits on the table).
- An *initial* begins producing the tension of the narrative arc (e.g., creature hand pulls the cookie bag down).
- One or more *prolongations* extend the narrative tension (e.g., cookie-eating noises are heard).
- A *peak* forms the height of narrative tension (e.g., the munching sounds reach maximum volume).
- A *release* dissipates the tension (e.g., the empty bag is tossed up on the table and a happy sigh is heard).

A video featuring a narrative arc as described above, however simple it may be, is more likely to engage viewers than a video with no discernible narrative structure. Thus, teachers who wish to create Three-Act Tasks for their students must understand the basics of video production and the creation of novel and appealing visual narratives that integrate seamlessly with the mathematics content.

5.6 Developing and Testing a Rubric for Evaluating and Creating Three-Act Tasks

The principles of effective storytelling, mathematics instruction, creativity, and multimedia design may not seem naturally interrelated. Based on the arguments set forth above, however, these four items are already integrated into a Three-Act Task, and therefore it is necessary to include all four constructs when evaluating this type of lesson. We developed a rubric that includes principles of effective mathematics instruction, creativity, and multimedia design within the framework of a three-act story to scaffold preservice teachers as they learn to evaluate or develop effective Three-Act Tasks.

Our rubric was developed through an iterative process, where we created a preliminary rubric, worked individually to apply it to sample Three-Act Tasks, compared our evaluations as a group, revised the rubric as needed, and tested again. We began by searching the *Math Forum* and selecting a rubric that appeared to

Table 5.1 Three-Act math lessons analyzed

Name	URL	Standards
Bucky the Badger	http://mrmeyer.com/threeacts/buckythebadger/	http://www.corestandards.org/Math/Content/3/OA/D/9/
The water boy	http://gfletchy.com/the-water-boy/	http://www.corestandards.org/Math/Content/3/NBT/#CCSS.Math.Content.3.NBT.A.2
The Juggler	http://gfletchy.com/the-juggler/	http://www.corestandards.org/Math/Content/1/NBT/A/1/

address some of our goals (http://mathforum.org/emc/three_act_task_rubric.pdf). The Math Forum rubric examined the three acts individually and included both "overall presentation" and "judge's discretion" categories. We used this rubric to analyze a Three-Act Task developed by a preservice teacher and quickly found we needed to revise it to explicitly address the essential feature of engaging video production and creativity to support mathematics learning to develop our first-draft rubric. We used this first-draft rubric to separately score selected Three-Act Tasks (Table 5.1) publicly available on two databases: Dan Meyer's *Three-Act Math Tasks* database (http://tinyurl.com/jx3fjvu) and Graham Fletcher's *Questioning My Metacognition* website (http://tinyurl.com/zq7mdal). Next, we came together to compare our scoring. We found that while we had general agreement on how to evaluate each example using the first-draft rubric, we had each written extensive notes about other important considerations that this draft did not capture. We discussed these additional observations at length, revised the rubric accordingly, and rated the examples independently again. We repeated this process until we each felt that the rubric was comprehensive.

As a result of the iterative process described above, we made the following changes to the original rubric:

- replaced the *judge's discretion* category with a theoretically grounded creativity assessment based on the NEW framework from Henriksen et al. (2015);
- added a *mathematical understanding* section which focused on developing pedagogical content knowledge;
- added wording throughout to highlight the story arc with a focus on the setting, protagonist, and problem;
- added technical quality assessment items such as hosting multimedia on the web instead of requiring a download;
- added comment boxes after each act in order to provide more guidance and feedback; and
- added wording to address relatability of protagonist and context (e.g. using children instead of adults as the protagonist if the intended audience is children).

The final rubric we created, the *Three-Act Math Rubric for Creativity and Multimedia Storytelling*, is presented in Fig. 5.4.

	Expert (4)	Practitioner (3.5)	Apprentice (3)	Novice (2)
Act One Exposition: Problem Posing:	The setting, protagonist, and problem of the story is clear and relatable to the intended age group. Unique or surprising approach. Leaves the viewer curious about what will happen next. Sets up a real-world math problem.	The setting, protagonist, and problem of the story is clear and relatable to the intended age group. Conventional or standard approach. Leaves the viewer somewhat curious about what will happen next. Sets up a real-world math problem	The setting, protagonist, and/or problem of the story is not clear and/or not relatable to the intended age group. Conventional or standard approach. Leaves the viewer confused or uninterested. Sets up a math problem that may or may not grounded in a real world context.	The story lacks setting, protagonist, and/or problem. Leaves the viewer confused or uninterested. No math problem is evident.
Act One Technical Video Quality	Video clips are exceptional quality. Images are clear. Framing is appropriate and adds to visual interest. Sound is crisp and clear. Titles and words are spelled correctly. Links to images and videos work properly and are easy to access. All video was immediately accessible without downloading.	Video clips are adequate. Video movement is smooth, though framing or sound may need to be adjusted to improve video. Titles and words are spelled correctly. Links to images and videos work properly and are easy to access, but some downloading was necessary.	Video clip movement is abrupt and bumpy Framing and sound need to be adjusted to improve video. Titles and words are sometimes spelled incorrectly. Links to images and videos are broken or user must download the images or video in order to view the task.	Still photos presented instead of a video clip, or video clip is bumpy, blurry and needs to be reshot to achieve clarity. Spelling errors, sound and/or framing errors, distract from the message. Links to images and videos are broken or missing.
Comments:				
	Expert (4)	**Practitioner (3.5)**	**Apprentice (3)**	**Novice (2)**
Act Two: Rising Action	Enough visual information is introduced in a provocative, or novel way. Gives enough new information to solve the problem presented in Act One in a variety of ways.	Enough visual information is introduced in a standard or conventional way. Gives enough new information to solve the problem presented in Act One.	Visual information is introduced in a standard or conventional way. Gives too much or too little new information to solve the problem presented in Act One.	Visual information gives too much or too little information and leaves the viewer confused or uninterested.

Fig. 5.4 Three-Act task rubric for creativity and multimedia storytelling

Act Two Technical Video/Photo Quality	Video clip/photos are exceptional quality. Images are clear. Framing is appropriate and adds to visual interest. Sound is crisp and clear. Titles and words are spelled correctly. Links to images and videos work properly and are easy to access. All video was immediately accessible without downloading.	Video clip/photos are adequate. Video movement is smooth, though framing or sound may need to be adjusted to improve video. Titles and words are spelled correctly. Links to images and videos work properly and are easy to access, but some downloading was necessary.	Video clip/photos are inadequate. Video movement is abrupt and bumpy / photos are blurry. Framing and sound need to be adjusted to improve video. Titles and words are sometimes spelled incorrectly. Links to images and videos are broken or user must download the images or video in order to view the task.	Video clip/photos are inadequate or missing. Video movement bumpy, blurry and needs to be reshot to achieve clarity. Spelling errors, sound and/or framing errors, distract from the message. Links to images and videos are broken or missing.
Comments:				
	Expert (4)	**Practitioner (3.5)**	**Apprentice (3)**	**Novice (2)**
Act Three Climax	The story arc brings the problem to resolution and sets up a sequel that generates further interest. Leaves the viewer curious about what ***could*** happen next.	The story arc brings the problem to resolution, and set up a sequel that does not generates further interest.	The story arc brings the problem to resolution, but does not set up a sequel.	The story arc does not bring the problem to resolution and does not set up a sequel.
Act Three Technical Video/Photo Quality	Video clip/photos are exceptional quality. Images are clear. Framing is appropriate and adds to visual interest. Sound is crisp and clear. Titles and words are spelled correctly. Links to images and videos work properly and are easy to access. All video was immediately accessible without downloading.	Video clip/photos are adequate. Video movement is smooth, though framing or sound may need to be adjusted to improve video. Titles and words are spelled correctly. Links to images and videos work properly and are easy to access, but some downloading was necessary.	Video clip/photos are inadequate. Video movement is abrupt and bumpy / photos are blurry. Framing and sound need to be adjusted to improve video. Titles and words are sometimes spelled incorrectly. Links to images and videos are broken or user must download the images or video in order to view the task.	Video clip/photos are inadequate or missing. Video movement bumpy, blurry and needs to be reshot to achieve clarity. Spelling errors, sound and/or framing errors, distract from the message. Links to images and videos are broken or missing.

Fig. 5.4 (continued)

Comments:				
	Expert (4)	**Practitioner (3.5)**	**Apprentice (3)**	**Novice (2)**
Mathematical Understanding	Demonstrates deep understanding of the mathematical concepts identified for this problem	Demonstrates adequate understanding of the mathematical concepts that they are trying to convey.	Demonstrates partial understanding of the mathematical concepts that they are trying to convey.	Demonstrates misconceptions of the mathematical concepts identified for this problem.
Comments:				
Standards	Standards are correctly chosen for the task and easily accessible. Standards are written out for the user.	Standards are correctly chosen for the task but aren't easily accessible. The user must search to find them or link to them from another website.	Standards are included but they do not fit with the goals and objectives of the three-act lesson.	Standards are not included.
Comments:				
	Expert (4)	**Practitioner (3.5)**	**Apprentice (3)**	**Novice (2)**
Creativity: Novelty	The lesson exhibits strong qualities of uniqueness and is exciting or interesting to learners. It takes a very novel approach to teaching of the subject matter in relative terms to similar lessons.	The lesson exhibits some qualities of uniqueness and is relatively interesting to learners. While aspects may bear certain similarities to standard teaching approaches to the subject matter, it also contains some interesting, fresh or novel qualities.	The lesson takes a relatively standard approach to the teaching of the subject matter. While there may be a few unique qualities, it does not necessarily stand out among other lessons. Average.	The lesson exhibits a complete lack of anything unique or novel, and/or a lack of content and substance to offer opportunities for novelty.

Fig. 5.4 (continued)

Creativity: Effectiveness	The lesson demonstrates an excellent and highly effective approach to teaching the subject matter. It makes the subject matter clear and comprehensible to most learnings and presents it in interesting and engaging ways that make the subject come alive.	The lesson shows an effective approach to teaching the subject matter. Most elements of the approach to or presentation of content work well to communicate the ideas clearly in interesting ways that should lead to solid understanding.	The lesson shows a fairly ineffective approach to or presentation of subject matter. It may have elements that are somewhat boring, confusing, dry, light on content, or do not sufficiently communicate the subject matter clearly to learners.	The lesson shows a complete lack of effectiveness and lack of content or substance to even offer opportunities for effective teaching. The approach is confusing, and/or the presentation of subject matter may lead to misconceptions or confusion for learners.
Creativity: Wholeness	The lesson exhibits excellent or exceptional aesthetic qualities. Flawless or near-perfect production values provide rich visual interest for learners. Provides aesthetically cohesive, or whole, learning that is exciting, thoughtful and stimulating to learners.	The lesson exhibits good aesthetic qualities and sharp or polished production values. The visual interest provided helps make the learning experience interesting and thought-provoking to learners.	The lesson shows weakness in aesthetic appeal or production values as well as clear flaws in the design of the learning experience.	The lesson has little or no aesthetic qualities. It shows poor or complete lack of production values.
Comments:				

Fig. 5.4 (continued)

5.7 Conclusions and Next Steps

As stated earlier, stories can transform learning into aesthetic, and thus engaging, experiences (Hobbs & Davis, 2012; Parrish, 2009), and the creative aspects of multimedia storytelling have the potential to amplify this effect (Muhtaris & Ziemke, 2015). Mishra and Koehler (2006) remind us that, "quality teaching requires developing a nuanced understanding of the complex relationships between technology, content, and pedagogy and using this understanding to develop appropriate, context-specific strategies and representations" (p. 1029). As teacher educators, we have a responsibility to model these relationships in our teacher-preparation classrooms by overtly demonstrating the effective integration of content knowledge and technology use within appropriate instructional strategies for a particular context. Additionally, Three-Act Tasks can incorporate all of the ten design qualities identified by Schlechty (2002) that support engagement: substantial content, well-organized information, affiliation (the ability to work with a group to solve problems), novelty and variety, and authenticity. The development of the Three-Act Task rubric addresses even more of these qualities: product focus, clear and compelling standards, choice (the story), and affirmation of a performance (the video). Because there can be more than one way to interpret the math story and multiple ways to solve the problems, Schlechty's freedom from fear of initial failure is also part of the design. These design qualities support creativity and the nuanced, integrated teacher knowledge identified by Mishra and Koehler (2006).

Three-Act Tasks, which can be created with tools as basic as those included in a smartphone, provide an engaging and satisfying mathematics encounter. This type of experience embodies the kind of mathematics Boaler (2008) contrasts to the often decontextualized, "boring" mathematics of school with the "interesting set of ideas that is the math of the world, and is curiously different and surprisingly engaging" (p. 5). Additionally, in order for students to be able to think and reason mathematically, teachers need to provide them with tasks that are worthwhile, challenging, and engaging (Henningsen & Stein, 1997).

In an era of rigorous content standards and high-stakes accountability, mathematics instruction is often reduced to step-by-step procedures that often do little to build conceptual knowledge and is stripped of its inherent potential for creativity in problem posing and problem solving (NCTM, 2014). Even though renowned mathematical thinkers are recognized as creative, mathematics in school is not typically thought of as a creative process (Pehkonen, 1997). Bonotto (2013) showed that problem posing could support flexible thinking skills, improve problem-solving skills, and equip learners to handle authentic situations in their out-of-school lives. The emphasis on STEM education continues to highlight the need for both creativity and strong conceptual knowledge (Lego Education, 2013). Additionally, international standards for teachers identify the need to design learning environments that incorporate digital tools and resources to promote student learning and creativity, communicate through a variety of digital-age media, and model digital-age citizenship and responsibility (ISTE, 2008). The Three-Act Task itself is a learning environment developed with digital tools, communicated through digital-age media, and designed to promote student mathematics learning and creativity in the context of a three-act story. The rubric presented in this chapter serves as a guide for teachers, ensuring the appropriate use of digital tools, the incorporation of story, and the application of the elements of creativity (novelty, effectiveness, and wholeness) to achieve the mathematics goal.

In this chapter we provided an introduction to Three-Act Tasks and the means for supporting teachers in creating and implementing these tasks, which have the potential to engage and challenge students in more authentic mathematical situations. Three-Act Tasks provide opportunities for both problem posing and problem solving in a creative digital format that leverages the strengths of the digital tools with the power of relevant, contextual stories to highlight the kind of mathematics situations students could find themselves in outside of school. These tasks provide options for teachers to create the kind of contextual lessons tailored to local contexts that make mathematics relative, authentic, and engaging in ways that textbooks cannot. Additionally, a quick Internet search reveals a myriad of Three-Act Tasks widely available for K-12 teachers to adopt that may fit their context or be easily adapted. The challenge exists in evaluating the quality of readily available Three-Act Tasks and identifying the design fundamentals of developing more locally relevant tasks. Using our respective strengths in mathematics and mathematics education, creativity, educational technology, and literacy, we approached the evaluation of the Three-Act Tasks as a multidisciplinary team. The resulting comprehensive rubric provides teachers with the support they need when creating, or when evaluating and selecting, appropriate Three-Act Tasks that meet their instructional goals. As with any other resource, it is critical for teachers to be supported in evaluating the quality of Three-Act Tasks prior to implementing it in their own instruction.

While Three-Act Tasks are becoming popular in the field of mathematics education (England, 2015; Meyer, 2015; Fletcher, 2016; Yenca, 2016), there is little existing research exploring how to develop teachers' ability to produce creative, engaging, and effective Three-Act Tasks. Also, there is no research to assess the effectiveness of the tasks in building and supporting conceptual knowledge of mathematics. The rubric presented here provides a starting place for a program of research centered on Three-Act Tasks. Our next steps include putting this rubric in the hands of teachers who will use it to critically analyze preexisting Three-Act Math Tasks. Additionally, we will use it as a tool to scaffold preservice teachers' creation of their own Three-Act Tasks. Other plans involve comparing student achievement between groups learning a concept through a Three-Act Task and groups learning the same concept through a more traditional lesson. In addition, we are interested in exploring the potential of the Three-Act Task to support learning in fields outside of mathematics. Science educators are beginning to adopt it, and we would also like to work with English and social studies educators. Our work developing this comprehensive rubric will support both the teaching of and the continued research on this innovative practice.

References

Allen, K. (2011). Mathematics as thinking: A response to "democracy and school math". *Democracy and Education, 19*(2).

Amabile, T. M. (1988). A model of creativity and innovation in organizations. *Research in Organizational Behavior, 10*(1), 123–167.

APA Work Group of the Board of Educational Affairs (APAWG). (1997). *Learner-centered psychological principles: A framework for school reform and redesign.* Washington, D.C.: Author. Retrieved from https://www.apa.org/ed/governance/bea/learner-centered.pdf.

Birth of Image (Producer). (2010, August 27, 2015). *Visual grammar: The 4 basic elements* [instructional video]. Retrieved from https://vimeo.com/17900961.

Boaler, J. (2008). *What's math go to do with it? Helping children fall in love with their least favorite subject—And why it's important for America.* New York, NY: Penguin.

Boaler, J. (2016). *Mathematical mindsets: Unleashing students' potential through creative math, inspiring messages, and innovative teaching.* San Francisco, CA: Jossey-Bass.

Boaler, J., & Staples, M. (2008). Creating mathematical futures through an equitable teaching approach: The case of Railside School. *Teachers College Record, 110*(3), 608–645.

Bonotto, C. (2013). Artifacts as sources for problem-posing activities. *Educational Studies in Mathematics, 83,* 37–55.

Brabazon, T. (2016). Let's talk about something important. Let's talk about me. Life, community and culture through digital storytelling. In *Play: A theory of learning and change* (pp. 205–223). New York, NY: Springer International Publishing.

Chandler, D. (2015). *The grammar of television and film.* Retrieved from http://visual-memory.co.uk/daniel/Documents/short/gramtv.html.

Clark-Wilson, A., Robutti, O., & Sinclair, N. (2014). Introduction. In A. Clark-Wilson, O. Robutti, & N. Sinclair (Eds.), *The mathematics teacher in the digital age: An international perspective on technology focused professional development.* Dordrecht, The Netherland: Springer Science +Business Media.

Cohn, N. (2013). Visual narrative structure. *Cognitive Science, 37*(3), 413–452. https://doi.org/10.1111/cogs.12016.

Cropley, A. J. (2001). *Creativity in education and learning: A guide for teachers and educators.* London, UK: Kogan Page.

England, L. (2015, June 22). *Engaging students in three-acts, Part 2.* Retrieved from http://www.nctm.org/Publications/Teaching-Children-Mathematics/Blog/Engaging-Students-in-Three-Acts,-Part-2/.

Fishman, B. J., Marx, R. W., Best, S., & Tal, R. T. (2003). Linking teachers and student learning to improve professional development in systemic reform. *Teaching and Teacher Education, 19*(6), 643–658.

Fletcher, G. (2016, April). Modeling with mathematics through three-act tasks. *Teaching Children Mathematics.* Retrieved from http://www.nctm.org/Publications/Teaching-Children-Mathematics/Blog/Modeling-with-Mathematics-through-Three-Act-Tasks/.

Fox, J. M., & Fox, R. L. (2010). *Exploring the nature of creativity.* Dubuque, IA: Kendall/Hunt Publishers.

Friedel, C., & Rudd, R. (2005). Creative thinking and learning styles in undergraduate agriculture students. In E. A. Moore, & D. Krueger (Eds.), *Proceedings of the American Association for Agricultural Education Conference* (pp. 199–211). San Antonio, Texas.

Greelish, D. (2013). An interview with computing pioneer Alan Kay. *Time.* Retrieved from http://techland.time.com/2013/04/02/an-interview-with-computing-pioneer-alan-kay/.

Henningsen, M., & Stein, M. K. (1997). Mathematical tasks and student cognition: Classroom-based factors that support and inhibit high-level mathematical thinking and reasoning. *Journal for Research in Mathematics Education, 28*(5), 524–549.

Henriksen, D., Mishra, P., & Mehta, R. (2015). Novel, effective, whole: Toward a NEW framework for evaluations of creative products. *Journal of Technology and Teacher Education, 23*(3), 455–478.

Hiebert, J., Morris, A. K., Berk, D., & Jansen, A. (2007). Preparing teachers to learn from teaching. *Journal of Teacher Education, 58*(1), 47–61.

Hiebert, J., & Wearne, D. (1993). Instructional tasks, classroom discourse, and students' learning in second-grade arithmetic. *American Educational Research Journal, 30*(2), 393–425.

Hobbs, L., & Davis, R. (2012). Narrative pedagogies in science, mathematics and technology. *Research in Science Education, 43*(3), 1289–1305. https://doi.org/10.1007/s11165-012-9302-5.

International Society for Technology in Education. (2008). *Standards for teachers*. Retrieved from http://www.iste.org/standards/iste-standards/standards-for-teachers.

Istenic Starčič, A., Cotic, M., Solomonides, I., & Volk, M. (2016). Engaging preservice primary and preprimary school teachers in digital storytelling for the teaching and learning of mathematics. *British Journal of Educational Technology, 47*(1), 29–50.

Jewitt, C. (2008). Multimodality and literacy in school classrooms. *Review of Research in Education, 32*(1), 241–267.

Kapur, M. (2014). Productive failure in learning math. *Cognitive Science, 38,* 1008–1022.

Kisa, M. T., & Stein, M. K. (2015). Learning to see teaching in new ways: A foundation for maintaining cognitive demand. *American Educational Research Journal, 52*(1), 105–136.

Koehler, M. J., & Mishra, P. (2008). Introducing TPCK. In *Handbook of technological pedagogical content knowledge (TPCK) for educators* (pp. 3–29).

Krulik, S., & Rudnik, J. A. (1999). Innovative tasks to improve critical- and creative-thinking skills. In I. V. Stiff (Ed.), *Developing mathematical reasoning in grades K–12* (pp. 138–145). Reston, VA: National Council of Teachers of Mathematics.

Lang, A. (2000). The limited capacity model of mediated message processing. *Journal of Communication, 50*(1), 46–70.

Lang, A. (2006). Using the limited capacity model of motivated mediated message processing to design effective cancer communication messages. *Journal of Communication, 56*(s1), S57–S80. https://doi.org/10.1111/j.1460-2466.2006.00283.x.

Lego Education. (2013, September). *Building future skills: Creativity and playful learning in the classroom*. Retrieved from http://www.bettshow.com/library_10/2728656_assocPDF.pdf.

Makel, M. C. (2009). Help us creativity researchers, you're our only hope. *Psychology of Aesthetics, Creativity, and the Arts, 3*(1), 38–42.

Malaguzzi, L. (1998). History, ideas, and basic philosophy: An interview with Lella Gandini. In Edwards, C., Gandini, L., & G. Forman (Eds.), *The hundred languages of children: The Reggio Emilia approach advanced reflections* (2nd ed.). Westport, CT: Ablex Publishing.

Mayer, R. E. (2002). Multimedia learning. *The Psychology of Learning and Motivation, 41,* 85–139.

Mayer, R. E. (2005). Cognitive theory of multimedia learning. In R. E. Mayer (Ed.), *The Cambridge handbook of multimedia learning* (pp. 31–48). New York, NY: Cambridge University Press.

Merriam-Webster. (2017). *Definition of technology*. Retrieved from https://www.merriam-webster.com/dictionary/technology.

Meyer, D. (2013a, May 14). *The three-acts of a mathematical story*. Retrieved from http://blog.mrmeyer.com/2011/the-three-acts-of-a-mathematical-story/.

Meyer, D. (2013b, May 14). *Teaching with three-act tasks: Act one*. Retrieved from http://blog.mrmeyer.com/2013/teaching-with-three-act-tasks-act-one/.

Meyer, D. (2013c, May 14). *Teaching with three-act tasks: Act two*. Retrieved from http://blog.mrmeyer.com/2013/teaching-with-three-act-tasks-act-two/.

Meyer, D. (2013d, May 14). *Teaching with three-act tasks: Act three & Sequel*. Retrieved from http://blog.mrmeyer.com/2013/teaching-with-three-act-tasks-act-three-sequel/.

Meyer, D. (2015). Missing the promise of mathematical modeling. *Mathematics Teacher, 108*(8). Retrieved from http://www.nctm.org/Publications/mathematics-teacher/2015/Vol108/Issue8/Missing-the-Promise-Of-Mathematical-Modeling/.

Mishra, P., & Koehler, M. (2006). Technological pedagogical content knowledge: A framework for teacher knowledge. *Teachers College Record, 108,* 1017–1054.

Muhtaris, K., & Ziemke, K. (2015). *Amplify: Digital teaching and learning in the K–6 classroom*. Portsmouth, NH: Heinemann.

Mullis, I., Martin, M., & Foy, P. (2008). *TIMSS 2007 international mathematics report*. TIMMS & PIRLS International Study Centre, Boston College.

National Centre for Excellence in Teaching Mathematics (NCETM). (2011). *ICT and digital technology used in mathematics teaching—Useful online resources*. Retrieved from https://www.ncetm.org.uk/resources/34752.

National Council of Teachers of Mathematics (NCTM). (2000). *Principles and standards for school mathematics*. Reston, VA.: NCTM.

National Council of Teachers of Mathematics (NCTM). (2014). *Principles to action: Ensuring mathematical success for all*. Reston, VA.: NCTM.

Ofsted. (2008). *Mathematics—Understanding the score*. London, UK: Ofsted.

Oldham, G. R., & Cummings, A. (1996). Employee creativity: Personal and contextual factors at work. *Academy of Management Journal, 39*(3), 607–634.

Ortiz, E. (2016). The problem-solving process in a mathematics classroom, *Transformations, 1*(1). Retrieved from http://nsuworks.nova.edu/transformations/vol1/iss1/1.

P21 Partnership for 21st Century Learning. (2015). *P21 framework definitions*. Retrieved from http://www.p21.org/storage/documents/docs/P21_Framework_Definitions_New_Logo_2015.pdf.

Padmavathy, R. D., & Mareesh, K. (2013). Effectiveness of problem based learning in mathematics. *International Multidisciplinary e-Journal, 2*(1). ISSN:2277-4262.

Parrish, P. E. (2009). Aesthetic principles for instructional design. *Educational Technology Research and Development, 57,* 511–528. https://doi.org/10.1007/1s1423-007-9060-7.

Pehkonen, E. (1997). The state-of-art in mathematical creativity. *ZDM Mathematics Education, 29*(3), 63–67.

Polly, D., & Hannafin, M. J. (2010). Reexamining technology's role in learner-centered professional development. *Education Technology Research Development, 58,* 557–571.

Plucker, J. A., Beghetto, R. A., & Dow, G. T. (2004). Why isn't creativity more important to educational psychologists? Potentials, pitfalls, and future directions in creativity research. *Educational Psychologist, 39*(2), 83–96.

Rubenstein, L. D., McCoach, D. B., & Siegle, D. (2013). Teaching for creativity scales: An instrument to examine teachers' perceptions of factors that allow for the teaching of creativity. *Creativity Research Journal, 25*(3), 324–334.

Schiro, M. S. (2004). *Oral storytelling and teaching mathematics: Pedagogical and multicultural perspectives*. Thousand Oaks, CA: Sage Publications.

Schlechty, P. (2002). *Working on the work: An action plan for teachers, principals, and superintendents*. San Francisco, CA: Jossey-Bass.

Smith, F. (1992). *To think*. New York, NY: Routledge.

Spector, J. M. (2012). Defining educational technology. In J. M. Spector (Ed.), *Foundations of educational technology: Integrative approaches and interdisciplinary perspectives* (pp. 3–15). New York, NY: Routledge.

Stansberry, S. L., Thompson, P., & Kymes, A. (2015). Teaching creativity in a master's level educational technology course. *Journal of Technology and Teacher Education, 23*(3), 433–453.

Stein, M. K., Smith, M. S., Henningsen, M., & Silver, E. A. (2009). *Implementing standards-based mathematics instruction: A casebook for professional development* (2nd ed.). New York, NY: Teachers College Press.

Turner, S. (2013, May). Teachers' and pupils' perceptions of creativity across different key stages. *Research in Education, 89*. https://doi.org/10.7227/rie.89.1.3.

Wells, G. (1987). Apprenticeship in literacy. *Interchange, 18*(1–2), 109–123.

Yenca, (2016, February). 3-Acts: Using digital tools to give every student a voice. *Mathematics Teaching in the Middle School*. Retrieved from http://www.nctm.org/Publications/Mathematics-Teaching-in-Middle-School/Blog/3-Acts_-Using-Digital-Tools-to-Give-Every-Student-a-Voice/.

Zhou, J., & George, J. (2001). When job dissatisfaction leads to creativity: Encouraging the expression of voice. *Academy of Management Journal, 44*(4), 682–696.

Part III
Creativity in Technology-Rich Mathematical Environments

Chapter 6
Interactive Technology to Foster Creativity in Future Mathematics Teachers

Alfinio Flores, Jungeun Park and Stephen A. Bernhardt

Abstract This chapter discusses ways in which the use of interactive technology in a problem-based course that integrates mathematics, science, and technology fosters creativity among future secondary mathematics teachers in their first year in college. The course was built on research-based principles to learn mathematics for understanding. We found that creativity is fostered naturally by teaching mathematics based on those principles. Creativity is fostered, promoted and developed when (a) learners themselves grapple with concepts and make concepts explicit; (b) learners actively build new understanding on previous knowledge; (c) learners engage with mathematics as a social process; (d) learners use multiple representations and connections to enhance their understanding; (e) learners pose and solve problems; and (f) learners exercise multiple modes of learning—when they read, talk, write, draw, analyze, apply, present, and reflect. We discuss the use of technology and issues related to future teachers' creativity as they solve problems; design experiments and collect, represent, and analyze data; develop mathematical models for phenomena in the physical, biological, and social sciences; and build and program their own robot.

Keywords Preservice mathematics teachers · Integrated mathematics, science, and technology · Teamwork · Communication · GeoGebra · Python Mathematical modeling

A. Flores (✉) · J. Park · S. A. Bernhardt
University of Delaware, Newark, DE, USA
e-mail: alfinio@udel.edu

J. Park
e-mail: jungeun@udel.edu

S. A. Bernhardt
e-mail: sab@udel.edu

© Springer International Publishing AG, part of Springer Nature 2018
V. Freiman and J. L. Tassell (eds.), *Creativity and Technology in Mathematics Education*, Mathematics Education in the Digital Era 10,
https://doi.org/10.1007/978-3-319-72381-5_6

6.1 Introduction

This chapter addresses the problem of providing future mathematics teachers the experience of learning mathematics integrated with science and technology, through a class in which students also had the opportunity to express their creativity by choosing their own strategies and methods to solve problems, conduct experiments, and model phenomena. Interactive technology played an active and essential role in providing students multiple opportunities to use and improve their creativity in their mathematical activities. The research reported here was conducted in a new problem-based course for future secondary mathematics teachers that integrates technology, mathematics, and science taught at the University of Delaware in the United States (Flores, 2014). This course was offered in fall 2013, spring 2015, and fall 2015. It serves as an alternative course to satisfy a technology requirement in the secondary mathematics teacher's preparation program. Students can meet this requirement also by taking an introductory computer science course. While not a methods course per se, the course was designed to offer students a rich learning experience that might suggest to them innovative paths in teaching.

The design and implementation of the course were guided by research-based principles to learn mathematics for conceptual understanding (Hiebert & Grouws, 2007): (a) learners themselves need to grapple with concepts and make concepts explicit; (b) learners actively build new understanding on previous knowledge; (c) learners benefit from engaging with mathematics as a social process; (d) learners' access to knowledge and understanding are enhanced when they use multiple representations and connections; (e) learners acquire significant mathematical content and know-how through problem solving and problem posing; and (f) students learn better when they exercise multiple modes of learning (Cakir & Stahl, 2013). That is, when students not only listen and take notes, but when they read, talk, write, draw, analyze, apply, present, and reflect. These research-based principles and this framework have been discussed in more detail elsewhere (Flores, Park, & Bernhardt, 2016). We also used the Technological Pedagogical Content Knowledge (TPACK) framework (Mishra & Koehler, 2006; Niess et al., 2009) to guide the design of the course. The course also was informed by professional recommendations about the dispositions and attributes that teachers need to have. For instance, the Interstate Teacher Assessment and Support Consortium (InTASC) developed a set of model core teaching standards in which they point out that "today's learners need … attributes and dispositions such as problem solving, curiosity, creativity, innovation, communication, interpersonal skills, the ability to synthesize across disciplines, …, and technological expertise" (InTASC, 2011, p. 4). Teachers themselves need to have these attributes and dispositions in order to help develop them in their students. With the purpose of putting students in control, we emphasized interactive technology, so students can would learn to solve specific modeling problems using computers tools and programs, such as GeoGebra and the programming language Python. We explored whether and how future teachers' creativity would be fostered as they used their own strategies and methods to solve problems; used invention to design experiments and decide how to collect, represent, and analyze data; developed their own mathematical models for phenomena in the physical, biological, and social sciences; and built and programmed their own robot to perform a task invented by them.

6.1.1 Overview of the Course

Originally we did not include fostering creativity in mathematics as an explicit goal for the course. However, the course was designed to incorporate ways of learning mathematics that the literature on creativity (Mason & Watson, 2008; Presmeg, 2003, Silver, 1997; Sheffield, 2009; Sriraman, 2004; Voica & Singer, 2013) had shown to be closely related to creativity in mathematics. In the current realization of the course, students work in cooperative groups to solve problems. They use interactive cognitive technologies in their inquiries. For example, they use GeoGebra (International GeoGebra Institute, 2017) to explore properties of linear, quadratic, and other functions. They use functions such as power, exponential, and periodic to model phenomena in the physical, biological, and social sciences. They use motion detectors to gather data and graphing calculators to transform and represent data. Students write their own programs in the freely available programming language Python (Enthought Scientific Computing Solutions, 2016) to simulate random phenomena. Students construct and program their own robot. They use interactive epsilon bands, where students can control the error bound epsilon and the threshold value *N*, in their re-invention of the definition of the limit of a sequence (Flores & Park, 2016).

Communication also plays a big role in the course. Students are asked to use written and verbal language to plan and modify their experiments and discuss problematic issues. For example, in the experiment with the pull-back car (see Sect. 6.1), they discuss in their small groups and write a full plan about what they are going to measure and how. After they collect their data they prepare a written report that includes computer generated tables and graphs, post it electronically, and share it verbally with other teams. They produce verbal and written mathematical signs by speaking, writing, or sketching to share their strategies, methods, and results with their peers in their small group or with the whole class.

As we will illustrate with specific examples later, such activities offer students opportunities to be creative in mathematics. In the next section we will draw from the literature on creativity in mathematics, first at a general level and then on the relation between creativity and specific aspects in the learning of mathematics that were emphasized in the course.

6.2 Literature Review

Creativity is a complex concept that can be viewed from multiple perspectives (Sriraman, 2004). Its multiple aspects have been defined variously by different researchers. For example, the National Teaching Fellows, a group of higher education instructors in the UK who have received recognition, include imagining, seeing unusual connections, combining ideas, and discovering original ideas (Fryer, 2006). Boden (2004) specifies combinatorial, exploratory, and transformational as different types of creativity. However, different characterizations of creativity also share some common features: novelty, value, and effectiveness (Aralas, 2008; Boden, 2004).

Although developing creativity is not often mentioned explicitly as one of the goals for mathematics courses, creativity is frequently mentioned when describing mathematical ability (Aiken, 1973; Krutetskii, 1976; Kattou et al., 2013) or gifted and talented students in mathematics (Sheffield, 1994; Sriraman, 2005). Creativity is an intrinsic component of *doing* mathematics (Halmos, 1968; Polya, 1962), of conducting accomplished work in mathematics, and of doing research in mathematics (Poincaré, 1920; Hadamard, 1945).

Creativity is also considered an important part of the process of creating mathematical statements such as proofs and definitions (Polya, 1957; Hanna & Winchester, 1990). For example, Winchester (1990) points out that creativity of thought is necessary for the production and enjoyment of mathematical proofs. The relationship between creativity and proofs is twofold. On one hand, finding a proof often requires creativity. On the other hand, proof is part of the social process through which the mathematical community validates the creative work of a mathematician (Hanna, 1991).

Of course, efforts to foster creativity in mathematics are not new (Gibb, 1970; Haylock, 1987; Leikin & Pitta-Pantazi, 2013). Researchers have identified ways to foster creativity by specifying features of the environment where students play an independent role and develop ownership of discussion (Jackson & Sinclair, 2006), where they are engaged in imaginative thinking and heuristic strategies, and where they develop risk-taking approaches and self-regulation (Fryer, 2006; Sternberg & Williams, 1996) without the threat of evaluation but with openness to unexpected responses (Torrance, 1979). Rather than trying to give a more comprehensive overview of the literature on fostering creativity in mathematics, we will focus next on the principles of fostering creativity from the research literature that are related to the design and implementation of a problem-based course for learning mathematics through technology.

6.2.1 Creativity in Independent Conceptual Understanding of Concepts and Relations

As mentioned earlier, students' independence in developing critical thinking is an important aspect of creativity (Jackson & Sinclair, 2006). In the context of learning mathematics, students' independence is required in their own meaning-making and knowledge construction, as they process mathematical concepts and the relationships between concepts (Von Glasersfeld, 1995). This process involves integrating the meanings and uses of previously learned (or partially learned) mathematical concepts into a new context or more complex system through which students "exercise their creative abilities and devise insightful ways to deal with mathematical topics and problems" (Hashimoto & Becker, 1999, p. 102). According to Freudenthal (1971), mathematics should be taught in "the order in which it could be invented by the student" (p. 416). This process engages several aspects closely related to creativity, such as inventiveness, guided re-invention, constructing knowledge, organizing matter, and mathematizing (Presmeg, 2003).

6.2.2 Creativity in Building New Knowledge on Previous Knowledge

Creativity is also involved when students build new knowledge from their existing knowledge (Jackson, 2006). Without existing knowledge, students would not be able to "generate," "hypothesize," "theorize," or "reflect," which are all creative activities that are crucial in building new knowledge (Biggs & Tang, 2007). Intense interest or involvement in a particular field accompanies creativity. At the end, new knowledge is produced as "a 'creative work' … comprising something new, a synthesis that did not exist quite like that before" (Biggs & Tang, 2007, p. 145). When such syntheses result in new concepts or systems in the field, these outcomes are also referred to as *invention* (Biggs & Tang, 2007, p. 145). Of course, novice students can frequently be inventive themselves, even if concepts or relationships are already known among experts. In fact, Mason and Watson (2008) take the view that mathematics:

> Can be presented and experienced as a constructive activity in which creativity and making choices are valued … in order to stimulate learners to use their own powers to make sense of phenomena mathematically (p. 192).

6.2.3 *Creativity in Engaging with Mathematics as a Social Process*

Although creativity has usually been studied at the individual level, the relation of students working in small groups to the development of their creativity has been the focus of research for quite some time (Banghart & Spraker, 1963). Because there is always a need to solve complex problems in our changing society, there has been recently increased interest in studying creativity at the group level (Van Oortmerssen, Van Woerkum, & Aarts, 2015).

Working in small groups may foster the creativity of students, but just putting students together to work in small groups does not mean that creativity will automatically flourish. As with other aspects of productive teamwork, this process requires learning. According to Meissner (2005), to further creative thinking in mathematics education, we need to further both individual and social abilities. Students need to learn how to avoid the negative factors that affect creativity: *cognitive interference*, which includes production blocking, task-irrelevant behavior, and cognitive overload; and *social inhibition*, which includes social anxiety, free riding, and illusion of productivity (Paulus, 2000). They also need to learn to recognize factors that can strengthen the potential of groups to generate ideas; for example, *social stimulation*, which includes both increased individual accountability and the development of shared standards for team performance; and *cognitive stimulation*, which includes stimulation of associations, attention to others' contributions, and opportunities to incubate ideas (Paulus, 2000).

6.2.4 *Creativity in Different Ways of Thinking, Learning, and Representing Ideas*

Creativity influences mathematical thinking and the representation of mathematical ideas (Sheffield, 1994). Components of mathematical creativity include various types of thinking, such as reversing the train of thought, solving problems in unique and unusual ways, reasoning clearly, and abstracting and generalizing mathematical content (Krutetskii, 1976; Sheffield, 2003). Mathematically, creative thinking also includes the ability to work with different representations and especially to adopt appropriate representations for the given problem or context. A crucial creative ability is to be able to switch among representations with flexibility, for example, "from computation to visual to symbolic to graphic representations" (Sheffield, 2003, p. 4).

6.2.5 *Creativity in Applying Knowledge and Reasoning in Problem Posing and Problem Solving*

It is widely recognized that creativity is involved while students apply their knowledge and reasoning in problem posing and problem solving, which are often considered two central goals of teaching and learning mathematics (Henderson & Pingry, 1953; Polya, 1962; Silver, 1997). The application and transfer of knowledge in posing and solving non-routine problems is considered a different and much more crucial mathematical ability than possession of information (Henderson & Pingry, 1953; Polya, 1962). The generative processes of problem posing and problem solving are central in creative activity (Silver, 1997). Polya (1954) points out that self-directed posing of problems to be solved is an important characteristic of engaging in the intellectual work of mathematics. Three of the factors mentioned by Mann (2006) to develop creativity in mathematics are problem solving, problem formulation, and open-ended problems. Problem solving often requires "some degree of independence, originality, creativity" (Polya, 1962, p. viii). The relationship between creativity and abilities in problem posing and problem solving became more explicit in Silver's work (1997), which showed "how mathematical problem posing and problem solving are connected to key aspects of the classic and contemporary conceptions of creativity and also to the assessment of creativity" (p. 75). Sheffield (2008) states that for students to become creative mathematicians, we need to cultivate and nurture their "abilities to recognize and define problems, generate multiple solutions or paths toward solution" (p. 370). Recent research (Voica & Singer, 2013) has shown the effectiveness of using problem posing as a tool to develop creativity among school children. Contreras (2013) fostered the creativity of students by using problem posing. Manuel (2009) summarizes findings in the literature by claiming that mathematical creativity happens when students have the opportunity to find different and original strategies and solutions to given problems, as they take risks and try to find new relationships between facts or ideas.

6.2.6 *Interactive Electronic Technology and Creativity*

In recent years there have been remarkable changes in access to interactive and communication technology and in what technology can do that has the potential to change the way mathematics is represented, learned and communicated (Hoyles & Lagrange, 2010; Bu & Schoen, 2011; Rivera, 2011; Abramovich, 2014). Consequently, technology should change the ways we prepare mathematics teachers (Clark-Wilson, Robutti, & Sinclair, 2014). Students who grew up having all sorts of information and communication technology available are portrayed by research as strong visual learners (Martinovic, Freiman, & Karadag, 2013). Mathematical activities suitable for these learners are informal, experimental, intuitive, and experiential, taking the form of multiple paths that coincide with the preferred learning activities for such students. Such students favor informal learning in exploring a concept or a process; they follow multiple paths when solving a problem, and they learn better when they are able to explore experientially and intuitively (Martinovic, Freiman, & Karadag, 2013, p. 211).

One of the ways to promote creativity in self-directed learning is through interactive educational technology. Groff (2013) mentions creativity, collaborative problem solving, and self-directed learning as three of the skills and capabilities for 21st century citizenship. According to Groff, on the one hand, advances in the learning sciences encourage educators to reconsider approaches to learning, instruction, and classroom environments; on the other hand, advances in educational technology inspire new ways for learners to engage with all kinds of content and activities in their own self-directed experiences. She mentions that technology is more than just part of the classroom resources. Technology can play a key, and at times a leading, role in all elements of the teaching and learning environment. Technology is integral to the organization component as it offers a critical mediating medium for the relationships of pedagogy and assessment. Sacristán et al. (2010) analyze how the affordances of digital technologies can shape learning trajectories. Abramovich (2014) illustrates how students in a technology-enabled mathematics pedagogy setting can use problem solving as a springboard into the domain of problem posing.

In some cases, creativity is an explicit goal of integrating technology into mathematics teaching. For instance, the approach Mathematics Integrating Computers and Applications (Buteau & Muller, 2006) has as one of its guiding principles to encourage student creativity and intellectual independence. Of course, this is not done at the expense of conceptual understanding. The other guiding principle of this approach is to develop mathematical concepts hand in hand with computers and applications. The main goal of the first Mathematics Integrating Computers and Applications course is to allow students to experience becoming the mediator through the design of original Learning Objects, which are instructional components that focus on one or two mathematical concepts and that are designed for another person. "These objects are interactive, engaging, easy to use, and are designed to mediate the user from information to understanding." (Buteau & Muller, 2006, p. 78). The role of the instructor in this course is radically different from traditional lecture based courses. The instructor acts as a mentor and encourages students' mathematical creativity as students design, program, and use their own interactive Exploratory Objects (Buteau & Muller, 2014).

Interactive, user-friendly technology can be used to adapt projects that foster creativity to make them more accessible to students. For example, a middle school teacher the authors have interacted with encouraged a strong sense of agency among her students and used technology to engage them in open-ended tasks. Students in her sixth grade honors math class, taught entirely in a computer lab, used a dynamical geometry program to design a miniature golf park. Students did this as a collaborative activity with each student in the group designing two to three holes. "The students really had fun with this and they were so creative" (A. Breitmeyer, personal communication, July 9, 2015). The teacher was inspired by an activity originally designed for high school students using traditional tools of geometry such as rulers, protractors, and compasses (Powell, Anderson, & Winterroth, 1994). By using interactive technology, the sixth grade students were able to explore and use mathematical concepts that are traditionally taught in high school.

Martinovic, Freiman and Karadag (2013) analyze the potential of technology to foster new ways of learning "to create an outcome that is collaborative, self-directed, democratic, co-constructed, coordinated, multimodal, sensuous, and empowering" (Martinovic, Freiman & Karadag, 2013, p. 216) from different perspectives, including (a) the visual aspects of different mathematics representations in software, (b) establishing connectivity, (c) dynamism and interactivity, and (d) processing power.

The use of technology also allows us to look at mathematical understanding not only from an individual perspective, but also as part of the group practices. Cakir and Stahl (2013) argue that "deep mathematical understanding can be located in the practices of multimodal reasoning displayed by groups of students through the sequential and spatial organization of their actions" (p. 91). By using technology, they were able to capture, document, and analyze the creative process of collaborative problem solving in mathematics.

As should be clear from this literature review, creativity is multifaceted and often plays a role in many innovative approaches in mathematics instruction. Creativity is invoked whenever students grapple with concepts, activate prior learning, make new connections, work collaboratively, and engage with technologies.

6.3 Research Questions

An innovative course for future mathematics teachers that integrates technology, science and mathematics can be studied from different points of view, some of which have been published elsewhere (Flores, 2014; Flores & Park, 2016; Flores, Park & Bernhardt, 2016). Consistent with the focus of this book, we will highlight in this chapter the relationship of students' use of interactive technology in the course with manifestations of their creativity. Two research questions help structure this chapter:

- In what ways does a problem-based course that integrates technology, mathematics, and science foster creativity in mathematics or related aspects of learning mathematics identified in the literature review (agency, problem solving, problem posing, imagination, inventiveness, guided re-invention, constructing knowledge, organizing matter, mathematizing)?
- What is the role of interactive electronic technology in fostering the creativity of future mathematics teachers in such a setting?

In the methods section (Sect. 6.4), we will make explicit how the principles that foster mathematics learning can shape the design and implementation of a course that fosters creativity. We will then discuss in the results section (Sect. 6.5) how students were provided with opportunities to display behaviors and approaches related to creativity in the activities of the course.

6.4 Theoretical Framework: Use of Guiding Principles in the Design and Implementation of the Course

This section will describe how, by using the research-based principles for learning mathematics with understanding that guided the design of the course, we incorporated into students' learning experiences elements identified in the literature that foster creativity in a natural way. We discuss the guiding principles and give examples of their implementation in the course. Although the implementation could be considered part of the methods, we think that coupling the principles with the corresponding examples would give readers a better picture.

Although the course was not originally designed with development of creativity as an explicit goal, we found that of the 25 strategies described by Sternberg and Williams (1996) to develop student creativity, we incorporated the first 15 basic strategies in a natural way. For instance, we allowed our students to define and redefine problems and projects. We allowed our students to choose their own random phenomenon for their programming project and choose their own ways of solving problems (Strategy 4, Sternberg & Williams, 1996). One of the main goals of the course was to help students to think across subjects and disciplines (Strategy 6, Sternberg & Williams, 1996).

With respect to the technology used in the course, emphasis was on interactive technology and on students telling the computer what to do rather than the other way around. Such adaptation of technology is closely related to creativity. The two main technologies used in the course, GeoGebra and Python, are technologies that were designed specifically with the purpose that users could adapt them for their own goals.

The design and implementation of the course were guided by research-based principles to learn mathematics for conceptual understanding: (a) learners themselves need to grapple with concepts and make concepts explicit; (b) learners actively build new understanding on previous knowledge; (c) learners benefit from engaging mathematics as a social process; (d) learners' access and understanding are enhanced when they use multiple representations and connections; (e) students learn significant mathematics as they engage in problem solving and problem posing, and (f) students learn better when they exercise multiple modes of learning—when they read, talk, write, draw, analyze, apply, present, and reflect. The research literature behind these principles was reviewed and the principles were discussed in more detail elsewhere (Flores, Park, & Bernhardt, 2016). Here we will make explicit how creativity is related to each of them and how each of the guiding principles was operationalized in the course.

6.4.1 Grappling with Mathematical Concepts and Making the Concepts Explicit

Creativity is fostered when learners themselves grapple with concepts and make concepts explicit. Rather than having the teacher organize the concepts for them and present them with finished and well-organized mathematical ideas, students need to grapple with the concepts themselves by experimenting, conjecturing, generalizing, testing, connecting concepts, and organizing their own thinking. All these activities offer opportunities for the students to be creative.

Equally important is that the mathematics concepts involved in the activity need to be made explicit during the learning process rather than remain implicit or tacit. Here are some examples. A student may describe the relation of two variables in qualitative terms, such as "the total distance increases as the pull-back distance increases." The student is guided to use this statement as a basis to re-state the relationship more precisely in terms of a constant ratio of the increments of the two variables and how this ratio is reflected by the slope of the line of best fit. Students are also guided to point to and make explicit what they mean when they use generic terms or references, such as "it" or "that," with the goal of full expression in well-formulated sentences. Students display creativity when they use language to frame problems and solutions in their own, original words so that they capture their understanding verbally. As Ernest (2005) points out, sign production or utterance involves primarily self-directed actions aimed at personal development or personally-chosen goals and is often a creative act.

Following the experimental and modeling activities, students are asked to read sections in the textbook (Gordon & Gordon, 2010) that provide technical and precise mathematics vocabulary about the shapes of the graphs of functions, such as *increasing*, *increasing at an increasing rate*, *concave*, and so on. Importantly, the class activities prime students to be receptive to the textbook content. Students are subsequently encouraged to use conceptual and precise mathematical terms explicitly when they discuss the shape of the graphs they were exploring, both in their small groups and whole class presentations. This integrated practice of talking, graphing, reading, and writing has been shown to help students develop conceptual understanding, while at the same time allowing the instructor to see where students are in their understanding and gauge the need for additional instruction or reinforcement (Sealey, Deshler, & Hazen, 2014).

6.4.2 Learners Actively Build New Understanding on Previous Knowledge

Creativity is supported by a smooth transition in knowledge building as learners actively build new understanding on previous knowledge. In general, in the course we work with mathematical concepts that students have already covered in prior coursework, and we draw widely from algebra, trigonometry, geometry, and calculus. Opportunities for creativity are provided as students need to cross borders between mathematics topics they learned separately, to think outside the boxes of separated mathematical topics to solve problems, and to rearrange and synthesize their knowledge. Because students build on their previous knowledge, instruction needs to address preconceptions, both true and false. In the course, we deliberately perturb existing schemas and allow time for reconstruction, moving forward when everyone on every team appears to have a solid understanding, often expressed in writing and presented orally.

6.4.3 Mathematics Learning as a Social Activity

Creativity is promoted when learners engage with mathematics as a social process. Rather than working on problems individually, students communicate with their teammates, explain their strategies and results, and work together to construct meaning. In a socially active classroom, the course fosters verbal and written communication. Communication in our course is not only between the instructor and the students but also, and even primarily, among the students. The activities are designed so that students work in small cooperative groups to stimulate the communication of mathematical ideas among students of all levels (Davidson, 1990). Of course, simply putting students in groups of four does not mean they are going to cooperate productively. Students need to develop multiple abilities to work in cooperative groups in mathematics (Artzt & Newman, 1990). We coach teams on process, provide peer facilitation, and use rubrics to observe and record the functioning of teams in the classroom. The four levels of student behavior, from more desirable to less desirable, of the innovation configuration map correspond to Table 6.1 and are provided to students. The students are asked to discuss advantages and disadvantages for each level.

Table 6.1 Rubric for working in cooperative groups

Working in cooperative groups			
(a) Students work productively with team members to solve problems. They consistently explain their ideas to the other members of the group. They listen carefully to each other's ideas, and they make sure everyone is heard and contributes to the joint effort	(b) Students attempt to work with team members, though not always productively. At times, they share solutions and processes with other members of the group once they have a solution	(c) Students work individually much of the time. One or more students try to dominate the process; other students follow	(d) Each student solves the problem or completes the task individually, and there is no sharing of results or processes

6.4.4 Making Connections and Using Multiple Representations

Creativity is also promoted when learners use multiple representations and connections to enhance their understanding. The course offers many opportunities for students to make connections across mathematics and to other disciplines, using multiple representations to access concepts. Students connect concepts within mathematics (algebra, geometry, trigonometry, calculus) to each other as well as to physical, biological, and social phenomena. The assessments in the course are geared to provide evidence that students are able to represent mathematics as an interconnected network of concepts and procedures and that they are able to establish connections between mathematics with other fields and everyday life. We design tasks that allow students to draw on the interplay of mathematics with the real world (Marrongelle, 2008). Students who deal with realistic problems can re-invent or develop mathematical concepts in a meaningful way (Gravemeijer, Cobb, Bowers, & Whitenack, 2000). Tasks are also designed to encourage students to connect algebra with geometry, geometry with modeling, and probability with statistics to encourage reasoning and sense making. We try to foster students' abilities to see connections and underlying themes in the mathematics they have learned.

One way we try to facilitate in the course connections in mathematics and enhance understanding is through the use of multiple representations of mathematical ideas. Creativity is promoted through different ways of representing ideas. We try to help students to develop efficient internal systems of representation that correspond coherently to, and interact with, the external systems of representation of mathematics that are used by mathematicians and schools. We want students to connect their internal representations of mathematical concepts to real phenomena so that they are more likely to retain the learning and be able to reconstruct their knowledge. Multiple external representations used in the course provide both opportunities to establish connections and additional points of entry to mathematical concepts. Tasks require students to use multiple representations of data (graphical/geometric/trigonometric, formulaic/algebraic, geometric/algebraic, graphical/analytic, symbolic/verbal) so they have more opportunities to use their creativity to help them develop deep understanding.

6.4.5 Learning Mathematics Through Problem Solving and Problem Posing

In addition to learning mathematical content and know-how, students develop creativity as they solve problems, modify them, and pose their own problems. We have adapted a model of problem-based learning (PBL), which relies on teams working collaboratively to solve real or realistic problems (Allen, Donham, & Bernhardt, 2011). Students thus have opportunities to try their creative approaches to solve the problems. Our classes typically involve a sequence of two or three short problems centered on modeling, graphing, or programming functions. Teams immediately dive into new problems with each class session. The instructor does not lecture or demonstrate at the board but sets tasks, monitors groups, maintains the pace, and offers questions to help teams stay on track. One or more peer facilitators perform a similar role. Students also pose their own problems when they identify situations that involve randomness and design their own programs to simulate the phenomenon with a probabilistic model, or when they set their own task for their robot to perform.

6.4.6 Multimodal Learning of Mathematics

With new technologies we can create multi-modal learning environments for students, where they can interact with mathematical objects and with each other in a variety of ways. “A multi-modal environment supports user interactions in more than one modality or communication channel (e.g., speech, gesture, writing) through perceptual, attentive or interactive interfaces” (Güçler, Hegedus, Robidoux, & Jackiw, 2013, p. 98).

Students develop creativity when they exercise multiple modes of learning. Students naturally adopt many different strategies as they attempt to solve problems. Crossing domains and strategies (speaking and writing, modeling, drawing, testing, measuring) supports the various learning and problem-solving strategies students naturally employ (Committee on Developments in the Science of Learning, 2000). Students also learn better using a multi-modal approach to learning. In a study of over 6500 students, Hake (1998) found that interactive engagement methods (broadly defined as heads-on, hands-on activities with immediate feedback) were strongly superior to lecture-centered instruction in improving performance on valid and reliable mechanics tests used to assess students' understanding of physics.

In our active classroom, students might read about and discuss a mathematical problem with their team members or decide what kind of data to collect from an experiment, discuss how to organize those data, draw a graph by hand that scales and represents the data, and then import and model those data more precisely using GeoGebra. At various stages, we ask a member of one team or another to explain their thinking to the class or use the document camera to project a solution. When student teams arrive at different solutions or different methods of describing a function, we ask them to compare their approaches. Students consolidate their understanding with either an individual or collaborative written description of the problem or modeling activity. Writing helps consolidate conceptual understanding on the part of students and teams, while allowing instructors insight in misconceptions or partial understanding (Habre, 2012). Incorporation of short, in-class writing assignments improves students' learning (Butler, Phillmann, & Smart, 2001; Davidson & Pearce, 1990; Drabick, Weisberg, Paul, & Bubier, 2007; Stewart, Myers, & Culley, 2010). When conceptual understanding is not evenly distributed across all individuals or teams, we return to a problem in a subsequent class to clear up misconceptions, or we direct the team to return to the problem and sort out their shared understanding. When we can, we compare earlier, faulty representations of data with later, better models, showing students that understanding is an approximate, iterative process.

6.5 Methods

A total of 45 students took the integrated math, science, technology course in fall 2013 (10 students), spring 2015 (19 students), and fall 2015 (17 students) at the University of Delaware. Of them 44 were secondary education majors, the majority of whom were either freshmen (23) or sophomores (16). Twenty-two were male and 23 female.

In the course students use four types of interactive electronic technology. First, they use GeoGebra to explore properties of functions of different types, including linear, quadratic, exponential, power, logarithmic, and logistic, and use these functions to model physical, biological and social phenomena in the physical. Second, they use motion detectors to collect data of experiments involving complex motion, such as a bouncing ball, or a parachute jump, and they use electronic devices to represent the data graphically. Third, they learn to code in Python with simple mathematical situations and later produce their own computer programs using Python to simulate random phenomena. Fourth, they build and program their own robot using an icon-based programming language.

We briefly outline some characteristics that would make the course valuable to the mathematics teacher education community as a whole. First, the course is hands-on, active, social, communicative, multimodal, and deeply collaborative and constructive. Second, it purposefully integrates representation and understanding across algebra, geometry, trigonometry, calculus, and science. Third, it uses a range of readily available technologies to gather data, model relationships and systems, and program actions, thus increasing computational reasoning. Fourth, it leads to students who can apply what they know, solve problems, work in teams, and apply technology. And fifth, the course is both fun and engaging for students and instructors.

6.6 Results

In this section, we address the implementation and outcomes of the course. We discuss how students' activities and responses in the course provide evidence that the course was effective in fostering creativity or related aspects of learning mathematics, and we highlight the central role of interactive technology in the process, illustrated by specific activities in the course. We try to make explicit how each activity discussed is connected to the principles of creativity identified in the literature review (Sect. 6.2) and to the creativity aspect connected with the research-based principles used to design the course.

6.6.1 Pull-Back Car

Students dealt with problem-based situations from the very first day because we wanted to immerse them in new models of teaching and learning designed to foster creative, team-based problem solving. In the first session of our class, students studied the motion of a small pull-back car (Bryan, 2014) and discussed in their groups how to represent the motion as a function (distance travelled vs. distance pulled back). They had to decide what experiments to perform, including what to measure and how to measure it. Students discussed in their cooperative groups how to organize and represent the data. They used technology to represent the data as a scatterplot, deciding what scales were appropriate, and they experimented to fit a function to the data visually. The problem had multiple solutions (Leikin, 2014). Most teams chose to fit a linear function, but one team decided to fit a quadratic function to their data. Students presented their data and analysis to the larger group. The higher level of agency for the students (Scardamalia & Bereiter, 2014) was noticed and appreciated by students, as one of them remarked in her "exit ticket," a short written reflection students turn in before leaving, for the session (Student exit ticket, February 10, 2015):

> I enjoyed being able to create our own experiment. Last semester I took physics and every lab was already very laid out for us. I enjoyed the freedom to run an experiment the way I thought it would work out.

This activity clearly highlighted the multimodal learning of mathematics supported by technology because students mathematized the movement of a physical object. These students' work reflected their creativity in the application of their mathematical knowledge to model physical movements and also in their different choices for their mathematical model. The first session set the tone for the semester. Students knew they would not watch the instructor solve problems at the board or listen to well-polished lectures, but instead they would need to become creative agents of their own learning.

6.6.2 Distance and Velocity for a Bouncing Ball

Collecting data from a bouncing ball with a motion detector attached to a graphing calculator (Cory, 2010) allowed students to call upon their understandings of key concepts that bridge physics and mathematics (see Fig. 6.1). They first designed an experimental approach to gather data and then analyzed the distance above the ground vs. time graph generated by the graphing calculator (Fig. 6.2), focusing on local maxima and minima; intervals during which velocity was decreasing, increasing, or increasing at a decreasing rate; concavity; and so on. Students were then asked to sketch individually a velocity vs. time graph that corresponded to their distance vs. time graph. Students needed to think carefully how the slope of the distance function changed dynamically. This activity highlighted (a) independent understanding of concepts and relations, (b) problem solving, (c) and use of technology in fostering creativity.

This activity fosters students' creativity by asking them to re-conceptualize the familiar mathematical concepts of functions, derivatives, and their relationships to explain a non-routine movement of a physical object. In this activity, the

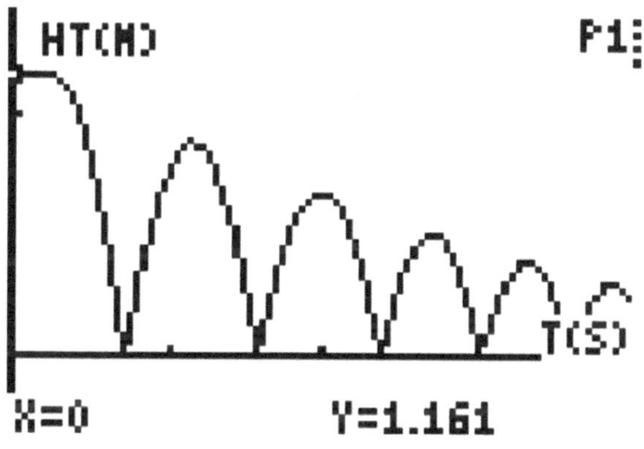

Fig. 6.1 Distance versus time for a bouncing ball

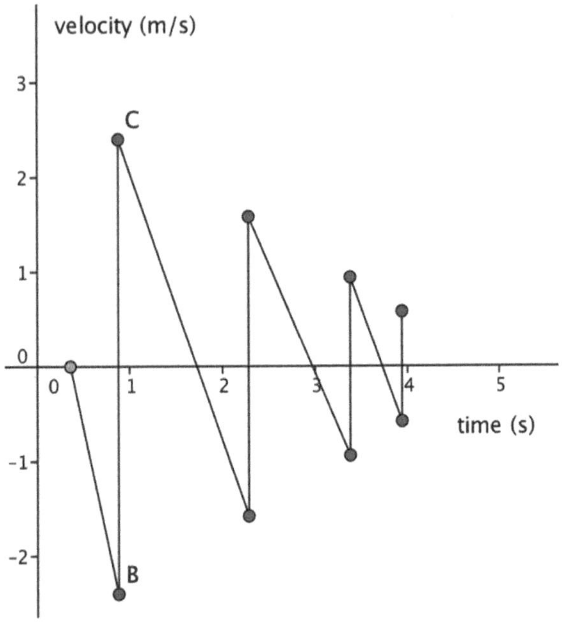

Fig. 6.2 One student's velocity versus time graph

mathematical objects of the object's movement cannot be explained by typical continuous version of the concepts listed above. Specifically, the data generated would not lead to a traditional symbolic representation for the derivative function. The movement of the physical object was described and graphed with technology. Specifically, in order to graph the velocity vs. time graph correctly, they also needed to make connections with their previous knowledge about physics (for example, that acceleration due to gravity is constant for a falling object), and about calculus (the derivative is positive if the function is increasing, the derivative is zero at local maxima and minima, and so on). Students had to decide what scale to use for their graph, whether the velocity graph would be continuous, and so on. Then they shared their graphs with each other in their cooperative groups.

Students grappled with concepts (is the velocity graph piecewise linear?) and helped each other in making sense of their sketches of velocity vs. time. Students demonstrated difficulties while representing the velocity of the ball for the very short interval when the ball hit the ground. Specifically, the individual sketches of the students were quite different from each other and often revealed misconceptions about the situation (for instance, velocity changing from negative to positive suddenly without being zero at some intermediate point). Even after talking with their peers, some groups reached a consensus that was not quite correct. In other cases, after discussing with their teammates, their sketches revealed a more precise understanding of the velocity of the ball at different parts of the experiment. However, students were still not quite confident and wanted the instructor to reveal a correct graph.

In order to foster students' independence in thinking, the instructor did not provide a sketch with the correct graph. Students had an additional opportunity to make sense of the velocity by using a GeoGebra sketch where they could manipulate the velocity values for some special points, connect the points they thought needed to be connected, and then work in small groups to represent the velocity vs. time graph (Fig. 6.2). Student groups were then asked to write an explanation of their graphs using their own words.

By comparing their initial sketches with their revised graphs, we can say that all groups showed a more solid understanding of the situation. The discussions in their small groups and in their reports used technical terms that reveal their understanding. They described the velocity when the ball was in the air using both their knowledge of derivatives (e.g., if the original function is decreasing, the derivative function is negative) and physics (e.g., constant negative acceleration due to gravity). They also showed better understanding for the brief interval when the ball was in contact with the ground, compressing and decompressing. They indicated, for instance, that the slope of the segment BC (Fig. 6.2) was a very large positive number but not infinite. They also were able to explain, both orally and in writing, why some of their initial individual or group graphs were not correct. Students felt confident with their answers even though the instructor never provided the correct graph. In this activity, being creative consisted of a set of productive behaviors: devising their own data-gathering methods using the appropriate electronic technology, modeling the data in various ways with interactive computer graphs, generating their own explanatory hypotheses, testing one explanation against competing explanations, expressing explanations both orally and in writing using their own language, arguing for one or another competing explanation, and coming to consensus on a best solution.

6.6.3 *Fitting Curves to Sets of Points*

In a mathematics classroom that fosters students' creativity, students are exposed to new ways of modeling data, including innovative approaches based on visualization of mathematical functions. In our class, students learned ways in which GeoGebra and the use of sliders within the software provided an alternative point of entry to the topic of fitting linear and quadratic curves to scatterplots of data (Fig. 6.3). Sliders allowed students to drag a point on a segment that represented values of the parameters of the function and thus had a visual and kinesthetic approach for the change of the parameters. Students also used sliders for fitting other types of functions (exponential and power functions) to sets of data. By adjusting the parameters on the slider, students developed a better understanding of the role and impact of the different parameters in the formula for a function. The feedback provided by the dynamic graph was immediate, and students could see coordinated changes across representations because the equation that described the function changed at the same time as the graph of the function. Students embraced and used

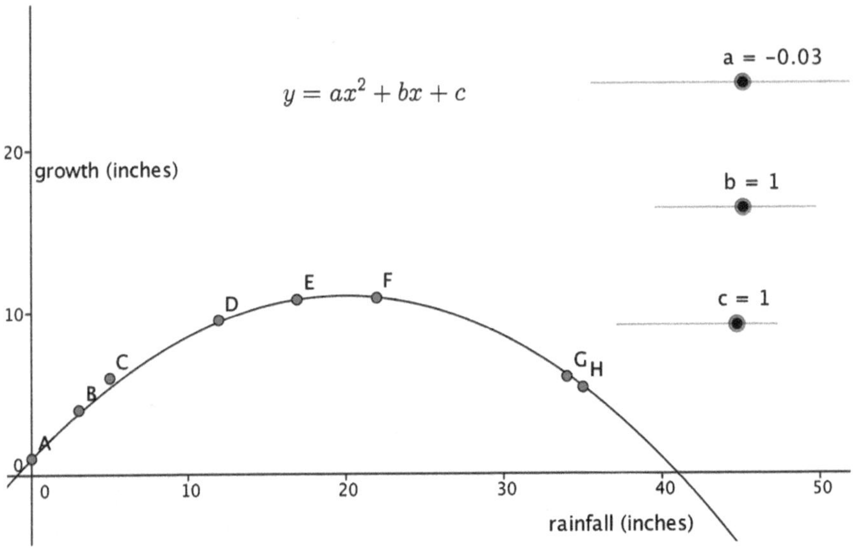

Fig. 6.3 Using sliders to fit a curve for a graph of tree growth vs. rainfall

to their advantage the capability of GeoGebra to use numeric, graphic and symbolic representations of a function, which changed simultaneously on the computer screen. Students quickly learned how to use sliders for the parameters of a variety of modeling functions with very little direct teaching. Team members took turns generating the graphs and adjusting them with the sliders with other members offering suggestions to increase or decrease the value of specific parameters.

Whenever possible, we attempted to present students with situations that called upon their understanding of mathematical functions as applied to phenomena in the physical world. Fitting functions to sets of points that represent real data encouraged students to connect mathematical functions and models to real-world phenomena. In the case represented in Fig. 6.3, students plotted data representing the growth of redwood trees in their first year as a function of rainfall. Students started to think about how tree growth might be measured, how much a tree might grow in a year, and why too little or too much rain (trees also need sunshine) might result in a concave curve. Using real growth data also caused students to think logically about why the graph for a given data set would or would not intersect one or both axes.

However, we observed that students did not automatically know how to "play" with the technology to model phenomena (Uribe-Zarain, 2015). Students needed time and guidance to develop their ability to explore mathematics using the technology. A few groups used the software in a more inquisitive manner by changing on their own the parameters of the functions to see how the graph changed. When students played with the software more, they gained better insights about the relation of the function and the situation they were modeling as revealed in their small group discussions and written reports. For example, one of our students pointed out how she benefitted from using GeoGebra in terms of finding best fit without limiting their possibility to the simple functions (Uribe-Zarain, 2015, p. 9):

> I think the visual part with the graph is really important. Because most people get okay linear, quadratic, but then it gets confusing beyond those. And I think playing with GeoGebra and showing how equations fit, like lines of best fit curves, stuff like that works well!

For instance, when studying sinusoidal functions of the form $y = A \sin(\omega x - \beta) + h$ in the beginning students did not have a clear understanding of the effect of changing each of the parameters on the graph of the function. After they had the opportunity to change the values of the parameters with the sliders and observe the corresponding change in the graph, students were able to see that the parameter h shifts the graph vertically, while the parameter A controls the amplitude, that changing the parameter ω affects the frequency, and so on.

6.6.4 Rowing Competition and Perspective

In the rowing competition activity (Flores, Bernhardt, & Shipman, 2015), students had a different opportunity to express their creativity in various ways. The context as presented was the challenge of gauging the progress of two rowing sculls at some distance when the viewer had only a single perspective. Human perception of moving objects on a level plane is often deceptive. To grapple with this problem, students first merged the use of two different types of technology, video and GeoGebra. By using screenshots of the video at different points in time and embedding them into GeoGebra files, students were able to compare the motion of sculls in more precise ways. Second, they made connections between the concept of the vanishing point, which is usually studied in static situations such as paintings, with the motion of sculls rowing in parallel. By doing this they extended their understanding of the vanishing point in perspective from a static to a dynamic situation. Third, the problem lent itself to multiple solutions. Some teams used the central vanishing point as the main mathematical tool to compare the motion of the sculls and used the lines connecting from the fronts of the sculls to the vanishing point in several snapshots to determine which scull was ahead and whether another scull was catching up. Other teams used the vanishing point corresponding to the fronts and ends of two sculls (Fig. 6.4) and determined, by observing the motion of the vanishing point across snapshots, whether one boat was moving faster than another. The students also had an opportunity to express their creativity in writing using a combination of genres. They were asked to write a persuasive letter to a fictitious TV producer to convince her that they had a way to use technology to allow viewers to determine more easily who was winning in the race and to compare speeds of sculls. The letter had to make explicit the mathematical principles used in their solution.

(a) Vanishing point at t = 34

(b) Vanishing point at t = 38

Fig. 6.4 Using GeoGebra with video snapshots to track the moving vanishing point

6.6.5 Student Construction of the Definition of Limit of a Sequence

In these activities, students participated in a cyclical process to re-invent their own definitions of the limit of a sequence for a particular value L = 5. The activity was preceded by having the whole class generate a set of qualitatively distinct examples and non-examples of sequence convergence. The instructor provided access to additional examples of sequences (Oehrtman, 2015). After this, students wrote their individual definitions of the limit of a sequence and then worked in their small groups for a first tentative definition. All of their first definitions included a dynamic language equivalent to "approaching" or "getting close to." Students shared with the whole group what problems they were having with their definitions. To develop students' understanding of the underlying structure of the convergence of sequences, the framework of approximations, errors, and error bounds was then introduced, and students wrote a second small group definition, tested it, and received feedback. The refining cycle started again. In total, students went through four refining cycles, and wrote four definitions as results of their refinement before they were asked to post a formal definition. This process of re-invention and the conceptual development students went through is described in detail in Flores and Park (2016). Here we focus on a couple of aspects related to creativity. One is that students experienced through this activity that creativity "is often associated with long periods of work and reflection rather than rapid, exceptional insight" (Silver, 1997, p. 75). For example, one student mentioned their refinement process as a valuable experience while starting from their informal conceptions of the limit which mainly included the qualitative behavior of the dots ("approaching" or "getting close to") and moving towards a definition equivalent to the formal definition in mathematics by identifying problems, fixing the problems, and revising their definitions in an end of course interview (Uribe-Zarain, 2015, p. 7):

> We had to learn the limits, we had to keep doing it over and towards the end I was like okay, I kind of just want the definition. But it was cool how we edit, and edit, and edit and we finally got it.

In the activity described in this quote students had to write tentative definitions for the limit of a sequence, contrast their definition with well-chosen examples and nonexamples, receive and give feedback to other teams, and after reflecting on the feedback, revise their definitions accordingly. As Biggs and Tang (2007) point out this kind of intense involvement in a particular field accompanies creativity. Some of the students appreciated that enough time was given to work in depth on the re-invention (Uribe-Zarain, 2015):

> I liked that it wasn't just a one class where we were focusing on just a definition, but it was like alright you write your definition, come back on Thursday and as a group we'll write them and alright on Tuesday now we're going to look at every other groups' definition. I liked that because it gave me more of an understanding whenever we would do that.

A second creative aspect relates to teamwork (Paulus, 2000). Students perceived that by working in small groups, they were able to re-invent something that would have been harder working individually. As one student expressed (Uribe-Zarain, 2015):

> It was a lot harder to work through it than I thought it would be. So, working with a group was helpful in that sense and then coming up with the definition on my own was always more complicated than when I was working with a group.

There is an important lesson here. Creativity is often iterative (Silver, 1997), as one design is tested, revised slightly, tested again, refined a bit more, at some point reaching a satisfactory design. Additionally, creativity in design often is enhanced by having more than one perspective.

6.6.6 Programmable Robots and Feedback Loops

The last activity of the course we will discuss is an open-ended project. Students built a robot working in small groups using Legos and a programmable brick. They controlled the robot using the programming language Mindstorms, which is based on icons. Students were asked to write their own programs that included a feedback loop, that is, that the robot would change its behavior in response to an outside input in a cyclical way. Students knew that we were "looking for them to demonstrate their knowledge, analytical and writing skills, and creativity" (Sternberg & Williams, 1996, p. 23). Although in most cases the teams chose to build a robot using the instructions for various examples available with the set or online, they showed their creativity in the way they programmed the robot. Often teams embarked on writing complex programs but ended up simplifying them because the sensors or the motors did not quite operate in the way they thought.

Students made formal presentations in which they were required to explain any feedback loops in their program, including how the robot changed behaviors depending on the signals received by the sensors, (Fig. 6.5), and also making explicit some of the concepts of mathematics and physics that could be illustrated in the functioning of the robot. The following are the mathematical concepts identified by students: ratio of gears, angle, rotation, rate of rotation, relation between the ratio of gears and ratio of rotation rate, multiplication of ratios, translation, reflection, symmetry, and inverse operation. Among the concepts related to physics, students identified speed, velocity, angular speed, centripetal acceleration, normal force, torque, momentum, angular momentum, mechanical advantage, and friction.

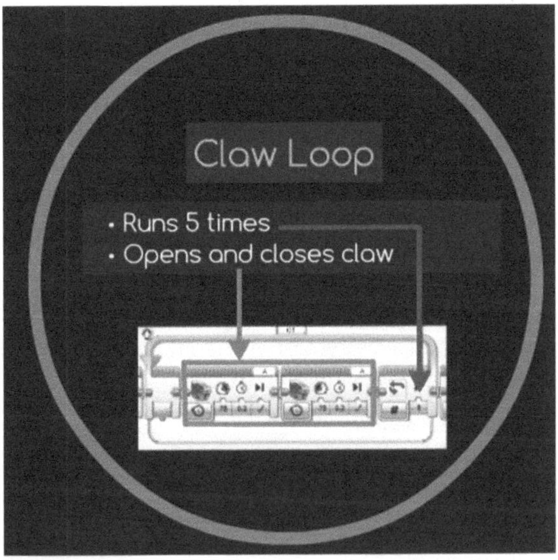

Fig. 6.5 A feedback loop programmed and explained by a team of students (Whiley & Tellup, 2015, used with permission)

The creativity of students was manifested through the specific examples they chose to illustrate the concepts and the connections they made between the behavior of the robot and feedback loops in nature depending on time, touch, light, or other inputs, as well as feedback loops in man-made devices, such as the automatic door at the supermarket, and the feedback loop when entering a password or PIN number. Students also showed creativity in the use of conveyance technologies for the formal presentation in the form of graphics or special effects. Frequently, students used a fictitious story to set the behavior of their robot in context. Although this type of creativity was not directly related to mathematics, it will be very handy in their future role as teachers.

6.7 Discussion

In this paper, we presented various ways that uses of interactive technology can foster creativity in the mathematics classroom. As we discussed in the literature review section, there have been multiple discussions on the role of creativity in learning of mathematics and the ways to promote them (e.g., Aiken, 1973; Hadamard, 1945; Krutetskii, 1976; Kattou, Kontoyianni, Pitta-Pantazi, & Christou, 2013; Poincaré, 1920; Sheffield, 1994; Sriraman, 2005). Considering the crucial role that creativity plays in creating mathematics (Polya, 1957; Hanna & Winchester, 1990), creativity should be considered in developing students' mathematical ability and deepening their mathematical understanding. Our results showed that students' creativity and related aspects of learning mathematics was enhanced through various activities involving the interactive electronic technology, which we designed using research-based principles for learning mathematics with understanding. Our detailed analysis in Sect. 6.6 showed that the use of interactive technology provided the environments where students express and promote their creativity in mathematical context by providing (a) tools to test, revise, and justify their conjectures and predictions multiple times (e.g., Sects. 6.6.1 and 6.6.3), (b) non-routine situations where they have to re-conceptualize their previous mathematical knowledge to explain the mathematical aspects of new situations (e.g., Sect. 6.6.2), (c) a new angle to look at physical objects with non-typical given data for students to think about a creative way to find out the mathematical information of the objects (e.g., Sect. 6.6.4), (d) the refinement cycle where students start with their initial idea and move towards mathematical formalization in which they had to condense and objectify their qualitative ideas to quantitative descriptions (e.g., Sect. 6.6.5), and (e) tools to build a physical robot that reflects and realizes what students plan to build from their own creativity (e.g. Sect. 6.6.6). While engaging these activities, other aspects of learning mathematics related to creativity were also promoted. For example, throughout each activity, students experienced learning mathematics in a social context by building shared understanding within their small group and the whole class (e.g., Sect. 6.6.5), how to express their mathematical idea in multiple ways but with connection among them (e.g., Sects. 6.6.2 and 6.6.4), and how multiple aspects of mathematics are affecting each other in various representations (e.g., Sect. 6.6.3).

Participating in creative experiences in mathematics is important for future teachers for several reasons. First, participation in such experiences can be very rewarding and fulfilling by itself. Second, it is unlikely that teachers who themselves have not experienced creativity in mathematics will be able to foster creativity in their own students. Teachers who have a rich repertoire of open-ended, problem-based tasks in which technology is used effectively can offer more opportunities to engage students in such activities. In turn, students who are expected to have more agency in their own learning will also be more creative. Furthermore, the importance of creativity goes beyond the realm of learning mathematics; creativity is important in learning to become a teacher. Teachers need to be creative to continue developing as teachers and grow professionally (Griffiths, 2014).

Students know from day one that our class is not a "traditional" mathematics class. They are surprised to immediately start working in teams, be presented with problems, and be expected to learn to apply new technologies without a lot of instruction. They are not used to an active class, where they spend very little time watching the instructor work at the board and where each class period engages them in new activities. They are not accustomed to mathematics classes where they are expected to draw upon prior understanding of algebra, trigonometry, geometry, statistics, and calculus, with the experience of moving across domains of knowledge on a daily basis, often applying what they understand to real-world phenomena. Nor do they expect to talk as much as they do, to argue as much, to stand in front and present as much, or especially to write as much as they do.

From the perspective of the instructor, teaching a course that fosters creativity of the students is also a very rewarding experience. So far, for each of the projects and most of the activities, students came up with approaches that enriched the experience of the instructor. For instance, the collection of student-generated examples of real world use of linear functions posted on the electronic forum constitutes a wealth of connections between mathematics and everyday life.

Of course, teaching such a course is quite labor intensive. It requires very careful preparation of the problems and activities, implementing them in class and observing their results, and keeping what works well and modifying what did not work so well.

Also, it is important to remember that students need guidance in terms of how to work productively in cooperative groups and how to deal with difficult and complex problems, but also that they need the opportunity to try their own strategies and solutions.

End of semester interviews with outside evaluators confirmed that students find the experiences of the course challenging and very valuable (Uribe-Zarain, 2015). Furthermore, two years after taking the course, one student reflected on how her participation in the integrated course helped her in her career at the university, and how the use of technology had an impact on her own creativity. She states that in the integrated course

> While utilizing technology, for example, using GeoGebra and the coding program, I learned myself connections within mathematics that would be of great benefit for my students conceptual learning. […] Ultimately this class, not only introduced me to new connections within mathematics but ways I can incorporate new and exciting topics to teach my students. These technology ideas will not only engage my students but build upon their conceptual learning. Before this class, I was unaware of the effects technology could have, but I have developed a much more creative mindset from not only doing the activities myself but also having many more resources. (Personal email communication with a student, 13/1/2017).

A student who took the course the year before highlights affective issues associated with the inherent creativity in experimenting with mathematical concepts and solving problems on their own.

> It was my first time using that much technology every day as a learning too and I think that it really benefited me. […] Most of the time in math classes, you are stuck taking notes from the board, or doing worksheets with a lot of problems, but this course was more fun because it was more hands on. It definitely was something that got me more excited to come to class each day. (Personal email communication with a student, 2/4/2017).

Three years after they had taken the course, two other former students also reflected on the impact of using interactive technology in their creativity. One of them is already a teacher.

> I think that the class opened me up to new ways to integrate programming into a mathematics course. Moreover, I think that the types of activities done in class work well in classes for problem based learning and project based leaning. The uses of technology in the course showed many possibilities and examples of how we could engage our own students in the future. (Personal email communication with a student, Jan 30, 2107).

In all cases former students remembered vividly the interactive technological tools used in the course and even specific activities. Their sense of excitement about using interactive technology in creative ways to foster mathematical understanding remains fresh in their minds.

Acknowledgements The course discussed in this chapter was funded in part by the National Science Foundation (TUES grant, award number 1140702): Bernhardt, S., Flores, A., Park, J., and Shipman, H. (2012–2016). *Integrated Science and Mathematics Education: A Model Course for Pre-Service Teachers*. The statements in this chapter are those of the authors and do not necessarily reflect the position of NSF.

References

Abramovich, S. (2014). *Computational experiment approach to advanced secondary mathematics curriculum*. New York, NY: Springer.

Aiken, L. R. (1973). Ability and creativity in mathematics. *Review of Educational Research, 43*, 405–432.

Allen, D. H., Donham, R. S., & Bernhardt, S. A. (2011). Problem-based learning. *New Directions for Teaching and Learning, 2011*(128), 21–29.

Aralas, D. (2008). Mathematical creativity and its connection with mathematical imagination. In *Proceedings of the Discussion Group 9: Promoting Creativity for all Students in Mathematics Education*, Section 1, 23–32. *The 11th International Congress on Mathematical Education Monterrey, Mexico, July 6–13, 2008*. http://www.dg.icme11.org/document/get/255.

Artzt, A. F., & Newman, C. M. (1990). *How to use cooperative learning in the mathematics class*. Reston, VA: National Council of Teachers of Mathematics.

Banghart, F. W., & Spraker, H. S. (1963). Group influence on creativity in mathematics. *The Journal of Experimental Education, 31,* 257–263.

Biggs, J., & Tang, C. (2007). *Teaching for quality learning at university* (3rd ed.). Berkshire, England: Society for Research into Higher Education & Open University Press. Retrieved from http://www.umweltbildung-noe.at/upload/files/OEKOLOG%202014/2_49657968-Teaching-for-Quality-Learning-at-University.pdf.

Boden, M. A. (2004). *The creative mind: Myths and mechanisms* (2nd ed.). London, UK: Routledge.

Bryan, J. (2014). *Back to school means back to STEM*. Retrieved from http://www.arborsci.com/wordpress/wp-content/uploads/2014/08/Pull-Back-Car.pdf.

Bu, L., & Schoen, R. (Eds.). (2011). *Model-centered learning: Pathways to mathematical understanding using GeoGebra*. Rotterdam, The Netherlands: Sense Publishers.

Buteau, C., & Muller, E. (2006). Evolving technologies integrated into undergraduate mathematics education. In C. Hoyles, J. Lagrange, L. H. Son, & N. Sinclair (Eds.), *Proceedings for the Seventeenth ICMI Study Conference: Technology Revisited, Part 2: Contributions* (pp. 74–81). Hanoi, Vietnam: Hanoi Institute of Technology. Retrieved from http://www.mathunion.org/fileadmin/ICMI/files/Digital_Library/icmi-study-17/ICMI17proceedingsPart2.pdf.

Buteau, C., & Muller, E. (2014). Teaching roles in a technology intensive core undergraduate mathematics course. In A. Clark-Wilson, O. Robutti, & N. Sinclair (Eds.), *The mathematics teacher in the digital era* (pp. 163–185). New York, NY: Springer.

Butler, A., Phillmann, K.-B., & Smart, L. (2001). Active learning within a lecture: Assessing the impact of short, in-class writing exercises. *Teaching of Psychology, 28,* 57–59.

Cakir, M. P., & Stahl, G. (2013). The integration of mathematics discourse, graphical reasoning and symbolic expression by a Virtual Math Team. In D. Martinovic, V. Freiman, & Z. Karadag (Eds.), *Visual mathematics and cyberlearning* (pp. 49–96), Dordrecht, The Netherlands, Springer Science+Business Media. https://doi.org/10.1007/978-94-007-2321-4.

Clark-Wilson, A., Robutti, O., & Sinclair, N. (2014). *The mathematics teacher in the digital era*. New York, NY: Springer.

Committee on Developments in the Science of Learning, National Research Council. (2000). *How people learn: Brain, mind, experience, and school (Expanded Edition)*. Washington, D.C.: The National Academies Press. E-book. Retrieved from http://www.nap.edu/read/9853/.

Contreras, J. N. (2013). Fostering mathematical creativity through problem posing and modeling using dynamic geometry: Viviani's problem in the classroom. *Journal of Mathematics Education at Teachers College, 42,* 66–72.

Cory, B. L. (2010). Bouncing balls and graphing derivatives. *Mathematics Teacher, 104*(3), 206–213.

Council of Chief State School Officers Interstate Teacher Assessment and Support Consortium. (2011). *InTASC model core teaching standards: A resource for state dialogue*. Washington, D. C.: Council of Chief State School Officers. Retrieved from http://www.ccsso.org/Documents/2011/InTASC_Model_Core_Teaching_Standards_2011.pdf.

Davidson, N. (Ed.). (1990). *Cooperative learning in mathematics*. Menlo Park, CA: Addison Wesley.

Davidson, D., & Pearce, D. (1990). Perspectives on writing activities in the mathematics classroom. *Mathematics Education Research Journal, 2,* 15–22.

Drabick, D. A. G., Weisberg, R., Paul, L., & Bubier, J. L. (2007). Keeping it short and sweet: Brief, ungraded writing assignments facilitate learning. *Teaching of Psychology, 34,* 172–176.

Enthought Scientific Computing Solutions. (2016). *Enthought canopy: Easy python deployment plus integrated analysis environment for scientific computing, data analysis and engineering*. Retrieved from https://store.enthought.com/downloads/#default.

Ernest, P. (2005). Agency and creativity in the semiotics of learning mathematics. In M. H. G. Hoffmann, K. Lenhard, & F. Seeger (Eds.), *Activity and sign: Grounding mathematics education* (pp. 23–34). New York, NY: Springer.

Flores, A. (2014). Integrating computers, science, and mathematics—A course for future mathematics teachers. In *Proceedings of the 6th International Conference on Computer Supported Education* (pp. 246–251). Setúbal, Portugal: Scitepress. https://doi.org/10.5220/0004942402460251.

Flores, A., Bernhardt, S. A., & Shipman, H. L. (2015). Rowing competitions and perspective. *International Journal of Mathematical Education in Science and Technology, 46,* 284–319.

Flores, A., & Park, J. (2016). Students' guided re-invention of definition of limit of sequence with interactive technology. In *Contemporary Issues in Technology and Teacher Education, 16(2)*. Retrieved from http://www.citejournal.org/volume-16/issue-2-16/mathematics/students-guided-reinvention-of-definition-of-limit-of-a-sequence-with-interactive-technology/.

Flores, A., Park, J., & Bernhardt, S. A. (2016). Learning mathematics and technology through inquiry, cooperation, and communication: A learning trajectory for future mathematics teachers. In M. L. Niess, S. Driskell, & K. Hollebrands (Eds.), *Handbook of research on transforming mathematics teacher education in the digital age* (pp. 324–352). Hershey, PA: Information Science Reference.

Freudenthal, H. (1971). Geometry between the devil and the deep sea. *Educational Studies in Mathematics, 3,* 413–435.

Fryer, M. (2006). Facilitating creativity in higher education: A brief account of National Teaching Fellows' views. In N. Jackson, M. Oliver, M. Shaw, & J. Wisdom (Eds.), *Developing creativity in higher education: An imaginative curriculum* (pp. 74–88). London, UK: Routledge.

Gibb, E. G. (1970). *Creative problem solving* (Preliminary edition). Austin, TX: University of Texas, Research and Development Center for Teacher Education. Retrieved from http://files.eric.ed.gov/fulltext/ED037367.pdf.

Gordon, S. P., & Gordon, F. S. (2010). *Functions, data, and models: An applied approach to college algebra*. Washington, D.C.: Mathematical Association of America.

Gravemeijer, K., Cobb, P., Bowers, J., & Whitenack, J. (2000). Symbolizing, modeling, and instructional design. In P. Cobb, E. Yackel, & K. McClain (Eds.), *Symbolizing and communicating in mathematics classrooms* (pp. 225–273). Mahwah, NJ: Lawrence Erlbaum Associates.

Groff, J. (2013). *Technology-rich innovative learning environments*. Paris, France: Organisation for Economic Co-operation and Development. Retrieved from http://www.oecd.org/edu/ceri/Technology-Rich%20Innovative%20Learning%20Environments%20by%20Jennifer%20Groff.pdf.

Griffiths, M. (2014). Encouraging imagination and creativity in the teaching profession. *European Educational Research Journal, 13*, 117–129. Retrieved from http://eer.sagepub.com/content/13/1/117.full.pdf+html.

Güçler, B., Hegedus, S., Robidoux, R., & Jackiw, N. (2013). Investigating the mathematical discourse of young learners involved in multi-modal mathematical investigations: The case of haptic technologies. In D. Martinovic, V. Freiman, & Z. Karadag (Eds.), *Visual mathematics and cyberlearning* (pp. 97–118). Dordrecht, The Netherlands: Springer Science+Business Media.

Habre, S. (2012). Improving understanding in ordinary differential equations through writing in a dynamical environment. *Teaching Mathematics and Its Applications, 31,* 153–166. https://doi.org/10.1093/teamat/hrs007.

Hadamard, J. (1945). *Essay on the psychology of invention in the mathematical field*. Princeton, NJ: Princeton University Press.

Hake, R. (1998). Interactive engagement versus traditional methods: A six thousand-student survey of mechanics test data for introductory physics courses. *American Journal of Physics, 66,* 64–74.

Halmos, P. R. (1968). Mathematics as a creative art. *American Scientist, 56,* 375–389.

Hanna, G. (1991). Mathematical proof. In D. Tall (Ed.), *Advanced mathematical thinking* (pp. 54–61). Dordrecht, The Netherlands: Kluwer.

Hanna, G., & Winchester, I. (Eds.). (1990). Creativity and mathematical proof. Special issue *Interchange, 21*(1).

Hashimoto, Y., & Becker, J. (1999). The open approach to teaching mathematics—Creating a culture of mathematics in the classroom: Japan. In L. J. Sheffield (Ed.), *Developing mathematically promising students* (pp. 101–119). Reston, VA: National Council of Teachers of Mathematics.

Haylock, D. W. (1987). A framework for assessing mathematical creativity in schoolchildren. *Educational Studies in Mathematics, 18*(1), 59–74.

Henderson, K. B., & Pingry, R. E. (1953). Problem-solving in mathematics. In H. F. Fehr (Ed.), *The learning of mathematics: Its theory and practice* (pp. 228–270). Washington, D.C.: National Council of Teachers of Mathematics.

Hiebert, J., & Grouws, D. A. (2007). The effects of classroom mathematics teaching on students' learning. In F. K. Lester (Ed.), *Second handbook of research on mathematics teaching and learning* (Vol. 1, pp. 371–404). Charlotte, NC: Information Age Publishing.

Hoyles, C., & Lagrange, J.-B. (Eds.). (2010). *Mathematics education and technology—Rethinking the terrain. The 17th ICMI Study.* New York, NY: Springer.

International GeoGebra Institute. (2017). *GeoGebra.* Retrieved from https://www.geogebra.org/cms/download.

Jackson, N. (2006). Imagining a different world. In N. Jackson, M. Oliver, M. Shaw, & J. Wisdom (Eds.), *Developing creativity in higher education: An imaginative curriculum* (pp. 1–9). London, UK: Routledge.

Jackson, N., & Sinclair, C. (2006). Developing students' creativity: Searching for an appropriate pedagogy. In N. Jackson, M. Oliver, M. Shaw, & J. Wisdom (Eds.), *Developing creativity in higher education: An imaginative curriculum* (pp. 118–141). London, UK: Routledge.

Kattou, M., Kontoyianni, K., Pitta-Pantazi, D., & Christou, C. (2013). Connecting mathematical creativity to mathematical ability. *ZDM—The International Journal on Mathematics Education, 45*(4), 167–181.

Krutetskii, V. A. (1976). *The Psychology of mathematical abilities in schoolchildren* (J. Teller, Trans.). Chicago, IL: University of Chicago Press.

Leikin, R. (2014). Challenging mathematics with multiple solution tasks and mathematical investigations in geometry. In Y. Li, E. A. Silver, & S. Li (Eds.), *Transforming mathematics instruction: Multiple approaches and practices* (pp. 59–80). Cham, Switzerland: Springer.

Leikin, R., & Pitta-Pantazi, D. (2013). Creativity and mathematics education: The state of the art. *ZDM—The International Journal on Mathematics Education, 45*(2), 159–166.

Mann, E. L. (2006). Creativity: The essence of mathematics. *Journal for the Education of the Gifted, 30*(2), 236–260. Retrieved from http://files.eric.ed.gov/fulltext/EJ750778.pdf.

Manuel, D. (2009). Does technology help building more creative mathematical environments? In B. Sriraman, V. Freiman, & N. Lirette-Pitre (Eds.), *Interdisciplinary, creativity, and learning: Mathematics with literature, paradoxes, history, technology and modeling* (pp. 233–247). Charlotte, NC: Information Age Publishing.

Marrongelle, K. A. (2008). Enhancing meaning in mathematics: Drawing on what students know about the physical world. In P. E. Elliott & C. M. Elliott Garnett (Eds.), *Getting into the mathematics conversation: Valuing communication in mathematics classrooms* (pp. 295–307). Reston, VA: National Council of Teachers of Mathematics.

Martinovic, D., Freiman, V., & Karadag, Z. (2013). Visual mathematics and cyberlearning in view of affordance and activity theories. In D. Martinovic, V. Freiman, & Z. Karadag (Eds.), *Visual mathematics and cyberlearning* (pp. 209–238). Dordrecht, The Netherlands: Springer Science +Business Media.

Mason, J., & Watson, A. (2008). Mathematics as a constructive activity: Exploiting dimensions of possible variation. In M. P. Carlson & C. Rasmussen (Eds.), *Making the connection: Research and teaching in undergraduate mathematics education* (pp. 191–204). Washington, D.C.: Mathematical Association of America.

Meissner, H. (2005). Challenges to provoke creativity. In *The Third East Asia Regional Conference on Mathematics Education.* Retrieved from www.math.uni-muenster.de/didaktik/u/meissne/WWW/mei135.doc.

Mishra, P., & Koehler, M. J. (2006). Technological pedagogical content knowledge: A new framework for teacher knowledge. *Teachers College Record, 108*(6), 1017–1054.

Niess, M. L., Ronau, R. N., Shafer, K. G., Driskell, S. O., Harper, S. R., Johnston, C., et al. (2009). Mathematics teacher TPACK standards and development model. *Contemporary Issues in Technology and Teacher Education, 9*(1), 4–24. Retrieved from http://www.citejournal.org/vol9/iss1/mathematics/article1.cfm.

Oehrtman, M. (2015). *Sequences that converge to 5. Sequences that don't converge to 5.* Retrieved from https://clearcalculus.okstate.edu/images/Guided%20Reinvention/Graphs%20-%20with%20Epsilon%20and%20N%20Lines/Sequence%20Graphs.html.

Paulus, P. B. (2000). Groups, teams, and creativity: The creative potential of idea-generating groups. *Applied Psychology: An International Review, 49*(2), 237–262.

Poincaré, H. (1920). *Science et méthode.* Paris, France: Flammarion. Retrieved from https://archive.org/stream/scienceetmthod00poin#page/42/mode/2up.

Polya, G. (1954). *Mathematics and plausible reasoning*. Princeton, NJ: Princeton University Press.

Polya, G. (1957). *How to solve it: A new aspect of mathematical method* (2nd ed.). Princeton, NJ: Princeton University Press.

Polya, G. (1962). *Mathematical discovery: On understanding, learning, and teaching problem solving* (Vol. 1). New York, NY: Wiley.

Powell, N. N., Anderson, M., & Winterroth, S. (1994). Reflections on miniature golf. *Mathematics Teacher, 87*(7), 490–495.

Presmeg, N. (2003). Creativity, mathematizing and didactizing: Leen Streefland's work continues. *Educational Studies in Mathematics, 54,* 127–137.

Rivera, F. D. (2011). *Toward a visually-oriented school mathematics curriculum.* Dordrecht, The Netherlands: Springer.

Sacristán, A. I., Calder, N., Rojano, T., Santos-Trigo, M., Friedlander, A., Meissner, H., et al. (2010). The influence and shaping of digital technologies on the learning—And learning trajectories—Of mathematical concepts. In C. Hoyles & J.-B. Lagrange (Eds.), *Mathematics education and technology—Rethinking the terrain* (pp. 179–226). New York, NY: Springer.

Scardamalia, M., & Bereiter, C. (2014). Smart technology for self-organizing processes. *Smart Learning Environments, 1*(1). Retrieved from http://www.slejournal.com/content/1/1/1, https://doi.org/10.1186/s40561-014-0001-8.

Sealey, V., Deshler, J. M., & Hazen, K. (2014). Strengthening student understanding of mathematical language through verbal and written representations of the intermediate value theorem. *PRIMUS: Problems, Resources, and Issues in Mathematics Undergraduate Studies, 24*(2), 175–190. https://doi.org/10.1080/10511970.2013.858282.

Sheffield, L. J. (1994). *The development of gifted and talented mathematics students and the National Council of Teachers of Mathematics Standards*. Storrs, CT: National Research Center on the Gifted and Talented.

Sheffield, L. J. (2003). *Extending the challenge in mathematics: Developing mathematical promise in K-8 children.* Thousand Oaks, CA: Corwin Press.

Sheffield, L. J. (2008). Promoting creativity for all students in mathematics education: An overview. In *Proceedings of the Discussing Group 9: Promoting Creativity for All Students in Mathematics Education, The 11th International Congress on Mathematical Education Monterrey, Mexico*, pp. 369–381. Retrieved from http://dg.icme11.org/tsg/show/10.

Sheffield, L. J. (2009). Developing mathematical creativity—Questions may be the answer. In R. Leikin, A. Berman, & B. Koichu (Eds.), *Creativity in mathematics and the education of gifted students* (pp. 87–100). Rotterdam, The Netherlands: Sense Publishers.

Silver, E. A. (1997). Fostering creativity through instruction rich in mathematical problem solving and problem posing. *ZDM—The International Journal on Mathematics Education, 29*(3), 75–80.

Sriraman, B. (2004). Characteristics of mathematical creativity. *The Mathematics Educator, 14*(1), 19–34. Retrieved from http://math.coe.uga.edu/tme/Issues/v14n1/v14n1.Sriraman.pdf.

Sriraman, B. (2005). Are giftedness and creativity synonyms in mathematics? An analysis of constructs within the professional and school realms. *The Journal of Secondary Gifted Education, 17,* 20–36.

Sternberg, R. J., & Williams, W. M. (1996). *How to develop student creativity.* Alexandria, VA: ASCD. Retrieved from http://www.ascd.org/publications/books/196073/chapters/Introduction@-Theory-of-Creativity.aspx.

Stewart, T. L., Myers, A. C., & Culley, M. R. (2010). Enhanced learning and retention through 'writing to learn' in the psychology classroom. *Teaching of Psychology, 37,* 46–49.

Torrance, E. P. (1979). A three-stage model for teaching for creative thinking. In A. E. Lawson (Ed.), *1980 AETS Yearbook: The psychology of teaching for thinking and creativity* (pp. 226–253). Columbus, OH: ERIC Clearinghouse for Science, Mathematics, and Environmental Education.

Uribe-Zarain, X. (2015). *TUES Math 267 Observations and interviews report.* Newark, DE: Delaware Education Research and Development Center.

Van Oortmerssen, L. A., Van Woerkum, C. M. J., & Aarts, N. (2015). When interaction flows: An exploration of collective creative processes on a collaborative governance board. *Group Organization & Management, 40*(4), 500–528.

Von Glasersfeld, E. (1995). *Radical constructivism: A way of knowing and learning.* New York, NY: Routledge/Falmer.

Voica, C., & Singer, F. M. (2013). Problem posing as a tool for the development of mathematical creativity in school children. *ZDM—The International Journal on Mathematics Education, 45*(4), 267–279.

Whiley, T., & Tellup, E. (2015). *The math and physics of Joshua.* Newark, DE: Final project presentation for Math267, University of Delaware.

Winchester, I. (1990). Introduction—Creativity, thought and mathematical proof. *Interchange, 21*(10), 1–6.

Chapter 7
Creativity and the Design of Music-Mathematics Activities in a Virtual Simulation Learning Environment

Trina J. Davis, Glenn Phillips and Gerald Kulm

Abstract Defined by digital age learning, the current education landscape offers unparalleled opportunities for creative and transformative experiences for students of all ages. Navigating the complexity of this new landscape means that students must be equipped with skills that foster creativity, and are poised to develop unique and innovative solutions. This requires educators to rethink what instructional design should look like and how students should be engaged. Mathematics classrooms, in particular, are fertile places for activities that integrate creativity. This chapter explores the role of creativity in mathematics learning and examines the intersection of mathematics, music, and virtual spaces. Built on Koestler (The concept of creativity in science and art. Springer, The Netherlands, pp. 1–17, 1981) work on creation and creativity, the chapter suggests how environmental (technology) and conceptual (music) frameworks can be juxtaposed to mathematics teaching to create more engaged and productive learning. It is in these unique collisions that new knowledge and new ways of knowing come to pass. A classroom simulation example involving practice teaching experiences in a virtual setting exhibits how technology and music can be incorporated into preservice teacher education. Implications of this work include an expanded idea of what contributes to feelings of efficacy and student success in the mathematics classroom as well as how music may help with challenging mathematical concepts like fractions and patterns.

Keywords Creativity · Mathematics · Technology · Music · Preservice teachers Second life

T. J. Davis (✉) · G. Phillips · G. Kulm
Texas A&M University, College Station, USA
e-mail: trinadavis@tamu.edu

G. Phillips
e-mail: glennallenphillips@gmail.com

G. Kulm
e-mail: gkulm123@yahoo.com

© Springer International Publishing AG, part of Springer Nature 2018

V. Freiman and J. L. Tassell (eds.), *Creativity and Technology in Mathematics Education*, Mathematics Education in the Digital Era 10,
https://doi.org/10.1007/978-3-319-72381-5_7

7.1 Introduction

We are now in a century that has been defined by digital age learning that offers unparalleled possibilities for creative and transformative learning experiences for students of all ages. Navigating the complexity of this new landscape means that students must be equipped with skills that foster creativity and problem solving and are poised to develop unique and innovative solutions (Mishra, Fahnoe, Henriksen, & The Deep Play Research Group, 2013; International Society for Technology in Education, 2007, 2016). Research suggests that these skill sets must be deeply ingrained and digital age learners must have strong disciplinary content knowledge, but also be able to see seamless connections across disciplines (Mishra et al., 2013). This project, part of a larger National Science Foundation-funded grant created to examine the intersection of technology, equity, and math content knowledge for preservice mathematics teachers, examines how technology can be used to link creativity and mathematics learning. As student avatars were placed into a virtual environment containing "musical play," researchers sought to understand what meaning these preservice teachers made of connections between technology, mathematics, and creativity. In short, this project seeks to understand what the experiences of preservice mathematics teachers in musical play environments tells us about creativity as pedagogy, and creativity as product. This piece uses Koestler (1981) notion of creativity as a theoretical frame to understand the context of the project and make meaning of the participants' responses. The perspectives that we share here are not exhaustive, but we hope help to illustrate the importance of creativity in mathematics learning. We will also highlight approaches that are situated in, as well as, leverage the affordances of using a simulated classroom and learning spaces in the virtual world of Second Life® (Davis, 2013). Dickey (2005) describes virtual worlds as networked desktop virtual reality where users or avatars (i.e., on screen customizable virtual personas), move around and engage in various three-dimensional (3-D) spaces. Users can communicate using text-based chat tools, or via audio (voice). Basic gesture functionality is also integrated into virtual worlds. Virtual world spaces will be elaborated on later in the chapter.

7.2 Creativity and Learning: Introducing a Creative Framework

The moment of creativity, as explained by Koestler (1981) occurs when two "mutually exclusive associative contexts… merge" (p. 5). The effect, leading to moments of comedy, discovery, or art is hinged upon the unexpected collision of two sets of rules or paradigms. Explained further,

> From Pythagoras, who combined arithmetic and geometry, to Einstein, who unified energy and matter in a single sinister equation, the pattern is always the same. The Latin word *cogito* comes from *cogitare*, "to shake together." The creative act does not create something

> out of nothing, like the God of the Old Testament; it combines, reshuffles, and relates already existing but hitherto separate ideas, facts, frames of perception, associative contexts (Koestler, 1981, p. 2).

Mishra et al. (2014) build on Koestler's argument insisting that "creativity is not a 'magical' process, rather it emerges from combining preexisting ideas and concepts in unique and novel ways" (p. 20). In short, it is at the nexus of two seemingly incongruent fields that we find creativity in business, in healthcare, in art, and in education. Therefore, the search for creativity is not as important as the search for an association that could create it.

The call for creativity in education is not new. The United States' National Advisory Committee on Creative and Cultural Education (NACCCE), in a 1999 report, claimed, "Education faces challenges that are without precedent. Meeting these challenges calls for new priorities in education, including a much stronger emphasis on creative and cultural education and a new balance in teaching and in the curriculum" (p. 5). Several books are devoted to both the creative pedagogy (Gregerson, Snyder, & Kaufman, 2012; Springer, Alexander, & Persiani, 2006; Woods & Jeffrey, 1996) and the practice of teaching creativity (Cropley, 2001; Craft, 2010). The implementation of creative pedagogy even contains parameters suggested by Cropley (2001). Three aspects that should guide how creativity is used in the classroom include novelty, effectiveness, and ethicality (Cropley, 2001, p. 6). Cropley (2001) suggests that creative education "departs from the familiar,…works, in the sense that it achieves some end," and is not "selfish or destructive behavior" (p. 6).

Moreover, creativity in education is not always an active process of infusion; it can also be an uncovering or unearthing of latent creativity. At the K–12 level, Runco (2008) argues that the job of educators is not to "start from scratch" as children often come with creative talents and ideas (p. 7). Instead, he continues, "[kids] already have the capacity for original interpretations and creative ideas. It is really more a matter of preventing loss of talent than it is the provision of new talents" (Runco, 2008, p. 7). And finally, creativity (even in education) is often difficult to define. Mitchell, Inouye, and Blumenthal (2003) provocatively suggest that "[c]reativity is a bit like pornography; it is hard to define, but we think we know it when we see it" (p. 7). Thirty years ago, Nicholson and Moran (1986) explored the ways that preschool teachers observe and measure creativity. They found that more often than not, teachers substituted or confused measures of intelligence for measures of creativity. While this flexibility should not discourage creativity work in education (PK–12), it should help us recognize the various ways that creativity can be both planted and dug up in the classroom context.

7.3 Creativity and Mathematics Learning: Contextualizing the Project

While the general push for creativity in education stems from broader concepts of creativity's value, it is important to recognize the importance of subject matter as a place of creativity. The ways that creativity can be used in a poetry lesson may vary greatly from how creativity would be infused into a lesson on the mitochondria, for example. Baer (2016) insists that "[t]here are simply no domain-general, decontextualized thinking skills, only domain- and content-specific thinking skills" (p. 16). If we follow Baer's argument (which we do), a logical step is to consider the particular relationship that creativity shares with mathematics education. What domain-specific thinking skills, practices, and projects can support creative learning within a mathematics classroom?

In his 1979 Presidential Address to the Mathematical Association of America, Tammadge builds a rationale for the inclusion of creativity in the mathematics classroom. He argues that "[w]hat makes the wise mathematician, now as [in the days of Archimedes], is experience and confidence, willingness to experiment, originality of approach" (Tammadge, 1979, p. 147). Critiquing the common models of classroom instruction, he writes:

> The "rational machine" model will do up to a point, but it is not enough. We create for pleasure as well as to meet needs or solve problems. Knowledge and understanding certainly develop in quanta, in jumps, not by continuous accretion (Tammadge, 1979, p. 147).

Tammadge's (1979) call was certainly heard. Today there are conferences devoted to creativity and mathematics, multiple websites that suggest creative mathematical pedagogy and projects, and numerous publications that both instruct on and consider the role of, creativity in mathematics education.

Researchers have explored the conceptions and executions of creativity in mathematics classrooms (Lev-Zamir & Leikin, 2013; Leikin, Subotnik, Pitta-Pantazi, Singer, & Pelczer, 2013; Henrickson & Mishra, 2015; Vale & Barbosa, 2015), evaluations of creative mathematicians (Mehta, Mishra, Henrickson, & The Deep-Play Research Group, 2016), and calls for future research on mathematics and creativity (Leikin & Pitta-Pantazi, 2013). In short, creativity in mathematics expands applications of concepts, engages students, and (in a Koestlerian way) "shakes" up the rote curriculum of record. Notably, creativity was introduced as a defensive tool against students with negative or disabling opinions on their own efficacy in mathematics (Boaler, 2015).

Kleinman (2008) suggests that while "there may be no single, 'hold-all' definition of creativity, there seems to be a general coalescing of agreement amongst creativity researchers that creativity involves notions of novelty and originality combined with notions of utility and value" (p. 209). It stands to reason then that creativity is often paired with those things to which we already assign "utility and value." This value can be understood in the recent evolution of STEM (Science,

Technology, Engineering, and Mathematics) to STEAM (Science, Technology, Engineering, Art + Design, and Mathematics). The incorporation of "Art + Design" both makes the popular acronymic field more inclusive, and it makes room for the creative interactions of art and design in traditional STEM fields. The movement (championed by the Rhode Island School of Design) has three main objectives:

- transform research policy to place Art + Design at the center of STEM.
- encourage integration of Art + Design in K–20 education.
- influence employers to hire artists and designers to drive innovation ("STEM to STEAM," 2016).

The lexical play both recognizes Art + Design as united with and central to STEM subjects.

It is at the intersection of art and mathematics that creativity, what Koestler (1981) referred to as the AHA (discovery), HAHA (humor), and AH…(emotive) moments, that the effects of creative "bisociation" are seen (p. 2). Methods for infusing creativity and mathematics include photography (Munakata & Vaidya, 2012; Furner & Marinas, 2014), painting and drawing, 3-dimensional art, computer technology, and performing arts, including music.

This final bisociation (mathematics and music) is most appropriate to our current project. Since Pythagoras (and perhaps before) the relationship of mathematics and music has been noted (Wright, 2009; Nisbet, 1991; Vaughn, 2000). At its base level, musical education is mathematics. Scales, meter, and rhythm are no more than sets, fractions, and integers. Even more complex ideas like chords are housed within a mathematical understanding of music. It is therefore unsurprising that music would make sense in a mathematics classroom. Vaughn (2000) explains it thusly:

> According to conventional wisdom, music and mathematics are related, and musical individuals are also mathematically inclined. After all, musical rhythm is based upon mathematical relations, and it is certainly reasonable to assume that an understanding of music requires some understanding of ratios (e.g., 3/4 time vs. 4/4 time) and repeating patterns. In addition, if music experiences [enhance] spatial-temporal reasoning, then music may also enhance understanding of those aspects of math that involve spatial-temporal reasoning, such as geometry and proportional reasoning (p. 149).

Introducing the arts into the mathematics classroom has multiple effects. Fiske (1999) suggests that engagement in the arts can (1) "reach students who are not otherwise being reached," (2) "reach students in ways that they are not otherwise being reached," (3) "connect students to themselves and others," (4) "transform the environment for learning," (5) "provide learning opportunities for the adults in the lives of young people," (6) "provide new challenges for those students already considered successful," and (7) "connect learning experiences to the world of real work." (pp. 12–13). Johnson and Edelson (2003) used music in the elementary classroom to increase spatial-temporal reasoning. Beal (2000) suggests that music can be used to help with functions and reasoning. Fernandez (1999) suggests creating math-based instruments upon which students can play songs. These creative practices lead to the confidence, connectedness, and challenge that Fiske (1999) reports.

In addition to the K–12 classroom, music can also be used in the university classroom to prepare future teachers. Though incorporation of music into traditional curriculum (as was done in this project) may positively challenge students and faculty, some teachers may feel that the time and energy requirements necessary to create these kinds of creative lessons are not feasible (Colwell, 2008). On the other hand, after incorporating music into a university classroom, An, Ma, and Capraro (2011) found that "preservice teachers' engagement, beliefs, motivation, and confidence toward mathematics teaching and learning were statistically significantly improved" (p. 240), indicating that the time and effort can be worthwhile.

7.4 Creativity and Technology: Exploring Intersections

Music and mathematics are separate paradigms whose intersection in the field of education creates space for new ideas, new ways of knowing, and creative footholds for both student and teacher growth. This pairing, however, must often be connected by a technology that can create a neutral space for interaction. Technology (from pianos and 3D graphics to virtual worlds) can act as this space. At its root, the question of creativity and education is not how to pair creativity with education, but instead, with what to pair education in an effort to produce creativity. Mishra and The Deep-Play Research Group (2012) suggest that "new tools, devices, and applications" have given birth to a "new world" of creative possibilities (p. 13). "Given this relationship between creativity and technology it is not surprising that educators (particularly those who are technically inclined) have argued that teaching and learning in this emerging world needs to emphasize these twin issues —technology and creativity" (Mishra & The Deep-Play Research Group, 2012, p. 13). However, Mishra and his research group are quick to clarify that technology is not limited to that digital or electronic "fad" that is currently *en vogue* (Mishra & The Deep-Play Research Group, 2012, p. 14). "Whether it's a stone-age tool, a Guttenburg Printing press, the simple crayon, or a high-tech digital simulation, any form of technology is a tool for living, working, teaching and learning" (Mishra & The Deep-Play Research Group, 2012, p. 14).

Technology can act as both a paradigm that interacts with education to promote creativity and an avenue or tool that makes the interaction of two unique paradigms possible. Technology offers what Mishra and The Deep-Play Research Group (2012) refer to as "(in)disciplined research" (p. 15). (In)disciplined research recognizes the importance of work within one discipline but suggests that equally important work is occurring across multiple disciplines (Mishra & The Deep-Play Research Group). Technology can serve as both a conduit and a catalyst for creativity. In Koestlerian terms, technology can be used to "shake" two paradigms together.

As one unpacks ways that technology can fuel or support creativity, a survey of 21st century knowledge frameworks can be instructive. Earlier works suggest that while being actively engaged in experiential and situated learning in virtual environments, learners can be afforded ubiquitous opportunities to authentically develop an array of skills: new literacies (Jenkins, Clinton, Purushotma, Robinson, & Weigel, 2006), problem-solving or mathematics skills (National Council of Teachers of Mathematics, 2000), scientific literacy skills, and computational thinking and information and communication technology (ICT) skills (Barr, Harrison, & Conery, 2011; ISTE, 2007). New literacies or skills outlined by various scholars and groups can include but are not limited to: creativity, collaboration, problem solving, simulation, critical thinking, and negotiation. More recently, Kereluick, Mishra, Fahnoe, and Terry (2013) in their critical review, identify common themes that converge to reveal three types of knowledge domains: foundational knowledge, meta-knowledge, and humanistic knowledge. The authors argue that little has changed in now, this 21st century related to the goals of education, rather they emphasize the need to critically express how technologies, in fact, change these knowledge types. The authors theorize that foundational knowledge gets at the core question "what do students need to know?" (Kereluick et al., 2013, p. 130). The frameworks that Kereluick et al. (2013) reviewed, summarized foundational knowledge by three key subcategories: core content knowledge, digital literacy, and cross-disciplinary knowledge. Meta knowledge, on the other hand, is delineated by the three subcategories: problem solving and critical thinking, communication and collaboration, and creativity and innovation. Kereluick et al. (2013) report that creativity was the most cited across multiple frameworks as an essential skill for success in the 21st century. They describe creativity and innovation thusly:

> Creativity and innovation involve applying a wide range of knowledge and skills to the generation of novel and worthwhile products (tangible or intangible) as well as the ability to evaluate, elaborate, and refine ideas and products.

Lastly, the authors describe humanistic knowledge as "a vision of the learner's self and its location in a broader social and global context" (Kereluick et al., 2013, p. 131). Three subcategories are delineated: life/job skills/leadership, cultural competence, and ethical/emotional awareness. It bares repeating, Kereluick et al. (2013) argue that little has changed related to our educational goals, however we note, one can assume that the technology skills that frame what students need to know and be able to do, are fluid over time.

The International Society for Technology in Education (ISTE) (2016) Standards for Students provide an illustrative example of this. The ISTE technology standards were recently updated and offer a framework for what creativity and student learning might look like in a technology-rich learning context. For example, the standards statement for the "Creative Communicator" focus area states: "[s]tudents communicate clearly and express themselves creatively for a variety of purposes using the platforms, tools, styles, formats and digital media appropriate to their goals. An example of an indicator for this standard is: "[s]tudents create original works or responsibly repurpose or remix digital resources into new creations." Similarly, in the ISTE (2008) Standards for Teachers, the "Facilitate and Inspire Student Learning and Creativity" standard states:" [t]eachers use their knowledge of subject matter, teaching and learning, and technology to facilitate experiences that advance student learning, creativity and innovation in both face-to-face and virtual environments. An illustrative indicator for this standard is: "[p]romote, support, and model creative and innovative thinking and inventiveness."

7.5 Technology Use and Affordances: Classroom Research

The results related to instructional effectiveness within Second Life® have been mixed across various studies (e.g., De Lucia, Francese, Passero, & Tortora, 2009; Wrzesien & Raya, 2010; Ho, Rappa, & Chee, 2009). Authors posit that a distinct advantage Second Life® offers is its presence, or *almost-like-being-there* feel. Across various contexts, SL participants express they have experienced a sense of being present in the space (e.g., De Lucia et al., 2009; Mikropoulos & Natsis, 2011; Witmer & Singer, 1998). The psychological construct of presence has been cited across numerous studies (Mikropoulos & Natsis, 2011) as a key to improving involvement [or engagement] and, by implication, outcomes.

In a computer mediated learning environment context, instructional time can be operationalized by the frequency or amount of time it takes to engage in virtual instruction, or complete exercises, or tasks. One can consider the amount of time participants spend in the environment or use the technology. Holt and Brockett (2012) suggest that a combination of pedagogies that foster time spent with technology and self-directed learning [or facilitated learning], help to improve technology use. In addition, several researchers have investigated self-efficacy and outcome expectancy as it relates to teaching (e.g., Enochs & Riggs, 1990; Enochs, Smith, & Huinker, 2000; Gibson & Dembo, 1984; Guskey, 1988). Design of virtual world learning activities that focus on modeling and working out mathematics problems are more challenging due to symbolic system use. With recent developments in SL such as extended use of displays that project an array of formats and now the availability of interactive pen displays (i.e., Smart Podium solutions) paired with streaming applications, mathematics concepts and problems can be presented/solved with greater ease (Davis, Chien, Brown, & Kulm, 2012).

Technology acceptance and use has also been linked to the comfort or ease of use, of the particular new technology. Usability can be defined by two subcomponents, perceived meaningfulness and perceived ease of use (Merchant et al., 2012). Davis (1989) conceptualized the well-known Technology Acceptance Model (TAM). Prominent factors that influence the decision of accepting a new technology are perceived meaningfulness and perceived ease of use (Davis, 1989). More recently, Holden and Rada (2011) found that technology self-efficacy (TSE) was more beneficial to the TAM than computer self-efficacy. Additionally, Wong (2015) in his study of preservice primary mathematics teachers found an overall positive attitude towards the use of technology, while perceived usefulness was more influential than perceived ease of use. Perceived ease of use was found to rely heavily on facilitating conditions rather than computer self-efficacy. Taken together, there are several factors that can impact a user's experiences with technology, we have highlighted just a few of these.

The remainder of the chapter provides an illustrative example of the Knowledge for Algebra Teaching for Equity (KATE) Project that utilizes a virtual environment to support creativity in the form of music-mathematics exercises. The research treatment of this classroom experience uses Koestler's (1981) notion of "creative *cogitare*" or a reshuffling of known contexts to produce new knowledge or creativity as framework to understand the effect of technology and music in student perceptions of creativity and mathematics learning.

7.6 An Environment Supporting Creativity and Music-Mathematics Learning

For a number of years, teacher educators have sought alternative methods for practical and clinical experiences for preservice teachers (Berliner, 1985; Metcalf, Hammer, & Kahlich, 1996). In recent years, emerging approaches in teacher education have evolved to include the use of virtual technologies to design and simulate authentic classroom teaching environments, or informal learning settings. For example, preservice teachers (PSTs) are able to set up avatars and engage in learning and teaching activities in novel computer-simulated environments. The KATE Project provides an illustrative example of exactly this. Funded by the National Science Foundation (Award #1020132), the KATE Project's research and design teams' efforts include the redesign of a required mathematics problem solving course at Texas A&M University. The course was designed by the KATE research team to include activities and assignments that address issues of diversity and culture in teaching algebra. The design of the course includes four primary, interrelated components: (1) math problem solving and problem posing, (2) math problem equity challenges, (3) readings and discussions on equity and diversity, and (4) Second Life® tutoring, teaching, and informal learning activities.

KATE team members designed a three-dimensional virtual classroom in Second Life® and creative learning spaces to provide a canvas for preservice teachers to engage in various instructional activities. We believe that there are specific approaches and areas of awareness about teaching for equity that preservice teachers must develop and practice early in their preparation (Darling-Hammond, 2000). We assert that providing early learning and teaching opportunities in a simulated learning environment offer a promising approach for experiential learning. We also posit that the provision of experiential learning activities for preservice teachers that include the use of creative spaces, may lead to both vision and utility for them to engage their future students in creative spaces or simulations.

Specifically, in the exploratory study described here, we sought answers to the following questions (1) To what extent did preservice mathematics teachers who participated in the music-mathematics activities find them (a) engaging, and (b) effective? (2) To what extent did preservice teachers' comfort level with using technology, efficacy about using Second Life® during course activities, and sense of presence, relate to them (a) finding the music-mathematics activities engaging, and (b) finding the music-mathematics activities effective? Through this exploratory investigation, we aim to better understand how students experience creativity at the nexus of technology, music, and mathematics. This work can give insight into both mathematics education and preservice teacher pedagogy.

7.7 Methods

7.7.1 Participants

The participants in the study included 24 preservice teachers enrolled during one semester of a required *Mathematics Problem Solving* (MASC 351) course at Texas A&M University. The course was part of the KATE NSF-funded project that focused on mathematical content knowledge and equity consciousness of middle grade preservice mathematics teachers. It was not required that any student had previous experience in Second Life® or virtual reality though some did. All virtual experiences referred to occurred in a 3-D virtual space (Glasscock Island), created and managed by the first author. The research project is one of many virtual world-based projects that occur on Glasscock Island.

7.7.2 Procedure

During the Spring semester, members of the research team designed an environment for music-mathematics activities (Davis, An, Cole, & Jett, 2012). The first author spearheaded the design of a playful creative space within Second Life®, *Music-Math Park.* This provided a multisensory setting for preservice teachers enrolled in the course, to engage in a series of music-mathematics activities (An, 2012) within this newly constructed virtual park. A KATE Project instructor led preservice teachers through various activities. All of the avatars (preservice teachers and instructor) walked around the virtual *Music-Math Park* and engaged in a number of simulation activities that included:

- Playing songs on a large virtual keyboard in the park, and solving related early algebra or mathematics problems based on grid representations or patterns of the notes in the songs (e.g., fractions, ratios, patterns),

- Composing their own songs and then listening to immediate playback of the songs using interactive music composition boards at stations in the park,
- Solving more advanced early algebra problems as they completed various problem solving activities in the park, and
- Fun extras included, listening to sample music selections across various genres (classical, jazz, blues, popular).

For example, Activity II began with the following introductory narrative (An, 2012):

> Unlike writing an essay that we write one sentence by one sentence, in [popular] music composing, it is more like building a house. We construct the frame first and then build and decorate every floor. The sequence of the chords is the frame of the music. We first put the chords in an order and then finish the music sentences under the chords. The most popular sequence of chords used by pop music composers is: I, V, VI, III, IV, I, II, V, I.

Preservice teachers then advanced to the next area in the learning station and completed compositions of their songs by entering notes using interactive music composition boards. Display screens at the various learning stations outlined detailed instructions (see Fig. 7.1).

The following Figs. 7.2, 7.3, 7.4, 7.5 and 7.6 highlight snapshots that were captured of the learning stations and of PSTs completing activities in *Music-Math Park* in Second Life®.

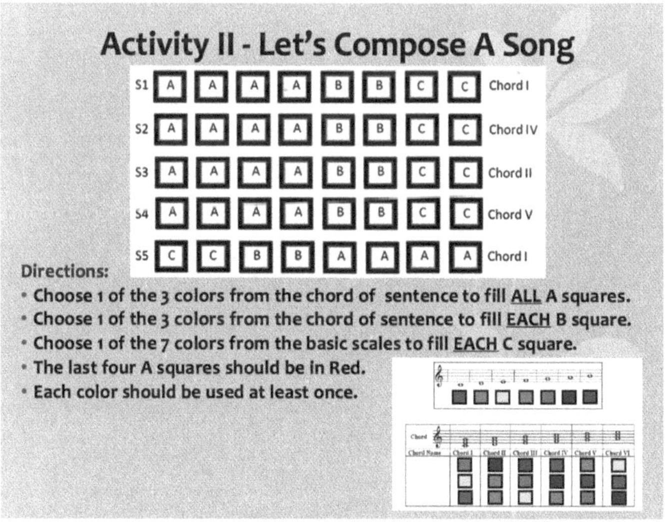

Fig. 7.1 Example of instructions for activity II displayed at a station in Music-Math Park

Fig. 7.2 Snapshot from an aerial view of Music-Math Park

Fig. 7.3 Preservice teacher avatars completing an introductory music-math activity after they played a children's song on the colorful piano keyboard

7.7.3 *Data Sources and Analyses*

The Preservice Teacher Second Life® Engagement (PTSLE) self-report measure, designed by the authors, was used to collect data for the study. Participants completed the PTSLE instrument following their participation in *Second Life®* activities in the problem solving course. The PTSLE instrument contained 32 items, 16 were Likert-type items with a 10-point scale spanning from strongly disagree to strongly agree. The current study focuses on the music-mathematics exercise related items

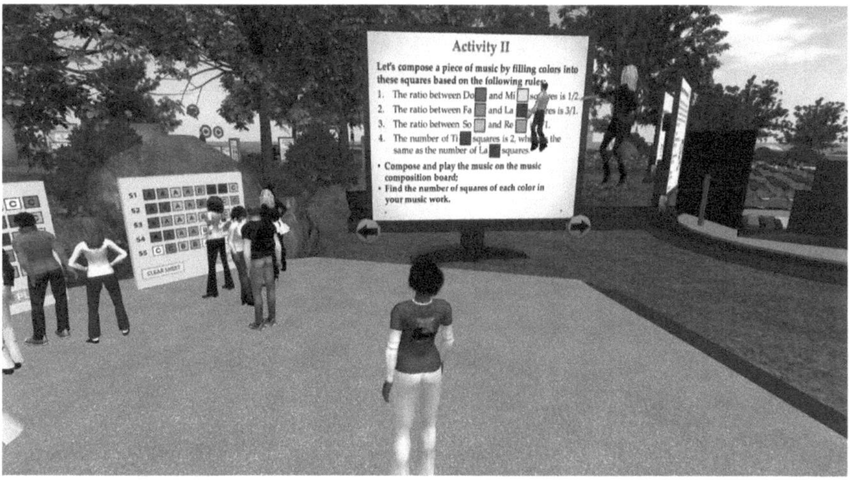

Fig. 7.4 Instructor and preservice teacher avatars completing activity II in Music-Math Park

Fig. 7.5 Preservice teacher avatars composing their songs using the interactive composition boards

(i.e., engagement, effectiveness) and other key foci (i.e., technology comfort/use, SL efficacy, presence). Additionally, all responses to an open-ended question are provided. While our sample size makes a traditional qualitative analysis of the open-ended responses inappropriate, the student responses offer more than just anecdotal commentary. Table 7.1 shows examples of these items.

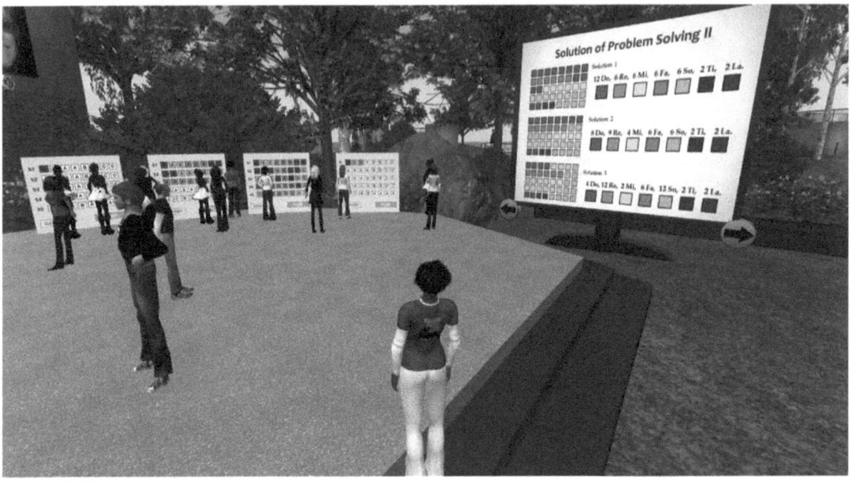

Fig. 7.6 Instructor and preservice teacher avatars finishing compositions and going over solutions to a music-math problem

Table 7.1 Sample items from the PTSLE

Component	Sample items
Technology comfort/use	The following best describes my comfort level with using technology in general
SL efficacy	At the beginning of the course, I felt that if I tried hard I would be successful using Second Life®
Presence	I had the sense of being present or being there in the virtual learning spaces
Engagement	I found the music-math SL exercises engaging
Effectiveness	I found the music-math SL exercises effective
Open Ended Question	Was there something you thought was good about the math and music SL exercise and homework challenge? Please explain and give one example

7.8 Results

The median and interquartile range (IQR) were calculated for technology comfort/use, Second Life® efficacy, presence, engagement, and effectiveness. The results of the analyses of the measures are provided in Table 7.2.

In general, preservice teachers' comfort level with using technology was high (Mdn = 8.0). They also indicated that the music-mathematics activities in Second Life® were engaging (Mdn = 7.0) and effective (Mdn = 7.0).

Nonparametric Spearman correlations (and Kendall Tau-B correlations) were calculated to examine the relationships between key variables, engagement, effectiveness, technology comfort/use, SL efficacy, and presence. Results show

Table 7.2 Medians and IQRs

Component	Median (IQR)
Technology comfort/use	8.0 (7.0–9.0)
SL efficacy	7.0 (6.0–9.0)
Presence	6.5 (4.5–8.75)
Engagement	7.0 (6.0–8.0)
Effectiveness	7.0 (5.25–8.0)

statistically significant relationships between several perceptual variables (at $\alpha = 0.05$): technology comfort/use and efficacy ($p = 0.04$), efficacy and presence ($p = 0.00$), effectiveness and engagement ($p = 0.00$), and effectiveness and presence ($p = 0.04$).

Responses from the Open Ended Question:

Was there something you thought was good about the Math and Music SL exercise and homework challenge? Please explain and give one example.

While there were not enough responses to warrant a rigorous or thorough qualitative analysis of the data, the responses gathered from the 24 participants hint towards student perceptions and the effect of including creativity in the virtual mathematics classroom. Three interacting themes can be drawn from data provided: enjoyment, engagement, and difference. These themes emerged through an in vivo coding process where student words were used as categories (Saldaña, 2015). These categories were then connected through axial coding which precipitated the three themes suggested. Future iterations of similar data would help determine if these themes were only representative of the unique group of participants we had or representative of a larger population of students engaged in a similar exercise.

What follows are brief discussions of each theme and some example quotes that contributed to that particular theme. It is important to remember that while limited, the ubiquity of these themes in the student responses suggest that they are good starting points for continued research.

7.8.1 Enjoyment

Many preservice teachers explained that the experience in the musical virtual environment was fun or enjoyable. The ubiquity of this response helps us to understand that the fun students had with the environment was often the gateway that led students to additional insights. "Fun" was often paired with explications of engagement, mathematical content, and discovery. One participant explained that while the experience of colliding mathematics and music was "fun," the additional manipulatives in the space were "icing on the cake." Koestler (1981) notion of wonder, exhibited by the "HA" moment. shows the integration of seemingly distinct things with which students have familiarity (in this case technology or virtual reality and music) can produce a creative *cogitare* that naturally leads to additional emotions like wonder, engagement, and interest. Illustrative quotes follow:

- I really enjoyed this activity.
- I had a lot of fun trying to figure out the missing notes and thought the activity was very beneficial.
- It connected math and music together which I think can engage more students.
- It is a good way to get students to engage in the lesson.
- I really enjoyed the Math and Music SL exercise. It was a lot of fun to see math and music come together in that way, and I thought that the music boards in SL that we could build these musical pieces we manipulated mathematically [and] were the icing on the cake.

7.8.2 Engagement

Preservice teachers reported that the project was "engaging for the class as a whole." Many preservice teachers who did not use the word "engaging," used language of engagement like "helpful," "interesting," and "spark." Preservice teachers saw engagement on a personal level, explaining that they "liked the hands-on engagement that [they] got" and that "it made [them] pay more attention knowing that [they] could actually play the notes and interact in the space". However, preservice teachers also saw the benefit of the project on a future professional level, exploring "how it is possible to integrate a core subject such as math with their future students' interest, in this case being music." Preservice teachers considered how future students would enjoy something like this activity in a future classroom. Engagement helped preservice teachers take the immediate reaction (fun) to an applicable future (pedagogy). Illustrative quotes follow:

- I thought this exercise was fun, I really liked the hands on engagement that we got. It made me pay more attention knowing that I could actually play the notes.
- I really thought it was engaging for the class as a whole.
- I thought it was really fun, I feel like this activity would be engaging for students and they would enjoy the lesson a lot.
- This homework challenge was very interesting. It was fun to get to use the Second Life® piano to help us solve the problem. It was a hands on assignment and it would definitely interest students.

7.8.3 Difference

Finally, preservice teachers found the activity to be novel. One of the hallmarks of creativity and Koestler's notion of creation, creative activities must be a new or unexpected collision. Preservice teachers shared that they not only appreciated the "difference" but the unusual or new nature of the activity was what caused them to appreciate the activity. "I liked it *because* it was different," one student admitted (emphasis added). Another student explained, "[The activity] offered a very unusual way to look at working with math so that made it more interesting because I hadn't ever thought of it that way." Another suggested, "It gave a new perspective on the content in a mathematical problem that I had never seen before." Difference became a conduit for experience, forcing students to react emotionally (enjoyment) and thoughtfully (engagement). Illustrative quotes follow:

- It is very interesting and puts new thoughts into math. I had never thought about incorporating music into math before this exercise.
- It gave a new perspective on the content in a mathematical problem that I had never seen before. The use of color and music helped to spark interests and catch students' attention.
- The SL exercise was different and was nice because it was more hands-on with the piano.
- I liked it because it was different.
- The exercise gave students a different way to look at math and its use in the real world and I feel that was very helpful.
- It offered a very unusual way to look at working with math so that made it more interesting because I hadn't ever thought of it that way.
- It allowed us to get more used to the Second Life® technology and showed us other ways to use it other than just sitting there and talking. The homework challenge was interesting, it was a different type of problem to solve.

7.9 Discussion

The exploratory work highlighted here was just a preliminary first step in tinkering around the edges of whether this kind of virtual space can be effective in engaging participants in creative mathematics activities. Our results show that preservice teachers found the music-mathematics exercises both engaging and effective. Responses to the open-ended question support this finding as well. Further, two important points emerge from the responses worth noting. First, more than a third of the twenty-four responses comment on approaches used in the music-mathematics activities as being "new, different, or novel." Second, preservice teachers reported little knowledge and experience integrating mathematics and music prior to the course. As one returns to the theoretical underpinnings described in the first part of this chapter, various scholars help to make the case for the importance of creativity in mathematics (Lev-Zamir & Leikin, 2013; Leikin, Subotnik, Pitta-Pantazi, Singer, & Pelczer, 2013; Henrickson & Mishra, 2015; Vale & Barbosa, 2015). Taken together, the theoretical framing and work reported here point to the need for introducing or [re]introducing creativity in mathematics and problem solving contexts, in an effort to expand applications of concepts, and engage learners at higher levels. If we are to be successful, creativity in mathematics must be introduced at both the preservice, as well as inservice stages. It is encouraging that the open-ended responses also reveal that preservice teachers are able to transfer their perceptions of this form of experiential learning, to perhaps see the value in designing similar activities for their future students. Simply put, the results suggest that at minimum, preservice teachers that were enrolled in the course are open to exploring these kinds of technology-enabled creative spaces from a mathematics learning and teaching lens. Notably, Wong (2015) in his study of preservice primary mathematics teachers found an overall positive attitude towards the use of technology, and perceived usefulness was more influential than perceived ease of use.

Additionally, preservice teachers' perceptions of Second Life® efficacy (technology efficacy) were moderately high. Our results also showed that the perceptual variables technology comfort/use and efficacy were related, and efficacy and presence were related. Holden and Rada (2011) found that technology self-efficacy was more beneficial to overall technology acceptance than computer self-efficacy. Moreover, the KATE music-math activities addressed the direct connections between music composition and fractions and patterns (Vaughn, 2000), an area of mathematics understanding that is traditionally difficult not only for middle grade students, but preservice teachers. The activities offered a creative and comfortable space for the participants to explore fractions and patterns in a new environment. One of the purposes of the music-math activities was to offer ideas to preservice teachers for approaches that might be meaningful for diverse middle grade students. The activities appeared to be successful in addressing this purpose. They provided experiences in some of Fiske (1999) criteria: perhaps reaching some students who are not reached by traditional instruction, connecting them with fellow students through transforming the learning environment. While these are all preliminary and observational conclusions, they offer options for further exploration.

In terms of future work, next steps for the project team might include exploring the utilization of *Music-Math Park* and similar spaces for virtual camps for secondary students or graduate students enrolled in STEM instructional design courses. These venues can provide both groups opportunities to design and explore various creative outlets or instructional development projects. Finally, one of the overarching objectives of the broader KATE project work was to provide preservice middle school mathematics teachers early experiences in teaching algebra for equity. The research and design teams went to great lengths to design engaging and immersive virtual learning spaces that could be used to structure mathematics teaching and learning experiences for preservice teachers. These preliminary results showed that presence was related to both SL efficacy and effectiveness. This suggests that the design team was reasonably successful in designing a virtual space that PSTs felt present in, and that presence was related to perceptions of the overall effectiveness of the music-mathematics exercises.

Exploring new ways to prepare teachers to engage all students in rich and effective learning experiences is critical. Investigations of the relationships of virtual learning and teaching experiences, the effects of creativity, engagement, learner characteristics and the affordances of using innovative technological resources like 3D virtual environments, all offer promise for redesigning select mathematics teacher preparation courses.

7.10 Conclusion and Future Research

This work has important implications for both preservice teacher pedagogy and math education in general. While technology and music are not new to mathematics curricula, purposefully fostering creativity as a product instead of a process is. If more educators engage in creating collisions or *cogitares* within mathematics classrooms, it could lead to patterns of student engagement and inquiry.

Future research can build on this and similar work by seeking to understand how preservice teachers experience creativity in classrooms and how they translate this experience into their own teaching and classroom planning. To change the pedagogy of future educators is to change the face of education. This project wedded technology and music to initiate new educational experiences. New research must proffer other creative juxtapositions and consider how they might shape mathematics education.

References

An, S. A., Ma, T., & Capraro, M. M. (2011). Preservice teachers' beliefs and attitude about teaching and learning mathematics through music: An intervention study. *School Science and Mathematics, 111*(5), 236–248.

An, S. (2012). *Music-Mathematics instructional activities* (unpublished).

Barr, D., Harrison, J., & Conery, L. (2011). Computational thinking: A digital age skill for everyone. *Learning & Leading with Technology, 38*(6), 20–52.

Baer, J. (2016). Creativity doesn't develop in a vacuum. *New Directions for Child and Adolescent Development, 2016*(151), 9–20.

Beal, S. (2000). *Functional melodies*. New York, NY: Curriculum Press.

Berliner, D. C. (1985). Laboratory settings and the study of teacher education. *Journal of Teacher Education, 36*(2), 2–8.

Boaler, J. (2015). *Mathematical mindsets: Unleashing students' potential through creative math, inspiring messages and innovative teaching*. New York, NY: Wiley.

Colwell, C. M. (2008). Integration of music and core academic objectives in the K-12 curriculum perceptions of music and classroom teachers. *Update: Applications of Research in Music Education, 26*(2), 33–41.

Craft, A. (2010). *Creativity and education futures: Learning in a digital age*. London, UK: Trentham Books.

Cropley, A. J. (2001). *Creativity in education & learning: A guide for teachers and educators*. Abingdon, VA: Psychology Press.

Darling-Hammond, L. (2000). How teacher education matters. *Journal of Teacher Education, 51*(3), 166–173.

Davis, F. D. (1989). Perceived usefulness, perceived ease of use, and user acceptance of information technology. *MIS Quarterly, 13*(3), 319–340.

Davis, T. J., An, S., Cole, M. & Jett, S. (2012). *Glasscock [Second Life Build]*. Retrieved from http://kate.tamu.edu/second-life.

Davis, T. (2013). Using three-dimensional virtual environments to prepare STEM teachers. In M. M. Capraro & R. M. Capraro (Eds.), *Improving urban schools: Equity and access in k-16 STEM education for all students* (pp. 125–141). Charlotte, NC: Information Age.

Davis, T., Chien, C., Brown, I., & Kulm, G. (2012b). Knowledge for algebra teaching for equity (KATE) project: An examination of virtual classroom simulation approaches. *National Forum of Multicultural Issues Journal, 9*(2), 67–87.

De Lucia, A., Francese, R., Passero, I., & Tortora, G. (2009). Development and evaluation of a virtual campus on Second Life: The case of second DMI. *Computers & Education, 52,* 220–233.

Dickey, M. D. (2005). Three-dimensional virtual worlds and distance learning: Two case studies of active worlds as a medium for distance education. *British Journal of Educational Technology, 36*(3), 439–451.

Enochs, L. G., & Riggs, I. M. (1990). Further development of an elementary science teaching efficacy instrument: A preservice elementary scale. *School Science and Mathematics, 90,* 695–706.

Enochs, L. G., Smith, P. L., & Huinker, D. (2000). Establishing factorial validity of the mathematics teaching efficacy beliefs instrument. *School Science and Mathematics, 100*(4), 194–203.

Fernandez, M. (1999). Making music with mathematics. *Mathematics Teacher, 92*(2), 90.

Fiske, E. B. (1999). *Champions of change: The impact of the arts on learning*. Washington D.C.: President's Committee on the Arts and the Humanities.

Furner, J. M., & Marinas, C. A. (2014). *Addressing math anxiety in teaching mathematics using photography and geogebra*. Presented at International Conference on Technology in Collegiate Mathematics.

Gibson, S., & Dembo, M. H. (1984). Teacher efficacy: A construct validation. *Journal of Educational Psychology, 76,* 503–511.

Gregerson, M. B., Snyder, H. T., & Kaufman, J. C. (Eds.). (2012). *Teaching creatively and teaching creativity*. Dordrecht, The Netherlands: Springer Science+Business Media.

Guskey, T. R. (1988). Teacher efficacy, self-concept, and attitudes toward the implementation of instructional innovation. *Teaching and Teacher Education, 4,* 63–69.

Henriksen, D., & Mishra, P. (2015). We teach who we are: Creativity in the lives and practices of accomplished teachers. *Teachers College Record, 117,* 070303.

Ho, C. M. L., Rappa, N. A., & Chee, Y. S. (2009). Designing and implementing virtual enactive role-play and structured argumentation: Promises and pitfalls. *Computer Assisted Language Learning, 22*(5), 381–408.

Holden, H., & Rada, R. (2011). Understanding the influence of perceived usability and technology self-efficacy on teachers' technology acceptance. *Journal of Research on Technology in Education, 43*(4), 343–367.

Holt, L., & Brockett, R. (2012). Self direction and factors influencing technology use: Examining the relationships for the 21st century workplace. *Computers in Human Behavior, 28*(6), 2075–2082.

International Society for Technology in Education. (2007). *National educational technology standards for students.* Retrieved from http://www.iste.org/docs/pdfs/20-14_ISTE_Standards-S_PDF.pdf.

International Society for Technology in Education. (2008). *ISTE standards for teachers.* Retrieved from http://www.iste.org/standards/standards/for-students-2016.

International Society for Technology in Education. (2016). *ISTE standards for students.* Retrieved from http://www.iste.org/standards/standards/for-students-2016.

Jenkins, H., Clinton, K., Purushotma, R., Robinson, A. J., & Weigel, M. (2006). *Confronting the challenges of participatory culture: Media education for the 21st century*. Chicago, IL: The MacArthur Foundation, with permission from Henry Jenkins.

Johnson, G. L., & Edelson, R. J. (2003). Integrating music and mathematics in the elementary classroom. *Teaching Children Mathematics, 9*(8), 474.

Kereluik, K., Mishra, P., Fahnoe, C., & Terry, L. (2013). What knowledge is of most worth: Teacher knowledge for 21st century learning. *Journal of Digital Learning in Teacher Education, 29*(4), 127–140.
Kleiman, P. (2008). Towards transformation: Conceptions of creativity in higher education. *Innovations in Education and Teaching International, 45*(3), 209–217.
Koestler, A. (1981). The three domains of creativity. In D. Dutton & M. Krausz (Eds.), *The concept of creativity in science and art* (pp. 1–17). Dordrecht, The Netherlands: Springer.
Leikin, R., & Pitta-Pantazi, D. (2013). Creativity and mathematics education: The state of the art. *ZDM Mathematics Education, 45*(2), 159–166.
Leikin, R., Subotnik, R., Pitta-Pantazi, D., Singer, F. M., & Pelczer, I. (2013). Teachers' views on creativity in mathematics education: An international survey. *ZDM Mathematics Education, 45*(2), 309–324.
Lev-Zamir, H., & Leikin, R. (2013). Saying versus doing: Teachers' conceptions of creativity in elementary mathematics teaching. *ZDM Mathematics Education, 45*(2), 295–308.
Mehta, R., Mishra, P., Henriksen, D., & The Deep-Play Research Group. (2016). Creativity in mathematics and beyond—Learning from fields medal winners. *Tech Trends, 60*(1), 14–18. http://doi.org/10.1007/s11528-015-0011-6.
Merchant, Z., Goetz, E. T., Keeney-Kennicutt, W., Kwok, O. M., Cifuentes, L., & Davis, T. J. (2012). The learner characteristics, features of desktop 3D virtual reality environments, and college chemistry instruction: A structural equation modeling analysis. *Computers & Education, 59*(2), 551–568.
Metcalf, K. K., Hammer, M. A., & Kahlich, P. A. (1996). Alternatives to field-based experiences: The comparative effects of on-campus laboratories. *Teaching and Teacher Education, 12*(3), 271–283.
Mishra, P., Henriksen, D., & The Deep-Play Research Group. (2014). Revisited and remixed: Creative variations and twisting knobs. *Tech Trends, 58*(1), 20–23.
Mishra, P., Fahnoe, C., Henriksen, D. & The Deep-Play Research Group. (2013). Creativity, self-directed learning, and the architecture of technology rich environments. *Tech Trends, 57*(1), 10–13.
Mishra, P., & The Deep-Play Research Group. (2012). Rethinking technology & creativity in the 21st century: Crayons are the future. *Tech Trends, 56*(5), 13–16.
Mikropoulos, T. A., & Natsis, A. (2011). Educational virtual environments: a ten-year review of empirical research (1999–2009). *Computers & Education, 56*(3), 769–778.
Mitchell, W., Inouye, A., & Blumenthal, M. (2003). *Beyond productivity: Information, technology, innovation, and creativity*. Washington, D.C.: National Academic Press.
Munakata, M., & Vaidya, A. (2012). Encouraging creativity in mathematics and science through photography. *Teaching Mathematics and its Applications, 31*(3), 121–132.
National Advisory Council on Creative and Cultural Education. (1999). *All our futures: Creativity, culture and education*. United Kingdom: Department of Education and Employment.
National Council of Teachers of Mathematics. (2000). *Principles and standards of school mathematics*. Reston, VA: Author.
Nicholson, M. W., & Moran, J. D. (1986). Teacher's judgements of preschoolers' creativity. *Perceptual and Motor Skills, 63,* 1211–1216.
Nisbet, S. (1991). Mathematics and music. *Australian Mathematics Teacher, 47*(4), 4–8.
Runco, M. A. (2008). Creativity and education. *New Horizons in Education, 56*(1), n1.
Saldaña, J. (2015). *The coding manual for qualitative researchers* (3rd ed.). Thousand Oaks, CA: Sage.
Springer, S., Alexander, B., & Persiani, K. (2006). *The creative teacher: An encyclopedia of ideas to energize your curriculum*. New York, NY: McGraw Hill Professional.
STEM to STEAM. (2016). *STEM to STEAM*. Retrieved from http://www.stemtosteam.org.
Tammadge, A. (1979). Creativity: Presidential address to the mathematical association at the annual conference, April 1979. *The Mathematical Gazette, 63*(425), 145–163.
Vale, I., & Barbosa, A. (2015). Mathematics creativity in elementary teacher training. *Journal of the European Teacher Education Network, 10,* 101–109.

Vaughn, K. (2000). Music and mathematics: Modest support for the oft-claimed relationship. *Journal of Aesthetic Education, 34*(3/4), 149–166.

Witmer, B. G., & Singer, M. J. (1998). Measuring presence in virtual environments: A presence questionnaire. *Presence, 7*(3), 225–240. https://doi.org/10.1162/105474698565686.

Wong, G. K. (2015). Understanding technology acceptance in pre-service teachers of primary mathematics in Hong Kong. *Australasian Journal of Educational Technology, 31*(6).

Woods, P., & Jeffrey, B. (1996). *Teachable moments: The art of creative teaching in primary schools*. Chicago, IL: Open University Press.

Wright, D. (2009). *Mathematics and music* (Vol. 28). Providence, RI, USA: American Mathematical Society.

Wrzesien, M., & Raya, M. A. (2010). Learning in serious virtual worlds: Evaluation of learning effectiveness and appeal to students in the E-Junior project. *Computers & Education, 55*(1), 178–187.

Chapter 8
Preparing Teachers to Use Excelets: Developing Creative Modeling Experiences for Secondary Mathematics Students

Ginger S. Watson and Mary C. Enderson

Abstract There are many challenges in preparing mathematics teachers for today's classrooms including content, pedagogy, technology, and creativity. This qualitative study was designed to examine how pre-service mathematics teachers solve modeling tasks using Excelets, an interactive form of an Excel spreadsheet that allows for the manipulation of data and the visualization of changes in numeric, graphic, and symbolic form (Sinex in Developer's guide to Excelets: dynamic and interactive visualization with "Javaless" applets or interactive Excel spreadsheets, 2005). Specific emphasis was given to how such experiences translate into providing creative learning environments in future teaching. The study focused on four specific participants and analyzed their Technological Pedagogical Content Knowledge (TPACK) scores (AMTE, 2009), "think-alouds", and written work to assess their understanding of modeling tasks that integrated technology as a tool for learning. Top-tier participants demonstrated abilities to recognize, accept, adapt, and explore mathematics creatively when using and integrating mathematical modeling tools while bottom-tier participants failed to exhibit these skills. Top-tier participants demonstrated high levels of creativity and TPACK yet rated themselves low in these skills while bottom-tier participants provided little creativity and TPACK yet rated themselves extremely high. These results indicate that there is more work to be done in preparing teachers to provide students with stimulating mathematics problems and explorations while scaffolding their integration of technological tools such as Excel.

Keywords Secondary preservice teachers · Modeling · Creativity
Excelets · TPACK

G. S. Watson (✉)
University of Virginia, Charlottesville, VA, USA
e-mail: ginger.watson@virginia.edu

M. C. Enderson
Old Dominion University, Norfolk, VA, USA
e-mail: menderso@odu.edu

© Springer International Publishing AG, part of Springer Nature 2018

V. Freiman and J. L. Tassell (eds.), *Creativity and Technology in Mathematics Education*, Mathematics Education in the Digital Era 10,
https://doi.org/10.1007/978-3-319-72381-5_8

8.1 Introduction

With adoption of the Common Core State Standards (National Governors Association Center for Best Practices, Council of Chief State School Officers, 2010) across the United States (U.S.), teachers are expected to engage students in "doing" mathematics, which involves exploring, making sense of, modeling, and using appropriate tools to solve problems. Such practices are to be embraced and promoted for all students, but how teachers develop such practices as learners and then progress towards implementation as teachers is often unclear. The Conference Board of Mathematical Sciences' (CBMS) *Mathematical Education of Teachers* (2001) professed that for many pre-service teachers, learning mathematics was equivalent to learning procedures and algorithms to solve problems. Many future teachers are quite successful in this style of learning although it provides a somewhat narrow view of what learning mathematics involves. With advancements in standards for teacher preparation and the Common Core movement taking hold in more U.S. classrooms, teachers are now confronted with novel practices and unfamiliar terrain in which to teach mathematics for all.

The more current publication, *Mathematical Education of Teachers II* (CBMS, 2012) addresses several critical experiences future mathematics teachers should encounter in their coursework in order to consider integrating mathematical processes and practices into teaching. Providing future mathematics teachers' opportunities to struggle with hard problems, discover their own solutions, reason and model mathematically, and develop mathematical habits of mind are essential in their preparation. It has been well established that early experiences, acquired through years as a student, influence the way new teachers think about learning and ultimately how they end up teaching (Borko & Putnam, 1996; Lortie, 1975). Thus, it is critical for such educational experiences to include experiential and inquiry-based learning rather than lecture and procedural learning. We believe that modeling scenarios can provide a rich environment for pre-service teachers to explore mathematics, be creative in their thinking and approach to problems, and witness that technological tools can benefit learning mathematics concepts in ways that are different from traditional paper/pencil learning. We also take the position that pre-service teachers must experience such learning activities themselves before they can transfer the creative ideas into future instruction. Such activities should "promote the development of both creative abilities of future teachers, and their skills to develop creative abilities of the pupils during school teaching in their future career" (Safuanov, 2008, p. 451). Encounters that support and promote creative development of ideas that allow one to explore and make sense of mathematics should be a component of every teacher preparation program.

This chapter briefly addresses mathematics teacher preparation as it relates to content, pedagogy, technology, and creativity. Further details are provided regarding the challenges associated with teachers' implementation of technology in mathematics classroom instruction and the role of creativity. In this particular study, we used an interactive spreadsheet tool programmed in Microsoft Excel called an

Excelet to study ways pre-service secondary mathematics teachers solve modeling problems. This digital modeling tool was selected because of the wide availability and use of Excel in the workplace and in classrooms across the U.S. While there has been research in the use of spreadsheets in mathematics, there is limited research using the specific tool of this study, Excelets. We situate this study within the context of spreadsheets, noting areas where they generalize to Excelets and ways that Excelets may offer some advantage. It has also been proposed that new or novel approaches to using technology in modeling real-world phenomena can help in the development of mathematical creativity (Arganbright, 2005). Thus, this study also examined how pre-service teachers transposed modeling experiences into future instruction for high school mathematics students, with an emphasis on connections to creativity.

8.2 Background

8.2.1 Mathematics Teacher Preparation

Preparing secondary (grades 6–12) mathematics teachers for today's classrooms involves developing well-rounded professionals with a strong knowledge base concentrated on content, pedagogy, and pedagogical content knowledge (PCK). The introduction of PCK launched teacher preparation programs to search for ways to find the "right" balance in order to teach content from a pedagogical content perspective (Shulman, 1986). As professed in *Mathematics Teaching Today* (Martin, 2007), teachers must be proficient in facilitating and guiding student learning, which includes:

- Designing and implementing mathematical experiences that stimulate student interests and intellect;
- Orchestrating classroom discourse in ways that promote the exploration and growth of mathematical ideas;
- Using, and helping students use, technology and other tools to pursue mathematical investigations; and
- Engaging in opportunities to deepen their own understanding of the mathematics being studied and its applications (pp. 5–6).

In addition, *Mathematics Teaching Today* (Martin, 2007) denotes the importance of teachers making improvements in the ways mathematics is taught and learned in schools. For many teachers, this equates to changing the ways they learned mathematics to include more student-centered learning rather than teacher demonstration of techniques. Such changes include active learners who are engaged in problem solving, reasoning, and making connections while having access to technology to study and solve real-life problems (Martin, 2007). These learning experiences promote multiple ways of solving problems, which in turn promote and nurture student creativity (Applebaum & Saul, 2009). While such practices appear reasonable, how pre-service teachers successfully develop these skills is often unclear or foreign and in many instances unnatural to the experiences they bring with them to university coursework. Exposing pre-service mathematics teachers to a different or creative view of mathematics and the tools one can use to investigate real problems is often the challenge to be met. Mathematics and pedagogy coursework should present cases where pre-service teachers can develop knowledge as learners of mathematics and then progress into thinking about creative ways to apply this knowledge as teachers of mathematics. Technology is one avenue to help in this process.

It has been well established that teaching mathematics requires more than knowledge and understanding of the content (Ball, Hill, & Bass, 2005; Hill, Ball, & Schilling, 2008; Hill, Rowan, & Ball, 2005). Development of mathematics knowledge should be evident and integrated seamlessly throughout one's undergraduate program in mathematics. The CBMS (2012) identifies two specific areas: (a) experience with reasoning and proof and (b) experience with technology as critical for future secondary teachers. The reasoning and proof assists pre-service teachers in making sense of the mathematics, which in turn, better prepares them to teach. Of more interest to this particular study, is the experience with technology. CBMS (2012) states:

> Teachers should become familiar with various software programs and technology platforms, learning how to use them to analyze data, to reduce computational overhead, to build computational models of mathematics objects, and to perform mathematical experiments. The experiences should include dynamic geometry environments, computer algebra systems, and statistical software used both to apply what students know and as tools to help them understand new mathematical ideas—in college and in high school. Not only can the proper use of technology make complex ideas tractable, it can also help one understand subtle mathematics concepts. At the same time, technology used in a superficial way, without connection to mathematical reasoning, can take up precious course time without advancing learning (p. 57).

Such technology experiences are not typical of what pre-service teachers have experienced in their own mathematics coursework at secondary and post-secondary levels. This presents challenges for mathematics teachers as they make attempts to integrate technology into their classroom instruction to solve real problems centered on modeling in a creative manner. We believe that lack of experience in technology integration for mathematical modeling scenarios or tasks hinders growth and development of using such tools in a valuable manner to promote creativity for learners. Arganbright (2005) also proposes that the use of spreadsheets to model real-life problems can aid in the development of mathematical creativity for those who interact with such tools.

8.2.2 Integrating Technology in the Teaching and Learning of Mathematics

Technology changes the way one investigates and makes sense of mathematics. Technology provides the learner with tools to study real problems with real numbers that are often "messy" and require a level of precision. Although this is a valuable position for teachers, preparing them to keep up with the technology revolution in planning and implementing instruction poses challenges. Future mathematics teachers should be equipped to implement technological tools in learning experiences for students and universities have a responsibility in preparing them (Association of Mathematics Teacher Educators (AMTE), 2006). Pre-service teachers should have opportunities to:

- Explore and learn mathematics using technology in ways that build confidence and understanding of the technology and mathematics.
- Model appropriate uses of a variety of established and new applications of technology as tools to develop a deep understanding of mathematics in varied contexts.
- Make informed decisions about appropriate and effective uses of technology in the teaching and learning of mathematics.
- Develop and practice teaching lessons that take advantage of the ability of technology to enrich and enhance the learning of mathematics (AMTE, 2009, p. 2).

In addition to AMTE's technology position, the mathematics education community is provided with a framework focused on TPACK (Technological Pedagogical Content Knowledge—AMTE, 2009) developed from the work of Mishra and Koehler (2006) and the ISTE (International Society for Technology in Education) standards (2008). This framework presents guidelines focused on four main areas: (1) designing and developing technology-enhanced learning experiences, (2) facilitating technology-integrated instruction, (3) evaluating technology-intensive environments, and (4) continuing to develop professional capacity in mathematics TPACK (AMTE, 2006).

Niess (2005), in her study of pre-service science and mathematics teachers found that there were numerous challenges in pre-service teachers' integration of technology into content instruction. One of the challenges brought forward was the fact that many pre-service teachers learn about technology outside the development of their content and pedagogical content knowledge bases. It often is addressed in a more generic form that is disconnected to the discipline and taught by a generalist. This does not bode well for exposing future teachers to technological tools that can be used in the study of mathematics as well as supporting a creative, more interpretive learning environment.

Niess, Sadri, and Lee (2007) and Niess et al. (2009) extensive work with teachers allowed them to develop a five-stage process they use when learning to integrate technology in the teaching and learning of mathematics. The focus of the technology was on use of interactive spreadsheets and included:

1. *Recognizing* (knowledge), where teachers are able to use the technology and recognize the alignment of the technology with mathematics content yet do not integrate the technology in teaching and learning of mathematics.
2. *Accepting* (persuasion), where teachers form a favorable or unfavorable attitude toward teaching and learning mathematics with an appropriate technology.
3. *Adapting* (decision), where teachers engage in activities that lead to a choice to adopt or reject teaching and learning mathematics with an appropriate technology.
4. *Exploring* (implementation), where teachers actively integrate teaching and learning of mathematics with an appropriate technology.
5. *Advancing* (confirmation), where teachers evaluate the results of the decision to integrate teaching and learning mathematics with an appropriate technology (Niess et al., 2009).

The five levels rely on integration of technology, pedagogy, and content (TPACK) and are not linear in progression but a more iterative process.

The study presented in this chapter was designed to observe where pre-service secondary mathematics teachers fell along this spectrum and how their own understanding of content with technology translated into future instruction. In addition to using Niess et al. (2009) five-stage process as a framework for the technology component of this study, identifying how creativity fit into the various levels—if at all—was of interest to us.

8.2.3 Developing Creative Learners and Teachers

Education is a complex process with various perspectives and techniques focused on crucial aspects of the teaching and learning processes. When one includes the element of creativity to these processes, it complicates the field even more. The early notion of creativity as a static trait one inherits has been replaced with the view that creativity is dynamic and should be developed (Silver, 1997; Milgram & Hong, 2009; Subotnik, Pillmeier, & Jarvin, 2009). The problem often lies in the fact that teachers are ill-trained to help students develop the practices that support creative thinking and learning. Evidence illustrates that many teachers have a limited understanding of creativity and what it means to think in a creative manner (Shriki, 2009; Bolden, Harries, & Newton, 2010). Thus, the need to develop student creativity requires teachers to have a better understanding of this field.

One way teachers can gain a better understanding of creativity and help students develop as creative thinkers is to become familiar with the types of problems and processes that are involved in this developmental progression. Davis and Rimm (2004) promote the concept that teachers should become familiar with processes and modes of creative behavior in order to develop learning tasks that make use of these skills. Such familiarity should be developed in teacher education programs in both content and pedagogy coursework. Pre-service teachers should be placed in learning situations that require them to think deeply about content, make sense of it, and communicate findings among peers. Without such experiences, teachers will lack the insight into how to provide stimulating and creative learning situations for their own students.

Defining creativity and how a teacher preparation program embraces it should be clearly articulated. There have been numerous interpretations of creativity in the field of education, which complicates the research and professional discourse in this field (Mann, 2006). For the purposes of this study, we embraced the perspective of "relative" creativity (Leikin, 2009), which is built upon one's own educational history and often compared to other students. Leikin (2009) proposed a focus on students' aptitude to produce solutions or details to new tasks or to present original work that supports previously learned concepts. This research examined pre-service teachers as creative learners and pre-service teachers as teachers promoting creative pedagogy both in a relative manner.

In preparing mathematics teachers for today's classrooms, it is essential to focus on both content and pedagogy as presented in Sect. 8.3.1. In order for teachers to embrace techniques of teaching that promote creative thinking, they must design unique methods to allow for various interpretations from students. Lin's work (2011) on creative thinking offers researchers a conceptual framework of creative pedagogy. He presents three main areas that have been recognized by previous researchers to foster creativity in the classroom: teaching, environment, and teacher ethos (Lin, 2011). Teaching is the act of providing students with creative and innovative practices, which in turn, promotes opportunities to explore and learn. The environment is critical in that it should stimulate and support learners' engagement in the activity—develop interest, excitement, and motivation to learn. Lastly, teacher ethos refers to teachers having an open position toward students who think and operate differently on given tasks. The teacher should support independent thinking and place value in different student approaches towards solving problems (Lin, 2011). We believe such creative thinking and learning can be nurtured and developed by exploring mathematical models that include the use of technological tools, from graphing utilities to various software packages, in the classroom. Such models have the potential to promote practices designed around problem solving and problem posing, which support student development of creative approaches to studying mathematics (Silver, 1997). In addition, incorporation of technology provides students with access to problems and applications to investigate, which allows them to engage in a level of creative discovery (Pead & Ralph, 2007).

Technology has the potential to provide learners with an environment that promotes creative mathematical thinking by exploring inquiry-based type problems. In order to develop creative thinking, access to technology is often the obstacle when in reality, there are many inexpensive and readily available tools that can be used in classrooms (Yerushalmy, 2009). Although Yerushalmy promotes use of inexpensive handheld technology in planning and developing unique scenarios for teaching and learning, we focus on computers that are pre-loaded with software packages that include spreadsheets. Pead and Ralph (2007) report that the "entry fee"—time and effort invested in learning the technological tools—can become obstacles to using the technology in valuable ways. Thus, generic applications like spreadsheets have a low entry fee and are often used by teachers for graphing and analyzing data (Pead & Ralph, 2007). It has also been reported that increases in the use of common spreadsheets (from 31.9% to 62.6%) over a period of time have real value in studying problems centered on statistics and real data (Thomas & Palmer, 2014). We propose that the low investment of spreadsheets, as well as the interactive environment where one can explore mathematics concepts in various ways, is an excellent tool to expose pre-service teachers to a creative element in the learning process. Brown and Gould (1987) state that spreadsheets are known for providing users with the ability to compare results in "what-if" scenarios by sorting out what remains fixed and what varies. This promotes a critical requirement for creativity in that it allows one to test or try out many different cases, which encourages exploration (Fischer & Nakakoji, 1994). This is exactly the type of environment we believed would benefit and support future teachers.

Preparing teachers to experience mathematics in a creative manner for themselves as learners and then transpose that learning as teachers is critical to developing an understanding of creative pedagogy. Lev-Zamir (2008) states, "To be able to encourage creativity in their students, teachers should experience themselves such kind of learning while being trained" (p. 444). This perspective is congruent with the focus of this research study on pre-service mathematics teachers' use of technology in approaching a modeling task as (1) learners and (2) teachers. In our search to understand these two roles, we were interested to identify how such environments nurtured the development of creative thinking and learning.

8.2.4 Technology Meets Creativity Through Mathematical Modeling Scenarios

With the standards movements in the U.S. centered on teaching and learning mathematics, we were interested in how mathematics teachers come to know, understand, and implement modeling into classroom instruction. In order for teachers to promote mathematics modeling in classroom instruction, they must have some experience in exploring what is involved in the process. Spreadsheets are tools that allow one to more fully explore problems and approaches as well as test questions that arise throughout the process (Pead & Ralph, 2007). Doerr (2007) raises the issue as to the knowledge that teachers need to possess in order to be effective in integrating applications and modeling into classroom instruction. She promotes the notion that pre-service teachers need to have experiences in modeling that consist of a range of contexts, tools, and analyses of the modeling task (2007). By gaining such experiences, it is anticipated that future teachers will come to recognize differences in exploring problems and become more open to unexpected approaches, connections, and strategies. It is these differences that open the door for creative thinking and learning processes.

The use of spreadsheets to explore and develop a deeper understanding of mathematics concepts has been one way of integrating technology into mathematics (Drier, 2001; Abramovich, 1995). Drier (2001) posits that one unique approach of using spreadsheets is the ability to interactively model and simulate mathematics concepts. She points out that spreadsheets can be used to create dynamic environments for discovering and making sense of mathematical ideas and relationships to build understanding. This description is very much in-line with the philosophy behind Excelets, interactive spreadsheets built in Excel, used in this study.

Excelets are interactive spreadsheets created in Microsoft Excel that allow for simulation, visualization, and exploration of mathematical models (Sinex, 2005). Recent research indicates that an increasing number of schools have access to spreadsheets such as Excel (Thomas & Palmer, 2014), making them a more feasible software tool for mathematical modeling. Unlike simple spreadsheets that require manual data entry or manipulation in individual cells with standard graphing capabilities, Excelets provide an interactive, graphical interface that facilitate a variety of input options and data displays. As proposed by Drier (2001), spreadsheet tools such as Excelets allow a learner to interact with the underlying mathematical model by entering data, manipulating variables, and viewing resulting displays of the model in numeric, graphical, and symbolic form. Excelets specifically do so without the need for extensive knowledge in spreadsheet use. Their interface provides for increased flexibility over traditional spreadsheets yet uses the same general functions reducing the "entry fee" time required for users who are not already familiar with spreadsheets (Pead & Ralph, 2007). Because Excelets provide a variety of input options including information boxes, input blocks, check boxes, and slider bars, they are often simpler to use than spreadsheets making it easier to manipulate the model to promote conceptual understanding. They also provide for a range of static visual overlays and dynamic graphic displays of the phenomena being modeled in order to visualize how changes in one variable affect other variables in the model. While Excelets have many benefits, they also have limitations. Their interface can easily become cluttered and difficult to read. Their display capabilities are limited to simple graphics and colors available in Excel. Finally, they are best used for modeling phenomena whose mathematics are easily calculated in spreadsheets. A more thorough description of the specific Excelets used in this study along with figures is provided in Sect. 8.4.2 Study Tasks.

The three elements of Lin's creativity framework (2011) were used in this study to present modeling scenarios and to interpret the creativity implemented by pre-service teachers when working through the models and developing ideas about implementation of such models in their future classroom instruction. The focus of this work was on how creativity relates to technological, pedagogical, and content knowledge in designing instruction (Niess, 2008; Thompson & Mishra, 2007) and in interpreting the creativity demonstrated by pre-service teachers in the current study.

8.3 Study

This study examined how pre-service mathematics teachers solve modeling tasks using Excelets and how such experiences translate into the ability to provide creative learning environments in their teaching. In this way, the study analyzed how pre-service teachers complete modeling tasks as students and as teachers (Lev-Zamir, 2008). The study sought to address the following research questions:

- How do pre-service mathematics teachers creatively solve problems when using Excelets as a mathematical modeling tool?
- How do pre-service mathematics teachers implement creativity in development of a task associated with a lesson that utilizes Excelets as a mathematical modeling tool?
- What is the self-reported TPACK of pre-service mathematics teachers who participated in this study and how does TPACK change after interaction with Excelet-based modeling tasks in a secondary mathematics teaching methods course?

Surveys, think-aloud protocols, and written work were collected and analyzed to answer these questions within the context of a qualitative study design. The study was limited to a single secondary (grades 6–12) mathematics methods course and bounded to a set of modeling tasks implemented over a two-week period. We independently analyzed and triangulated a variety of data sources both within and across modeling tasks, conferring to verify observations and themes.

8.3.1 Participants

Four participants, members of a larger group of fourteen pre-service mathematics teachers, are the focus of this chapter. All students were enrolled in a secondary mathematics methods course where they were encouraged to think about teaching and learning mathematics in a variety of ways—many of which were different than what they had experienced as learners. Two participants, classified in the top tier of the group along with two in the bottom tier, proved to be of great interest to this study and we believe provide evidence of creativity—or lack of creativity—in making sense of mathematical modeling using technology. Participants were identified independently by us based on their written work demonstrating their understanding of modeling tasks that integrated technology as a tool for learning and independent observations and scoring of their "think-alouds" when using the Excelets during modeling tasks. We present a brief overview of the participants.

Ian, in the graduate master's program, was seeking licensure to teach middle school mathematics. He classified himself as an above average mathematics student and carried a 4.0 grade point average (GPA), which included mathematics coursework. He was quite methodical and organized in his approach toward attacking and/or completing problems and elaborated on processes involved in such approaches. Ian had completed the instructional technology course for all education majors, which is designed to expose teachers to technological tools that support them in their classroom. He had already passed the Praxis content test required for mathematics licensure so was in good standing for the forthcoming student teaching experience.

Valerie, an undergraduate mathematics education major, began her education at the community college prior to transferring to this 4-year university. She was a strong mathematics student and had completed up through Calculus 2 at the community college level. Although her GPA was a 3.35, she exemplified strength in understanding mathematics from a conceptual orientation. One of the points that she shared with us was over-committing to too many things and trying to balance it all—work, academics/studying, and personal. She was often overly critical of her responses and work to support problems or tasks, which was very detailed and quite insightful. Valerie was active in class discussions and often looked at concepts differently while remaining open to sharing these insights. She had also completed the general technology course and had passed the Praxis content test required for student teaching.

Anna, also a graduate student pursuing licensure to teach mathematics at the middle grades, had a 3.74 GPA and was confident in her mathematics abilities but not always articulate in her communication of mathematics concepts. When it came to presenting mathematics concepts, she often was vague and did not provide many details to really get at making meanings in mathematics. She was procedural in her presentation and explanation of mathematics concepts, which did not allow for much creative thought or elaboration. Anna had taken the instructional technology course for education majors as well as passing the Praxis content test.

Natalie, the final participant, was an undergraduate mathematics education major who had a 3.21 GPA. As she progressed in her program of study, her GPA dropped due to average and below average grades in many mathematics courses. She did not pass the required Praxis content test and will have to re-take the test before she can student teach. She was not an active participant in the course and often did not engage in valuable discourse related to mathematics concepts covered in the course. Her responses were often low-level, not very thought-provoking, and more procedural. Similar to the other participants, she had completed the technology course for education.

8.3.2 Study Tasks

All tasks for this study occurred within the context of a secondary (grades 6–12) mathematics methods course during a single semester. Participants completed all explorations and surveys as part of the class but were given the option to opt out of having their data analyzed for study purposes. Study tasks occurred midway through the course.

At the beginning of the study, participants completed a demographic survey reporting their year in school, academic major, prior coursework, and grades. At that same time, participants completed a 22-item survey assessing their self-reported secondary mathematics pre-service teachers' TPACK (Zelkowski, Gleason, Cox, & Bismarck, 2013). For each item, participants rated their knowledge and skills on a five-point Likert scale ranging from strongly agree (5) to strongly disagree (1). The TPACK instrument provided four subscale scores, one each for technological, pedagogical, content, and integrated TPACK knowledge. Scores were derived by calculating the mean response for all items in each scale resulting in scores that ranged from high (5) to low (1) for each subscale. The first administration of this survey served as a baseline of participants' self-reported TPACK knowledge and skill. This survey was completed again after the final modeling task to measure any changes to TPACK that may have occurred at the end of this experience.

Participants completed two modeling tasks using Excelets. The specific Excelets used in this study were chosen based on their interactive coverage of content, scaffolding of the content from simple to complex, simplicity of their user interface, and effective visualization of concepts. Participants were given access to a handout and online tutorial for Excel and another for Excelets at the beginning of the task, to facilitate review if needed.

In the first modeling task, participants solved problems using the "cookie stack" Excelet (Sinex, 2011a). This Excelet had four worksheets with an overview and objectives on the left-most worksheet (see Fig. 8.1) and progressively more complex, interactive worksheets to the right where participants could explore the linear relationship between the number of cookies and the height of the stack of cookies (see Fig. 8.2), variations in the height of cookies (see Fig. 8.3), and error in the ruler used to measure the cookies (see Fig. 8.4). The primary graphical display was a linear plot of the number of cookies and stack height. Each interactive worksheet contained an interactive graph and questions to guide exploration. Participants were provided with links to the Excelet, a worksheet to accompany the Excelet ("Investigating the height of a stack of cookies" Sinex, 2011b), instructions for recording the exploration, and a self-reflection survey for completion following the task. Participants were asked to video record their interaction with the Excelet while "thinking aloud" expressing their thought process, understanding, challenges, and strategies used during problem solving. The recordings included synchronized voice and captures of the screen.

After finishing the modeling task, participants completed a reflection where they rated their experience on a ten-item survey measuring their understanding, motivation, and perceived effectiveness on a five-point Likert scale ranging from low (1) to high (5). A mean reflection score was calculated for each participant based on their responses to the ten items. The resulting mean score ranged from low (1) indicating little motivation and perceived effectiveness to high (5) indicating extremely high perceptions on of their understanding, motivation, and skills. Participants also responded to five open-ended questions soliciting details of their understanding, comfort, interests, and barriers in completion of the modeling task.

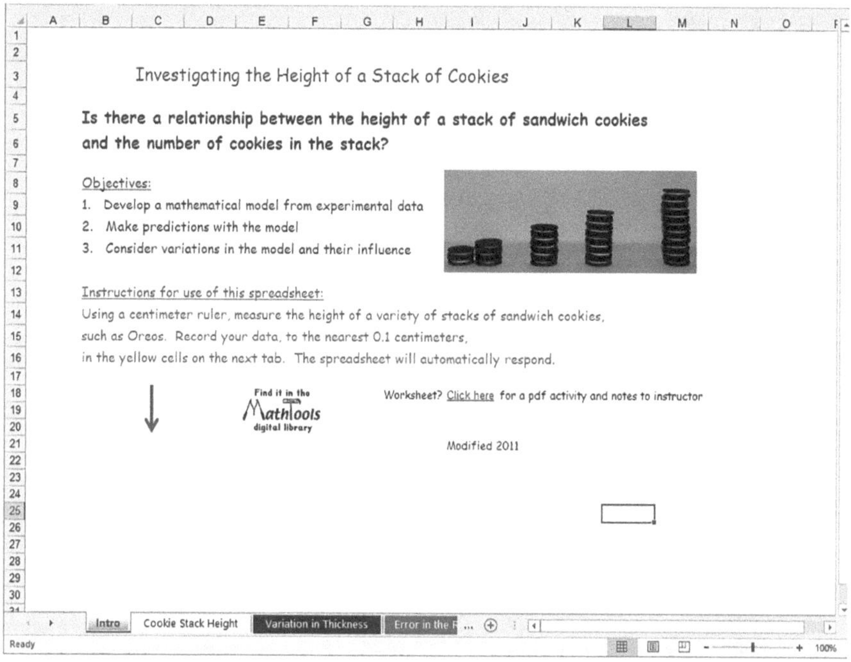

Fig. 8.1 Introduction for the investigating the height of a stack of cookies Excelet (Sinex, 2011b)

A second modeling task focused on developing a lesson that utilized Excelets as a mathematical modeling tool. For this task, participants were asked to explore one of three different Excelets on topics including exploring the nature of cyclic data in nature (Sinex, 2006), kinetics of cancer cell growth (Sinex, 2011c), or probability in mathematical models (Sinex, 2012). Figures 8.5, 8.6 and 8.7 provide an example of a primary modeling task within each of these Excelets. They were similar in design to the Excelet used in the first modeling task of this study. All three were comprised of multiple worksheets with an overview and objectives on the left-most worksheet and progressively more complex, interactive worksheets to the right where participants could enter data, view resulting line graphs, and ponder increasing complex principles such as error, exponential growth, curvilinear relationship, and fitting of data. For this task, participants were asked to locate a Common Core or state standard that their Excelet would address and to create a handout to guide the exploration for students in a secondary mathematics class. Following this task, participants responded to open-ended questions framing their reflections on the modeling task. The handout generated by participants for students and the descriptive reflection were the primary outcome measures for this task.

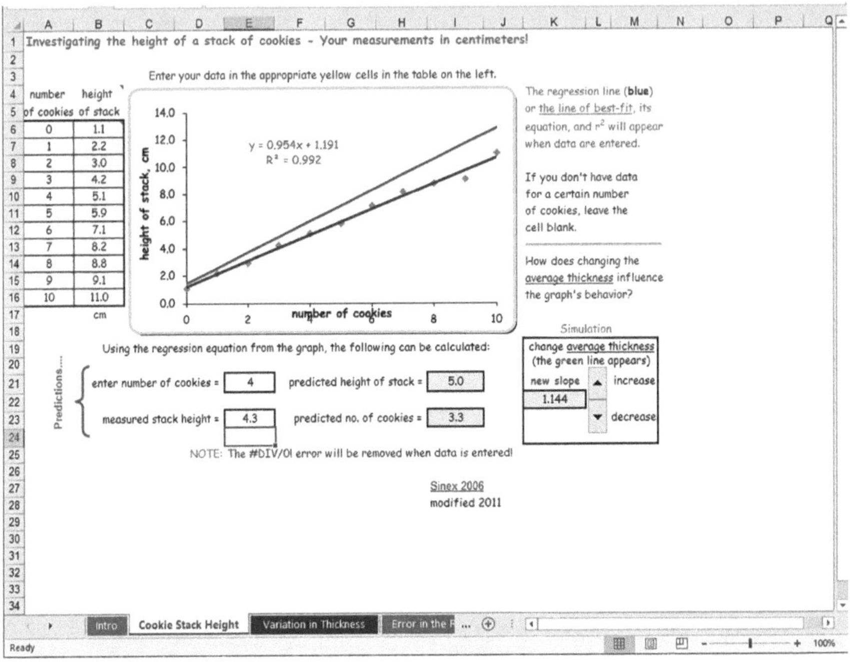

Fig. 8.2 Entering cookie stack data within the investigating the height of a stack of cookies Excelet (Sinex, 2011b)

8.3.3 *Analysis*

A framework combining TPACK (Niess et al., 2009) and creative pedagogy (Lin, 2011) was developed to analyze and interpret levels of TPACK integration and creativity demonstrated by participants in our study. The resulting framework was used to assess the level and extent to which participants displayed each of the following attributes during modeling tasks:

- Recognized and demonstrated knowledgeable and appropriate use of Excelets to promote learning of mathematics.
- Displayed positive, consistent attitudes toward the teaching and learning of mathematics with Excelets.
- Adapted appropriate learning strategies using Excelets as a modeling tool.
- Creatively explored while taking advantage of the many technological affordances of Excelets as an interactive modeling tool.
- Demonstrated a higher-order and creative "ethos" that embraced innovative and sound approaches to mathematical modeling and problem solving.

The think-aloud recordings were independently reviewed, transcribed, and analyzed by both researchers. Works and self-reflections from the second modeling

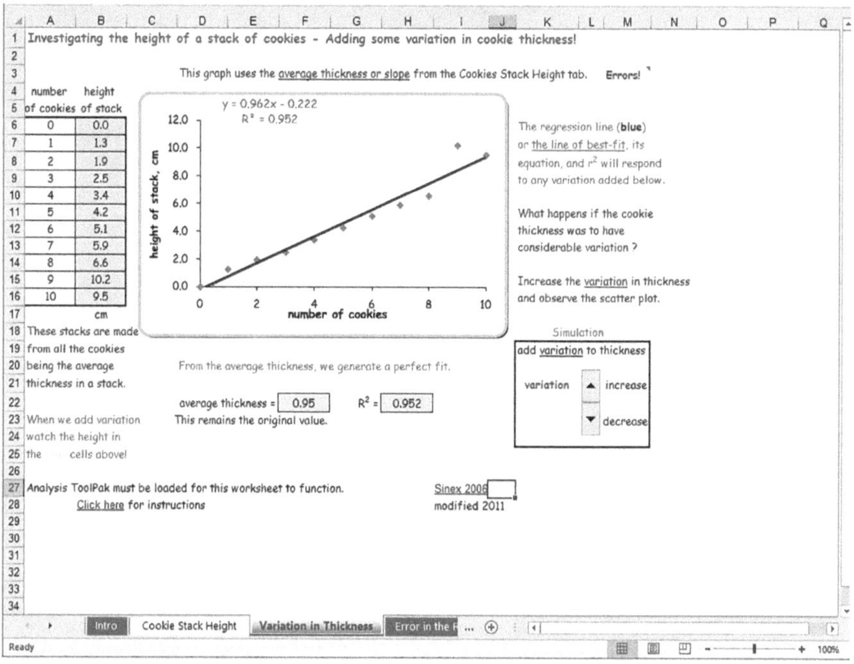

Fig. 8.3 Assessing variations in thickness within the investigating the height of a stack of cookies Excelet (Sinex, 2011b)

tasks were also independently reviewed. The framework above was used to describe and interpret TPACK and creativity with the tasks. Attention was given to their use of the tools, demonstrated understanding and connections of mathematical concepts, and overall creativity on both tasks.

After all descriptive analysis was complete, we individually ranked participants from highest to lowest. Rankings were compared and top-tier and bottom-tier participants were selected for in-depth, descriptive analysis.

TPACK and reflection scores were analyzed for the select participants in order to understand how top-tier and bottom-tier participants perceived their knowledge of effective technology use in the teaching and learning of mathematics. These scores were aligned with each of Lin's (2011) three aspects of pedagogy: teaching, environment, and teacher ethos, and used as a framework for the study. In analysis of data, it was determined that there was interplay between the designated areas where creative teaching was connected with the environment in that a teacher must know what it means to teach in a way that promotes a learning system for students to be excited about their own mathematics learning. We believe that participant "think alouds" along with the development of a handout to guide through a selected Excelet activity, allowed us to see the overlap in Lin's three areas. In addition, pre-service teachers' reflections in using the Excelets to promote mathematical

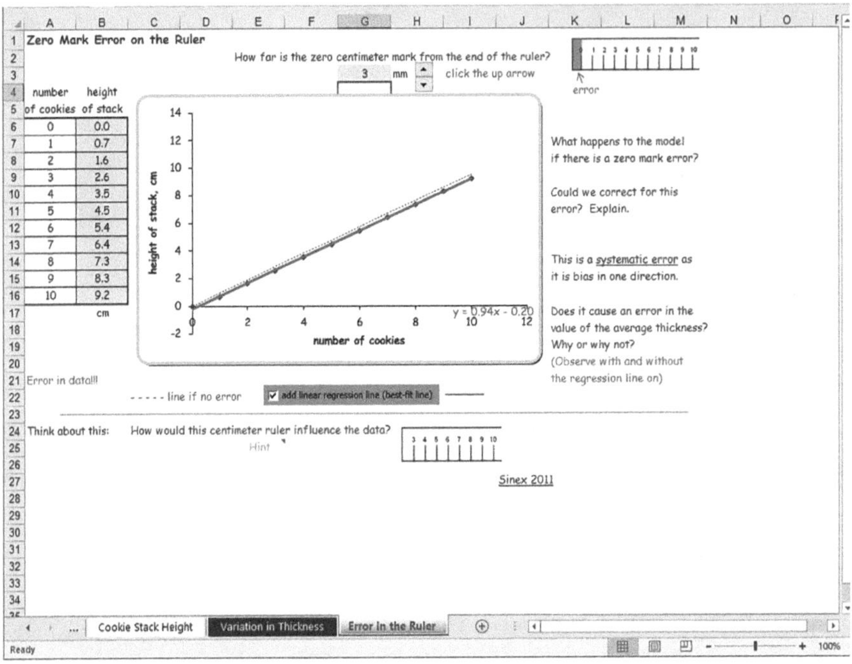

Fig. 8.4 Determining error in the ruler within the investigating the height of a stack of cookies Excelet (Sinex, 2011b)

creativity emerged by focusing on pedagogy related to teaching real-life mathematics concepts.

8.4 Findings

The goals of this study were to assess how pre-service mathematics teachers solved problems using Excelets and how they developed tasks associated with a modeling lesson for students in their future classroom. The study also explored the relationship between the self-reported TPACK of participants, their creativity, and reflections of their effectiveness when completing these tasks. Results are presented through the lens of the two top-tier and two bottom-tier participants identified during analysis of modeling tasks.

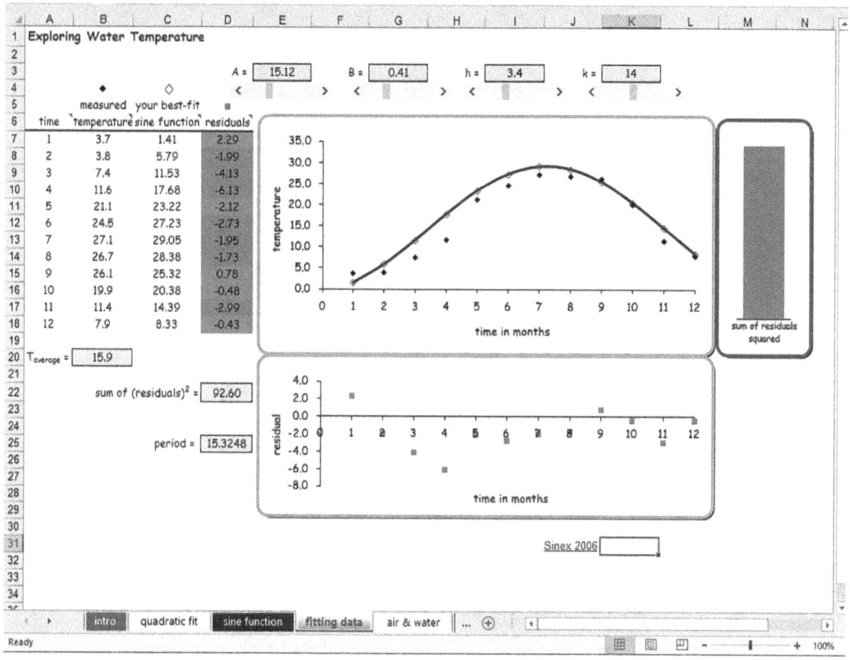

Fig. 8.5 Exploring cyclic data in nature Excelet (Sinex, 2006)

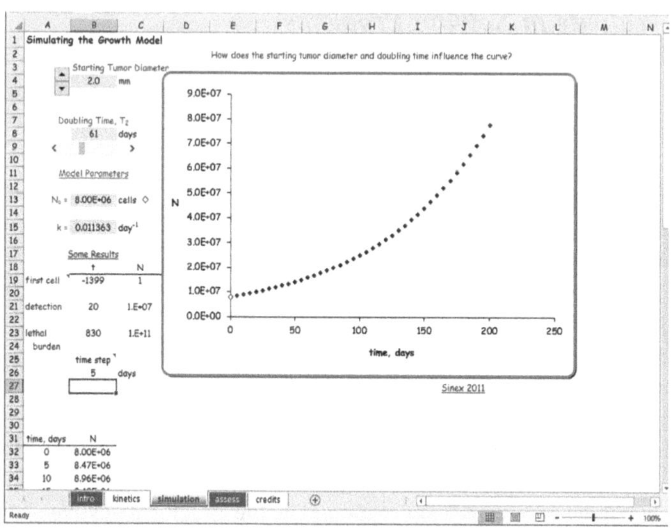

Fig. 8.6 Kinetics of cancer cell growth Excelet (Sinex, 2011c)

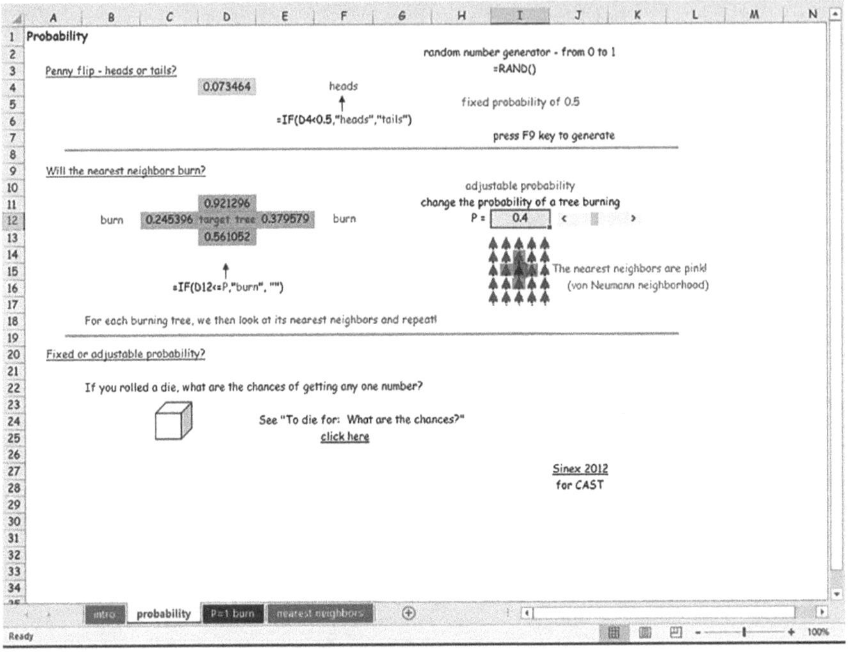

Fig. 8.7 What are the odds? Using probability in mathematical models Excelet (Sinex, 2012)

8.4.1 Creativity in Solving Problems

In order to assess how pre-service mathematics teachers solve problems when using Excelets as a mathematical modeling tool, we analyzed video recordings of participants completing the first modeling task where they acted as students using the Excelet for learning. This task required participants to measure a stack of cookies, enter the data for the height of the stack of cookies, and note how the height of the stack linearly increased with each cookie. A number of probing questions and conceptual variations were available on each Excelet worksheet. A separate and unique paper worksheet was also provided to guide learners through use of the Excelet and interpretation of content.

Ian, a high-tier performer took almost 19 minutes to complete this task. In doing so, he very methodologically read each page of the Excelet, verbally answering the question prompts and generating his own questions during the process. His answers were accurate and detailed. He made important connections between the prompts, general mathematics concepts, and changes in the graphical displays. He rigidly followed the Excelet flow and was a proficient user of the interface, locating and using various input areas, highlighting changes in other variables and associated displays, and easily maneuvering between worksheet pages. He demonstrated appropriate use of the technology, positive attitudes toward the tool and learning strategies, while exploring. His use of the Excelet for higher-order thinking was clear and apparent. Ian was methodical yet creative in his connections.

Valerie, another high-tier performer, completed the task in approximately seven minutes. She demonstrated fluency with both the Excelet interface and the mathematics tasks. She was also very methodical but quick, as she read through the statements and question prompts in the Excelet covering more material than Ian but at a more rapid pace. She slowed on a few occasions to process more complex reasoning and to resolve an input error. Like Ian, she covered all Excelet pages but with a rapid pace. She explored more extreme variations in mathematical inputs and made fluid connections of the concepts from prior pages. Her proficiency with technology was very apparent both in the use of the Excelet and in her recording which consisted of an annotated Microsoft PowerPoint presentation. She was the most fluid and innovative user of both the mathematical content and modeling tool. Her exploration was extensive, innovative, and creative.

Anna was a lower-tier performer who completed the task in 3.5 min. At the beginning, she noticed that there were various pages in the Excelet, articulating that it would be best to go through these pages in order from left to right. She proceeded to access pages in this order but failed to develop a systematic process for exploring the content and answering the questions prompts presented on each page. She spent little time reading and failed to interact with the more advanced concepts, visualization, and variables. She glanced at the separate worksheet and noted that the questions would be "good to go through" but she neither read nor answered them. While she noted that the Excelet could help visualize and make connections between mathematical concepts, she failed to explore the content and affordances of the tool. She did make a few low-level connections but did not demonstrate complete, systematic, or innovative connections during the task. Anna demonstrated limited use of the technology and only surface-level exploration of mathematical concepts presented. She demonstrated no creativity or higher-order integration.

Natalie, another low-tier performer, demonstrated the lowest creativity and use of the technology. She completed the task in less than three minutes. Like Anna, she was neither thorough nor systematic, ignoring the prompts and guidance presented on each page of the Excelet. She struggled with the Excelet and encountered several errors. One error occurred when she incorrectly entered information in cells intended for output. While she noted the error, she incorrectly stated that the error was due to her measurement of the cookies, failing to connect the problem to data entry in the interface. When encountering subsequent error with the interface, she scarcely acknowledged the problem and did not attempt to resolve or isolate the source. She also made at least two mistakes in connections between mathematical concepts, graphical representations, and the Excelet interface. Instead of processing, she stated that she was "not really sure what this means" and moved quickly on. Her attitude toward the Excelet was dismissive and negative. This was expressed in her concluding comment: "I really don't like Excel. It doesn't really help with anything besides like being an accountant or like a gradebook." Natalie was not a proficient user of the technology, demonstrating few, if any, connections to mathematics. She made several errors that led to her progressively restricted exploration. Her inability to resolve the errors in tool use and conceptual understanding limited her efficient, effective, and creative use of Excelets for learning.

The ways that participants solved problems when using Excelets was very different. High-tier performers were systematic and detailed. They completed all tasks, utilizing the probing questions on each worksheet of the Excelet. They made connections between the variables, the graphic and numeric displays, and more difficult tasks of prediction and error. High-tier performers were proficient and creative users of the technology, the mathematics content covered, and the instructional strategies presented as questions and prompts in the interface. They displayed positive and innovative attitudes while using Excelets to complete the modeling task. Low-tier performers lacked a detailed or systematic approach. They presented surface-level understanding and skills, never reaching the point of effective or creative TPACK integration. In the case of the lowest performer, unresolved error and ineffective tool use led to a progressive decline in attitude and little learning.

8.4.2 Creativity in Developing Tasks

To assess how pre-service mathematics teachers develop a task associated with a modeling lesson, participants were asked to select one of three Excelets and to create a handout to guide the exploration of the Excelet for students in their future class. They were asked to connect this exploration to a Common Core or state standard. The handout and reflection they created were analyzed to assess creativity and ability of the task to promote learning. Particular attention was given to appropriate use of the technology and the pedagogy to promote exploration and integration of content knowledge within the context of creativity. High levels of creativity were those that promoted opportunities for students to explore and learn, while encouraging independent problem solving (Lin, 2011) within the mathematics context presented in the Excelet. Low levels of creativity were those that sought routine responses, promoted little exploration or independent problem solving for the learner.

Ian, a high-tier performer, created a two-page handout that asked broad and high-level questions associated with fundamental concepts of material presented in the Excelet. The handout presented a new dataset for entry, followed by questions regarding associated changes to the graph. Ian's handout provided a scaffold for his students taking them from simple interpretation to explanation, and ending with a more complex scenario and associated interpretations. In his reflection, he provided a thorough analysis of his thought process and a clear rationale for his handout. His handout was deemed effective and his reflection, accurate. Ian demonstrated creative TPACK with appropriate levels of guidance for his students.

Valerie, another high-tier performer, provided the most intricate handout. It was nicely formatted with a title and the associated state standard at the top. The handout included a well-structured goal for the activity and prompted the student to carefully read the introduction provided in the Excelet, drawing attention to the importance of this information. Then, the handout asked the student to define the variables used in the calculation, followed by a rich example with data for the student to enter and interpret. The handout included an appropriate balance of information and probing questions. The second page of the handout articulated a broader exploration with less but sufficient scaffolding to prompt and support students to extend their understanding while manipulating various pages of the Excelet. Valerie's handout was excellent and her reflection of the process accurate. She demonstrated creative guidance and integration of TPACK to promote learning.

Anna, a low-tier performer, produced a simple and logical handout for her students. It included a title and abbreviated version of the associated standard for learning. The handout also included a table for recording data but was unclear exactly how and where this activity linked to the Excelet. The handout included three probing questions and two extending questions but did not directly link these to the Excelet interface or associated content, making it potentially difficult for the learner to follow. In her reflection, Anna simply expressed her process and articulated the difficulty she had creating "good" questions that promoted critical thinking. The exploration task that she generated in the handout was logical but would have benefited significantly from additional and more exploratory prompts. Her use of TPACK was acceptable and slightly creative. To her credit, she expressed concerns that the task she created was not straightforward for students, but she did not resolve this in her final handout.

Natalie, a low-tier performer, provided related standards and explained the context for how high school students might use this Excelet, but she did not generate a handout with activities to guide the process. In her explanation, she recommended asking students to measure tumors and record them in Excel, then she recommended having students measure, chart, and calculate "without the help of Excel." The rationale provided for this approach was that it would "assess if the students actually understand the equations and different processes they will be using." Ironically, Natalie did not recommend any exploration using the Excelet or probing questions to facilitate conceptual understanding with the tool. Regrettably, she demonstrated no creative TPACK for her students.

The handouts that participants prepared for their students varied significantly in quality and approach. High-tier performers provided clear instructions, context, and well-formulated activities for learning. Their reflections were detailed and accurate, indicating that they were aware of their decisions and difficulties they encountered. High-tier performers promoted proficient and creative activities for their students. They found solutions to the difficulties that they encountered during development. Low-tier performers lacked a detailed or systematic approach. One of the low-tier performers produced a better quality product than the other. She demonstrated a simple solution. The other failed to produce an effective or creative handout. She demonstrated a comprehensive lack of TPACK integration for the content of Excelet.

8.4.3 Creativity and TPACK

This analysis assessed the relationship between participants' self-reported TPACK, reflection scores, and their performance in using and creating modeling tasks. Additionally, self-reported TPACK scores of pre-service mathematics teachers collected before and after modeling tasks were compared to assess the impact of Excelet-based modeling on TPACK awareness. These scores provide insight into how participants assessed their technological, pedagogical, and content knowledge and reflected on their creative teaching of real-life mathematics concepts both as students and as teachers (Lin, 2011; Lev-Zamir, 2008).

Ian, a high-tier performer, rated himself as a moderately proficient user of TPACK with scores hovering around 3.5. He appraised himself as slightly higher on pedagogical knowledge and lowest on content knowledge. There were no major changes in his self-reported TPACK when comparing scores collected before and after the modeling tasks, although he dropped his content knowledge score by approximately half a point. He expressed generally positive ratings of his abilities to generate tasks as evidenced by his mean reflection score of 3.9. Figure 8.8 provides an overall display of his results. Ian was a very proficient performer on both modeling tasks and as such, he accurately or slightly underestimated his abilities.

Valerie, a high-tier performer, rated herself as a moderately proficient user of TPACK. She rated her abilities in the technological, pedagogical, and content areas higher than those of integrated TPACK. Of particular interest was her shift in scores after completing the modeling tasks. Her technological mean increased by over one point and her mean integrated TPACK score decreased by one point. The change in her technological score seemed appropriate given her proficient use of the Excelet in the first modeling task, signifying that this task reinforced her self-awareness of technological skill. The decrease in her integrated TPACK score was evidenced in her struggle to determine the appropriate level of support when designing exploration activities in the second modeling task. A further indicator of this struggle was

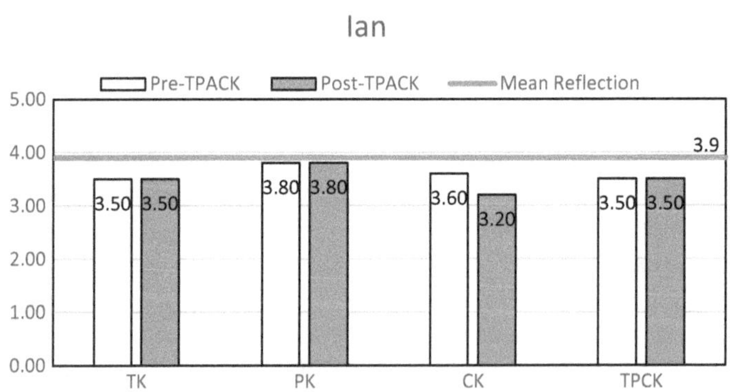

Fig. 8.8 TPACK and mean reflection for Ian, a high-tier performer

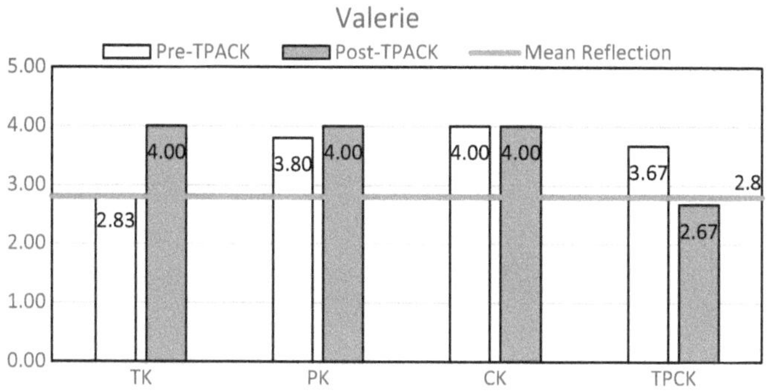

Fig. 8.9 TPACK and mean reflection for Valerie, a high-tier performer

her slightly negative mean reflection score of 2.8. Figure 8.9 provides a graph of her results. Valerie was a proficient user of modeling but was aware of the difficulties in creating effective exploration for students, even though her explorations were very elaborate and deemed highly effective. She slightly underestimated her abilities to create modeling tasks and was very thoughtful in her reflection of the difficulty of this task.

Anna, a low-tier performer, rated herself as a moderately proficient user of TPACK as seen in her scores before completing the modeling tasks. She rated herself slightly lower in technological skills and slightly higher in pedagogical skills. Anna made several meaningful changes to her TPACK scores following the modeling tasks, dropping her technological score by 0.83, her pedagogical score by 0.8, and her content score by 0.6. Her integrated TPACK score remained the same. She expressed a generally positive rating of her abilities to generate tasks as evidenced by her mean reflection score of 3.7. Figure 8.10 provides a numerical overview of Anna's results. Her qualitative reflection articulated the difficulty she had creating "good" questions that promoted critical thinking. The exploration task that she generated for students was logical but would have benefited significantly from additional and more exploratory prompts. Anna was a relatively low-performer on both modeling tasks and, while the changes to TPACK were in the right direction, it is believed that she still overestimated her abilities after both modeling tasks.

Natalie was a low-tier performer who rated herself as a highly proficient user of TPACK with most scores reported as perfect fives. Specifically, she rated herself as a five on technological, content, and integrated TPACK skills. Interesting, she reported lower scores in the area of pedagogical knowledge yet still rated herself with a top score on integrated TPACK. Natalie appeared to have a flawed understanding of the importance of technological, pedagogical, and content knowledge to be an effective integrator of TPACK. Such disconnects permeate Natalie's self-rating. The only change to her self-reported TPACK when comparing scores collected before and after the modeling tasks, was an increase in of 0.8 in pedagogical score. Given that she was ranked lowest on both modeling tasks, her mean

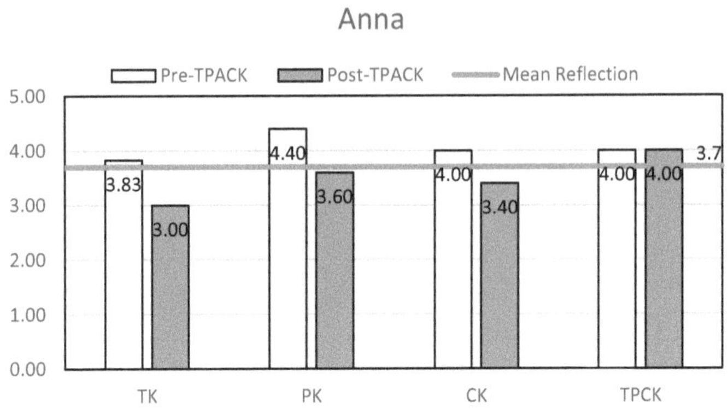

Fig. 8.10 TPACK and mean reflection for Anna, a low-tier performer

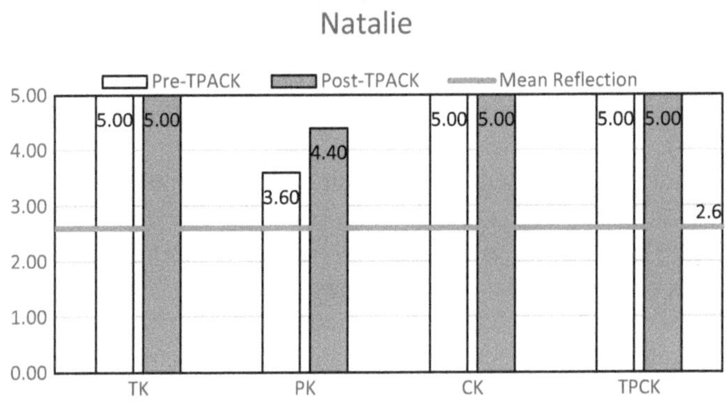

Fig. 8.11 TPACK and mean reflection for Natalie, a low-tier performer

reflection score of 2.7 was rather accurate indicating that she was aware of deficiencies in completing the actual modeling tasks. Figure 8.11 provides a graph of Natalie's results. While she accurately reflected on her poor performance on these tasks, she failed to connect the relationship of these tasks to her TPACK abilities, as evidenced by her increase in post-TPACK scores. Natalie was a low performer who reflected accurately on her performance yet overestimated her abilities, failing to make connections between the two.

In order to understand the magnitude of these misperceptions, TPACK scores of all pre-service mathematics teachers enrolled in the course were ranked and the rankings of the four participants in this study were compared. When two individuals had the same scores, the average ranking between those individuals was used (e.g., two individuals ranked 12th and 13th who report the same mean score have an adjusted rank of 12.5). The TPACK rankings for individuals in this study are provided in Table 8.1. Top-tier participants demonstrated high levels of creativity and TPACK yet rated themselves low in these skills while bottom-tier participants

Table 8.1 Self-reported TPACK rank relative to classroom peers

Participant	Pre-TPACK (n = 13)	Post-TPACK (n = 12.5)
Ian	11	12.5
Val	12	11
Anna	5	12.5
Natalie	3	2

1 = Highest TPACK. 13 or 12.5 = Lowest TPACK

provided little creativity and TPACK yet rated themselves extremely high. This highlights the issues associated with low-tier performers who do not accurately assess their knowledge and skills.

8.5 Summary

Findings from this study indicate that high-tier and low-tier performers use and produce modeling activities of very different quality and substance. While all participants were challenged by the activities, the high-tier performers were able to think creatively, articulate issues, solve problems, and develop meaningful explorations for students. They displayed a consistent and positive attitude toward the technology, and demonstrated a respect for the role of these tools in learning. Low-tier performers struggled with the technology, content, and pedagogy yet they rated themselves higher in skill than their high-tier peers. The low-tier performers need more awareness of their true skills and opportunities to resolve conflicts. This comes through prolonged, strategic development of TPACK.

Teacher preparation programs have much work to do. A mathematics methods course is often the first time that a pre-service teacher may be challenged to explore mathematics content creatively. The technology is a tool that opens up many opportunities for learners to explore yet it is also a challenge since many pre-service teachers lack experience with this creative thinking. There are still many challenges in educating pre-service and in-service teachers to become proficient users of technology for the classroom. Future instruction should consider ways to scaffold instruction, particularly for those pre-service teachers who struggle with the integration of TPACK.

8.6 Implications

Based on the evidence presented, many points arise that are in need of future research and study. First, as was presented in research (Beck & Wynn, 1998; Niess, 2005; Polly, Mims, Shepherd, & Inan, 2010), many pre-service teacher education programs have a general technology course that really does not serve the specific disciplines well. This was evident in this study by the fact that all participants identified they had taken such a course and were familiar with Excel, but not all were able to use it to the full capacity. It was apparent that the high-tier participants were able to use the technology as a tool to help them understand the mathematics, but the low-tier participants only used it to get answers. There was a wide gap in how the high-level users explored with Excelets compared to the low-level users. This has strong ties to Thomas and Palmer's work identifying obstacles and opportunities for teaching with technology (2014). They brought out that teachers who are most successful in integrating technology into class instruction were those who let the mathematics rather than the technology be the focus. We believe this phenomenon appears to be the case with the high-level users in this study.

This study also brings to light that pre-service teachers need more modeling experiences in mathematics and thinking about concepts in a more open manner. This openness allows for more creative thinking and understanding, which is what learning is all about. If teachers do not experience such practices for themselves as learners, they will not be able to act on them as teachers. Teachers must provide the "right kind of environment" for learners to be creative in thinking about how to solve real problems. This just does not happen—it happens when one experiences it and has the vision to promote individuality and creativity in learning tasks for students. We believe that use of spreadsheets allows one to take risks and to make use of visual cues to make sense of mathematics. Thomas and Palmer (2014) bring out the point that a major challenge of teachers is the time and effort required of both students and teachers to become familiar with technology and make sense of concepts.

We also found that reflective students often develop the skills to be creative in what they see, say, and do. It was determined that Ian and Valerie were great at "thinking aloud" and freely shared what they were thinking which in turn related to creativity. They often were unique in how they said things as well as what they used to describe their thinking and processes. On the other hand, the lower-level performers, Anna and Natalie, did not demonstrate much thought in their comments and actions and as a result were not creative in how they approached or discussed mathematics concepts. This leads us to consider how lower-level learners develop traits of creative responses and insight into tasks. More specifically, how can creativity be developed in pre-service teachers who are weak in their mathematics foundation? This would require greater discussion and investigation into what occurs in mathematics coursework pre-service teachers complete.

Preparing future teachers is an important task for colleges and universities. With expectations to prepare teachers around content, pedagogy, and technology, how to include a creative lens for such work is critical to advance learning. Teachers must know what to look for and how to provide a nurturing environment for such learning to occur. This takes time and effort on the part of the teacher (Thomas & Palmer, 2014). Pre-service teachers gain insight by experiencing technology-rich scenarios steeped in creative interpretation as learners prior to acting on them as teachers. As Applebaum and Saul (2009) state, "Teachers can practice the skill of analyzing creative responses. They can provide students with situations which leave room for the creative response, and they can provide classroom environments where the unusual is welcomed" (p. 282). Teacher preparation programs need to educate creative teachers so they in turn produce unusual but creative students!

References

Abramovich, S. (1995). Technology-motivated teaching of advanced topics in discrete mathematics. *Journal of Computers in Mathematics and Science Teaching, 14*(3), 391–418.

Applebaum, M., & Saul, M. (2009). Anecdotes and assertions about creativity in the working mathematics classroom. In R. Leikin et al. (Eds.), *Creativity in mathematics and the education of gifted students* (pp. 271–284). Rotterdam, The Netherlands: Sense Publishers.

Arganbright, D. (2005). Developing mathematics creativity with spreadsheets. *Journal of the Korea Society of Mathematical Education Series D: Research in Mathematical Education, 9*(3), 187–201.

Association of Mathematics Teacher Educators. (2006). *Preparing teachers to use technology to enhance the learning of mathematics: A position of the association of mathematics teacher educators*. San Diego, CA: Author.

Association of Mathematics Teacher Educators. (2009). *Mathematics TPACK (Technological Pedagogical Content Knowledge) Framework*. San Diego, CA: Author.

Ball, D., Hill, H. C., & Bass, H. (2005). Knowing mathematics for teaching: Who knows mathematics well enough to teach third grade, and how can we decide? *American Educator, 29*(3), 14–22.

Beck, J. A., & Wynn, H. C. (1998). *Technology in teacher education: Progress along the continuum*. ERIC Document Reproduction No. ED 424 212.

Bolden, D. S., Harries, A. V., & Newton, D. P. (2010). Pre-service primary teachers' conceptions of creativity in mathematics. *Educational Studies in Mathematics, 73*(2), 143–157.

Borko, H., & Putnam, R. T. (1996). Learning to teach. In D. C. Berliner & R. C. Calfee (Eds.), *Handbook of educational psychology* (pp. 673–708). New York, NY: Macmillan.

Brown, P. S., & Gould, J. D. (1987). An experimental study of people creating spreadsheets. *ACM Transactions on Office Information Systems, 5*(3), 258–272.

Conference Board of the Mathematical Sciences. (2001). *The mathematical education of teachers*. Providence, RI American Mathematical Society. Washington D.C.: Mathematical Association of America.

Conference Board of the Mathematical Sciences. (2012). *The mathematical education of teachers II*. Providence, RI: American Mathematical Society. Washington D.C.: Mathematical Association of America.

Davis, G. A., & Rimm, S. B. (2004). *Education of the gifted and talented* (5th ed.). Boston, MA: Pearson Education Press.

Doerr, H. M. (2007). What knowledge do teachers need for teaching mathematics through applications and modelling? In W. Blum et al. (Eds.), *Modelling and applications in mathematics education: The 14th ICMI study* (pp. 69–78). New York, NY: Springer.

Drier, H. S. (2001). Teaching and learning mathematics with interactive spreadsheets. *School Science and Mathematics, 10*(4), 170–179.

Fischer, G., & Nakakoji, K. (1994). Amplifying designers' creativity with domain-oriented design environments. In T. Dartnall (Ed.), *Artificial intelligence and creativity, Part V* (pp. 343–364). Dordrecht, The Netherlands: Kluwer Academic Publishers.

Hill, H. C., Ball, D. L., & Schilling, S. G. (2008). Unpacking "pedagogical content knowledge": Conceptualizing and measuring teachers' topic-specific knowledge of students. *Journal for Research in Mathematics Education, 39*(4), 372–400.

Hill, H. C., Rowan, B., & Ball, D. L. (2005). Effects of teachers' mathematical knowledge for teaching on student achievement. *American Educational Research Journal, 42*(2), 371–406.

International Society for Technology in Education. (2008). *National educational technology standards for teachers.* Retrieved from http://www.iste.org/standards/standards/standards-for-teachers.

Leikin, R. (2009). Exploring mathematical creativity using multiple solution tasks. In R. Leikin, A. Berman, & B. Koichu (Eds.), *Creativity in mathematics and the education of gifted students* (pp. 129–145). Rotterdam, The Netherlands: Sense Publishers.

Lev-Zamir, H. (2008). Can an algorithm be a way to elevate the development of creativity in mathematics teachers? In R. Leikin (Ed.), *Proceedings of the 5th International Conference on Creativity in Mathematics and the Education of Gifted Students* (pp. 444–446). Haifa, Israel.

Lin, Y. S. (2011). Fostering creativity through education: A conceptual framework of creative pedagogy. *Creative Education, 2*(3), 149–155.

Lortie, D. C. (1975). *School teacher: A sociological study.* Chicago, IL: University of Chicago Press.

Mann, E. (2006). Creativity: The essence of mathematics. *Journal for the Education of the Gifted, 30*(2), 236–260.

Martin, T. S. (Ed.). (2007). *Mathematics teaching today: Improving practice, improving student learning*. Reston, VA: National Council of Teachers of Mathematics.

Milgram, R., & Hong, E. (2009). Talent loss in mathematics: Causes and solutions. In R. Leikin, A. Berman, & B. Koichu (Eds.), *Creativity in mathematics and the education of gifted students* (pp. 149–163). Rotterdam, The Netherlands: Sense Publishers.

Mishra, P., & Koehler, M. J. (2006). Technological pedagogical content knowledge: A framework for teacher knowledge. *Teachers College Record, 108*(6), 1017–1054.

National Governors Association Center for Best Practices, Council of Chief State School Officers. (2010). *Common core state standards (Mathematics).* Washington, D.C.: National Governors Association Center for Best Practices, Council of Chief State School Officers. Retrieved from http://www.corestandards.org/Math/

Niess, M. L. (2005). Preparing teachers to teach science and mathematics with technology: Developing a technology pedagogical content knowledge. *Teaching and Teacher Education, 21,* 509–523.

Niess, M. L. (2008). Knowledge needed for teaching with technologies—Call it TPACK. *AMTE Connections, 17*(2), 9–10.

Niess, M. L., Sadri, P., & Lee, K. H. (2007). *Dynamic spreadsheets as learning technology tools: Developing teachers' technology pedagogical content knowledge (TPCK).* Paper Presented at the Meeting of the American Educational Research Association Annual Conference, Chicago, IL.

Niess, M. L., Ronau, R. N., Shafer, K. G., Driskell, S. O., Harper, S. R., Johnston, C., et al. (2009). Mathematics teacher TPACK standards and development model. *Contemporary Issues in Technology and Teacher Education, 9*(1), 4–24.

Pead, D., & Ralph, B. (with Muller, E.). (2007). Uses of technologies in learning mathematics through modelling. In W. Blum et al. (Eds.), *Modelling and applications in mathematics education: The 14th ICMI study* (pp. 309–318). New York, NY: Springer.

Polly, D., Mims, C., Shepherd, C. E., & Inan, F. (2010). Evidence of impact: Transforming teacher education with preparing tomorrow's teachers to teach with technology (PT3) grants. *Teaching and Teacher Education, 26,* 863–870.

Safuanov, I. S. (2008). Fostering creativity of prospective teachers in Tatarstan. In R. Leikin (Ed.), *Proceedings of the 5th international conference on creativity in mathematics and the education of gifted students* (pp. 449–451). Haifa, Israel.

Shriki, A. (2009). Developing prospective teachers' awareness of mathematical creativity through generating new concepts. *Educational Studies in Mathematics, 73*(2), 159–179.

Shulman, L. S. (1986). Those who understand: Knowledge growth in teaching. *Educational Researcher, 15*(2), 4–14.

Silver, E. A. (1997). Fostering creativity through instruction rich in mathematical problem solving and problem posing. *ZDM—The International Journal on Mathematics Education, 29*(3), 75–80.

Sinex, S. A. (2005). *Developer's guide to Excelets: Dynamic and interactive visualization with "Javaless" applets or interactive excel spreadsheets*. Retrieved from http://academic.pgcc.edu/~ssinex/excelets.

Sinex, S. A. (2006). *Exploring cyclic data in nature*. Retrieved from http://academic.pgcc.edu/~ssinex/excelets/periodic_data.xls.

Sinex, S. A. (2011a). *Cookies stack*. Retrieved from http://academic.pgcc.edu/~ssinex/cookies_stack.xls.

Sinex, S. A. (2011b). *Investigating the height of a stack of cookies*. Retrieved from http://academic.pgcc.edu/~ssinex/cast/cookies_stack.pdf.

Sinex, S. A. (2011c). *Kinetics of cancer cell growth*. Retrieved from http://academic.pgcc.edu/~ssinex/excelets/Kinetics_cancer_cells.xls.

Sinex, S. A. (2012). *What are the odds? Using probably in mathematical models*. Retrieved from http://academic.pgcc.edu/~ssinex/excelets/Probability.xls.

Subotnik, R. F., Pillmeier, E., & Jarvin, L. (2009). The psychosocial dimensions of creativity in mathematics: Implications for gifted education policy. In R. Leikin, A. Berman, & B. Koichu (Eds.), *Creativity in mathematics and the education of gifted students* (pp. 165–179). Rotterdam, The Netherlands: Sense Publishers.

Thomas, M. O. J., & Palmer, J. M. (2014). Teaching with digital technology: Obstacles and opportunities. In A. Clark-Wilson, O. Robutti, & N. Sinclair (Eds.), *The mathematics teacher in the digital era* (pp. 71–89). Dordrecht, the Netherlands: Springer.

Thompson, A. D., & Mishra, P. (2007). Breaking news: TPCK becomes TPACK! *Journal of Computing in Teacher Education, 24*(2), 38–64.

Yerushalmy, M. (2009). Educational technology and curricular design: Promoting mathematical creativity for all students. In R. Leikin, A. Berman, & B. Koichu (Eds.), *Creativity in mathematics and the education of gifted students* (pp. 101–113). Rotterdam, The Netherlands: Sense Publishers.

Zelkowski, J., Gleason, J., Cox, D. C., & Bismarck, S. (2013). Developing and validating a reliable TPACK instrument for secondary mathematics preservice teachers. *Journal of Research on Technology in Education, 46*(2), 173–206.

Chapter 9
Creativity in Question and Answer Digital Spaces for Mathematics Education: A Case Study of the Water Triangle for Proportional Reasoning

Benjamin Dickman

Abstract As digital spaces evolve, mathematics educators must develop an awareness of the ways in which these environments can facilitate discussion and foster creativity. Question and Answer (Q&A) sites such as Mathematics Educators Stack Exchange (MESE) provide a platform through which those interested in the teaching and learning of mathematics can harness new technologies to address novel queries, and engage collaboratively with others who share their interests. This chapter aims to trace one example of a question-answer combination on MESE as situated in the broader context of technology and creativity in mathematics education, and to utilize the example as a lens through which we can critically examine the current state of digital environments and reflect on their potential use by mathematics educators.

Keywords Collaborative emergence · Mathematical creativity · Online spaces
Participatory model of creativity · Q&A sites

9.1 Introduction

The ideas outlined in this chapter coincide with an evolution of the digital spaces that can foster mathematical creativity. Geographical barriers no longer pose the same sort of hindrance to collaboration among education researchers and practitioners. Today, mathematics educators come together through social media such as the MathTwitterBlogosphere (MTBoS), deftly navigate vast repositories of mathematical information such as the arXiv, and communicate directly through various web forums.

B. Dickman (✉)
New York, USA
e-mail: benjamindickman@gmail.com

© Springer International Publishing AG, part of Springer Nature 2018

V. Freiman and J. L. Tassell (eds.), *Creativity and Technology in Mathematics Education*, Mathematics Education in the Digital Era 10,
https://doi.org/10.1007/978-3-319-72381-5_9

Question and Answer (Q&A) sites for mathematics are not new to the decade, or even the millennium. The online *Geometry Forum*, which is now NCTM's *Math Forum*,[1] traces the history of its Q&A component *Ask Dr. Math* back to 1994.[2] Two years later, in 1996, the similarly titled *Math Doctor* began at Nicholls State University, but changed its name shortly thereafter to *Math Nerds*, which had "about one hundred volunteers and answer[ed] about 1500 questions per month" as measured about a decade after its creation (De Angelis et al., 2008, p. 28). Writing out of the Technical Education Research Centers (TERC), Rubin's (1999) *Technology Meets Math Education: Envisioning a Practical Future Forum on the Future of Technology in Education* "insists that, rather than looking at math education from the perspective of the computer, we must look at computers from the perspective of mathematics education" (p. 1). In doing so, Rubin identifies five powerful uses of technology in mathematics education, among which *Resource-Rich Mathematical Communities* includes the *Math Forum* as "the best known" resource site at the time (Rubin, 1999, p. 8). This subsection on resource sites concludes by remarking that the forum "has served as an important portal for mathematics educators and as a kind of social center for the mathematics education community" (p. 9).

The social centers for mathematics education communities have continued to exist on the Internet, but have evidently changed over the past two decades. One feature is the inclusion of sites that allow anyone, not just those who are confirmed as experts, to answer questions about mathematics and, in some cases, mathematics education. For example, the *reddit* community dedicated to socializing around mathematics[3] has subscribers in the hundreds of thousands, and allows anyone to sign up and post or comment about mathematical links and questions.

In this chapter, we look at a particular Q&A digital space, Mathematics Educators Stack Exchange (MESE), which fits within the broader Stack Exchange network. Unlike many of its predecessors, MESE is organized around *mathematics education* rather than mathematics proper. To gain insight into how MESE fits into the landscape of technology and creativity in mathematics education, we proceed as follows: First, we articulate the three key ideas that will be covered throughout the chapter, after which we look to the literature as concerns the breadth of definitions that have arisen over the years in investigations of 'technology' and 'creativity'. Next, we provide brief remarks around the connections between our specific subject of study and the broader topics of this text: mathematics and mathematics education, technology, and creativity. The third section summarizes our specific subject —a question posted to MESE about a tool used in proportional reasoning—and then investigates the ways in which responding to a reference request is an act of creative collaboration. The fourth and final section provides avenues for further research by proposing three open questions related to the initial key ideas, before closing with our conclusion.

---

[1](http://mathforum.org/).

[2](http://mathforum.org/dr.math/abt.drmath.html).

[3](https://www.reddit.com/r/math).

9.1.1 Key Ideas: Q&A Sites, Definitions, and Creative Collaboration

We non-exhaustively list here three key ideas for the chapter, which relate, respectively, to Q&A sites for mathematics education, the importance of defining terms and scope when discussing technology and creativity, and the situating of everyday ideas as examples of creative collaboration within a participatory model of creativity.

Key Idea 1 Question and Answer (Q&A) sites specific to mathematics education are a recent phenomenon, and have emerged along with a collaborative paradigm in which users frequently serve as both askers and answerers of questions. Earlier precursors include Q&A sites specific to mathematics, which have existed for over two decades, yet for which the askers and answerers have sometimes constituted disjoint, or nearly disjoint, groups.

Key Idea 2 The many definitions for 'technology' and 'creativity' require a certain amount of specificity in any discussion for which they play prominent roles. We advocate for an interpretation of 'technology' that admits both digital technologies (such as online Q&A forums) and domain-specific tools (such as the water triangle for proportional reasoning). Moreover, we advocate for an interpretation of 'creativity' that coincides with Hanson's (2015a, b) description of a participatory model: Rather than focusing on single ideas or identifying individual creators, we look at how creative collaboration (e.g., through a Q&A forum) is distributed among actors and objects.

Key Idea 3 Our particular example of a tool (the *water rectangle*) paired with a necessarily incomplete account of its history does not constitute a watershed, domain-shifting moment in mathematics education; rather, the collaborative creativity exemplified by the reference request described in this chapter contributes to an ongoing conversation about proportional reasoning, in particular, and mathematics education, in general. It is a conversation that began before the modern language of proportional reasoning existed, has continued with the predecessors for this tool and the tool itself, was furthered by the satisfied reference request, and which will continue far beyond the everyday ideas put forth in this chapter.

9.1.2 Definitions for 'Technology' and 'Creativity' Over Time

Definitions for 'technology' and 'creativity' abound. As contemporary conceptions of technology undergo rapid change, we begin by looking back to Hansen and Froelich's (1994) early attempt at articulating the variety of definitions for 'technology' in their aptly-titled *Defining Technology and Technological Education*, in which they remark that "philosophers, anthropologists, sociologists, historians, and teacher educators continue to study the subject [of technology], yet a widely accepted definition remains obscure" (p. 179). The authors continue in exploring definitions of 'technology' by looking to dictionaries and considering its etymology; by looking to individual scholars from a variety of domains; by considering, among other conceptions, 'technology' with regard to products and processes; and by examining technology as relates to feminism and the evolution of women's roles in society (Franklin, 1992). Analogously, there are elsewhere discussions about the emergence of 'creativity' in English dictionaries (e.g., Mason, 2003); debates about whether creativity is domain-specific (e.g., Baer, 1998; Plucker, 1998); conceptions that include creativity with respect to products and processes (Rhodes, 1961); and discussions around creativity as relates to the evolution of women's roles in society (Bateson, 2001, 2004). Beyond these parallels, in Treffinger et al. (2002) the authors remark that "Treffinger (1996) reviewed and presented more than 100 different definitions [of creativity] from the literature" (p. 5), and Sawyer (2011) goes so far as to contend that "defining creativity may be one of the most difficult tasks facing the social sciences" (p. 11). Defining either term is certainly no easy task.

There is a school of thought within creativity research, originating with work by Amabile and Hennessey (e.g., Amabile, 1983, 1996; Hennessey, 1994; Hennessey & Amabile, 1999), in which one operationalizes subjective agreement on that which constitutes 'creativity' with respect to particular products, rather than providing a formal catch-all definition for the term. There are also schools of thought, more process-oriented, that essentially identify creativity with problem solving; for example, Weisberg (2006) writes that "it seems reasonable to adopt as a working assumption the premise that creative thinking is an example of problem solving" (p. 581). In a similar vein, others associate creativity with problem *posing*; for example, Getzels (1975) quotes Einstein as stating that, "The formulation of a problem is often more essential than its solution, which may be merely a matter of mathematical or experimental skill. To raise new questions, new possibilities, to regard old questions from a new angle, requires creative imagination and marks real advances in science" (p. 12). Such an approach (see also: Getzels & Csikszentmihalyi, 1976; Runco, 1994) continues the line of thought associated with Guilford's (1950) trait of "sensitivity to problems" as relates to creativity (p. 454). This work of Guilford appeared in his APA presidential address, and also touched upon the ability to reorganize, or redefine, in the sense of Gestalt psychology; relatedly, one finds Wertheimer (1959) remarking that "often in great discoveries the most important thing is that a certain question is found. Envisaging, putting the productive question is often more important, often a greater achievement than solution of a set question" (p. 141).

9.1.3 Perspectives Adopted for Investigating the Creative Use of Technology

In this chapter, we will adopt a combination of perspectives on creativity: We will use the Question and Answer (Q&A) format of web-based platforms to frame the creative use of technology. The notion that both questions and answers are important is established within the discipline of mathematics education. We may associate 'answers' with the process of finding solutions, and look to the vast literature on mathematical problem solving (e.g., Polya, 1945; Schoenfeld, 1985, 2010) and, similarly, associate 'questions' with the process of problem formulation, and look to the vast literature on problem posing (e.g., Brown & Walter, 2005; Duncker & Lees, 1945; NCTM, 1989, 1991; Kilpatrick, 1987; Silver, 1994). Furthermore, we assume of the reader a familiarity with the domain of mathematics education, and, therefore, the capability to apply the subjective commendation of 'creative' to products (cf. Baer & McKool, 2009). The sheer breadth among conceptions of technology and creativity will make our own study intractable without first limiting our scope; we use here a case study of one, which is an approach foreign neither to creativity research (e.g., Gruber & Davis, 1988; Gruber & Wallace, 1999) nor mathematics education (e.g., Brizuela, 1997; Erlwanger, 1973). Specifically, we will trace a single example of a question-answer combination posted on the Mathematics Educators Stack Exchange (MESE) website, and unpack from a seemingly straightforward reference request the ways in which technology and creativity collide in a present-day digital space designed for those interested in the teaching and learning of mathematics.

9.2 Brief Connections to Mathematics, Technology and Creativity

In this section, we situate our subject of investigation by connecting it to the three broad topics of mathematics and mathematics education, technology, and creativity.

9.2.1 Brief Connections to Mathematics and Mathematics Education

The Stack Exchange network includes over 150 Q&A communities; among these are MathOverflow (MO), which is designed for those engaged in research level mathematics, as well as Mathematics Educator Stack Exchange (MESE), which is designed for those interested in the teaching and learning of mathematics. Earlier work by Tausczik et al. (2014) explored the collaborative problem solving that takes place on MO, and the five "collaborative acts" of providing information, clarifying the question, critiquing an answer, revising an answer, and extending an answer identified through a process of open coding (p. 359). Though not explicitly connected there, the authors' research fits well with the notion of *collaborative emergence* in the creativity literature (Sawyer, 2011; Sawyer & DeZutter, 2009). At present, there appear not to have been any investigations of MESE, which was proposed as a site in 2014, and currently holds over 8000 combined questions and answers in the domain of mathematics education.

MESE is specific to mathematics education, and the example traced here is no exception: "The 'water triangle' proportional reasoning task"[4] is a reference request about the origin of a tool previously depicted on Wikipedia's *proportional reasoning* page.[5] Proportional reasoning is a fundamental component of early mathematics education, and relates to work with such topics as ratios, fractions, rational numbers, and rates (Tourniaire & Pulos, 1985; Lobato et al., 2010). The asker suggests the tool may have been created by mathematics educator Robert Karplus in the 1970s, but is otherwise unaware of its history. This tool inspired the construction of a *water rectangle* in the asker's dissertation on mathematics education, as well as subsequent investigations presented at the ICMI-East Asia Regional Conference on Mathematics Education (Noche, 2013; Noche & Vistro-Yu, 2015).

9.2.2 Brief Connections to Technology and Online Forums

The movement to incorporate technology into learning trajectories can be seen by the growing presence of online classes, MOOCs, sites such as Coursera and MIT OpenCourseWare, and web-based platforms such as Moodle and Blackboard to supplement classroom-based courses. There are also digital spaces associated with post-secondary programs in mathematics education, such as *The Math Forum* (Drexel University, mathforum.org) and *The Mathematics Teaching Community* (University of Georgia, mathematicsteachingcommunity.math.uga.edu). *Mathematics Educators Stack Exchange* (MESE, mateducators.stackexchange.com) is not associated with an academic institution, and instead fits within the Stack Exchange (SE) network; the network includes an additional site specifically for mathematics questions at the research level (MathOverflow.net) and another for general mathematical queries (math.stackexchange.com). Although SE contains over 150 different Q&A communities, MESE is, at present, the only one concerned specifically with education.

In addition to the technology involved in interacting through a digital environment, both the question and answer connect to technology, as well. The question explored is about a particular form of technology: although it is not *digital* technology, the water triangle is itself a tool for investigating proportional reasoning (Kurtz, 1976). With regard to digital technology, the answer emerged from a

[4]http://matheducators.stackexchange.com/q/29.

[5]https://en.wikipedia.org/wiki/Proportional_reasoning.

confluence of sources: MESE, Wikipedia, ProQuest, e-mail and more. To unpack the power of a modern technological tool such as a Q&A digital space, we will explore both the 'Q' and 'A' part of the given example; more precisely, we must remain cognizant of the types of technology that exist outside of the computer-based forms commonly associated with contemporary conceptions of 'tech'—a deep understanding of connections to technology emerges most prominently when the digital requirement is dropped, and a broader toolbox conception is adopted.

9.2.3 *Loose Ends: A Couple of Additional Connections to Creativity*

In addition to the already mentioned problem solving, problem posing, and collaborative emergence, we consider two more important connections to the literature on creativity. First, Stokes (2005, 2010) discusses the development of creativity through *constraints*. The digital space under discussion is designed specifically for questions about mathematics education; this *precluding constraint* ensures that questions that are deemed off-topic by other site users are either refined, migrated to another site, or closed entirely. Moreover, there exists an additional *promoting constraint* with regard to novelty; namely, that new questions be distinct from earlier ones: If the question already exists on the network, then site users may choose either to close the new version or encourage its modification so as to prevent repetition. Second, Rhodes' (1961) classical framework around situating creativity concerns the person, process, product, and environmental press. These are only a few of the many conceptions of creativity, and, although this chapter contains a portion narrativized as a personal recollection, our ultimate goal is to consider creativity from a variety of perspectives; as is the case with connections to technology, a deeper understanding of creativity emerges when we adopt a broader toolbox conception.

9.3 Collaborative Creativity Through a Reference Request

In this first sub-section, we detail the history of a single example of a routine reference request, which will provide us with a lens through which, in the subsequent sub-section, we may examine the current state of question and answer digital environments as we reflect on their potential use by mathematics educators.

9.3.1 Reference Request: 'Water Triangle' for Proportional Reasoning

"The 'water triangle' proportional reasoning task" was initially posted on the *Mathematics Teaching Community* in 2012[6] where it remained unresolved. The question was modified by its creator and re-posted to MESE in 2014. The question essentially asked about the source of the 'water triangle' depicted on Wikipedia's proportional reasoning page; the illustration under discussion can be seen in Fig. 9.1.

The author of this chapter ultimately located the original source of the water triangle, and provided an accepted answer to the query; in recounting how this answer came about, we change voice here to the first-person for the sake of clarity:

I began by investigating the Wikipage for proportional reasoning, and also looked through its history to see if there was relevant information to be found in earlier versions of the page. Earlier incarnations of the Wikipage included an additional photograph of a physical water triangle being used by students (Fig. 9.2) and the image had the same credited uploader as the illustrated version already shown in Fig. 9.1.

The original question on MESE included a mention of mathematics educator Robert Karplus; however, both of the images were credited to Barry L. Kurtz, whose e-mail address was included, as well. I wrote to Professor Kurtz to ask whether he was aware of the water triangle's origins; his response message was as follows:

> I completed my Ph.D. under Bob Karplus at UC Berkeley. I was his last Ph.D. student. My dissertation dealt with teaching for proportional reasoning. I invented the idea of a "water triangle" to teach inverse proportions. There were all made by the workshop at the Lawrence Hall of Science; they were not a commercial item. I doubt any exist today; I certainly don't have any. Thanks for your interest. You did a good job tracking me down!

I followed up on this lead by using ProQuest[7] to find Kurtz's doctoral dissertation, where the water triangle can be found on page 34; an image of the dissertation (Kurtz, 1976) is displayed in Fig. 9.3.

In a follow-up message, Kurtz pointed to an article based on his dissertation (Kurtz & Karplus, 1979) and noted that it was later reprinted in Fuller's (2002) *A Love of Discovery: Science Education—The Second Career of Robert Karplus*.

[6]https://mathematicsteachingcommunity.math.uga.edu/index.php/685/the-water-triangle-proportional-reasoning-task.

[7]http://www.proquest.com/.

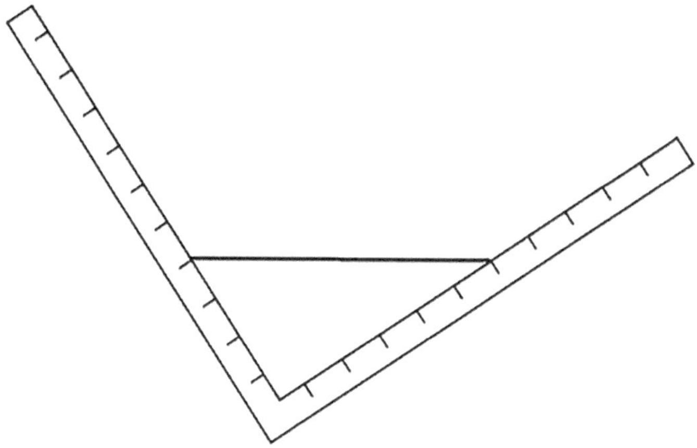

Fig. 9.1 The water triangle for proportional reasoning (released by its author into public domain for any purpose, and without any terms or conditions; cf. https://commons.wikimedia.org/wiki/File:Water-triangle.JPG)

Fig. 9.2 Students using a physical version of the water triangle (released by its author into public domain for any purpose, and without any terms or conditions; cf. https://commons.wikimedia.org/wiki/File:Constant-product.png)

An apparatus called a "Water Triangle," shown in figure 2.3, was used to demonstrate a constant product relationship. The variables were the height of the water level on the left side and the right side of the triangle.

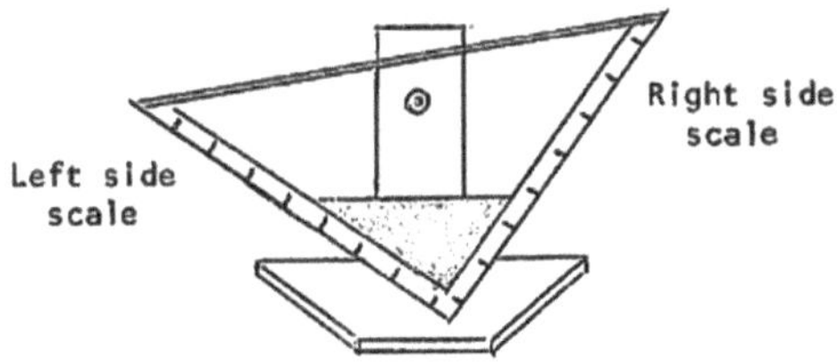

Figure 2.3 The Water Triangle

Fig. 9.3 An excerpt from the doctoral dissertation of Barry L. Kurtz

9.3.2 Collaborative Creativity: What Do We Mean by 'Creative'?

The previous sub-section briefly detailed how a question on Math Educators Stack Exchange, which asked about the origin of a water triangle for proportional reasoning, was investigated and resolved. This section aims to answer the natural follow-up question with respect to inclusion in a textbook on creativity and technology in mathematics education; succinctly: *Okay, so what?*

We begin by observing the manifold ways in which technology allowed for such a question-answer combination. Specifically, we have an educator in The Philippines posing a question about mathematics education, for which a mathematics educator in the East Coast of the United States was able to respond by tracking down graduate research completed in the West Coast of the United States some four decades earlier. Without a confluence of technological means—Wikipedia, which allows users to participate by uploading original content; e-mail, which allows individuals to communicate without the geographical barriers that would have hindered such an interaction in the recent past; and ProQuest, which includes doctoral dissertations uploaded to a searchable database—such a query would have been essentially intractable. Moreover, there needed to exist a digital space that could facilitate such interactions between individuals with the relevant curiosity and expertise: for the asker, this meant the curiosity and expertise to formulate the question after utilizing a domain-specific tool based on the water triangle in their own work on mathematics education; for the answerer, this meant the curiosity and expertise to understand and investigate the question, and to use a combination of digital technologies in order to resolve it.

Creating spaces in which members of a field can come together to interact meaningfully is a nontrivial endeavor. As mentioned earlier, the reference request at the heart of this chapter was already posted to another digital space for mathematics education two years prior to its re-post on MESE. Internet-based Q&A digital spaces provide both a platform and a technological tool that can be used to foster creativity within the domain of mathematics education; such an observation points to the requirement not only for tools and technologies, but also to *users* who can intentionally and capably operate them. For the question itself, we needed an environment that not only could house such a query, but could also do so in an accessible manner: for example, using again the language of Stokes (2005, 2010), we needed the search space to be both broad enough to allow for a variety of questions (e.g., by instituting promoting constraints around novelty such as instituting a *reputation-based* system to award credit to those who formulate well-received questions and answers) and narrow enough to appeal to members of a particular field (e.g., by instituting precluding constraints such as closing duplicated questions, or, more generally, having a site specific not to mathematics or to education but rather to *mathematics education*).

Similar observations hold outside of the digital-interpretation of 'technology': The original water triangle, depicted as an illustrated diagram in Fig. 9.1, is itself an example of a tool and technology; however, its educational relevance is perhaps more clearly depicted in Fig. 9.2, where we observe individuals interacting with a physical model. Even still, the images themselves require additional interpretation and exploration; to this end, we require members of a field to push concepts further by creating web-based content such as a Wikipage or a response on MESE, by developing the language of *proportional reasoning* and explaining its domain-relevance in the earlier dissertative work and publications, by continuing with new ideas around proportional reasoning in more recent dissertative work and presentations, and, self-referentially, by summarizing and connecting these various contributions in our present account of matters as they stand today. None of this is accomplished by individuals working in a vacuum, just as none of the ideas has emerged ex nihilo; rather, we have the *collaborative emergence* (Sawyer, 2011; Sawyer & DeZutter, 2009) that is enabled by the collision of technologies: physical and digital, old and new.

At this point in our discussion, it is hopefully clear that Q&A digital spaces exemplify an environment that has the potential to facilitate discussion among those with domain-relevant expertise and interest. But our argument here is that these spaces can foster *creativity*, and it is to this claim that we must now attend. What do we really mean by 'creativity'? In Hanson's (2015b) *Worldmaking: Psychology and the Ideology of Creativity* there are a variety of conceptions around creativity put forth and described. Specifically, Hanson (2015a) writes that:

> …the concept of creativity provides a site to explore important issues within a framework of often unquestioned assumptions. Beyond claims of the specific theories, the amalgam of theories have contributed to an underlying ideology. This ideology is important because it concerns one of the most salient characteristics of our times: change. It is also a fascinating ideology because, in keeping with the values it represents, the ideology changes over time.

The belief espoused in our own account is not that satisfying a reference request constitutes a sort *eminent* or 'big-C' Creative (e.g., Kaufman & Sternberg, 2009) achievement that shifts an entire domain (Csikszentmihalyi, 1999). Instead, we wish not to leave the underlying assumptions around creativity unquestioned, and adopt a view of creativity as a *participatory model*. Quoting Hanson (2015a) again:

> Instead of focusing almost entirely on how to get people to think of new ideas, the participatory models situate ideation within individual development, group dynamics and historical settings. The support roles and the field (gatekeeper) roles that people take up as they integrate novelty become more central. Choosing, supporting, interpreting and refining ideas are as important as 'having' an idea. Indeed, on close examination, distinctions between field roles and the 'creator' role begin to disappear… Creator as curator [of ideas] emphasizes the tasks of selecting, emphasizing, and powerfully presenting ideas that – as always – derive from historical domains, broader culture ('commonsense'), the artifacts of culture and other people's ideas.

The creativity in our discussion is distributed among the many actors and objects involved, ranging from the research of Karplus and students on proportional reasoning, to the question-answer combination on MESE, to the understanding reached (extended, challenged, etc.) by the reader of the work at hand. In most any direction we look, there is more to be unpacked and asked about (e.g., How did Wikipedia, and the wiki-paradigm more generally, support these interactions? How did the availability of resources at the UC-Berkeley Lawrence Hall of Science contribute to Kurtz's work?). Technologies such as Q&A digital spaces allow educators to expand and develop their own *network of enterprises* (Gruber & Wallace, 1999) as they participate in projects that allow for new ways of organizing the self—what *needs* to be done, what *can* I do, and what *must* I do (cf. Gruber & Barrett, 1974)—in relation to one's own ongoing work and the creative endeavors of others.

Our attention to the role of Q&A digital spaces in fostering creativity is not restricted to Rhodes' (1961) framework around the *person*, *process*, *product*, or *press*; nor is it with respect to some sort of *self-actualization* (e.g., Maslow, 1943) in becoming "Creative Educators," or by conflating creativity writ large with the separate construct of *divergent thinking* within *out-of-the-box* models (Runco, 2010), Gestaltist views on creative insight (Wertheimer, 1959), or eminence-based theories of creativity (Csikszentmihalyi, 1999). Rather, we focus on continuous participation in the evolving *ideology* of creativity (Hanson, 2015b). As mathematics education develops as a domain, the corresponding field of mathematics educators is confronted by challenges that require the use of tools and technologies both new and old. Even, or especially, as central concepts in mathematics education, such as proportional reasoning, have changed over the past several decades, we must be prepared to exchange and modify ideas and understandings within spaces that allow us to engage in the sort of dynamic interactions that best support our work as individuals and as groups. Neither a routine reference request nor the novel pathway carved out in resolving it constitute paradigm-shifts in creativity within mathematics education. Instead, these questions and answers, the digital spaces that provide platforms for their formulation, and the many people whose work is inextricable from the technologies and tools that are built and used to resolve them are all parts of a conversation around the teaching and learning of mathematics, and all contribute to the creativity necessary to face the continuous change found in the ever-evolving world of education.

9.4 Looking Ahead and Wrapping Up

We conclude our chapter by posing three follow-up questions that correspond to possible ways in which our initial three key ideas can be further explored.

9.4.1 Looking Ahead: Three Open Questions

We conclude with three open questions for future investigation, consideration, and research. Each of the three questions is intended to extend, or challenge, components of the respective key ideas from the beginning of the chapter. None of these questions is intended to be answered succinctly, or even directly; instead, they are posed as questions to guide, or influence, those who wish to think further around the subject matter contained in, or related to, this chapter.

Open Question 1 As the distinction between asker and answerer is blurred within Q&A sites, and forums become more inclusive of participants at various positions along a novice-to-expert continuum, how can we better ensure that creative collaboration will ultimately be productive both within the confined digital space, and within broader conversations throughout the domain of mathematics education?

Open Question 2 For those who adhere to, or advocate for, interpretations of 'technology' and 'creativity' different from ours, what conclusions can be drawn about the impact of evolving Q&A sites for mathematics education? For example, for those whose definitions of technology are restricted to the digital, and for whom the participatory model of creativity is rejected in favor of an out-of-the-box divergent thinking model, are digital spaces such as Math Educators Stack Exchange well-positioned to support and foster creativity?

Open Question 3 Given a conception of creativity in which it is viewed as a feature of an ongoing conversation, and the creative collaboration is distributed across many individual actors to the extent that supporter roles and creator roles fade away, what impact does the adoption of such a perspective have on students, teachers, and other stakeholders in mathematics education?

9.4.2 Conclusion

We have advocated in this chapter for a view of creativity that is participatory: rather than looking to identify eminent individuals or singular insights, we chose to examine everyday creativity as exemplified by a reference request. Our view of technology was similarly inclusive, as it admitted not only the digital sort, but also tools in a more general sense, up to and including a physical water triangle for proportional reasoning. As we did not concern ourselves with finding the one person, or one moment, or one idea, to which the commendation of 'creativity' can be applied, we focused instead on the distribution across multiple actors, and the ways in which they fulfilled their roles as facilitated by a particular Q&A site. In doing so, we actively push back against the *myth* (cf. Weisberg, 1986) that creativity is strictly the work of geniuses. And in promulgating a participatory model of creativity in which a hierarchy of creator and supporter roles ceases to exist, we hope to shift the focus away from precisely who or what can be considered creative, and instead to think more inclusively about the ways in which a diverse array of individuals can productively work together, whether this collaboration occurs in the Q&A digital spaces of today, or in yet-to-be-created spaces of the future.

References

Amabile, T. M. (1983). A consensual technique for creativity assessment. In *The social psychology of creativity* (pp. 37–63). New York, NY: Springer.

Amabile, T. M. (1996). *Creativity in context: Update to the social psychology of creativity*. Boulder, CO: Westview.

Baer, J. (1998). The case for domain specificity of creativity. *Creativity Research Journal, 11*(2), 173–177.

Baer, J., & McKool, S. (2009). Assessing creativity using the consensual assessment. In *Handbook of assessment technologies, methods, and applications in higher education*. Hershey, PA: IGI Global.

Bateson, M. C. (2001). *Composing a life*. New York, NY: Grove Press.

Bateson, M. C. (2004). Composing a life story. In *Willing to learn: Passages of personal discovery* (pp. 66–76). Hanover, NH: Steerford Press.

Brizuela, B. (1997). Inventions and conventions: A story about capital numbers. *For the Learning of Mathematics, 17*(1), 2–6.

Brown, S. I., & Walter, M. I. (2005). *The art of problem posing* (3rd ed.). London, UK: Psychology Press. https://books.google.com/books?hl=en&lr=&id=xn95AgAAQBAJ&oi=fnd&pg=PP1.

Csikszentmihalyi, M. (1999). Implications of a systems perspective for the study of creativity. In R. J. Sternberg (Ed.), *Handbook of creativity* (pp. 313–338). Cambridge, UK: Cambridge University Press.

De Angelis, V., Kang, K., Mahavier, W. T., & Stenger, A. (2008). Nerd is the word. *Math Horizons, 16*(1), 28–29.

Duncker, K., & Lees, L. S. (1945). On problem-solving. *Psychological Monographs, 58*(5).

Erlwanger, S. H. (1973). Benny's conception of rules and answers in IPI mathematics. *Journal of Children's Mathematical Behavior, 1*(2), 7–26.

Franklin, U. (1992). *The real world of technology*. Concord, ON: House of Anansi Press Ltd.

Getzels, J. W. (1975). Problem-finding and the inventiveness of solutions. *The Journal of Creative Behavior, 9*(1), 12–18.

Getzels, J. W., & Csikszentmihalyi, M. (1976). *The creative vision: A longitudinal study of problem finding in art*. New York, NY: Wiley.

Gruber, H. E., & Barrett, P. H. (1974). *Darwin on man: A psychological study of scientific creativity*. New York, NY: EP Dutton.

Gruber, H. E., & Davis, S. N. (1988). 10 inching our way up Mount Olympus: The evolving-systems approach to creative thinking. *The Nature of Creativity: Contemporary Psychological Perspectives, 243*.

Gruber, H. E., & Wallace, D. B. (1999). The case study method and evolving systems approach for understanding unique creative people at work. *Handbook of Creativity, 93,* 115.

Guilford, J. P. (1950). Creativity. *American Psychologist, 5*(9), 444–454.

Hansen, R., & Froelich, M. (1994). Defining technology and technological education: A crisis, or cause for celebration? *International Journal of Technology and Design Education, 4*(2), 179–207.

Hanson, M. H. (2015a). *Worldmaking: Psychology and the ideology of creativity*. New York, NY: Palgrave Macmillan.

Hanson, M. H. (2015b). The ideology of creativity and challenges of participation. *Europe's Journal of Psychology, 11*(3), 369–378.

Hennessey, B. A. (1994). The consensual assessment technique: An examination of the relationship between ratings of product and process creativity. *Creativity Research Journal,* 7(2), 193–208.

Hennessey, B. A., & Amabile, T. M. (1999). Consensual assessment. *Encyclopedia of Creativity, 1,* 347–359.

Karplus, R., & Fuller, R. G. (2002). *A love of discovery: Science education—The second career of Robert Karplus*. New York, NY: Springer Science+Business Media.

Kaufman, J. C., & Beghetto, R. A. (2009). Beyond big and little: The four c model of creativity. *Review of General Psychology, 13*(1).

Kilpatrick, J. (1987). Formulating the problem: Where do good problems come from? In A. H. Schoenfeld (Ed.), *Cognitive science and mathematics education* (pp. 123–147). Hillsdale, NJ: Lawrence Erlbaum Associates.

Kurtz, B. L. (1976). *A study of teaching for proportional reasoning* (Unpublished doctoral dissertation). Berkeley, CA: University of California.

Kurtz, B., & Karplus, R. (1979). Intellectual development beyond elementary school VII: Teaching for proportional reasoning. *School Science and Mathematics, 79*(5), 387–398.

Lobato, J., Ellis, A., & Zbiek, R. M. (2010). *Developing essential understanding of ratios, proportions, and proportional reasoning for teaching mathematics: Grades 6–8*. Reston, VA: National Council of Teachers of Mathematics.

Maslow, A. H. (1943). A theory of human motivation. *Psychological Review, 50*(4), 370.

Mason, J. H. (2003). *The value of creativity: The origins and emergence of a modern belief*. Farnham, UK: Ashgate Publishing.

National Council of Teachers of Mathematics. (1989). *Curriculum and evaluation standards for school mathematics*. Reston, VA: National Council of Teachers of Mathematics.

National Council of Teachers of Mathematics. (1991). *Professional standards for teaching mathematics*. Reston, VA: National Council of Teachers of Mathematics.

Noche, J. R. (2013). *Conceptual and procedural knowledge in proportional reasoning of undergraduate students* (Unpublished doctoral dissertation). University of the Philippines Open University.

Noche, J. R., & Vistro-Yu, C. P. (2015). Teaching proportional reasoning concepts and procedures using repetition with variation. In *Paper Presented at the 7th ICMI—East Asia Regional Conference on Mathematics Education*, Cebu City, Philippines.

Plucker, J. A. (1998). Beware of simple conclusions: The case for content generality of creativity. *Creativity Research Journal, 11*(2), 179–182.

Pólya, G. (1945). *How to solve it*. Princeton, NJ: Princeton University Press.

Rhodes, M. (1961). An analysis of creativity. *Phi Delta Kappan, 42*, 305–310.
Rubin, A. (1999). *Technology meets math education: Envisioning a practical future forum on the future of technology in education*. Retrieved June 11, 2017 from http://files.eric.ed.gov/fulltext/ED453055.pdf.
Runco, M. A. (Ed.). (1994). *Problem finding, problem solving, and creativity*. Westport, CT: Greenwood Publishing Group.
Runco, M. A. (2010). Divergent thinking, creativity, and ideation. In J. C. Kaufman & R. J. Sternberg (Eds.), *The cambridge handbook of creativity* (pp. 413–446). Cambridge, UK: Cambridge University Press.
Sawyer, R. K. (2011). *Structure and improvisation in creative teaching*. New York, NY: Cambridge University Press.
Sawyer, R. K., & DeZutter, S. (2009). Distributed creativity: How collective creations emerge from collaboration. *Psychology of Aesthetics, Creativity, and the Arts, 3*(2), 81.
Schoenfeld, A. H. (1985). *Mathematical problem solving*. New York, NY: Academic Press.
Schoenfeld, A. H. (2010). *How we think: A theory of goal-oriented decision making and its educational applications*. New York, NY: Routledge.
Silver, E. A. (1994). On mathematical problem posing. *For the Learning of Mathematics, 14*(1), 19–28.
Stokes, P. D. (2005). *Creativity from constraints: The psychology of breakthrough*. New York, NY: Springer.
Stokes, P. D. (2010). Using constraints to develop creativity in the classroom. In R. A. Beghetto & J. C. Kaugman (Eds.), *Nurturing creativity in the classroom* (pp. 88–112). Cambridge, UK: Cambridge University Press.
Tausczik, Y. R., Kittur, A., & Kraut, R. E. (2014, February). Collaborative problem solving: A study of mathoverflow. In *Proceedings of the 17th ACM Conference on Computer Supported Cooperative Work & Social Computing* (pp. 355–367). New York, NY: ACM.
Tourniaire, F., & Pulos, S. (1985). Proportional reasoning: A review of the literature. *Educational Studies in Mathematics, 16*(2), 181–204.
Treffinger, D. J. (1996). *Creativity, creative thinking, and critical thinking: In search of Definitions*. Sarasota, FL: Center for Creative Learning.
Treffinger, D. J., Young, G. C., Selby, E. C., & Shepardson, C. (2002). *Assessing creativity: A guide for education*. Sarasota, FL: The National Research Center on the gifted and talented.
Weisberg, R. (1986). *Creativity: Genius and other myths*. New York, NY: WH Freeman/Times Books/Henry Holt & Co.
Weisberg, R. W. (2006). *Creativity: Understanding innovation in problem solving, science, invention, and the arts*. Hoboken, NJ: Wiley.
Wertheimer, M. (1959). *Productive thinking: Enlarged edition*. New York, NY: Harper.

Part IV
New Learning Paths and Creative Teaching Approaches

Chapter 10
Nurturing Creativity in Future Mathematics Teachers Through Embracing Technology and Failure

Marina Milner-Bolotin

Abstract This chapter discusses how modern educational technologies open new opportunities for educating creative and engaging mathematics teachers. In particular, the focus is on using technology to engage mathematics teacher-candidates in exploring how technology can facilitate productive mathematical thinking. The chapter emphasizes the need for viewing mathematics learning as a creative, collaborative and constructive process that sometimes is fraught with inevitable challenges and productive failures, and at other times filled with exhilarating discoveries and new insights. The chapter suggests various ways of implementing digital technologies, such as data collection and analysis tools, electronic response systems, PeerWise, computer simulations, dynamic mathematical software, and Collaborative Learning Annotation System in mathematics teacher education courses in order to inspire teacher-candidates to embrace technology-enhanced creative mathematical thinking. In addition, the importance of technology in scaffolding teacher-candidates and consequently mathematics learners in experiencing and overcoming productive mathematics learning failures is emphasized. The challenge of the implementation of these technologies in mathematics teacher education and the opportunities they offer for embracing creative mathematical thinking are also discussed.

Keywords Educational technology · Technology-enhanced collaboration
STEM teacher education · Student engagement

M. Milner-Bolotin (✉)
Faculty of Education, University of British Columbia,
2125 Main Mall, Vancouver V6T 1Z4, Canada
e-mail: marina.milner-bolotin@ubc.ca

© Springer International Publishing AG, part of Springer Nature 2018

V. Freiman and J. L. Tassell (eds.), *Creativity and Technology in Mathematics Education*, Mathematics Education in the Digital Era 10,
https://doi.org/10.1007/978-3-319-72381-5_10

10.1 Introduction

Success is not final, failure is not fatal: it is the courage to continue that counts.

Winston Churchill

For too many people in North America *mathematics* is synonymous with *Failure* with an upper-case F. Paraphrasing Winston Churchill's famous quote, this failure to do mathematics is often fatal—it projects on our identity ("math is not for me"); it defines who we are ("I am just not good at math"); it demoralizes us ("It doesn't matter how much I try, I will never be good at math"); and most importantly, it takes away our courage to continue ("I should avoid math-heavy disciplines at all costs"). Considering the high value western societies place on innovations and technological development, the Failure to do mathematics has significant social and economic implications on students' lives and on our society as a whole (Let's Talk Science, 2013). In the 21st century, mathematics has become a roadblock to success for millions of North American students, which has also been reflected in their low performance in the international assessments of mathematics and science learning (National Center for Education Statistics, 2011; OECD, 2014) and in their lack of interest in science, technology, engineering and mathematics (STEM) careers (Chachashvili-Bolotin, Milner-Bolotin, & Lissitsa, 2016; Garforth & Stockelova, 2012; UMass Donahue Institute Research and Evaluation Group, 2011).

In the last fifty years in North America, students' Failure to understand, engage with, and appreciate mathematics has become a widely accepted phenomenon by educators, students and parents alike (Adams, 2001; Tobias, 1993). We, as a society, have become complacent with our children's Failure to master basic mathematical problem solving skills, get comfortable with fractions, percentages, or exponents, and be able to apply mathematics they learn at school to their lives. This also explains why declaring their dislike and anxiety about doing mathematics is rather common both among the North American general public and K-12 (non-mathematics or science) teachers (Adams, 2001; Bursal & Paznokas, 2006; Ma, 1999). A quick Google search for "*failure and mathematics*" produced 95,600,000 results in 0.44 s! Considering that a failure in the North American K-12 education system is often perceived as fatal and not as an opportunity to "summon the courage and continue" learning, associating Failure with mathematics has significant educational and societal implications.

The failure to engage with mathematics in a meaningful way is exacerbated by the widely acceptable and profoundly erroneous beliefs about secondary school mathematics: (a) After all, unlike language arts or fine arts disciplines, it is a "hard subject" that not everybody can master; (b) Students who are "innately good at math" need little practice and can solve problems fast without making any mistakes; (c) being "good at math" means being able to solve problems almost instantaneously without making mistakes along the way, and (d) K-12 mathematics is not a subject that requires or nurtures creativity, as the rigor of mathematics means that there is always only one right answer and one right way to get to it (Boaler, 2010; Chinn, 2012; National Numeracy, 2016). One can contest these false and very harmful "truisms" on multiple grounds, not the least of which is that there are no "easy fields" if one strives to achieve mastery (Dweck, 2016). Most famous mathematicians, scientists, musicians, and artists have overcome significant difficulties before producing the work they have become famous for. Achieving mastery in secondary school mathematics requires time, dedication and effort on behalf of the students, as much as becoming a member of the city youth orchestra or a varsity team. Excelling in any area, be it mathematics, music, medicine, sport, or business requires a significant investment of time and effort (Gladwell, 2008). Moreover, it requires perseverance in the face of failure, multiple opportunities to reflect on this failure, while continuing to practice and perfect the desired skills (Lewis, 2014). "Lower-case *failure*", which we define as an opportunity to approach the task differently, is an inherent part of learning any new subject, skill or craft, and achieving mastery in it. Not only that failure is not fatal, but it is extremely important for productive learning. Therefore, it is crucial to create learning environments where students are supported in overcoming failures by having multiple opportunities to master knowledge and skills while receiving formative feedback during the learning process. A low-case failure is also an inherent part of developing one's creativity and confidence. Being creative means being able to think of new ideas, conceive new approaches, travel off the beaten path and eventually take risks and overcome obstacles. This can only be achieved if one feels safe to fail and is given ample opportunities to try again and again. For example, in the Finland K-12 education system, students are not expected to succeed instantaneously and failure is considered to be an important and inherent part of learning. This encourages students to take academic risks and continue learning in the face of temporary and inevitable setbacks (Ripley, 2013).

In the early decades of the 21st century mathematics educators have many innovative opportunities to turn students' fatal Failure to learn mathematics into an opportunity to help them build problem-solving skills, confidence and interest in STEM subjects in general and in mathematics in particular (Ge, Ifenthaler, & Spector, 2015). We can use modern technologies to provide students with ample formative assessment and guide them on the way to achieving mastery of mathematical concepts, acquiring positive attitudes about mathematics, and developing creative mathematical thinking. However, to seize this opportunity, we have to change how we educate mathematics teachers. In particular, we call on educators to reconsider how they incorporate STEM teacher education (Milner-Bolotin, 2015). The goal of this chapter is to unpack some of the possibilities of using technology in mathematics teacher education, so that future mathematics teachers will be willing and capable of fostering creativity and risk taking in their classrooms. In order to do that we first have to take a deeper look at the theoretical underpinnings of technology-rich mathematics learning environments and the knowledge mathematics teachers need to possess in order to be able to implement them into practice.

10.2 Technological Pedagogical and Content Knowledge in the 21st Century

It is not surprising that with the rapidly growing proliferation of digital technologies, access to information, as well as with the increased availability of new visualization tools that can help learners make sense of factual and conceptual knowledge, the goals of K-12 and post-secondary education have changed dramatically (Milner-Bolotin & Nashon, 2012; USA National Research Council, 2013). This can be seen from the U.S. and Canadian national policy documents, newly developed curricular materials, and increased emphasis on core cross-curricular competencies, such as creative and critical thinking, problem solving, dealing with open-ended problems and ambiguous information, interpreting and analyzing rich data, communicating specialized technical information, and connecting the mathematics and science learned at school to everyday life (British Columbia Ministry of Education, 2013, 2015; OECD, 2014; Schmidt et al., 2011; USA National Research Council, 2013). Many of these competencies have a direct relationship to mathematics education. However, while new technological tools, such as computers, tablets, iPads®, smartphones and other personal mobile devices have a lot of potential to change *how* students learn both inside and outside of the classroom, teachers need a lot of support in learning *how* to use these technologies to enhance student productive engagement with mathematics (British Columbia Ministry of Education, 2015; Guerrero, 2010; O'Grady, Deussing, Scerbina, Fung, & Muhe, 2016; OECD, 2016a).

We have to pay more attention to *how* teachers use these tools to promote meaningful learning and how teachers acquire the necessary knowledge needed to take a full advantage of these new tools in a mathematics classroom.

In the 21st century, many countries are working on developing a new vision about *what* we want mathematics teachers to know, *how* we want them to educate our children, and *what* mathematics we would like our children to engage with (National Governors Association Center for Best Practices, & Council of Chief State School Officers, 2010). This new vision coupled with the increased emphasis on improving student mathematics performance on international assessments such as PISA influence parental and student expectations from modern schools, making the teaching of mathematics more difficult (O'Grady et al., 2016; OECD, 2016a, b). Some scholars even referred to the challenging situation faced by contemporary teachers and the lack of resources to support them, as a *crisis in the teaching profession* (Troen & Boles, 2003). The expectations of teachers to adapt their pedagogical approaches to the new reality are also reflected in the governmental documents, but it is unclear how teachers are supposed to achieve that. For example, British Columbia's newly developed K-12 curriculum document states on its opening page:

> The world is changing – and we have to change too. Technology and innovation are reshaping society – and the future. That's why it's critical we refine our education system, designed in the last century, so students can succeed in the 21st century (British Columbia Ministry of Education, 2015, p. 1).

The refinement of the education system should begin with the refinement of how we educate teachers, as it is hardly possible to find a factor that has a larger impact on student learning than teachers (Schmidt et al., 2011; Troen & Boles, 2003).

If teacher education matters so much, then what is the theoretical framework that we can utilize to analyze, evaluate, and improve teachers' *knowledge for teaching*? In this chapter we will use the Technological Pedagogical and Content Knowledge (TPCK or TPACK) framework proposed by Koehler and Mishra as an extension of the original Pedagogical Content Knowledge (PCK) framework suggested by Shulman two decades earlier (Koehler & Mishra, 2009, 2015; Shulman, 1986). While introducing the TPACK framework, Koehler and Mishra write:

> TPACK is an emergent form of knowledge that goes beyond all three "core" components (content, pedagogy, and technology). Technological pedagogical and content knowledge is an understanding that emerges from interaction among content, pedagogy, and technology knowledge. Underlying truly meaningful and deeply skilled teaching with technology, TPACK is different from knowledge of all three concepts individually. Instead, TPACK is the basis of effective teaching with technology, requiring an understanding of the representation of concepts using technologies; pedagogical techniques that use technologies in constructive ways to teach content; knowledge of what makes concepts difficult or easy to learn and how technology can help redress some of the problems that students face; knowledge of students' prior knowledge and theories of epistemology; and knowledge of how technologies can be used to build on existing knowledge to develop new epistemologies or strengthen old ones (Koehler & Mishra, 2009, p. 66).

Teachers who acquire necessary TPACK and who hold the positive attitudes about nurturing student engagement and independent thinking will be more likely to implement learning environments where mathematical creativity blossoms and failure is viewed as a stepping stone on the road to conceptual understanding (Koehler & Mishra, 2005; Milner-Bolotin, 2016b). However, TPACK is a very sophisticated and complex form of knowledge, unfamiliar to many practicing and especially new teachers. Acquiring it will require mathematics teacher-candidates to take significant personal and pedagogical risks and inevitably experience failures. At the same time, this process is an opportunity to experience pedagogical success and an exhilarating feeling of figuring things out (Feynman, 1999). And what a better place to learn about new pedagogical approaches, and experience them both as learners and as future teachers, than a mathematics methods course in a teacher education program (Milner-Bolotin, 2015; Milner-Bolotin, Fisher, & MacDonald, 2013). Contemporary mathematics teacher-candidates should be offered ample opportunities to develop their own mathematical pedagogical creativity through *deliberate engagement* with subject-specific technologies in their teacher education programs (Milner-Bolotin, 2016a). In this context, by mathematical pedagogical creativity we mean teacher's ability to expand their pedagogical repertoire through identifying student learning difficulties and devising (novel) pedagogical approaches that support students in addressing these challenges. Some of these pedagogies might employ new digital tools but technology should be used deliberately to enhance pedagogy and not for the sake of using it. In order to assure that mathematics teacher-candidates learn how to use technology deliberately in order to support mathematics creativity and meaningful learning, we suggest that mathematics methods courses in teacher education programs provide them with the following opportunities:

(a) To experience mathematics education that promotes future teachers' conceptual understanding, authentic learning, and mathematical creativity. This can be achieved by proposing and evaluating various problem-solving approaches to the same problem, evaluating pros and cons of using multiple representations in problem solving, or designing new problems that illustrate specific mathematical concepts.
(b) To utilize new technologies as invitations for new ways of mathematical thinking as opposed to using new educational technologies to support old pedagogies.
(c) To feel supported and safe in taking pedagogical risks in designing their lessons, in experimenting with new technologies, in reflecting on their own learning and teaching, as well as on the teaching of their peers, and trying again and again.
(d) To learn from failures, to accept feedback as an opportunity for improvement, and to learn that creative educational solutions rarely come from big technological breakthroughs, but from a new vision about what mathematics education is all about. In order for mathematics teachers to design and implement novel and creative pedagogical approaches to mathematics learning, the teachers have to believe that there are many ways to learn and experience mathematics. The teachers have to possess a deep understanding of the mathematical concepts, students' potential difficulties in understanding them, and available tools to scaffold student learning.

Fortunately, we have a number of modern technologies that can help teacher educators to create learning environments that open these opportunities for mathematics teacher-candidates. While there are a vast number of educational technologies that could be relevant to mathematics education, in this chapter we decided to limit ourselves to four big clusters of these technologies: data acquisition and analysis systems; Classroom Response Systems and collaborative online systems for creating multiple-choice questions; computer simulations and mathematical modeling software; and online systems for collaborative analysis of videos (Table 10.1). These tools were chosen for five interrelated reasons: (a) They have been widely used in post-secondary classrooms, thus we have a lot of knowledge about how to implement them; (b) There is ample research evidence about their pedagogical effectiveness in promoting student engagement and conceptual understanding; (c) With the proliferation of technology, the access to these tools has increased significantly in K-12 classrooms; (d) These tools open unprecedented opportunities for supporting student-driven investigations; and lastly (e) They "reduce the cost of failure", thus opening doors to creative experimentation: allowing students to experiment, make mistakes, reflect on these lower-case failures, make changes and do it again. Yet, it is always important to remember that technologies are only tools and should be used to support pedagogies that promote meaningful engagement with mathematics. Paraphrasing a famous quote by one of the pioneers in the use of educational technologies, Alan Kay, we can say that a technological tail should not be wagging a pedagogical dog (Kay, 1987). We will discuss four examples of deliberate use of educational technologies in order to increase learners' engagement with mathematics in the following section.

10.3 Fostering Technology-Enhanced Creativity in Mathematics Education

This section explores how four different technology-enhanced pedagogies (Table 10.1) can be implemented in mathematics methods courses (or in-service professional development courses for mathematics educators) in order to support teacher-candidates (or practicing mathematics teachers) in acquiring meaningful TPACK, taking pedagogical risks, learning from failures and eventually developing creative pedagogical mathematical thinking (Campbell et al., 2014). While this section focuses on specific examples of technologies, it is important to emphasize that what matters is not the tools themselves (which are likely to change in the future), but their affordances and the way they are used in the context of teacher education, teaching and learning.

Table 10.1 Examples of four clusters of educational technologies used for engaging mathematics teacher-candidates in creative mathematical thinking

	Technology cluster	Selected technology examples
1	Data acquisition and analysis systems	Logger Pro, Pasco, smartphone apps
2	Classroom response systems and online systems for collaborative creation of multiple-choice questions	Clickers (hand-held or virtual), PeerWise collaborative system
3	Computer simulations and mathematical modeling software	PhET computer simulations, GeoGebra, Cabri geometry, etc
4	Online systems for collaborative analysis of videos	The collaborative learning annotation system (CLAS)

10.3.1 Pushing Mathematical Boundaries Through Authentic Data Acquisition and Analysis

In order to inspire future teachers to promote creative technology-enhanced mathematics education in their own classrooms, they have to experience the power of authentic technology-enhanced learning environments as students (Lewis, 2014; Schoenfeld, 2014). Thus one of the goals of the methods courses is to create these opportunities while encouraging teacher-candidates to reflect on their experiences as learners and as future teachers (Harris & Hofer, 2011; Milner-Bolotin, 2016a, b; Milner-Bolotin et al., 2013). Technology, such as the Logger Pro data acquisition system, can be a great asset for supporting creativity, reducing the cost of risk-taking and allowing learners to experiment with abstract mathematical ideas through connecting them to everyday life (Eijck & Roth, 2009; Milner-Bolotin, 2012; Vernier-Technology, 2016).

For example, it is widely known that students experience significant difficulties in creating and interpreting graphs representing mathematical relationships, as well as building bridges between algebraic, graphical and physical representations (Eshach, 2014; Lingefjärd & Ghosh, 2016). Learners also find it challenging distinguishing temporal [i.e., $x(t)$] versus spatial [i.e., $y(x)$] representations of motion and connecting abstract graphs to physical scenarios (McDemott, Rosenquist, & Zee, 1987). This prompted researchers to suggest the use of graphing calculators (Kastberg & Leatham, 2005; Ruthven, Deaney, & Hennessy, 2009). However, for many students graphing calculators still remain "black boxes" that connect two abstract representations (graphical and algebraic) without necessarily relating these concepts to their lives.

An alternative solution to helping students connect real-life phenomena with their mathematical representations is using data collection and analysis tools (sensors). For instance, a Logger Pro motion detector (a *sonic ranger*) can generate a real-time position-time, $x(t)$, and velocity-time, $v(t)$, graphs of a 1-D motion, such as a student moving along a straight line in front of the detector (Vernier-Technology, 2016) (Fig. 10.1). This experience can prompt an all-class discussion of the physical meaning of abstract concepts, such as a slope of a graph, a y-intercept, an area under the graph. Moreover, as collecting data with a motion detector is fast and easy, students can repeat and adjust the experiment multiple times. If they "fail" to produce the desired graph they can do it again till they succeed. The lower-case failure in this case becomes a stepping stone to a meaningful conceptual understanding. The data represented by the graph provides a continuous formative assessment, such as the students can adjust their own thinking (Schuster, Undreiu, Adams, Brookes, & Milner-Bolotin, 2009). This process opens doors to new questions thus inviting creative thinking. For example: What is the relationship between $x(t)$ and $v(t)$ graphs? What is the physical meaning of different features of these graphs (i.e., slopes, areas under the graphs)? How will relevant graphs of an accelerating object look like? What is the meaning of the term *acceleration*? How will the graphs transform if we reverse the direction of the x-axis?

Fig. 10.1 An author helps a novice mathematics teacher to experience data collection technology in action. The Logger Pro motion detector is located on the instructor's desk and the display shows $x(t)$ and $v(t)$ graphs of teacher's motion. She first moves in front of the motion detector and observes the generated graphs, then she moves back and forth and observes how motion is reflected in the graphs. She tries to generate graphs of a given shape. During her motion, the rest of the group observes her and comments on her motion and corresponding $x(t)$ and $v(t)$ graphs

Answering these questions without having a motion detector would be much more difficult. Moreover, having this technology allows teachers to differentiate instruction, thus challenging students of different levels of abilities during the same in-class activity.

Using data collection tools in methods courses invites teacher-candidates to think creatively about student engagement. It also encourages them to design creative mathematical tasks for the school practicum. We have observed that more than half of the teacher-candidates who were exposed to these tools in their methods courses designed lessons during their practicum that implemented similar tools. These authentic technology-supported activities are relatively easy to implement, yet they can be much more cognitively engaging than traditional paper-based activities, because technology provides unique opportunities for student experimentation—trying different ideas, getting continuous feedback, modifying their solutions and trying again and again. These technologies also encourage student mathematical collaboration and communication, thus building their confidence in being able to do mathematics and to communicate their mathematical understanding to others.

An example of such an activity can be a graph-matching exercise where students produce motion that matches a given *x(t)* graph (the top graph in Fig. 10.2). In this activity a student is asked to walk in front of the motion detector in order to reproduce a pre-existing top graph with a bottom graph produced by her. After the students figure it out, they can be asked to match *v(t)* graphs or to produce graphs to be matched by their peers.

Data collection technology also allows data analysis that can be conducted either in or out of class. After the real-life data has been collected, the students can generate and analyze relevant graphs using the software tools, such as graph fitting, slopes, areas under the graph, average values, etc. Thus, the activity can also focus on an algebraic description of motion, connecting physical, graphical and algebraic representations, or constructing the conceptual understanding of calculus, such as the meaning of derivatives or integrals.

Lastly, we would like to point out that in addition to Logger Pro and similar dedicated data collection tools, many modern smartphone or tablet apps allow rather sophisticated live data collection and analysis (Maciel, 2015). Creative teachers can use these 21st century tools to enable students to ask and answer their own authentic mathematical questions using the devices located literally at their fingertips. However, this will not happen unless teachers have acquired TPACK necessary for facilitating meaningful student engagement and until they feel comfortable with the affordances of these technologies (Carr, 2012). Without it, these expensive tools will remain underused gadgets and missed opportunities, while the goal of fostering student engagement and mathematical creativity will remain an

Fig. 10.2 A novice mathematics teacher looks at the *x(t)* graph she produced (the bottom graph) in order to match a given graph (the top graph). This was her third attempt as she was not satisfied with her previous attempts that according to her were unsuccessful. So she, the rest of the group, and the author were very happy about the final result

unattainable dream (Cuban, 2001). Unfortunately, in many cases, educational administrators see technology and not the teachers (and their TPACK) as the driving force behind innovation. Thus, teacher professional development and support in using technology deliberately to promote meaningful learning are still often lacking (Burridge & Carpenter, 2013; Luft & Hewson, 2014; OECD, 2016c). Therefore, it is not surprising that Cuban's almost two-decade old lament that in many schools computers and new technologies "are oversold and underused" is still relevant (2001).

10.3.2 Developing Mathematics Argumentation Through Technology-Supported Questioning

Question-driven pedagogy has been used by educators for centuries. Yet, not all questioning strategies are equally effective (Shahrill, 2013). One of the most common and pedagogically proven question-driven pedagogies in postsecondary STEM classrooms is Peer Instruction (PI) developed by Eric Mazur at Harvard University in the 1980s (Lasry, Mazur, & Watkins, 2008; Mazur, 1997b; Vickrey, Rosploch, Rahmanian, Pilarz, & Stains, 2015). It utilizes Classroom Response Systems (clickers, mobile devices, or even flashcards) to engage students in discussions through responding to conceptual multiple-choice questions that target common student difficulties, often referred to as misconceptions (Milner-Bolotin, 2015; Milner-Bolotin et al., 2013). The key element of Peer Instruction is student small group discussions of alternative answers to multiple-choice questions. Since effective Peer Instruction questions deliberately use common student misconceptions as distractors (Fig. 10.3), the students are asked to articulate not only the reasons behind the presumably correct answer they voted for, but also the reasons for why the alternatives given in the question are wrong.

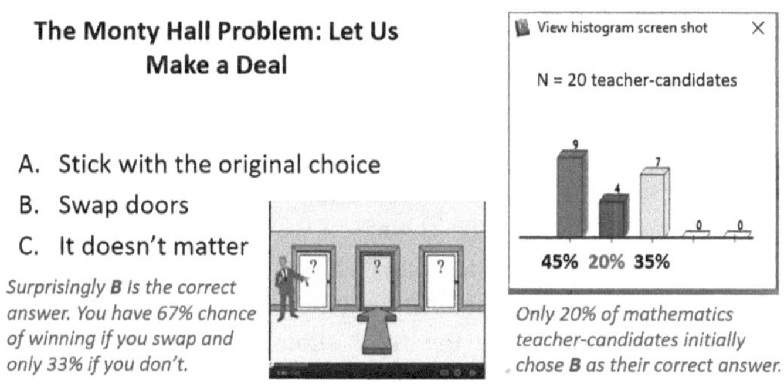

Fig. 10.3 An example of a conceptual multiple-choice question called the Monty Hall problem and the distribution of mathematics teacher-candidates' responses. The correct answer B was chosen by 4 (20%) out of 20 teacher-candidates

Initially implemented in large undergraduate STEM college courses, Peer Instruction has been found to be effective in actively engaging students when either implemented with clickers (Hake, 1998; Milner-Bolotin, Antimirova, & Petrov, 2010) or with flashcards (Lasry, 2008). With the advent of new cost-effective models for its implementations (such as smartphones, tablets or iPads®), it is becoming more popular in secondary schools. There is extensive research evidence that the success of Peer Instruction or any other clicker-enhanced pedagogy is not in the technology itself, but in the pedagogical skills of the teachers and in the quality of the multiple-choice questions (Milner-Bolotin et al., 2013; Vickrey et al., 2015). These findings highlight the importance of developing teacher-candidates' TPACK, so they will be ready to utilize this technology in their own classrooms (Milner-Bolotin, 2014).

Peer Instruction is especially valuable for helping teacher-candidates develop conceptual understanding crucial for creative mathematical thinking (Milner-Bolotin, 2016b). There are at least five reasons for that. First, mathematics teacher-candidates, despite earning relevant degrees, often possess insufficient content knowledge relevant for teaching secondary mathematics. This might come as a surprise to them, as they have successfully passed upper level courses to complete their B.Sc. degrees. For example, probability and statistics are included in most secondary mathematics curricula. A famous counterintuitive reasoning problem that requires understanding of conditional probability included in most secondary mathematics curricula, called the Monty Hall Problem, does not involve advanced mathematical knowledge, but requires critical thinking and logical reasoning. It is modeled after a famous game show that involves three doors behind which two goats and a car are hidden. A game participant picks a door (hoping to win a car). Then the show host (Monty Hall, who knows behind which door the car is hidden) opens one of the other two doors that has a goat behind it. The challenge for the game participant (who wants to win a car) is to decide if she should stick with the original choice or to switch to the third unopened door. The rules of the game are very simple, yet it challenges our intuition, as surprisingly switching will help the participant to win 2/3 of the time, versus sticking with the original choice that will bring success only 1/3 of the time.

The reason for this apparent paradox is that as the host opens the door that has a goat behind it, he introduces new information, thus breaking the symmetry between the two doors that are left. There are many creative ways to think about this problem and how you can convince a peer in the validity of your solution. For example, one might think of having 100 doors instead of three. Then if a participant were to choose a door randomly, the chance to win a car would have been 1/100. However, if the host were to open the other 98 doors that have goats behind them, the chance that the car is hidden behind the other door, would have been 99/100. Thus switching would have been the right strategy in 99 cases out of a 100. Understanding why switching is a winning strategy and being able to convince others that they should switch requires a conceptual understanding of conditional probability, which many mathematics teacher-candidates are lacking (Rosenhouse, 2009).

Second, Peer Instruction opens opportunities for teacher-candidates to make an individual choice safely without being embarrassed for making a mistake (voting with clickers is anonymous, so nobody else knows what choices others have made). In other words, Peer Instruction turns capital-case Failure into a lower-case failure. This is very important for teacher-candidates who might be afraid to expose the lack of basic mathematical knowledge in front of their peers and the instructor. As novice teachers, they do not yet realize that it doesn't matter how much formal post-secondary education one has acquired, there is always much left to learn at the secondary mathematics level and making mistakes is expected in this learning process. Moreover, after teacher-candidates make their own choice and then can see the aggregate of choices made by their peers, they realize that their peers encounter similar difficulties.

Third, having an opportunity to discuss the problem with their peers helps teacher-candidates to learn how to listen and how to communicate their ideas to others. It also opens doors to creativity through discussing multiple solutions to mathematical problems. Listening, being open to multiple ways of thinking about the problem, and being able to communicate your ideas in multiple ways are crucial qualities for mathematics teachers. Therefore, Peer Instruction can become a process of turning a Failure with an upper-case F into a lower-case failure that is integral to the process of learning.

Fourth, using Peer Instruction helps instill the importance of high quality conceptual questions in mathematics learning as opposed to factual (memory-recall) questions. Unlike the latter, conceptual questions address higher levels of Bloom's taxonomy (Bloom, 1956). For example, conceptual multiple-choice questions that use students' common misconceptions as distractors, such as the Monty Hall problem mentioned earlier. This is what makes these questions pedagogically valuable and mathematically challenging.

Fifth, using Peer Instruction in methods courses opens doors for the development of mathematics knowledge *for* teaching or TPACK. The choice of questions, the choice of distractors and the pedagogies teachers employ to address students' difficulties are very important for helping to foster teachers' creativity. For example, in the case of Monty Hall problem, the use of Peer Instructor can reveal the issue—student disagreements about the best strategy for winning the game. After this disagreement is pointed out, a teacher can use a computer simulation, such as the one found here (http://www.grand-illusions.com/simulator/montysim.htm) or some other strategy to help students figure it out (Franco-Watkins, Derks, & Dougherty, 2010). Another option is asking students to repeat the experiment themselves, collect data, and then analyze it using a tool such as a spreadsheet. After all, Peer Instruction can help students to focus on asking interesting questions and the teacher can guide students in using available technologies to figure out the answers (Hake, 2012; Paul & Elder, 2007). The key for success here is using instructional design principles that support the development of student creativity and critical thinking, such as questions that ask students to organize concepts, compare, categorize, contrast, and apply concepts outside of the original context they were initially introduced (Chin, 2007; Dickinson, 2011; Kuo & Wieman, 2016).

Another technology that can supplement the use of Peer Instruction in teacher education and help teacher-candidates develop questioning skills is called PeerWise (Denny, 2016). It is a free online collaborative tool that allows students (teacher-candidates in our case) to design and share their own multiple-choice questions, comment and respond to the questions of their peers, and eventually create a shared database of conceptual questions, explanations, comments and solutions (Milner-Bolotin, 2014). PeerWise has been used extensively in undergraduate STEM education in order to support students in asking novel and meaningful questions and providing feedback on questions asked by their peers (Bates & Galloway, 2013; Denny, Luxton-Reilly, & Simon, 2009; McQueen, Shields, Finnegan, Higham, & Simmen, 2014). Thus, PeerWise can be especially beneficial for teacher education, since it equips future teachers with the skills needed to come up with their own questions rather than always relying on the questions asked by others.

10.3.3 Developing Creative Mathematical Thinking Through Computer-Supported Modeling

Data collection and analysis are powerful tools for unleashing student creativity and helping them to connect abstract mathematical ideas to their lives (Erickson & Cooley, 2006). However, not every mathematical concept can be easily demonstrated through a hands-on activity that students can engage in. In addition, many mathematical concepts are rather complex and include multiple abstract features that might manifest themselves very differently in a variety of cases. As a result, novice learners often fail to see the "prototypical" examples, the underlying relationships behind the plethora of abstract mathematical objects. For example, while to an expert, all quadratic functions can be represented by a common generalized expression: $y(x) = ax^2 + bx + c(provided\ a \neq 0)$, a novice might fail to internalize the roles played by different coefficients, or the graphical representation of this relationship (the parabola) depending on the values of these coefficients. These features might be very obvious to an expert, but might remain unnoticed by novice learners unless they have multiple opportunities to manipulate these features, ask and answer their own "what-if questions" and figure out for themselves what each one of these features represents (Bransford, Brown, & Cocking, 2002).

Therefore, it is not surprising that a concept of a mathematical function in general proves to be an obstacle to many students (Clement, 2001; Mesa, 2008; Watson & Harel, 2013). Educational research suggests that in order to help students build a more accurate and "functional" understanding of the function concept (pun intended), teachers should offer students different "prototypes" as well as multiple opportunities to manipulate functions and represent them graphically, algebraically and even verbally (Gagatsis & Shiakalli, 2004; Mesa, 2008). Highlighting examples of functions that do not fit the types of functions most often seen in textbooks and helping students notice the features of the functions that might be critical or atypical, is another pedagogically effective strategy.

Modern technologies, such as computer simulations, offer learners and opportunity to meaningfully engage with mathematics. For example, a suite of freely available online computer simulations designed by the Physics Education Technology (PhET) research group at the University of Colorado (https://phet.colorado.edu/) (Wieman, Adams, Loeblein, & Perkins, 2010), offers a wide range of activities built on solid educational research evidence. These simulations help students not only to understand mathematical concepts, such as functions, the quadratic equation coefficients, and the roots of a quadratic equation shown in Fig. 10.4, but also conduct independent investigations in these virtual environments. Without having graphing technology such as the PhET Equation Grapher simulation, conducting such an investigation would have been if not impossible then very time consuming. Thus, this tool offers mathematics educators an unprecedented creativity carte blanche to invite students to ask "*what if*" questions. However, as in the previous examples, mathematics teachers should possess the necessary TPACK in order to take full advantage of this powerful tool.

In addition, computer simulations offer students an opportunity to develop mathematical intuition that is still uncommon in traditional mathematics learning. Yet, this intuition, based on experience of "mindful playing with mathematical objects", is crucial for achieving creative thinking. Simulations like PhET put a very low "price" on failure, emphasizing the important of the process of figuring things out. Therefore, the learners are provided with ample learning opportunities (such as activities of graduated difficulty levels), instantaneous feedback (the results of manipulating various parameters become known instantaneously), as well as deliberately chosen number of variables the students can manipulate in the

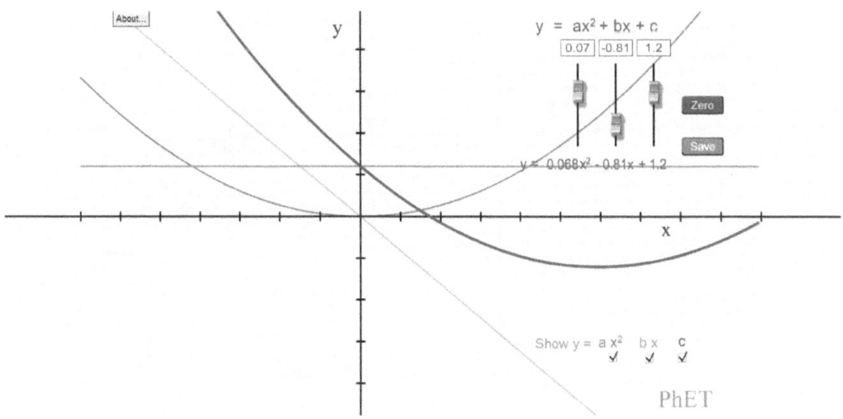

Fig. 10.4 PhET Computer simulation "Equation Grapher" allows students to plot various quadratic functions and observe how changing coefficients *a*, *b*, and *c* can affect the graphical representation of the function (https://phet.colorado.edu/en/simulation/equation-grapher). They can also save their work, thus allowing long-term investigations and progress tracking

simulation (Perkins et al., 2006; Wieman et al., 2010). Interacting with such a simulation shows students that mathematical knowledge can be acquired and not knowing the correct answer right away and making wrong turns (lower-case failures) is natural in the learning process.

Mathematics teachers should develop this intuition first and have confidence in their own ability to think creatively. At the beginning of the chapter, we mentioned that one of the common misconceptions about mathematics is that it is all about finding "the right answer right away". Yet, in reality, mathematics is about posing big problems and finding ways of thinking about them which often take a very long time (Sfard, 2012). For example, many mathematics problems deal with multiple dimensions, transformations, and the concept of scaling (Milner-Bolotin, 2009). The concept of scaling has very important applications for many STEM fields, yet it is rarely acquired by students. From our experience of working with mathematics teacher-candidates, few of them feel comfortable with this concept as well. However, a computer simulation can offer students an opportunity to experiment with scaling through a game, where students are provided with instantaneous feedback and where making a mistake, such as predicting a wrong outcome, is just a step in the game (Fig. 10.5). In this PhET computer simulation, the students as asked to estimate how many small 1-D, 2-D or 3-D objects can fit into a large object. For example, in Fig. 10.5, a player has to estimate how many small cubes can fit into a large one. In order to do that, they have to understand the concept of scaling: the volume of an object changes as the cube of the scaling factor (Barnes,

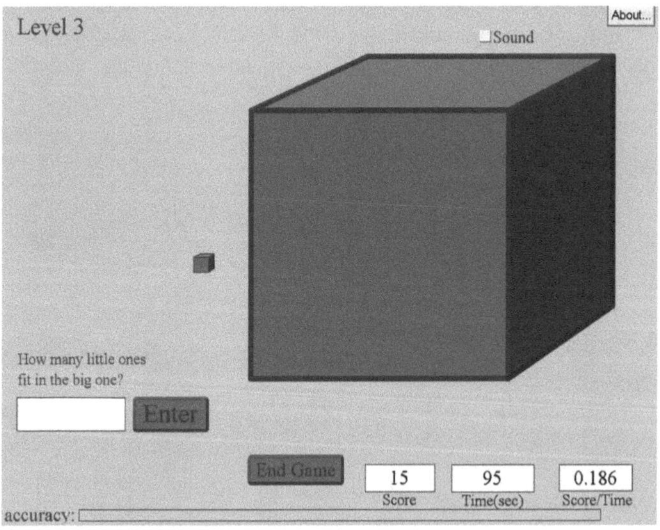

Fig. 10.5 PhET Computer simulation "Estimation" invites students to estimate how many little cubes can fit in the big one. Though playing the game, the students get instantaneous feedback while building an intuitive understanding of scaling of 3-D shapes. The game progresses from 1-D to 3-D shapes and allows a player to repeat it as many times as they wish. If a mistake is made, a student is provided with relevant feedback and multiple opportunities to try again

1989; Milner-Bolotin, 2009). This virtual activity opens doors to asking many creative real-life questions, including the relationships between surface areas and volumes of biological systems, such as plants and animals. These are very important questions as in most cases surface area of a living organism is responsible for its heat exchange with the environment, while its volume is in charge of the energy generation. In order to ask and answer these questions, students have to acquire mathematical knowledge and to apply it to real life.

Since PhET simulations are free and are designed to operate on multiple platforms, they can be used by students both in school and at home. Moreover, PhET simulations are also a hub for a STEM teaching community of K-12 and post-secondary teachers and educational researchers who share ideas about how these resources can be implemented into practice and how students learn with simulations. This makes PhET simulations especially valuable for developing creative mathematical thinking by the new teachers.

For example, the concept of scaling mentioned above penetrates all areas of science, engineering and mathematics (Goth, 2009; Milner-Bolotin, 2009). The idea that when all dimensions of an object increase by the same factor its surface area and volume change as a square and as a cube of this scaling factor respectively has long-ranging implications for many biological systems and creative human endeavors, such as architecture and design (Salvadori, 1980). However, students often have few opportunities to experience this phenomenon first hand and appreciate its breathtaking power. PhET computer simulation "Estimation" provides students with this opportunity by asking them to estimate what happens to the volume of the object when its dimensions increase in the same proportion (Fig. 10.5).

While PhET simulations open many opportunities for developing creative and critical mathematical and scientific thinking, simulations are purposefully rather structured and closed tools (Wieman et al., 2010). Their designers have decided in advance what parameters the user will be able to manipulate. In other words, the decision about the model embedded in the simulation has been made by the designers and the learner has no input into it. Computer simulations, where those decisions have been made based on education research (such as PhET) might be powerful for beginners (Finkelstein et al., 2005). However, when the learners have acquired more advanced subject knowledge and they are ready to ask more sophisticated questions and test their own models, the simulations might prove to be somewhat limited.

A higher level of mathematical creativity is reached when the students are not only invited to engage with the simulations designed by others, but when they have an opportunity to create and test their own mathematical models (Hohenwarter, Hohenwarter, & Lavicza, 2008; Martinovic, Karadag, & McDougall, 2014; Martinovic & Manizade, 2014). An example of modeling technology that has the potential to support an even higher level of students' mathematical creativity and empower them to experience mathematics in a very personal way is dynamic modeling software, such as GeoGebra (Hohenwarter, 2014). Unlike traditional paper and pencil geometrical or algebraic constructions, where a construction or a graphical representation cannot be changed or manipulated easily, or unlike computer simulations where mathematical models have been chosen by the designers, GeoGebra allows students to develop a mathematical language, mathematical models, dynamically test their understanding, visualize abstract mathematical relationships and eventually foster their mathematical intuition (Martinovic et al., 2014). In addition, GeoGebra's dynamic features invite students to manipulate geometrical and algebraic objects, visualize abstract mathematical concepts and search for mathematical patterns and relationships behind artistic artifacts, everyday life or natural phenomena. In summary, GeoGebra allows students to experience mathematical construction that was not possible before. For example, students can use GeoGebra to explore regular and semi-regular tessellations, mosaics and geometrical patterns, and their use in art and architecture (many of these activities can be found on GeoGebraTube—www.geogebratube.org) (Fig. 10.6).

In the hands of knowledgeable mathematics educators, GeoGebra offers ample opportunities for mathematical creativity, but to unleash it, this dynamic mathematics software should be explored in mathematics teacher education. GeoGebra is freely available to teachers and students, and its educational community offers extensive pedagogical resources (Fenyvesi, Budinski, & Lavicza, 2014; Hohenwarter et al., 2008). However, to support teacher-candidates in appreciating the affordance of GeoGebra for mathematics learning, it is advisable that they

Fig. 10.6 An example of a construction of a tessellation created in GeoGebra software (http://www.geogebra.org/material/simple/id/2101729)

explore GeoGebra under the guidance of more experienced educators, who can model creative ways of using this software. For instance, when creating tessellations with GeoGebra, such as a Pythagorean tiling (Fig. 10.6), a knowledgeable teacher can help students to experience and visualize the Pythagorean original theorem describing the arrangement of quadrilateral polygons, as well as appreciate its implications to design and architecture (Burger & Starbird, 2000). Then the students can create their own mathematically inspired tessellations using GeoGebra.

As mentioned earlier, technology provides new opportunities to explicitly connect mathematics and the arts. For example, University of Alberta mathematics professor, Gerda de Vries, creates mathematically inspired quilts (Black, 2017). She "sews mathematics" into her quilts using the concepts such as tessellations, symmetry and transformations thus uncovering the mathematics behind the arts. GeoGebra or other modeling software can become a useful tool in this process.

In the previous sections, we discussed various educational technologies that have potential to nurture creativity in mathematics teaching and learning. Throughout the chapter, we kept emphasizing that the key to creative technology-enhanced mathematics education is in the hands of mathematics teachers who have acquired necessary TPACK in order to take full advantage of these technologies to promote meaningful mathematics learning. Raising a generation of mathematics teachers open to creative use of technology is a long and laborious process that requires teachers to continuously reflect on their teaching practices. Once again technology can play an important role here. In the following section, we discuss a special kind of technology that can be used to support mathematics teacher-candidates in acquiring TPACK and developing their own pedagogical creativity and confidence. This technology is called the Collaborative Learning Annotation System or CLAS (Dang, 2016) and it will be described in details below.

10.3.4 Becoming Creative Mathematics Teachers Through Collaborative Reflection on Mini-lessons

We began this chapter by discussing the importance of turning an upper-case Failure in mathematics learning into a learning opportunity, or a lower-case failure. The same applies to the process of becoming a mathematics teacher. It is impossible to become a mathematics teacher without experiencing temporary setbacks—unsuccessful lessons, failed activities, disengaged students, or negative feedback from parents. Those setbacks, however, should not deter teachers from reflecting on their practice and trying again. Mathematics methods courses are promising opportunities for novice mathematics teachers to practice their teaching, to try new activities, to experiment with new technology-enhanced pedagogies, and most importantly, to learn how to learn from mistakes and temporary failures. In this section, we will discuss a technology that we have found to be extremely useful in this process.

The Collaborative Learning Annotation System (CLAS, https://clas.sites.olt.ubc.ca/) has been created at the University of British Columbia and is freely available to students and teachers (Fig. 10.7). It is a media player used to record, share, and comment on videos uploaded onto it by the students or by the teachers (Dang, 2016). The videos are stored on a secure server at the University of British Columbia and are only visible to students enrolled in the course. Every member of the course, including the instructor, can post timely annotations on the videos that can be visible only to the person who uploaded the video or to the entire class.

We have used CLAS widely with STEM teacher-candidates during their methods courses. In the past, during these courses teacher-candidates were required to do microteaching (teach 12–15 minute lessons) to their peers. This was a traditional practice, where they had to teach a mini-lesson and then reflect on it. Very often these mini-lessons were recorded such as teacher-candidates could reflect on them at home. Yet educational research indicates that teachers learn not only from their own lessons, but also from observing and reflecting on the lessons taught by their peers (Cole & Knowles, 2000; Kemmis, 2011; Ma, 1999; Stigler & Hiebert, 1999).

Giving feedback on peers' lessons is one of the most powerful ways of improving one's own teaching (Stigler & Hiebert, 1999). This was the main reason why we introduced the peer feedback assignment using CLAS into the methods courses: we wanted teacher-candidates not only to experience teaching mini-lessons, but also to engage in the peer feedback. This process teaches teacher-candidates to observe carefully, to provide constructive feedback and to

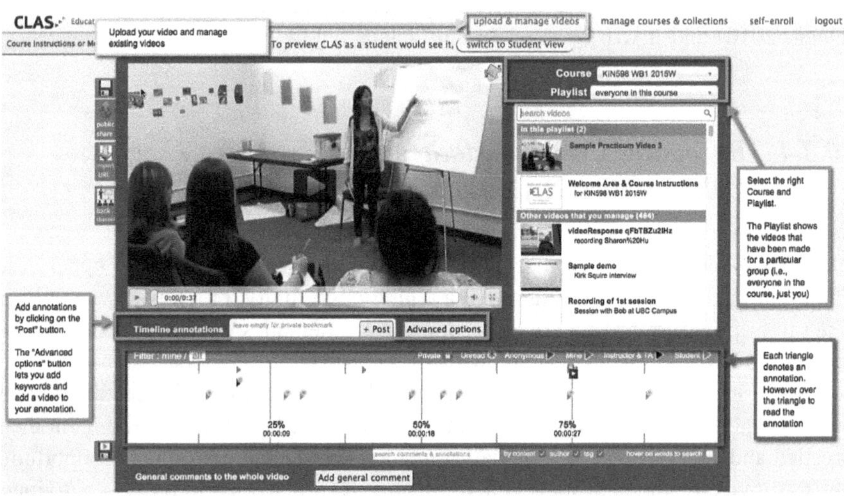

Fig. 10.7 A screenshot of CLAS with the description of its functionality. The playlist includes all the videos available to the students in the course. Timeline annotations below the video show annotations made by different students and by the instructor. By clicking on the annotation one can see its content, respond to it in writing or post a video response. All the communication can be made public or private. For more information see: http://ets.educ.ubc.ca/clas/

accept peer feedback in a positive way. It also helps them notice what they want to improve and what they want to work on. For example, as a result of this reflection, teacher-candidates might decide to re-teach a lesson. This was a very common experience in our methods courses. After watching their own mini-lessons and seeing the feedback from the instructor and from their peers, several teacher-candidates decided to re-teach their lessons. Since both mini-lessons were recorded and uploaded on CLAS, teacher-candidates could see the difference between their original and improved lessons. It was an empowering learning experience that gave the teacher-candidates the much-needed confidence that they can figure things out and if a lesson doesn't go as well as they wanted the first time around, they have a second chance.

Learning to remove yourself from your own lesson and to reflect on it in order to teach it better in the future is extremely important. As mathematics educators, we will always be looking for new teaching ideas, educational technologies, and ways to engage students. Not all of these ideas will work for us and it will take time to tweak them to fit our learning environments. Thus, accepting our own teaching practice as a work in progress is a very important attribute of a creative teacher. Using CLAS to reflect on their own lessons will allow novice teachers to view their own teaching failures as opportunities for learning. Lastly, having an opportunity to observe recorded lessons allows us to slow down and to see where potential student difficulties might come from. While we recorded mini-lessons where the students were also teacher-candidates, nobody (as long as we satisfy ethical requirements) precludes us from recording and analyzing real lessons. For example, teacher-candidates might want to record their own lessons during the practicum and then analyze them at home. Coming back to an earlier discussion of creativity as the freedom to explore, challenge other people's ideas, be challenged by others, learn, fail, re-learn, and move on (Lewis, 2014), we believe that CLAS can be used to support the development of the creative mathematics teaching. While we only used CLAS in a pilot study, we believe that it can become a very useful tool for helping mathematics teacher-candidates develop creative pedagogies in a safe and supportive learning environment.

The examples discussed above illustrate how new digital tools, such as sensors and software for data collection and analysis (e.g., Logger Pro), collaborative technology for designing and responding to multiple-choice questions (e.g., PeerWise and clickers), computer simulations and modeling software (e.g., PhET, GeoGebra), as well as collaborative tools for reflection (e.g., CLAS) can inspire deliberate pedagogical thinking with technology by future mathematics teachers. However, exposing teacher-candidates to the opportunities these tools offer is a necessary but insufficient condition for changing how mathematics will be taught in our classrooms. Implementing novel pedagogies is a venture inevitably fraught with perils. Teachers need to be supported and have to have tolerance for lower-case failure, such as it doesn't turn into Failure of using new technologies to promote creativity in mathematics education. Finding ways to support teachers on the road to designing these learning environments is one of the key challenges in modern mathematics teacher education.

10.4 Conclusions

There are many approaches to define the notion of *mathematical creativity* and ample studies exploring its different facets and definitions (Aralas, 2008). As discussed by Aralas, while there might be several disagreements about the most encompassing definition of mathematical creativity, there is an agreement about its core elements, such as novelty, originality and relevance of ideas and approaches for solving mathematical problems, effectiveness and usefulness of thinking, independence and originality while appropriate use of available knowledge and resources. What is often neglected in defining mathematical creativity is learner's ability to take necessary risks and steer off the beaten path, while building on the prior knowledge. Akin to Newton, who in his famous letter to his rival, Robert Hook, wrote that he could see further than his contemporaries because he was standing on the shoulders of giants, mathematics students cannot be expected to produce significant creative insights if they do not have multiple opportunities to acquire and challenge the knowledge produced by others. Creativity is built on prior knowledge and on the freedom to explore, challenge other people's ideas, be challenged by others, learn, fail, re-learn, and move on (Lewis, 2014). In this chapter, we emphasized the aspects of mathematical creativity that we believe can benefit directly from the use of educational technologies in mathematics learning, such as the ability to connect multiple representations of abstract concepts, the ability to ask novel questions, the ability to communicate and critique mathematical ideas, the ability to connect abstract mathematical concepts to everyday life and apply these concepts to novel contexts.

The goal of this chapter was to discuss how technology can support mathematics teacher-candidates and practicing teachers who want to create opportunities for their students to engage with mathematics in a creative way. This takes courage and willingness to take risks as new ways of teaching need time for the teachers to master and for the students to get used to and to accept. Eric Mazur, a pioneer of Peer Instruction, compared this process to moving mountains and emphasized that the key to success in this process is teachers' and students' attitude about STEM learning (Mazur, 1997a). In order to succeed in mathematics learning, students should see mathematics not as a collection of facts and procedures, but as a process of figuring things out, thinking creatively and making sense of the world around them with the newly acquired mathematical concepts. If we want our students and our society to change their views about mathematics, we have to educate mathematics teachers who see mathematics as a highly creative pursuit (Lockhart, 2009). For these teachers, modern technologies will unleash the opportunities for authentic and meaningful learning. These teachers will have mastery of TPACK that they began acquiring during their teacher education program and will continue building during their entire careers. These teachers will have experienced successes and failures and they will know firsthand that failure in mathematics learning or teaching is a temporary setback that can and should be overcome. These teachers will remember how they came to understand new ideas and will have intuitive knowledge of why some mathematical concepts are difficult to master (Goodstein, 2000; Milner-Bolotin et al., 2013).

The discussions and arguments brought in this chapter also open new venues for educational research in the area of technology use to promote creativity in mathematics teacher-education and in mathematics teaching and learning. While there are many questions one can ask on the topic, we list a few we find especially interesting. How do we support the development of deliberate pedagogical thinking with technology in future mathematics teachers? How do we use technology to promote mathematical creativity in teacher education and in mathematics classrooms? How do we make sure that new technologies open new pedagogical opportunities for engaging students in meaningful and creative mathematics learning? What are new creative ways of using technology in mathematics education? How do we support practicing mathematics teachers and help them to uncover the potential of new technologies to promote mathematical creativity in their classrooms? How do we evaluate the effectiveness of technology use in a mathematics classroom in the context of developing mathematical creativity? How do we use technology to promote student mathematical collaboration on non-standard open-ended problems? How do we use technology to promote student creative thinking in and out of class that help them bridge mathematics to their lives?

The main argument of this chapter is that in order to prepare creative mathematics teachers for a successful teaching career in the 21st century, teacher-candidates have to experience mathematical creativity in their own teacher education program (Milner-Bolotin, 2015). It is not enough to acquire TPACK, mathematics teachers should be open and willing to implement these novel pedagogical strategies in their classrooms (Milner-Bolotin, 2016b). Therefore, mathematics methods courses should challenge their creative thinking, help them remember what it means not to know a "simple" mathematical concept and what it means to come to understand it (Goodstein, 2000). Teacher-candidates should experience the transient nature of understanding—there is no limit to understanding—one can always dig deeper, find new connections, and challenge previous understandings. Most importantly, in addition to experiencing technology-enhanced pedagogies as students and as teachers, teacher-candidates should acquire pedagogical values congruent with creative mathematical thinking (Aralas, 2008). This means they should be open to being vulnerable (not knowing something) and being proven wrong. They should realize that lower-case failure is instructive and inevitable when one learns something new, including mathematics. But upper-case (fatal) Failure has no room in the 21st century mathematics classroom. This is where mathematics teachers should summon all their courage, TPACK and creative abilities in order to continue with what matters—engaging their students in meaningful and creative mathematics learning.

Acknowledgements The use of educational technologies in the courses described in this chapter became possible thanks to the support of the Teaching and Learning Enhancement Fund and the Faculty of Education at the University of British Columbia. We also would like to thank Davor

Egersdorfer (a Teaching and Research Assistant for the project) for his contributions to the project and for providing valuable feedback on this chapter. In addition, we would like to acknowledge the book editors for their valuable, thoughtful and productive feedback on this chapter. Lastly, we would like to express our gratitude to the mathematics and science teacher-candidates at the University of British Columbia for their participation, feedback and encouragement for this project. Their enthusiasm for creative use of technology in mathematics and science education served the continuous motivation and inspiration for us.

References

Adams, C. (2001). Overcoming math anxiety. *The Mathematical Intelligencer, 23*(1), 49–50.

Aralas, D. (2008, July 6–13). *Mathematical creativity and its connection with mathematical imagination.* Paper presented at the 11th International Congress on Mathematical Education Monterey, Mexico.

Barnes, G. (1989). Physics and size in biological systems. *The Physics Teacher, 27*(4), 234–253.

Bates, S., & Galloway, R. (2013). *PeerWise: Student-generated content for enhanced engagement and learning*. Paper presented at the Western Conference on Science Education, London, Ontario. http://ir.lib.uwo.ca/wcse/WCSEThirteen/july11/5/.

Black, K. (2017). *The calculated art of quilting*. Retrieved from https://www.ualberta.ca/newtrail/featurestories/the-calculated-art-of-quilting.

Bloom, B. S. (1956). *Taxonomy of educational objectives: Cognitive domain* (Vol. 1). New York, NY: Longman.

Boaler, J. (2010). *The elephant in the classroom: Helping children learn and love maths*. London, UK: Souvenir Press Ltd.

Bransford, J. D., Brown, A. L., & Cocking, R. R. (2002). *How people learn: Brain, mind, experience, and school*. Washington, D.C.: The National Academies Press.

British Columbia Ministry of Education. (2013). *Defining cross-curricular competencies: Transforming curriculum and assessment*. Retrieved from https://curriculum.gov.bc.ca/.

British Columbia Ministry of Education. (2015). *Building students success: BC's new curriculum*. Retrieved from https://curriculum.gov.bc.ca/.

Burger, E. B., & Starbird, M. (2000). *The heart of mathematics: An invitation to effective thinking*. Emeryville, CA: Key College Publishing.

Burridge, P., & Carpenter, C. (2013). Expanding pedagogical horizons: A case study of teacher professional development. *Australian Journal of Teacher Education, 38*(9), 10–24.

Bursal, M., & Paznokas, L. (2006). Mathematics anxiety and preservice elementary teachers' confidence to teach mathematics and science. *School Science & Mathematics, 106*(4), 173–180. https://doi.org/10.2307/749455.

Campbell, P. F., Nishio, M., Smith, T. M., Clark, L. M., Conant, D. L., Rust, A. H. .... Choi, Y. (2014). The relationship between teachers' mathematical content and pedagogical knowledge, teachers' perceptions, and student achievement. *Journal of Research in Mathematics Education, 45*(4).

Carr, J. M. (2012). Does math achievement h'APP'en when iPads and game-based learning are incorporated into fifth-grade mathematics instruction? *Journal of Information Technology Education: Research, 11,* 269–286.

Chachashvili-Bolotin, S., Milner-Bolotin, M., & Lissitsa, S. (2016). Examination of factors predicting secondary students' interest in tertiary STEM education. *International Journal of Science Education, 38*(2), 366–390. http://dx.doi.org/10.1080/09500693.2016.1143137.

Chin, C. (2007). Teacher questioning in science classrooms: Approaches that stimulate productive thinking. *Journal of Research in Science Teaching, 44*(6), 815–843.

Chinn, S. (2012). Beliefs, anxiety and avoiding failure in mathematics. *Child Development Research,* vol. 2012, Article ID 396071. http://dx.doi.org/10.1155/2012/396071.

Clement, L. L. (2001). What do students really know about functions? *Mathematics Teacher, 94* (9), 745–748.

Cole, A. L., & Knowles, J. G. (2000). *Researching teaching: Exploring teacher development through reflexive inquiry*. Boston, MA: Allyn and Bacon.

Cuban, L. (2001). *Oversold and underused: Computers in the classroom*. Cambridge, MA: Harvard University Press.

Dang, T (Producer). (2016). *Collaborative learning annotation system [Media player]*. Retrieved from http://ets.educ.ubc.ca/clas/.

Denny, P. (2016). *PeerWise*. Retrieved from http://peerwise.cs.auckland.ac.nz/.

Denny, P., Luxton-Reilly, A., & Simon, B. (2009). *Quality of student contributed questions using PeerWise*. Retrieved from Enschede http://doc.utwente.nl/fid/1445.

Dickinson, M. (2011). Writing multiple-choice questions for higher-level thinking. *Learning Solutions*. Retrieved from https://www.learningsolutionsmag.com/articles/804/writing-multiple-choice-questions-for-higher-level-thinking.

Dweck, C. S. (2016). *Mindset: The new psychology of success*. New York, NY: Penguin Random House LLC.

Eijck, M., & Roth, W. M. (2009). Authentic science experiences as a vehicle to change students' orientations towards science and scientific career choices: Learning from the path followed by Brad. *The Culture of Science Education, 4,* 611–638.

Erickson, T., & Cooley, B. (2006). *A den of inquiry* (Vol. 1). Beaverton, OR: Vernier Technology.

Eshach, H. (2014). The use of intuitive rules in interpreting students' difficulties in reading and creating kinematic graphs. *Canadian Journal of Physics, 92*(1), 1–8.

Fenyvesi, K., Budinski, N., & Lavicza, Z. (2014). *Two solutions to an unsolvable problem: Connecting origami and GeoGebra in a Serbian high school*. Paper Presented at the Bridges 2014: Mathematics, Music, Art, Architecture, Culture, Seoul, South Korea.

Feynman, R. P. (1999). *The pleasure of finding things out: The best short works of Richard P. Feynman*. New York, NY: Helix Books; London, UK: Perseus Books.

Finkelstein, N. D., Adams, W. K., Keller, C. J., Kohl, P. B., Perkins, K. K., Podolefsky, N. S., et al. (2005). When learning about the real world is better done virtually: A study of substituting computer simulations for laboratory equipment. *Physical Review Special Topics—Physics Education Research, 1*(1), 010103.

Franco-Watkins, A., Derks, P., & Dougherty, M. (2010). Reasoning in the monty hall problem: Examining choice behaviour and probability judgements. *Thinking & Reasoning, 9*(1), 67–90.

Gagatsis, A., & Shiakalli, M. (2004). Ability to translate from one representation of the concept of function to another and mathematical problem solving. *Educational Psychology, 24*(5), 645–657. https://doi.org/10.1080/0144341042000262953.

Garforth, L., & Stockelova, T. (2012). Science policy and STS from other epistemic places. *Science, Technology and Human Values, 37*(2), 226–240.

Ge, X., Ifenthaler, D., & Spector, J. M. (Eds.). (2015). *Emerging technologies for STEAM education: Full STEAM ahead*. New York, NY: Springer.

Gladwell, M. (2008). *Outliers: The story of success* (Vol. Spring). New York, NY: Little, Brown and Company.

Goodstein, D. (2000). The coming revolution in physics education. *APS News: The American Physical Society, 9*(6), 8.

Goth, G. (2009). *Physics and size in biological systems*. Retrieved from http://www.smccd.net/accounts/goth/MainPages/Scaling/Scaling.htm.

Guerrero, S. (2010). Technological pedagogical content knowledge in the mathematics classroom. *Journal of Digital Learning in Teacher Education, 26*(4), 132–139.

Hake, R. R. (1998). Interactive-engagement versus traditional methods: A six-thousand-student survey of mechanics test data for introductory physics courses. *American Journal of Physics, 66*(1), 64–74.

Hake, R. R. (2012). Helping students to think like scientists in socratic dialogue-inducing labs. *The Physics Teacher, 50*(1), 48–52. https://doi.org/10.1119/1.3670087.

Harris, J., & Hofer, M. (2011). Technological pedagogical content knowledge (TPACK) in action: A descriptive study of secondary teachers' curriculum-based, technology-related instructional planning. *Journal of Research on Technology in Education, 43*(3), 211–229.
Hohenwarter, J., Hohenwarter, M., & Lavicza, Z. (2008). Introducing dynamic mathematics software to secondary school teachers: The case of geogebra. *Journal of Computers in Mathematics and Science Teaching, 28*(2), 135–146.
Hohenwarter, M. (2014). *GeoGebra*. www.geogebra.org.
Kastberg, S., & Leatham, K. (2005). Research on graphing calculators at the secondary level: Implications for mathematics teacher education. *Contemporary Issues in Technology and Teacher Education, Communication and Information, 5*(1), 25–37.
Kay, A. (Producer). (1987). Doing with images makes symbols. *The Distinguished Lecture Series: Industry Leaders in Computer Science*. Retrieved from https://www.youtube.com/watch?v=0oonXT-gYjU.
Kemmis, S. (2011). A self-reflective practitioner and a new definition of critical participatory action research. In N. Mockler & J. Sachs (Eds.), *Rethinking educational practice through reflexive inquiry* (pp. 11–29). New York, NY: Springer.
Koehler, M. J., & Mishra, P. (2005). What happens when teachers design educational technology? The development of technological pedagogical content knowledge. *Journal of Educational Computing Research, 32*(2), 131–152.
Koehler, M. J., & Mishra, P. (2009). What is technological pedagogical content knowledge? *Contemporary Issues in Technology and Teacher Education, 9*(1), 60–70.
Koehler, M. J., & Mishra, P. (2015). Technological pedagogical content knowledge. In M. J. Spector (Ed.), *The SAGE encyclopedia of educational technology* (Vol. 2, pp. 782–785). Los Angeles, CA: SAGE.
Kuo, E., & Wieman, C. E. (2016). Toward instructional design principles: Inducing Faraday's law with contrasting cases. *Physical Review Physics Education Research, 12,* 0101128.
Lasry, N. (2008). Clickers or flashcards: Is there really a difference? *The Physics Teacher, 46*(5), 242–244.
Lasry, N., Mazur, E., & Watkins, J. (2008). Peer instruction: From harvard to the two-year college. *American Journal of Physics, 76*(11), 1066–1069.
Let's Talk Science. (2013). *Spotlight on science learning: The high cost of dropping science and math*. Retrieved from http://www.letstalkscience.ca/our-research/spotlight2013.html.
Lewis, S. (2014). *The rise: Creativity, the gift of failure, and the search for mastery*. New York, NY: Simon & Schuster.
Lingefjärd, T., & Ghosh, J. B. (2016). Learning mathematics as an interplay between internal and external representations. *Far East Journal of Mathematical Education, 16*(3), 271–297. http://dx.doi.org/10.17654/ME016030271.
Lockhart, P. (2009). *A mathematician's lament: How school cheats us out of our most fascinating and imaginative art form*. New York, NY: Bellevue Literary Press.
Luft, J. A., & Hewson, P. W. (2014). Research on teacher professional development knowledge in science. In N. G. Lederman & S. K. Abel (Eds.), *Handbook of research on science education* (Vol. 2, pp. 889–909). New York, NY: Routledge.
Ma, L. (1999). *Knowing and teaching elementary mathematics: Teachers' understanding of fundamental mathematics in China and in the United States*. New York, NY : Lawrence Erlbaum Associates, Inc.
Maciel, T. (2015). Smartphones in the classroom help students see inside the black box. *APS News, 24*(3), 5–6.
Martinovic, D., Karadag, Z., & McDougall, D. (Eds.). (2014). *Proceedings of the fifth north american geogebra conference: Explorative learning with technology: GeoGebra-NA 2014*. Toronto, Canada: University of Toronto.
Martinovic, D., & Manizade, A. G. (2014). Technology as a partner in geometry classrooms. *The Electronic Journal of Mathematics and Technology, 8*(2), 69–87.
Mazur, E. (1997a). Moving the mountain: Impediments to change. *The Physics Teacher, 35,* 1–4.
Mazur, E. (1997b). *Peer instruction: User's MANUAL*. Upper Saddle River, NJ: Prentice Hall.

McDemott, L., Rosenquist, M. L., & van Zee, E. H. (1987). Student difficulties in connecting graphs and physics: Examples from kinematics. *Americal Journal of Physics, 55*(6), 503–513.
McQueen, H. A., Shields, C., Finnegan, D. J., Higham, J., & Simmen, M. W. (2014). PeerWise provides significant academic benefits to biological science students across diverse learning tasks, but with minimal instructor intervention. *Biochemistry and Molecular Biology Education, 42*(5), 371–381. https://doi.org/10.1002/bmb.20806.
Mesa, V. (2008). Solving problems on functions: Role of the graphing calculator. *PNA, 2*(3), 109–135.
Milner-Bolotin, M. (2009). Exploring scaling: From concept to applications. *Science Education Review, 8*, 70–77.
Milner-Bolotin, M. (2012). Increasing interactivity and authenticity of chemistry instruction through data acquisition systems and other technologies. *Journal of Chemical Education, 89*(4), 477–481.
Milner-Bolotin, M. (2014). Using PeerWise to promote student collaboration on design of conceptual multiple-choice questions. *Physics in Canada, 70*(3), 149–150.
Milner-Bolotin, M. (2015). Technology-enhanced teacher education for 21st century: Challenges and possibilities. In X. Ge, D. Ifenthaler, & J. M. Spector (Eds.), *Emerging technologies for STEAM education* (pp. 137–156). Basel, Switzerland: Springer International Publishing.
Milner-Bolotin, M. (2016a). Promoting deliberate pedagogical thinking with technology in physics teacher education: A teacher-educator's journey. In T. G. Ryan & K. A. McLeod (Eds.), *The physics educator: Tacit praxes and untold stories* (pp. 112–141). Champaign, IL: Common Ground and The Learner.
Milner-Bolotin, M. (2016b). Rethinking technology-enhanced physics teacher education: From theory to practice. *Canadian Journal of Science, Mathematics and Technology Education, 16*(3), 284–295. https://doi.org/10.1080/14926156.2015.1119334.
Milner-Bolotin, M., Antimirova, T., & Petrov, A. (2010). Clickers beyond the first year science classroom. *Journal of College Science Teaching, 40*(2), 18–22.
Milner-Bolotin, M., Fisher, H., & MacDonald, A. (2013). Modeling active engagement pedagogy through classroom response systems in a physics teacher education course. *LUMAT: Research and Practice in Math, Science and Technology Education, 1*(5), 523–542.
Milner-Bolotin, M., & Nashon, S. (2012). The essence of student visual–spatial literacy and higher order thinking skills in undergraduate biology. *Protoplasma, 249*(1), 25–30. https://doi.org/10.1007/s00709-011-0346-6.
National Center for Education Statistics. (2011). *Trends in international mathematics and science study*. Retrieved from Washington, D.C. http://nces.ed.gov/timss/.
National Governors Association Center for Best Practices, & Council of Chief State School Officers. (2010). *Common core state standards for mathematics*. Retrieved from Washington, D.C. http://www.corestandards.org/wp-content/uploads/Math_Standards1.pdf.
National Numeracy, U. K. (2016). *Attitudes towards maths—Research and approach overview*. Lewes, UK: National Numeracy.
O'Grady, K., Deussing, M.-A., Scerbina, T., Fung, K., & Muhe, N. (2016). *Measuring up: Canadian results of the OECD PISA study: The performance of Canada's youth in science, reading and mathematics (2015 First Results for Canadians Aged 15)*. Toronto, Canada: Council of Ministers of Education.
OECD. (2014). *PISA 2012 Results: What students know and can do—Student performance in mathematics reading and science*. Retrieved from http://www.oecd.org/pisa/keyfindings/pisa-2012-results.htm.
OECD. (2016a). *PISA 2015 results: Excellence and equity in education* (Vol. I). Paris, France: OECD Publishing.
OECD. (2016b). *PISA 2015 results: Policies and practices for successful schools* (Vol. II). Paris, France: OECD Publishing.
OECD. (2016c). *School leadership for developing professional learning communities*. Paris, France: OECD Publishing.
Paul, R., & Elder, L. (2007). Critical thinking: The art of socratic questioning. *Journal of Developmental Education, 31*(1), 36–37.

Perkins, K., Adams, W., Dubson, M., Finkelstein, N., Reid, S., Wieman, C., et al. (2006). PhET: Interactive simulations for teaching and learning physics. *The Physics Teacher, 44*(January), 18–23.

Ripley, A. (2013). *The smartest kids in the world: And how they got that way*. New York, NY: Simon and Schuster.

Rosenhouse, J. (2009). *The monty hall problem: The remarkable story of math's most contentious brain teaser*. Oxford, UK: Oxford University Press.

Ruthven, K., Deaney, R., & Hennessy, S. (2009). Using graphing software to teach about algebraic forms: A study of technology-supported practice in secondary-school mathematics. *Educational Studies in Mathematics, 71*, 279–297.

Salvadori, M. (1980). *Why buildings stand up: The strength of architecture*. New York, NY: Norton & Company.

Schmidt, W. H., Blömeke, S., Tatto, M. T., Hsieh, F.-J., Cogan, L. S., Houang, R. T., et al. (2011). *Teacher education matters: A study of middle school mathematics teacher preparation in six countries*. New York, NY: Teachers College Press.

Schoenfeld, A. H. (2014). What makes for powerful classrooms, and how can we support teachers in creating them? A story of research and practice, productively intertwined. *Educational Researcher, 43*(8), 404–412.

Schuster, D., Undreiu, A., Adams, B., Brookes, D., & Milner-Bolotin, M. (2009). Motion-matching: A challenge game to generate motion concepts. *The Physics Teacher, 47*(7), 381–385.

Sfard, A. (2012). New clothes—and no emperor. *For the Learning of Mathematics, 32*(2, July), 2–4.

Shahrill, M. (2013). Review of effective teacher questioning in mathematics classrooms. *International Journal of Humanities and Social Science, 3*(17), 224–231.

Shulman, L. S. (1986). Those who understand: Knowledge growth in teaching. *Educational Researcher, 15*(2), 4–14.

Stigler, J. W., & Hiebert, J. (1999). *The teaching gap: Best ideas from world's teachers for improving education in the classroom*. New-York, NY: The Free Press.

Tobias, S. (1993). *Overcoming math anxiety: Revised and expanded*. New York, NY: W.W. Norton and Company.

Troen, V., & Boles, K. C. (2003). *Who's teaching your children?: Why the teacher crisis is worse than you think and what can be done about it*. New Haven, CT: Yale University Press.

UMass Donahue Institute Research and Evaluation Group. (2011). *Increasing student interest in science, technology, engineering, and math (STEM): Massachusetts STEM pipeline fund programs using promising practices* (M. D. o. H. Education, Trans.). Boston, MA: University of Massachusetts Donahue Institute. Retrieved from http://www.mass.edu/stem/documents/Student%20Interest%20Summary%20Report.pdf

USA National Research Council. (2013). *Next generation science standards: For states, by states*. Washington, D.C.: The National Academies Press.

Vernier-Technology. (2016). *Logger pro (Version 3.6.1)*. Portland, OR: Vernier Technology. Retrieved from www.vernier.com.

Vickrey, T., Rosploch, K., Rahmanian, R., Pilarz, M., & Stains, M. (2015). Research-based implementation of peer instruction: A literature review. *CBE—Life Science Education, 14*(Spring), 1–11.

Watson, A., & Harel, G. (2013). The role of teachers' knowledge of functions in their teaching: A conceptual approach with illustrations from two cases. *Canadian Journal of Science, Mathematics and Technology Education, 13*(2), 154–168.

Wieman, C. E., Adams, W. K., Loeblein, P., & Perkins, K. K. (2010). Teaching physics using PhET simulations. *The Physics Teacher, 48*(April), 225–227.

Chapter 11
Harnessing Early Spatial Learning Using Technological and Traditional Tools at Home

Joanne Lee, Ariel Ho and Eileen Wood

Abstract Parents and early childhood educators share a unique role in scaffolding the acquisition of foundational mathematical concepts in young children. Targeting early skill development is critical as differences in children's early mathematical competence emerge as young as four years old, and these differences persist into formal schooling (e.g., Duncan et al. in Dev Psychol, 43(6):1428–1446, 2007). Skills in geometry and spatial sense represent one of the mathematical strands recommended by the National Council of Teachers of Mathematics (NCTM) in the United States that can be acquired by young children prior to formal schooling. This chapter introduces important differences in spatial talk and activities elicited during play by parents and early childhood educators both in the context of traditional 3-dimensional play (e.g., blocks and puzzles) environments and virtual 2-dimensional digital formats (e.g., iPads® and computers). Substantial literature reveals the important array of creative and educational experiences afforded through play and particularly manipulatives. This chapter reviews previous research and extends findings to digital contexts involving our youngest learners and discusses ways to capitalize on the affordances offered by both digital applications and traditional manipulatives to harness children's spatial learning. We also examine the benefits and concerns about educational software programs (e.g., what makes educational software programs more or less effective) in general and in the context of mathematics education.

Keywords Early learning software · Parents and math development
Children's spatial knowledge · Spatial play · Scaffolding children's learning

J. Lee (✉) · A. Ho · E. Wood
Department of Psychology, Wilfrid Laurier University,
Waterloo, Ontario, Canada
e-mail: jlee@wlu.ca

A. Ho
e-mail: arielwh.ho@gmail.com

E. Wood
e-mail: ewood@wlu.ca

© Springer International Publishing AG, part of Springer Nature 2018
V. Freiman and J. L. Tassell (eds.), *Creativity and Technology in Mathematics Education*, Mathematics Education in the Digital Era 10,
https://doi.org/10.1007/978-3-319-72381-5_11

11.1 Identifying the "Big Picture": Why Should We Design and Support Early Play Contexts that Target Spatial Skill Development?

Early acquisition of foundational spatial concepts can occur in formal or informal learning contexts. Formal contexts include early childhood education programs and explicit instruction provided by parents and other knowledgeable individuals in young children's lives. However, informal learning contexts, especially play, provide some of the richest early spatial learning experiences (e.g., Ginsburg, 2006; Jirout & Newcombe, 2015; Pollman, 2010; Verdine, Golinkoff, Hirsh-Pasek, & Newcombe, 2014a). Play encourages exploration, manipulation, construction and deconstruction (e.g., Reifel, 1984; Stiles-Davies, 1988; Stiles & Stern, 2009; Vygotsky, 1978). Through creatively constructing, juxtapositioning, and deconstructing objects, children can derive critical information that will aid them in understanding spatial properties and spatial relations (e.g., Casey & Bobb, 2003; Casey et al., 2008; Levine, Ratliff, Huttenlocher, & Cannon, 2012; Moyer-Packenham & Bolyard, 2016). In this review, we will demonstrate how children's spatial knowledge is facilitated when creative play is augmented by appropriate spatial language. For example, although many parents might provide the label for a triangle or a square and might further define these objects by identifying the number of "sides" each has, children's full understanding of the concepts such as straight lines, angles, orientation in space, and relative size can be realized when children explore the outline of the objects, extend "sides", align objects with other like or different objects, create patterns and build structures—some that will stand and some that will fall. These important creative play opportunities allow children to experience spatial properties first hand, elaborate on the prior knowledge and information provided by parents, and become more fluent in their understanding. This creative discovery can occur in many contexts and through two dimensional and three dimensional representations (e.g., Clements & Sarama, 2007; Martin, Lukong, & Reaves, 2007; Moyer-Packenham & Westenskow, 2013; Verdine, Golinkoff, Hirsh-Pasek, & Newcombe, 2014b). The current review explores spatial development in young children. We ask whether different types of representations promote "better" or "different" learning than others. In addition, we explore the impact of spatial language for the acquisition of spatial concepts. The review begins by summarizing important outcomes associated with spatial play, language and representation and then continues with a consideration of the particular contributions afforded by each.

During the early childhood years, hearing spatial language (e.g., Ferrara, Hirsh-Pasek, Newcombe, Golinkoff, & Lam, 2011; Foster & Hund, 2012; Pruden, Levine, & Huttenlocher, 2011) and engaging in spatial activities with three-dimensional (3D) blocks or puzzles (e.g., Casey, Ceder, Erkut, & Mercer Young, 2008; Jirout & Newcombe, 2014; Levine et al., 2012) facilitates spatial learning. Interestingly, the amount of spatial talk a child is exposed to is related to the types of play in which a child-adult dyad are engaged. For example, parents use more spatial language when they play with their young children using blocks and puzzles compared to other non-spatial activities (Hermer-Vazquez, Moffet, & Munkholm, 2001; Ferarra et al., 2011; Levine et al., 2012; Pruden et al., 2011). However, the amount of spatial language provided by parents is also influenced by the amount of information available through the toys themselves. For example, parents use less spatial language when they and their child play with "talking" electronic shape sorters that provide labels for shapes as well as other sounds than when they play with traditional 3D shape sorters that do not "talk" (Zosh et al., 2015). Both the amount of spatial talk and the diversity of spatial talk impacts learning. Spatial words—describing the features of an object (e.g., corner, edge), the location of an object (e.g., next to, left of), or the spatial relations between objects to build a castle (e.g., top, under, between, in the middle)—are important as they help to direct children's attention to and encoding of spatial concepts (Dessalegn & Landau, 2008; Gentner, 2003; Loewenstein & Gentner, 2005; Plumert & Nichols-Whitehead, 1996).

Existing research on early spatial learning has mainly focused on spatial play with 3D objects such as tangible blocks and jigsaw puzzles, that permit children to engage in tactile-kinesthetic experiences, including touching, holding, rotating, and manipulating in their hands (e.g., Levine et al., 2012; Needham, 2009). Today, traditional modes of play have expanded to include technology based platforms. Despite the growing use of interactive technological devices (e.g., iPad® & smartphones) by very young children (e.g., Kabali et al., 2015; Common Sense Media, 2013), little is known about the quality of mathematical instruction available through these technologies, the impact of these technologies on early skill development, or the contributions care providers can offer in these contexts (e.g., Kurcirkora, 2014). Yet, early introduction of mobile digital technologies is increasingly evident in younger age groups and for longer periods of play time (Rideout, 2014). In addition, software development has seen an increase in programs explicitly geared towards educational content in early learning (e.g., Eagle, 2012; Mueller et al., 2011). For example, over 80,000 apps on iTune store are classified as educational (Apple, 2015). Although this chapter reviews considerations applicable to software design and use in general, it specifically targets the impact of a popular, interactive and contingent responsive electronic device—the iPad® and its apps featuring games using blocks and puzzles—that 58% of parents in the United States have downloaded for their young children (Common Sense Media, 2013).

A significant body of research supports both learning gains and positive social outcomes when young children use well-designed instructional software (e.g., McKenny & Voogt, 2010; Savage et al., 2013; Thorell, Lindquist, Bergman, Bohlin, & Klingberg, 2009; Willoughby et al., 2009). In addition, a growing literature describes and supports the importance of the interaction of parent-child dyads in technology-based settings (Berkowitz et al., 2015; Flynn & Richert, 2015; Wood et al., 2016). However, the quality of software programs and their associated learning outcomes have largely been examined in the context of literacy (e.g., Grant et al., 2012; Neumann, 2016). Recently, research has shifted focus to early mathematical learning contexts, in particular, number sense (Berkowitz et al., 2015).

One growing area of interest is the development of spatial skills (e.g., Larkin, 2016; Verdine et al., 2014b). Specifically, affordances available in software design can manipulate representations of objects to provide two dimensional (2D) and/or three dimensional (3D) perspectives (e.g., Pan, Cheok, Yang, Zhu, & Shi, 2006; Rosen & Hoffman, 2009). Rotation of objects in the available screen space can also provide information about relative size and dimensionality. These representations differ from information children may gain from 3D objects manipulated in real world contexts. For example, a child manipulating a building block on screen may have access to a 2D representation only, or the software could provide the opportunity to rotate the block to allow investigation of front, back, top and bottom (e.g., Clements & Sarama, 2007; Moyer-Packenham & Bolyard, 2016). However, in all these contexts the block is understood within the context of the available screen space. In real-world block play contexts, the child can physically manipulate, rotate and examine the block in conjunction with other real-world information including the relationship of the object to the child's personal size, the size and contours of other familiar objects in the room, as well as relative to the movement of other objects and people in the environment. Affordances of 2D screen-based and 3D real-world based manipulatives may contribute to differences in the nature of spatial input provided by adults during interactions with preschoolers (e.g., Clements & Sarama, 2016; Ho et al., 2017). Such differences could, in turn, provide opportunities for children to construct multiple representations (i.e., 2D vs. 3D) of information in specific contexts and to apply these in new ways and new contexts, thus nurturing the key dimensions of creativity including flexibility, elaboration and fluency (Guilford, 1967).

Additional affordances such as immediate and interactive feedback, and swipe and touch interfaces offered by these responsive and easy-to-use devices can complement or even leverage affordances traditionally associated with tangible 3D toys (Cooper, 2005; Geist, 2014; Guernsey & Levine, 2015; National Association for the Education of Young Children (NAEYC) & Fred Rogers Centre for Early Learning and Children's Media, 2012). It is important, therefore, to explore potential differences between tangible 3D toy play and play involving digital devices such as iPads®.

This chapter reviews extant literature in an effort to identify and describe key ideas and research findings relevant to the use of physical and digital/virtual manipulatives for promoting early spatial development. This context serves as the foundation for exploring the importance of supporting spatial-visual skill development in early childhood years through the complementary use of 2D and 3D spatial play activities. Particular attention is given to the use of 2D spatial-visual iPad® applications to nurture creative thinking afforded by technology and the importance of the design of applications to support positive learning outcomes. We introduce a novel idea that playing with *both* 2D (i.e., iPad® apps) and 3D manipulatives may maximize learning opportunities to foster creative and flexible spatial thinking. In summary, our goal is to provide a foundation for understanding early spatial development in the home and in the context of a technologically rich learning environment. These two contexts provide opportunities for creative expression, discovery and exploration. A key objective is to identify how these aspects of creativity intersect in the current literature and may be important for future study—especially in the critical early years where home influences are most likely to establish fundamental skills and where touchscreen technologies are increasingly prevalent.

11.2 Laying Good Foundations: Providing Early Spatial Activities Fosters Development of Spatial Skills

Strong spatial knowledge—such as geometric knowledge, spatial visualization, spatial perception, spatial orientation, and mental rotation—in young children facilitates growth in mathematics (e.g., Casey, Nuttall, Pezaris, & Benbow, 1995; Gunderson, Ramirez, Beilock, & Levine, 2012; Kamii, Miyakawa, & Kato, 2004; Verdine, Golinkoff, Hirsh-Pasek, & Newcombe, 2014), science such as chemistry in post-secondary school (Stieff, 2007), geometry in high school (Delgado & Prieto, 2004) and creativity (Kell, Lubinski, Benbow, & Steiger, 2013). In addition, geometry and spatial skills are highly correlated to engagement and success in science, technology, engineering, and mathematics (STEM) occupations in later adulthood, even after accounting for verbal and mathematical abilities (Shea et al., 2001; Wai, Lubinski, & Benbow, 2009). A strong correlation between spatial and numeracy competencies has also been noted. For example, 12- to 14-year-olds who were gifted in mathematics also had stronger spatial abilities compared to their peers functioning at their own grade level (Hermelin & O'Connor, 1986). Thus, the National Research Council Committee on Early Childhood Mathematics in the United States (Cross, Woods, & Schweingruber, 2009) strongly recommends that children between three and six years of age learn both geometry and numeracy, to provide the foundations needed to engage in developmentally appropriate early childhood mathematical activities.

Spatial activities such as 3D block and puzzle play have been linked to spatial development. For example, block play promotes preschoolers' spatial visualisation and visual-motor coordination skills (Caldera et al., 1999) and spatial skills and geometric knowledge in 5- to 7-year-olds (Casey et al., 2008). Nine-year-olds who could build a Lego model performed better in visual spatial tasks such as mental rotation than their peers who could not (Brosnan, 1998). In addition, engaging in more complex block play by preschoolers (e.g., building a tower versus stacking blocks vertically) is predictive of other mathematical concepts and skills such as symmetry, counting, and patterning in middle and high school (e.g., MacDonald, 2001; Park, Chae, & Boyd, 2008; Wolfgang, Stannard, & Jones, 2001). Peterson and Levine (2014) also demonstrated an advantage for children who engaged in block play as part of their daily home routines. These researchers observed children for 90 min every four months when the children were between 26 and 46 months old. Children who engaged in more block play with their parents performed better on mathematical (Math Equivalence problems) and geometry (KeyMath Geometry) tasks at grade three compared to their peers. Furthermore, Verdine, Irwin, Golinkoff, and Hirsh-Pasek (2014) found that block constructing ability at three years of age three-year-olds' was predictive of math problem solving task performance at age four even after accounting for number knowledge and executive function. In a different task, 26 to 46 month old children who played with jigsaw puzzles more frequently with their parents performed better at a spatial task involving mental transformation of 2-dimensional shapes than children who did not (Levine et al., 2012). Even in older children between six and eight years of age, activities such as identifying the shape after the pieces have been mentally rotated were found to improve their math calculation, mental rotation and spatial relation skills (Cheng & Mix, 2014).

Overall, much of the extant research focuses on spatial activities involving traditional 3D toys (i.e., physical blocks and puzzles) and subsequent mathematical development. However, over the past few years attention has begun to include activities involving digital media (e.g., Hirsh-Pasek et al., 2015; Moyer-Packenham & Bolyard, 2016). It is argued that playing with virtual 2D manipulatives extends opportunities for young learners in two key ways. First, exposure to two representations—3D representations which are typically sensory and concrete and 2D representations which are typically pictorial and abstract/symbolic—allows children to understand that key concepts can be represented in more than one way. Second, access to more than one form of representation permits learners to make observations and connections about differences in representations that can help facilitate sophisticated mathematical understanding (e.g., Martin, 2009; Moyer-Packenham & Westenskow, 2013; Sarama & Clements, 2016). For example, touch-screen tablets permit creative exploration where young children are able to easily experiment with attributes such as the aspect ratio (i.e., the ratio of height to base) of perpendicularity of shapes to understand what makes a rhombus a rhombus and not a square. Such flexible manipulation and capability to merely isolate a specific attribute would not be feasible with traditional 3D toys (see example, Fig. 11.1). The ease with which touch screen software can allow learners to isolate, add, subtract and augment attributes may also lead to more efficient acquisition of mathematical concepts. For example, composition and decomposition of shapes could be readily executed in a 2D touch screen context compared to a 3D context. Research supports both the acquisition and efficiency of mathematical knowledge through 2D technology-driven learning experiences. For example, among fifth graders, spatial abilities—in terms of spatial perception, mental rotation, and spatial visualization—improved significantly after children played a computer game involving tangram puzzles and pentominoes for even just one hour (Yang & Chen, 2010)! Together the research with tangible 3D activities and research examining 2D digital representations indicate that early exposure to mathematical concepts through play can have immediate and long-lasting impact on development.

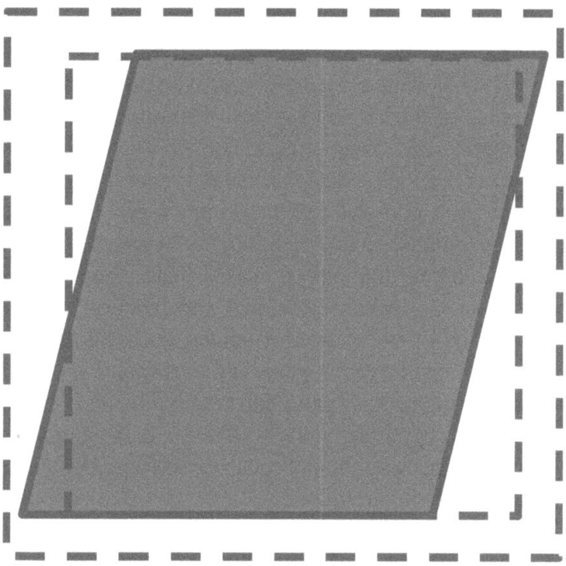

Fig. 11.1 Using technology, a child can easily manipulate a rhombus to become a square by changing the aspect ratio using the zooming and shrinking features on a touchscreen device

11.3 Fostering Creative Spatial Learning Opportunities: Contributions from Technology

Acquiring early mathematical skills inherently relies on creative discovery as well as explicit instruction (e.g., Alfieri, Brooks, Aldrich, & Tenenbaum, 2011; Hirsh-Pasek, Berk, Singer, & Golinkoff, 2008). Considerable literature in early mathematics education supports the need for children to explore and mathematize key concepts through application to diverse novel contexts (e.g., Newcombe, 2010; Uttal, 2000). This would require mastery of spatial concepts embodied physically as in the real world and represented abstractly as in visual images and representations such as maps and models.

Physical manipulatives are static but concrete, thus multiple manipulatives might be necessary for a child to discover or realize key spatial concepts In addition, providing both prototypical and non-prototypical examples (Satlow & Newcombe, 1998) or creating model representations (Siegal & Schadler, 1977) may further understanding. For example, understanding the concept of a cylinder can involve exposure to multiple examples evident in children's everyday lives (e.g., tinned goods, cookie jars) and also by having children manipulate various cylindrical shapes in block building, puzzle solving or other contexts. Virtual 2D manipulatives, however, afford learners an opportunity to dynamically manipulate objects on screen using screen tools such as swiping and touching. For example, given that the visual prototype of a triangle for young children between four and six years old is an isosceles triangle, identifying triangles (other than an isosceles triangle) is difficult for most young children (Aslan & Aktas-Arnas, 2007; Clarke, 2004; Clements, Swaminathan, Hannibal, & Sarama, 1999; Yin, 2003). However, software could make salient features of different triangles quickly apparent to young children by systematically resizing defining attributes such as aspect ratios of width and height as well as skewness (or lack of symmetry) as a child expands or shrinks the image of a triangle or series of triangles on screen. As such, an extensive and "manipulateable" representation of various types of triangles—isosceles, equilateral, right angled—could be made available through the use of technology. Such dynamic and non-static representations of multiple variations of triangles are not easily accomplished in single integrated presentations using physical manipulatives. Dynamic representations of spatial features and relationships in a virtual medium may, therefore, enhance children's acquisition of abstract spatial concepts.

Abstract representation of spatial concepts underlies learners' spatial thinking and reasoning. To make sense and to respond appropriately in our environment, we organize and process spatial information using two frames of reference—egocentric and exocentric (Klatzky, 1998; Shelton & Mcnamara, 2001). Spatial information in the environment in an egocentric or self-based frame of reference is processed and organized from an individual's perspective, bearing or orientation such as the distance of an object from oneself. This frame of reference allows us, for example, to extend our arm to an appropriate length to pick up a mug on the table by judging the distance between oneself and the table. However, representation of spatial relations among objects in the environment varies accordingly to the individual's bearing. In our mug example, the distance between oneself and the table would differ if one were to stand at the corner of the table versus the side of the table. On the other hand, spatial information in the environment in an exocentric or external-based frame of reference is organized based on salient features or landmarks in that environment. Hence, representation of spatial relations among objects (object-to-object relations) in the environment is invariant to an individual's bearing/perspective, which is foundational in forming a cognitive map for one to think, reason and visualize spatial information.

Both frames of reference are fostered using physical manipulatives; for example, the egocentric or self-based frame of reference could be used to judge the distance of the blocks between a tower children are building and themselves to ensure each block is stacked on top of each other without toppling them, while at the same time, the exocentric/external-based frame of reference could be used to decide the distance among the different towers to build a fort. However, technology may be introduced to foster the use of exocentric/external-based frame of reference, and ultimately the abstract representation of spatial concepts, by young children more effectively. If an egocentric/self-based frame of reference were adopted while children are stacking blocks to build a fort on an iPad®, they would realize very quickly that the spatial information they have is between themselves and the iPad®. Furthermore, the lack of sensory-concrete-tactile information in virtual 2D presentation makes it more difficult for learners to adopt an egocentric/self-based frame of reference to organize spatial information in the virtual environment. Similar to maps and models, the virtual 2D platform offers opportunities for children to explore and mathematize key concepts by developing abstract/symbolic spatial representations of actual representations in the environment (e.g., Moyer-Packenham & Westernskow, 2013; Uribe-Florez & Wilkins, 2010), and ultimately acquiring the ability to switch intermodally between concrete and abstract spatial representations (e.g., Clements, 1999; Sarama & Clements, 2016). Such mastery in switching between the two types of spatial representations helps to foster young children's creative and flexible thinking. Thus, play contexts that involve both virtual 2D and physical 3D representations may provide the mix of play opportunities that will encourage greater creative exploration and subsequently greater learning (e.g., Moyer-Packenham et al., 2015; Musawi, 2011; Yelland, 2002). Interestingly, we currently know very little about how virtual 2D media are used in the home or even in the classroom to support creative thinking or play in early spatial development. Emerging research in literacy suggests that the dual learning opportunities may indeed foster different yet meaningful creative play and learning (e.g., Beschorner & Hutchison, 2013; Patchan & Puranik, 2016; Price, Jewitt, & Crescenzi, 2015).

The use of technology can be effectively introduced into play contexts to elicit enriched and spatially diverse parental input at home. In our recent study, 34 parents and their preschoolers engaged in 30-min of 3D play using blocks and puzzles and virtual 2D play using an iPad® in two separate home visits (Ho et al., 2017). Our findings reveal that parents did not differ in the amount of spatial talk and the number of spatial categories in both the 3D and 2D play contexts. Specifically, parents provided the same amount of talk regarding: spatial dimensions (e.g., large, small), spatial features and properties (e.g., straight, angles), shapes (e.g., circle, triangle), location and directions (e.g., top, bottom), orientation and transformation (e.g., turn, rotate), continuous amount (e.g., half, part), and deitics (e.g., here, there). In addition, all spatial categories were used by parents in both play contexts. However, significant differences did arise in the specific types of spatial categories that were used in the traditional 3D versus 2D virtual play contexts. Specifically, in the 3D play contexts (for example, see Fig. 11.2), parents produced more words related to spatial dimensions, location and directions, and continuous amount than in the 2D play contexts. In contrast, in the 2D play contexts (for example, see Fig. 11.3), they produced more words associated with orientations and transformations as well as deictics than in the 3D play contexts. The use of these two types of spatial categories in the 2D play contexts could be due to the function of the game design, which requires parent-child dyads to complete targeted learning objectives in order to progress by *rotating* a right-angled triangle clockwise to fit into the existing square space.

This study provides evidence that, using both 2D and 3D manipulatives may maximize learning opportunities to foster creative and flexible thinking. Our findings suggest that differential parental input in specific types of spatial categories was elicited during the two play contexts. Spatial language serves as a representational tool to help children encode and represent spatial concepts (Gentner, 2003; Gentner & Lowenstein, 2002; Kuhn, 2000). Without such a symbolic, linguistic system that embodies our thoughts and concepts, one would not have the foundational skill to assimilate and accommodate (in Piaget's terms) existing spatial

Fig. 11.2 A child building a castle with 3D blocks would require an understanding of an egocentric frame of reference (i.e., spatial relations are between oneself and the objects in his environment). The spatial representation using 3D manipulatives is a sensory-concrete one

concepts to generate new knowledge (e.g., Casasola & Bhagwat, 2007; Casasola, Bhagwat, & Burke, 2009; Piaget & Inhelder, 1969). Additionally, each of these play contexts enhances exposure to different representations—abstract versus sensory-concrete—which in turn may promote differential development of various types of spatial thinking such as spatial navigation, orientation, and transformation. For example, building a castle on a 2D tablet versus using actual 3D blocks would require cognitive flexibility in mental representations of spatial concepts such as relations, dimensions, scaling (Zbiek, Heid, Blume, & Dick, 2007). Thus, the ability to manipulate and switch between sensory-concrete and abstract/symbolic spatial representations from both frames of reference helps young children in becoming creative and flexible learners to mathematize spatial concepts. The relative advantages of providing children with both sensory-concrete and abstract/symbolic representations require access to appropriate manipulatives and software. Although manipulatives may be readily accessible, facilitating children's ability to make the desired connections from technology-based contexts requires high-quality instructional software.

Fig. 11.3 A child building a castle out of blocks on an iPad® would require an understanding of an exocentric frame of reference [i.e., spatial relations are among objects (i.e., object-to-object relations) in the virtual iPad® environment]. The spatial representation using virtual 2D manipulatives is an abstract/symbolic

11.4 Importance of Instructional Software Design in Early Spatial Learning

Potential learning and social gains associated with the use of well-designed instructional software has been documented consistently for learners across ages and for learning that occurs in the classroom and, informally, at home (Archer et al., 2014; Flynn & Richert, 2015; Mayer, 2005; McKenny & Voogt, 2010; Savage et al., 2013; Thorell et al., 2009; Willoughby et al., 2009). Although high quality, pedagogically appropriate software is a critical feature in determining learning outcomes, learner interest and persistence are also key considerations when understanding the impact of technology-based instruction (e.g., Grant et al., 2012; Wood, Hui, & Willoughby, 2008). Many children are highly motivated to engage with computer-based learning contexts and to persist longer even when working on challenging tasks when engaged with interactive technologies (e.g., Karemaker, Pitchford, & O'Malley, 2010; Swing & Anderson, 2008). A key contributor to extended engagement and attraction for digitally-based delivery systems is the "game-like format" (e.g., Gee, 2008; Vogel et al., 2006). Most children's learning software also employs an element of competition where children can compete against themselves or others to reach goals, obtain rewards or tokens or advance to more advanced levels of play (Lucas & Sherry 2004). Software programs also provide opportunities for children to navigate, explore and manipulate novel environments (e.g., Abdul Jabbar & Felicia, 2015). Exploration can take different perspectives including first person and third person. Children are often accompanied on their 'learning adventures' by supportive animated characters which advise, reinforce, and interact with the learner. Interactivity between the learner and the software environment further promotes engagement and learning (e.g., Abdul Jabbar & Felicia, 2015).

Software design is a key element in maximizing learning. Within the literature, discussions have examined how software design—especially instructive ones that afford only static visuo-spatial information (e.g., naming shapes) versus manipulable or constructive ones that afford dynamic visuo-spatial information (e.g., composing, decomposing shapes)—may promote, inhibit or even constrain learning (e.g., Goldwin & Highfield, 2013; Hirsh-Pasek et al., 2015; McQuiggan, Kosturko, McQuiggan, & Sabourin, 2015; Travers & More, 2013). Given that there are different types of spatial skills that can be present as early as the first year of life (Lauer & Lourenco, 2016; Uttal et al., 2013), the pressing concern is how to design high quality software applications (apps). Such software must use sound pedagogical and learning principles to present developmentally appropriate content, including appropriate graphic and user-interface design, and formats that capitalize on the unique affordances (e.g., manipulatability and scalability) offered by technology. This constellation of requirements in design maximizes the potential to support creative, constructive and flexible learning.

In addition to appropriate content pedagogy that would be expected in any instructional delivery system, multimedia formats offer affordances known to support learning (e.g., Clark & Mayer, 2008; Mayer, 2005; Takacs, Swart, & Bus, 2015). For example, multimedia formats, by definition, present information through more than one modality. This means learners can experience information visually, verbally and through touch, often simultaneously. Redundancy of information through differing modalities (e.g., visual and verbal) can enhance learning (Takacs, Swart & Bus, 2015). Instructional supports such as levelled activities comprised of hierarchically arranged sub-goals, organize and structure the learning experience. Similarly, scaffolds such as hints and following responses with immediate, elaborated and accurate feedback allow opportunities to acquire skills and correct errors (e.g., Van der Kleij, Feskens, & Eggen, 2015). These instructional features support individual learning needs and may be especially relevant for very young or inexperienced learners.

Clearly, digitally delivered instruction has the promise to attract and engage learners. Hence, well-conceived and designed programs offer the potential to facilitate learning by minimizing the cognitive resources needed to navigate the software. However, striking the right balance between providing all the relevant and desirable affordances and providing too little or too much is challenging. For example, software that depicts excessive perceptual richness or that requires high interactivity between users and the program have been found to hinder learning (e.g., Levinson, Weaver, Garside, McGinn, & Norman, 2007; Song et al., 2014; Stull & Mayer, 2007). Although perceptual richness or interactivity is generally considered a positive feature of software design, it is clear that design considerations need to be aligned with developmental, educational and cognitive capabilities of the target audience to maximize learning opportunities.

The above requirements have yet to be fully translated to apps to benefit early childhood learning, especially with respect to the visual-spatial strand for parents to engage with their children. Despite a dramatic increase in the use of mobile touch-screen devices (e.g., iPad® and smartphones) from 52% in 2011 to 75% in 2013 among children under eight years old at home (Common Sense Media, 2013), there are fewer educational apps on spatial skills compared to language- and literacy-focused apps geared to preschool aged children on the iTunes store. For example, our survey of educational apps on iTunes store in January 2016 revealed that there were only 47 free and paid apps featuring blocks and puzzles for children between three and five years of age. Moreover, there is concern that learning goals of available educational apps are often not grounded in evidence-based early childhood mathematics curriculum (Falloon, 2013; Hirsh-Pasek et al., 2015). A similar conclusion has been drawn from language- and literacy-focused apps (Guernsey & Levine, 2015). Appropriate diversity in skills taught and high quality instruction are fundamental software design features for promoting early development (Grant et al., 2012; Wood, Grant, Gottardo, Savage, & Evans, 2017).

The importance of pedagogically appropriate content on optimal children's learning outcomes is underscored by emerging research demonstrating learning gains when relevant iPad® apps were used at home and in school settings (Berkowitz et al., 2015; Pitchford, 2015). For example, Berkowitz et al. (2015) reported significant numeracy gains in first-grade children at the end of their academic year after their primary caregivers read short numerical story problems to them using an iPad® app several times a week for a year. Similarly, students from first to third grades were reported to achieve significant numeracy skills after eight weeks having their math lessons delivered 30 min per day using four iPad® apps (Pitchford, 2015). In this study, significant learning gains were observed only in students in the tablet group with the math apps, but not in students who used the tablet without the math-focused apps or in a group of students receiving math lessons only from their teachers.

Software design that targets key learning goals is especially important. For example, Falloon (2013) found that apps that are free of distractions such as additional graphics/images helped 5-year-olds' achieve the learning objectives of the apps. Similarly, the numeracy-focused story app used in Berkowitz et al. (2015) consisted of minimal sounds, graphics and animation to permit greater attention to target information. Inclusion of non-target or additional superfluous features used in story e-books distracts young children and decreases learning (Parish-Morris, Mahajan, Hirsh-Pasek, Golinkoff, & Collins, 2013). Thus, effective program design must attract attention to key elements and information while permitting creative exploration and discovery. For example, software with programming features such as "turtle" geometry or Logo (Clements & Meredith, 1993; Yelland, 1994) and Lego WeDo and Bee-bots (Bers, 2010) have been found to promote active, hands-on learning and exploration of mathematical concepts in young children between 3 and 7 years old. Clearly, well-designed software offers significant potential as a learning tool. In fact, Highfield (2015) argues that incorporating technology in early childhood mathematics education could support both academic STEM goals such as measuring and visualizing as well as intellectual STEM goals such as problem-solving and reflecting (Katz, 2010). Fostering cognitive growth when using technology requires examination of specific learning contexts in supporting creative mathematical constructs.

11.5 Considering Both the Benefits and Pitfalls of Technology as a Tool to Support Creativity in Spatial Development

Acquiring competence in mathematics begins early in life. Children's play environments provide a natural and important context for children to discover, create and explore important precursor mathematical knowledge that will persist throughout their school and, potentially, occupational lives. Providing a rich learning environment that engages children, reinforces important concepts, and encourages creative mathematical play requires knowledge of tools that can best facilitate learning. In this chapter, we have outlined how technology can be introduced as a complementary learning tool to traditional physical manipulatives in early spatial learning at home. The use of mobile touchscreen devices such as the iPad® has the potential to foster early spatial learning at home. The addition of technology can foster learning gains through two avenues—increased parental involvement and direct learning experiences for the child. The use of iPad® has been shown to help parents, especially those with high anxiety in mathematics, to increase their engagement in mathematical talk through bedtime math stories (Berkowitz et al., 2015). Their increased math engagement—in terms of amount of time and input—helped their children improve significantly in their numeracy competence. In addition, parents and early childhood educators may not know how best to engage in mathematics activities with young children (Cannon & Ginsburg, 2008; Lee, Kotsopoulos, & Zambrzycka, 2012). Thus, well-designed apps could offer appropriate content pedagogy to help parents and early childhood educators to scaffold children's spatial learning through diverse, relevant and engaging creative learning opportunities.

However, to harness the full potential of the creative use of technology in fostering early spatial development, a few concerns should be addressed. First, readily available resources offering reliable formal review of educational apps is necessary for appropriate selection of resources for parents and educators. At present, such comprehensive reviews are not available. There is no government body that screens the educational quality or rates the educational value of apps for children to ensure the veracity of the claims in the packaging (Guernsey & Levine, 2015; Willoughby & Wood, 2008). Such claims of apps traditionally were left to parents, teachers, and a handful of web-based consumer concern groups (e.g., Common Sense Media, Moms with Apps, BestAppsforKids) to evaluate (Guernsey & Levine, 2015), often with parents and teachers using the websites as their source for information and evaluation. Although these websites serve as a useful resource guide, not all of them screen or rate the educational quality of apps. For example, review websites such as Moms with Apps and BestAppsforKids provide the description of what the app does while others such as Common Sense Media provide both the description and rating of the educational quality of apps.

Fortunately, comprehensive evaluations of apps involving mathematical constructs have emerged recently. Unfortunately, the findings suggest shortcomings in many apps currently available. For example, Larkin (2016) evaluated the quality of 53 Geometry apps targeted for children between 5 and 12 years old based on three criteria: content, pedagogy (e.g., ease of use without instruction), and facilitation of learner's thought process. Only 7 of the 53 apps received a rating of 6 or higher on their 10 point scale. Similarly, evaluation of 19 apps targeting foundational geometric and visual-spatial skills designed for preschoolers aged 3 to 5 years old, yielded only 4 apps that featured at least 4 out of the 5 spatial concepts in a taxonomy designed to reflect developmental progression of key concepts (Lee, Douglas, Wood, & Andrade, 2017). In addition, when assessed for instructional quality (i.e., levelling of difficulty and feedback), all but one app received a rating of 2 or lower out of 5. These findings underscore the need for well-conceived and well-designed apps if technology is to effectively foster creative and flexible spatial thinking. Building on these findings suggests that developing and standardizing a taxonomy that depicts the developmental progression of foundational mathematical skills is an important priority. In addition, program design, ideally, should include instructional materials that would provide support for parents and teachers on how use the app effectively (e.g., Lysenko et al., 2016). These supports should reinforce elements within the program and extend beyond the elements in the software to examples in everyday life.

Second, parents and caregivers should be cognizant of the fact that technology use should not replace or reduce their level of engagement or the opportunities for them to provide scaffolding to support children's learning. Thus, the importance of adult engagement in joint media attention with children in technology-based contexts needs to be enhanced. For example, parents have been observed to reduce their language input during computer-based game context with older preschoolers (Flynn & Richert, 2015) and their spatial language input during an iPad® play context (Ho et al., 2017). Some parents may believe that software programs are intact, comprehensive and complete, thus requiring little if any parental contributions. As such, software developers and accompanying advertisements for mathematics-based software should incorporate parents as active agents in media learning contexts to promote enriched spatial adult input. The inclusion of parents as key contributors in the learning context will avoid what has been called the "pass-back" effect where parents hand their mobile touchscreen devices to their young children to play by themselves (Chiong & Shuler, 2010).

Third, spatial learning may not be optimized in a 2D context alone. For example, some emerging evidence with third graders suggests that although the use of either a physical 3D or virtual abacus helped students to learn basic identification of numbers on the abacus, only students using the physical 3D abacus were successfully able to apply their knowledge to new and advanced questions (Flanagan, 2013). Ongoing research is required to examine the cognitive processes involved in spatial learning in abstract and sensory-tactile modes.

Fourth, our spatial perception about the world processed by the visual system relies on multiple perceptual cues and measurement scales (e.g., absolute and relative size, depth) (e.g., Gogel & Tietz, 1980). However, little is known regarding whether images presented on a 2D iPad® are perceived in the same manner spatially as those of 3D objects in the real world—also known as perceptual fidelity. This is a concern especially for young learners who are less familiar with the technology as well as spatial concepts. Before launching programs that rely heavily on 2D representations, we need to know more about how children understand the information that is being presented to them.

Finally, both 2D and 3D forms of play offer the opportunity for creative play. Although the opportunities for tangible physical manipulation and exploration are immediately apparent, it is also important to note that the touchable feature of the iPad® affords physical manipulation and exploration (Plowman & Stephen, 2003). However, these opportunities can be easily constrained by less flexible learning contexts. It is important to ensure that opportunities to encourage mathematical development in either 2D or 3D environments include sufficient flexibility to allow children to create knowledge and to use mathematical concepts and structures in creative ways (e.g., NCTM, 2007; Newcombe & Frick, 2010).

11.6 Conclusions

In summary, emerging research suggests that technology could be introduced into play contexts at home to complement the types of spatial input and engagement elicited in the traditional 3D play contexts (Ho et al., 2017). However, effective introduction of technology into play would require a multi-facet approach among researchers and application developers to create evidence-based educational apps to foster early spatial learning (Chiong & Shuler, 2010; Falloon, 2013; Hirsh-Pasek et al., 2015). Specifically, careful evaluation of the impact technology has on play, for both the parent and the child, must be conducted to ensure that play is augmented, fostered and open to creative exploration. For example, understanding that the "talking" features in an electronic shape sorter actually reduced parental spatial input (Zosh et al., 2015) informs us that the intention underlying some design features may not yield desired outcomes. In addition, it is important to identify and embrace different experiences in 2D versus 3D learning environments.

For example, in the Ho et al., study (2017), play with traditional 3D spatial toys elicited more words related to spatial dimensions and locations whereas play with iPad® apps elicited more words related to spatial orientations and transformations by parents with their preschoolers. Given the important influence adults can have in early play contexts, parents and educators should be supported with best practices and resources to help them in terms of ways to engage in technology with their preschoolers (e.g., asking questions related to or elaborating the content of the games) and with the selection of developmentally appropriate software. Encouraging children and those who care for them to jointly engage in creative mathematical play that is aligned with early childhood mathematics curriculum may provide a richer context for discovery and development.

References

Abdul Jabbar, A. I., & Felicia, P. (2015). Gameplay engagement and game-based learning: A systematic review. *Review of Educational Research, 85*(4), 740–779. https://doi.org/10.3102/00346543155772210.

Alfieri, L., Brooks, P. J., Aldrich, N. J., & Tenenbaum, H. R. (2011). Does discovery-based instruction enhance learning? *Journal of Educational Psychology, 103,* 1–18.

Apple. (2015). *iPad in education* [website]. Retrieved from http://www.apple.com/education/ipad/apps-books-and-more/.

Archer, K., Savage, R., Sanghera-Sidhu, S., Wood, E., Gottardo, A., & Chen, V. (2014). Examining the effectiveness of technology use in classrooms: A tertiary meta-analysis. *Computers & Education, 78,* 140–149.

Aslan, D., & Aktas-Arnas, Y. (2007). Three-to six-year-old children's recognition of geometric shapes. *International Journal of Early Years Education, 15*(1), 83–104.

Berkowitz, T., Schaeffer, M. W., Maloney, E., Peterson, L., Gregor, C., Levine, S. C., et al. (2015). Math at home adds up to achievement in school. *Science, 350*(6257), 196–198.

Bers, M. U. (2010). The Tangible Robotics program: Applied computational thinking for young children. *Early Childhood Research and Practice, 12*(2), 1 20.

Beschorner, B., & Hutchison, A. (2013). iPads as a literacy teaching tool in early childhood. *International Journal of Education in Mathematics, Science and Technology, 1*(1), 16–24.

Brosnan, M. J. (1998). Spatial ability in children's play with Lego blocks. *Perceptual and Motor Skills, 87,* 19–28.

Caldera, Y. M., Culp, A. M., O'Brien, M., Truglio, R. T., Alvarez, M., & Huston, A. C. (1999). Children's play preferences, construction play with blocks and visual-spatial skills. Are they related? *International Journal of Behavioral Development, 23,* 855–872.

Cannon, J., & Ginsburg, H. P. (2008). "Doing the math": Maternal beliefs about early mathematics versus language learning. *Early Education & Development, 19,* 238–260.

Casasola, M., & Bhagwat, J. (2007). Does a novel word facilitate 18-month-olds' categorization of a spatial relation? *Child Development, 78,* 1818–1829.

Casasola, M., Bhagwat, J., & Burke, A. S. (2009). Learning to form a spatial category of tight-fit relations: How experience with a label can give a boost. *Developmental Psychology, 45,* 711–723.

Casey, B. M., Andrews, N., Schindler, H., Kersh, J. E., Samper, A., & Copley, J. (2008a). The development of spatial skills through interventions involving block building activities. *Cognition and Instruction, 26,* 269–309.

Casey, B. M., & Bobb, B. (2003). The power of block building. *Teaching Children Mathematics, 10,* 98–102.

Casey, B. M., Erkut, S., Ceder, I., & Mercer Young, J. (2008b). Use of a storytelling context to improve girls' and boy's geometry skills in kindergarten. *Journal of Applied Developmental Psychology, 29,* 29–48.

Casey, B. M., Nuttall, R., Pezaris, E., & Benbow, C. P. (1995). The influence of spatial ability on gender differences in mathematics college entrance test scores across diverse samples. *Developmental Psychology, 31*(4), 697–705. https://doi.org/10.1037/0012-1649.31.4.697.

Cheng, Y., & Mix, K. S. (2014). Spatial training improves children's mathematics ability. *Journal of Cognition and Development, 15*(1), 2–11.

Chiong, C., & Shuler, C. (2010). *Learning: Is there an app for that? Investigations of young children's usage and learning with mobile devices and apps.* New York: The Joan Ganz Cooney Center at Sesame Workshop.

Clark, R. C., & Mayer, R. E. (2008). Learning by viewing versus learning by doing: Evidence-based guidelines for principled learning environments. *Performance Improvement, 47*(9), 5–13.

Clarke, B. A. (2004). A shape is not defined by its shape: Developing young children's geometric understanding. *Journal of Australian Research in Early Childhood Education, 11*(2), 110–127.

Clements, D. H. (1999). 'Concrete' manipulates, concrete ideas. *Contemporary Issues in Early Childhood, 1*(1), 45–60.

Clements, D. H., & Meredith, J. S. (1993). Research on logo: Effects and efficacy. *Journal of Computing in Childhood Education, 4,* 263–290.

Clements, D. H., & Sarama, J. (2007). Effects of a preschool mathematics curriculum: Summative research on the building blocks project. *Journal for Research in Mathematics Education, 38,* 136–163.

Clements, D. H., Swaminathan, S., Hannibal, M. A., & Sarama, J. (1999). Young children's concepts of shape. *Journal of Research in Mathematics Education, 30,* 192–212.

Common Sense Media. (2013). *Zero to eight: Children's media use in America: A common sense research study.* Retrieved from https://www.commonsensemedia.org/research/zero-to-eight-childrens-media-use-in-america-2013.

Cooper, L. Z. (2005). Developmentally appropriate digital environments for young children. *Library Trends, 54,* 286–302.

Cross, C. T., Woods, T. A., & Schweingruber, H. (Eds.). (2009). *Mathematics learning in early childhood.* Washington, D.C.: National Academies Press.

Delgado, A. R., & Prieto, G. (2004). Cognitive mediators and sex related differences in mathematics. *Intelligence, 32*(1), 25–32. https://doi.org/10.1016/s0160-2896(03)00061-8.

Dessalegn, B., & Landau, B. (2008). The role of language in binding and maintaining feature conjunctions. *Psychological Science, 19*(2), 189–195. https://doi.org/10.1111/j.1467-9280.2008.02066.x.

Duncan, G. J., Dowsett, C., Claessens, A., Magnuson. K., Huston, A., Klebanov, P., … Japel, C. (2007). School readiness and later achievement. *Developmental Psychology, 43*(6), 1428–1446.

Eagle, S. (2012). Learning in the early years: Social interactions around picturebooks and digital technologies. *Computers & Education, 59,* 38–49.

Falloon, G. (2013). Young children using iPads: App design and content influences on their learning pathways. *Computers & Education, 68,* 505–521.

Ferrara, K., Hirsh-Pasek, K., Newcombe, N. S., Golinkoff, R. M., & Lam, W. S. (2011). Blocktalk: Spatial language during block play. *Mind, Brain, and Education, 5*(3), 143–151.

Flanagan, R. (2013). Effects of learning from interaction with physical or mediated devices. *Cognitive Processes, 14,* 213–215.

Flynn, R. M., & Richert, R. A. (2015). Parents support preschoolers' use of a novel interactive device. *Infant and Child Development.* https://doi.org/10.1002/icd.1911.

Foster, E. K., & Hund, A. M. (2012). The impact of scaffolding and overhearing on young children's use of the spatial terms between and middle. *Journal of Child Language, 39*(2), 338–364.

Gee, J. (2008). Good videogames, the human mind, and good learning. In E. Wood, & T. Willoughby (Eds.), *Children learning in a digital world* (pp. 40–63). Oxford, UK: Blackwell Publishing.

Geist, E. (2014). Using tablet computers with toddlers and young preschoolers. *Young Children, 69,* 58–63.

Gentner, D. (2003). Why we're so smart. In D. Gentner & S. Goldin-Meadow (Eds.), *Language in mind: Advances in the study of language and cognition.* Cambridge, MA: MIT Press.

Gentner, D., & Lowenstein, J. (2002). Relational language and relational thought. In E. Amsel & P. Brynes (Eds.), *Language, literacy, and cognitive development: The development and consequences of symbolic communication* (pp. 87–120). Mahwah, NJ: Erlbaum.

Ginsburg, H. P. (2006). Mathematical play and playful mathematics: A guide for early education. In D. G. Singer, R. M. Golinkoff, & K. Hirsh-Pasek (Eds.), *Play = Learning: How play motivates and enhances children's cognitive and social-emotional growth.* New York, NY: Oxford University Press.

Gogel, W. C., & Tietz, J. D. (1980). Relative cues and absolute distance perception. *Perception and Psychophysics, 28,* 321–328.

Goldwin, K., & Highfield, K. (2013). A framework for examining technologies and early mathematics learning. In L. D. English & J. T. Mulligan (Eds.), *Reconceptualizing early mathematics learning* (pp. 205–226). New York, NY: Springer.

Grant, A., Wood, E., Gottardo, A., Evans, M., Phillips, L., & Savage, R. (2012). Assessing the content and quality of commercially available reading software programs: Do they have the fundamental structures to promote the development of early reading skills in children? *NHSA: Dialogue: A Research to Practice Journal for the Early Intervention Field, 15*(4), 319–342.

Guernsey, L., & Levine, M. H. (2015). *Tap, click, read: Growing readers in a world of screens.* San Francisco, CA: Wiley/Jossey-Bass.

Guildford, J. P. (1967). *The nature of human intelligence.* New York, NY: McGraw-Hill.

Gunderson, E. A., Ramirez, G., Beilock, S. L., & Levine, S. C. (2012). The relation between spatial skill and early number knowledge: The role of the linear number line. *Developmental Psychology, 48*(5), 1229–1241. https://doi.org/10.1037/a0027433.

Hermelin, B., & O'Connor, N. (1986). Spatial representations in mathematically and in artistically gifted children. *British Journal of Educational Psychology, 56,* 150–157.

Hermer-Vazquez, L., Moffet, A., & Munkholm. (2001). Language, space, and the development of cognitive flexibility in humans: The case of two spatial memory tasks. *Cognition, 79,* 263–299.

Highfield, K. (2015). Stepping into STEM with young children: Simple robotics and programming as catalysts for early learning. In C. Donohue (Ed.), *Technology and digital media in the early years* (pp. 150–161). New York, NY: Routledge.

Hirsh-Pasek, K., Berk, L. E., Singer, D. G., & Golinkoff, R. M. (2008). *A mandate for playful learning in preschool: Presenting the evidence.* New York, NY: Oxford University Press.

Hirsh-Pasek, K., Zosh, J. M., Golinkoff, R. M., Gray, J. H., Robb, M. B., & Kaufman, J. (2015). Putting education in "Educational" apps: Lessons from science of learning. *Psychological Science in the Public Interest, 16*(1).

Ho, A., Lee, J., Wood, E., Kassies, S., & Heinbuck, C. (2017). Tap, swipe & build: Parental spatial input during iPad® and toy play. *Infant and Child Development.* https://doi.org/10.1002/icd.2061.

Jirout, J., & Newcombe, N. (2014). Mazes and maps: Can young children find their way? *Mind, Brain, and Education, 8,* 89–96.

Jirout, J., & Newcombe, N. (2015). Building blocks for developing spatial skills: Evidence from a large, representative U.S. sample. *Psychological Science, 26*(3), 302–308.

Kabali, H. K., Irigoyen, M. M., Nunez-Davis, R., Budacki, J. G., Mohanty, S. H., Leister, K. P. … Bonner, R. L. (2015). Exposure and use of mobile media devices by young children. *Pediatrics.* https://doi.org/10.1542/peds.2015-2151.

Kamii, C., Miyakawa, Y., & Kato, Y. (2004). The development of logico-mathematical knowledge in a block-building activity at ages 1–4. *Journal of Research in Childhood Education, 19,* 13–26.

Karemaker, A., Pitchford, N. J., & O'Malley, C. (2010). Enhanced recognition of written words and enjoyment of reading in struggling beginner readers through whole-word multimedia software. *Computers & Education, 54*(1), 199–208.

Katz, L. G. (2010). STEM in early years. *Early childhood research and practice*. Collected Papers from the SEED (STEM in Early Education and Development) Conference. Retrieved from http://ecrp.uiuc.edu/beyond/seed/katz.html.

Kell, H. J., Lubinski, D., Benbow, C. P., & Steiger, J. H. (2013). Creativity and technical innovation: Spatial ability's unique role. *Psychological Science, 24,* 1831–1836.

Klatzky, R. L. (1998). Allocentric and egocentric spatial representations: Definitions, distinctions, and interconnections. In C. Freksa, C. Habel, & W. F. Heidelberg (Eds.), *Spatial cognition*. LNCS (LNAI), (Vol. 1404, pp. 1–17). Heidelberg, Germany: Springer.

Kuhn, D. (2000). Metacognitive development. *Current Directions in Psychological Science, 9*(5), 178–181.

Kurcirkora, N. (2014). iPads in early education: Separating assumptions and evidence. *Frontiers in Psychology, 5*(715). https://doi.org/10.3389/fpsyg.2014.00715.

Larkin, K. (2016). Geometry and iPads in primary schools: Does their usefulness extend beyond tracing an oblong? In P. S. Moyer-Packenham (Ed.), *Mathematics Education in the Digital Era: International Perspectives on Teaching and Learning Mathematics with Virtual Manipulatives* (pp. 247–274). Basel, Switzerland: Springer International Publishing.

Lauer, J. E., & Lourenco, S. F. (2016). Spatial processing in infancy predicts both spatial and mathematical aptitude in childhood. *Psychological Science, 27*(10), 1291–1298.

Lee, J., Douglas, E., Wood, E., & Andrade, S. (2017). *How educational are math apps for preschoolers?* Poster presented at the 2017 Society for the Research of Child Development Biennial Meeting, Austin, TX.

Lee, J., Kotsopoulos, D., & Zambrzycka, J. (2012). Does block play support children's numeracy development? In L. R. Van Zoest, J. J. Lo, & J. K. Kraty (Eds.), *Proceedings of the 34th Annual Meeting of the North American Chapter of the International Group for the Psychology of Mathematics Education* (pp. 1028–1031). Kalamazoo, MI: Western Michigan University.

Levine, S. C., Ratliff, K. R., Huttenlocher, J., & Cannon, J. (2012). Early puzzle play: A predictor of preschoolers' spatial transformation skill. *Developmental Psychology, 48*(2), 530–542.

Levinson, A. J., Weaver, B., Garside, S., McGinn, H., & Norman, G. R. (2007). Virtual reality and brain anatomy: A randomized trial of e-learning instructional designs. *Medical Education, 41,* 495–501.

Loewenstein, J., & Gentner, D. (2005). Relational language and the development of relational mapping. *Cognitive Psychology, 50,* 315–353.

Lucas, K., & Sherry, J. L. (2004). Sex differences in video game play: A communication based explanation. *Communication Research, 31,* 499–523.

Lysenko, L., Rosenfield, S., Dedic, H., Savard, A., Idan, E., Abrami, P. C., et al. (2016). Using interactive software to teach foundational mathematical skills. *Journal of Information Technology Education: Innovations in Practice, 15,* 19–34.

MacDonald, S. (2001). *Blockplay: The complete guide to learning and playing with blocks*. Beltsville, MD: Gryphon House.

Martin, T. (2009). A theory of physically distributed learning: How external environments and internal states interact in mathematics learning. *Child Development Perspectives, 3*(3), 140–144.

Martin, T., Lukong, A., & Reaves, R. (2007). The role of manipulatives in arithmetic and geometry tasks. *Journal of Education and Human Development, 1*(1) (Electronic version).

Mayer, R. (2005). *The Cambridge handbook of multimedia learning*. New York, NY: Cambridge University Press.

McKenney, S., & Voogt, J. (2010). Technology and young children: How 4–7 year olds perceive their own use of computers. *Computers in Human Behavior, 26*(4), 656–664.

McQuiggan, S., Kosturko, L., McQuiggan, J., & Sabourin, J. (2015). *Mobile learning: A handbook for developers, educators, and learners*. Hoboken, NJ: John Wiley & Sons Inc.

Moyer-Packenham, P. S., & Bolyard, J. J. (2016). Revisiting the definition of a virtual manipulative. In P. S. Moyer-Packenham (Ed.), *Mathematics Education in the Digital Era: International Perspectives on Teaching and Learning Mathematics with Virtual Manipulatives* (pp. 3–23). Basel, Switzerland: Springer International Publishing.

Moyer-Packenham, P. S., Shumwa, J. F., Bullock, E., Tucker, S. I., Anderson-Pence, K., Westernskow, A., et al. (2015). Young children's learning performance and efficiency when using virtual manipulative mathematics iPad apps. *Journal of Computers in Mathematics and Science Teaching, 34*(1), 41–69.

Moyer-Packenham, P. S., & Westenskow, A. (2013). Effects of virtual manipulatives on student achievement and mathematics learning. *International Journal of Virtual and Personal Learning Environments, 4*(3), 35–50.

Mueller, J., Wood, E., De Pasquale, D., & Archer, K. (2011). Students learning with mobile technologies in and out of the classroom. In A. Mendez-Vilas (Ed.), *Education in a technological world: Communicating current and emerging research and technological efforts*. Badajoz, Spain: Formatex Research Centre.

Musawi, A. (2011). Redefining technology role in education. *Creative Education, 2,* 130–135.

National Association for the Education of Young Children (NAEYC) and the Fred Rogers Center for Early Learning and Children's Media at Saint Vincent College. (2012). *Technology and Interactive Media as Tools in Early Childhood Programs Serving Children from Birth through Age 8*. Joint Position Statement. Washington, D.C: NAEYC.

National Council of Teachers of Mathematics. (2007). *Second handbook of research on mathematics teaching and learning*. Washington, D.C: Council of Teachers of Mathematics.

Needham, A. (2009). Learning in infants' object perception, object-directed action, and tool use. In A. Woodward & A. Needham (Eds.), *Learning and the infant mind* (pp. 208–226). New York, NY: Oxford University Press.

Neumann, M. M. (2016). Young children's use of touch screen tablets for writing and reading at home: Relationships with emergent literacy. *Computers & Education, 97,* 61–68.

Newcombe, N. S. (2010). Picture this: Increasing math and science learning by improving spatial thinking. *American Educator*, Summer, 29–43.

Newcombe, N. S., & Frick, A. (2010). Early education for spatial intelligence: Why, what, and how. *Mind, Brain, & Education, 4*(3), 102–111.

Pan, Z., Cheok, A. D., Yang, H., Zhu, J., & Shi, J. (2006). Virtual reality and mixed reality for virtual learning environments. *Computers and Graphics, 30,* 20–28.

Parish-Morris, J., Mahajan, N., Hirsh-Pasek, K., Golinkoff, R. M., & Collins, M. F. (2013). Once upon a time: Parent–child dialogue and storybook reading in the electronic era. *Mind, Brain, and Education, 7*(3), 200–211.

Park, B., Chae, J.-L., & Boyd, B. F. (2008). Young children's block play and mathematical learning. *Journal of Research in Childhood Education, 23*(2), 157–162.

Patchan, M. M., & Puranik, C. S. (2016). Using tablet computers to teach preschool children to write letters: Exploring the impact of extrinsic and intrinsic feedback. *Computers & Education, 102,* 128–137.

Petersen, L., & Levine, S. (2014, September). Early block play predicts conceptual understanding of geometry and mathematical equivalence in elementary school. *SILC Showcase*. http://bit.ly/1nwuS4Q.

Piaget, J., & Inhelder, B. (1969). *The psychology of the child*. New York, NY: Basic Books.

Pitchford, N. J. (2015). Development of early mathematical skills with a tablet intervention: a randomized control trial in Malawi. *Frontiers in Psychology, 6*, 485. https://doi.org/10.3389/psyg.2015.00485.

Pollman, M. J. (2010). *Blocks and beyond: Strengthening early math and science skills through spatial learning*. Baltimore, MD: Paul Brookes Publishing Co.

Price, S., Jewitt, C., & Crescenzi, L. (2015). The role of iPads in pre-school children's mark making development. *Computers & Education, 87,* 131–141.

Pruden, S. M., Levine, S. C., & Huttenlocher, J. (2011). Children's spatial thinking: Does talk about the spatial world matter? *Developmental Science, 14*(6), 1417–1430.

Plowman, L., & Stephen, C. (2003). 'A benign addition'? Research on ICT and pre-school children. *Journal of Computer Assisted learning, 19,* 149–164.

Plumert, J. M., & Nichols-Whitehead, P. (1996). Parental scaffolding of young children's spatial communication. *Developmental Psychology, 32*(3), 523–532.

Reifel, S. (1984). Block construction: Children's developmental landmarks in representation of space. *Young Children, 40,* 61–67.

Rideout, V. J. (2014). *Learning at home: Families' educational media use in America. A report of the families and media project.* New York, NY: The Joan Ganz Cooney Center at Sesame Workshop.

Rosen, D., & Hoffman, J. (2009). Integrating concrete and virtual manipulatives in early childhood mathematics. *Young Children, 64*(3), 26–29, 31–33.

Sarama, J., & Clements, D. H. (2016). Physical and virtual manipulatives: What is "Concrete"? In P. S. Moyer-Packenham (Ed.), *Mathematics education in the digital era: International perspectives on teaching and learning mathematics with virtual manipulatives* (pp. 71–93). Basel, Switzerland: Springer International Publishing.

Satlow, E., & Newcombe, N. S. (1998). When is a triangle not a triangle? Young children's developing concepts of geometric shape. *Cognitive Development, 13*(4), 547–559.

Savage, R., Abrami, P. C., Piquette, N., Wood, E., Deleveaux, G., Sanghera-Sidhu, S., et al. (2013). A (pan-Canadian) cluster randomized control effectiveness trial of the ABRACADABRA web-based literacy program. *Journal of Educational Psychology, 105*(2), 310–328.

Shea, D. L., Lubinski, D., & Benbow, C. (2001). Importance of assessing spatial ability in intellectually talented young adolescents: A 20-year longitudinal study. *Journal of Educational Psychology, 93*(3), 604–614. http://dx.doi.org/10.1037/0022-0663.93.3.604.

Shelton, A., & Mcnamara, T. (2001). Systems of spatial reference in human memory. *Cognitive Psychology, 43,* 274–310.

Sigel, A., & Schadler, M. (1977). The development of young children's spatial representations of their classrooms. *Child Development, 48,* 388–394.

Song, H. S., Pusic, M., Nick, M. W., Sarpel, U., Plass, J. L., & Kalet, A. L. (2014). The cognitive impact of interactive design features for learning complex materials in medical education. *Computers & Education, 71,* 198–205.

Stieff, M. (2007). Mental rotation and diagrammatic reasoning in science. *Learning and Instruction, 17*(2), 219–234. https://doi.org/10.1016/j.learninstruc.2007.01.012.

Stiles, J., & Stern, C. (2009). Developmental change in spatial cognitive processing: Complexity effects and block construction performance in preschool children. *Journal of Cognition and Development, 2,* 157–187.

Stiles-Davis, J. (1988). Developmental change in young children's spatial grouping activity. *Developmental Psychology, 24,* 522–531.

Stull, A. T., & Mayer, R. E. (2007). Learning by doing versus learning by viewing: Three experimental comparisons of learner-generated versus author-provided graphic organizers. *Journal of Educational Psychology, 99,* 808–820.

Swing, E. L., & Anderson, C. A. (2008). How and what do video games teach? In T. Willoughby & E. Wood (Eds.), *Children's learning in a digital world* (pp. 64–84). Oxford, UK: Blackwell.

Takacs, Z., Swart, E., & Bus, A. (2015). Benefits and pitfalls of multimedia and interactive features in technology-enhanced story books: A meta-analysis. *Review of Educational Research, 85(4),* 698–739. https://doi.org/10.3102/0034654314566989.

Thorell, L. B., Lindqvist, S., Bergman Nutley, S., Bohlin, G., & Klingberg, T. (2009). Training and transfer effects of executive functions in preschool children. *Developmental Science, 12*(1), 106–113.

Travers, J. C., & More, C. M. (2013). TECH IT: Obtaining, evaluating, and using instructional technology innovations in early childhood. In H. P. Parette., & C. H. Blum (Eds.), *Instructional technology in early childhood: Teaching in the digital age.* Baltimore, MD: Brookes Publishing Co.

Uribe-Florez, L. J., & Wilkins, J. L. M. (2010). Elementary school teachers' manipulative use. *School Science & Mathematics, 110*(7), 363–371.

Uttal, D. H. (2000). Seeing the big picture: Map use and the development of spatial cognition. *Developmental Science, 3*(3), 247–264.

Uttal, D. H., Meadow, N. G., Tipton, E., Hand, L. L., Alden, A. R., Warren, C., et al. (2013). The malleability of spatial skills: A meta-analysis of training studies. *Psychological Bulletin, 139,* 352–402.

Van der Kleij, F. M., Feskens, R. C., & Eggen, T. J. (2015). Effects of feedback in a computer-based learning environment on students' learning outcomes: A meta-analysis. *Review of Educational Research, 85*(4), 475–511. https://doi.org/10.3102/0034654314564881.

Verdine, B. N., Golinkoff, R. M., Hirsh-Pasek, K., & Newcombe, N. S. (2014a). Deconstructing building blocks: Preschoolers' spatial assembly performance relates to early mathematical skills. *Child Development, 85*(3), 1062–1076.

Verdine, B. N., Golinkoff, R. M., Hirsh-Pasek, K., & Newcombe, N. S. (2014b). Finding the missing piece: Blocks, puzzles, and shapes fuel school readiness. *Trends in Neuroscience and Education, 3,* 7–13.

Verdine, B. N., Irwin, C., Golinkoff, R. M., & Hirsh-Pasek, K. (2014c). Contributions of executive function and spatial skills to preschool mathematics achievement. *Journal of Experimental Child Psychology, 126,* 37–51.

Vygotaksy, L. S. (1978). *Mind in society: The development of higher mental processes.* Cambridge, MA: Harvard University Press.

Vogel, J. J., Vogel, D. S., Cannon-Bowers, J., Bowers, C. A., Muse, K., & Wright, M. (2006). Computer gaming and interactive simulations for learning: A meta-analysis. *Journal of Educational Computing Research, 34*(3), 229–243.

Wai, J., Lubinski, D., & Benbow, C. P. (2009). Spatial ability for STEM domains: Aligning over 50 years of cumulative psychological knowledge solidifies its importance. *Journal of Educational Psychology, 101*(4), 817–835. http://dx.doi.org/10.1037/a0016127.

Willoughby, T., & Wood, E. (Eds.). (2008). *Children learning in a digital world: Opportunities and challenges.* Oxford, UK: Blackwell.

Willoughby, T., Wood, E., Desjarlais, M., Williams, L., Leacy, K., & Sedore, L. (2009). Social interaction during computer-based activities: Comparisons by number of sessions, gender, school-level, gender composition of the group, and computer-child ratio. *Sex Roles, 61*(11–12), 864–878.

Wolfgang, C. H., Stannard, L. L., & Jones, I. (2001). Block play performance among preschoolers as a predictor of later school achievement in mathematics. *Journal of Research in Childhood Education, 15,* 173–180.

Wood, E., Hui, B., & Willoughby, T. (2008). Introduction to formal learning with technologies: Exploring the role of digital technologies. In T. Willoughby & E. Wood (Eds.), *Children Learning in a Digital World.* Oxford, UK: Blackwell.

Wood, E., Grant, A., Gottardo, A., Savage, R., & Evans, M. A. (2017). Software to promote young children's growth in literacy: A comparison of online and offline formats. *Early Childhood Education Journal, 45*(2), 207–217.

Wood, E., Petkovski, M., De Pasquale, D., Gottardo, A., Evans M. A., & Savage, R. (2016). Parent scaffolding of young children when engaged with mobile technology. *Frontiers in Psychology, 7.* https://doi.org/10.3389/fpsyg.2016.00690.

Yang, J. C., & Chen, S. Y. (2010). Effects of gender differences and spatial abilities within a digital pentominoes game. *Computers & Education, 55,* 1220–1233.

Yelland, N. J. (1994). The strategies and interactions of young children in LOGO tasks. *Journal of Computer Assisted learning, 10,* 33–49.

Yelland, N. J. (2002). Playing with ideas and games in early mathematics. *Contemporary Issues in Early Childhood, 3,* 197–215.

Yin, S. H. (2003). Young children's concept of shape: Van Hiele visualization level of geometric thinking. *The Mathematics Educator, 7*(2), 71–85.

Zbeik, R. M., Heid, M. K., Blume, G. W., & Dick, T. P. (2007). Research on technology in mathematics education: A perspective of constructs. In F. K. Lester (Ed.), *Second handbook of research on mathematics teaching and learning* (Vol. 2, pp. 1169–1207). Charlotte, NC: Information Age Publishing Incorporated.

Zosh, J. M., Verdine, B. N., Filipowicz, A., Golinkoff, R. M., Hirsh-Pasek, K., & Newcombe, N. S. (2015). Talking shape: Parental language with electronic versus traditional shape sorters. *Mind, Brain, and Education, 9*(3), 136–144.

Chapter 12
Video Game Play, Mathematics, Spatial Skills, and Creativity—A Study of the Impact on Teacher Candidates

Janet Lynne Tassell, Elena Novak and Mengjiao Wu

Abstract This chapter explores the relationships among video gaming, spatial skills and creativity in mathematics education. Specifically, it highlights the importance of spatial abilities for pre-service elementary teachers, and suggests video games as a teaching approach for potentially enhancing creativity, spatial abilities, and mathematics performance. We argue that spatial abilities deserve more attention in mathematics education, as a major predictor of achievements in science, technology, engineering, and mathematical fields. To support this notion, we describe an experimental study that examined the effects of playing the Angry Birds and Action Video recreational video games on education majors' math problem-solving and perceptions, math anxiety, working memory, and spatial skills. Individuals with high spatial abilities had significantly higher confidence in learning mathematics, ACT mathematics, science, composite scores, as well as geometry, word, and non-word math problem solving than individuals with low spatial abilities. In addition, students with low spatial abilities had significantly higher math anxiety. After ten hours of playing, both video game intervention groups significantly improved their spatial skills, working memory, and geometry performance from pre- to post-test. These findings suggest potential impact of video gaming in mathematics education and open new horizons for future research that explores how schools and homes working together with strategic gaming plans can help students improve their spatial reasoning and problem solving. The chapter concludes with future research suggestions on spatial abilities and creativity in mathematics education.

Keywords Action video games · Mathematics education · Spatial skills
Creativity · Preservice education

J. L. Tassell (✉)
Western Kentucky University, Bowling Green, KY, USA
e-mail: janet.tassell@wku.edu

E. Novak · M. Wu
Kent State University, Kent, OH, USA
e-mail: enovak6@kent.edu

M. Wu
e-mail: Mwu8@kent.edu

© Springer International Publishing AG, part of Springer Nature 2018

V. Freiman and J. L. Tassell (eds.), *Creativity and Technology in Mathematics Education*, Mathematics Education in the Digital Era 10,
https://doi.org/10.1007/978-3-319-72381-5_12

12.1 Introduction

How can mathematics learning be impacted by video game play? This is a question that many people in our society, both past and present, might find as ridiculous or inappropriate to even consider. Often, video game play is perceived as a waste of time by parents. Teachers tend not to look toward creative uses for video game play in light of today's focused climate of the standards-based world in education. However, research studies are finding that playing certain recreational video games can improve various cognitive and perceptual abilities, such as attention, working memory, and reaction times (Boot, Kramer, Simons, Fabiani, & Gratton, 2008; Green & Bavalier, 2012). A considerable number of studies also documented improved performance on tasks, such as to rehabilitate the elderly, hone surgical skills, and sharpen flight performance (Green & Bavalier, 2006).

In light of the growing literature body on the impact of video games on learning and cognition, we were interested in exploring whether casual video games are making an impact with video-game player's spatial and mathematics abilities as well and recommend future research directions that can connect these findings to creativity. This question drove us to examine the effects of playing *Unreal Tournament,* an intense first-person shooter video game, and *Angry Birds*, a popular low-stress video game, on spatial abilities and mathematics proficiency and attitudes toward mathematics. The results have led us to another question: How might video-gaming be used as a creative instructional approach to increase student's skills in mathematics problem solving and spatial reasoning?

We assert that teaching approaches aiming at improving cognitive abilities that are essential for successful mathematics learning have a potential of better preparing students for learning mathematics and increasing student interest toward mathematics. When considering what population to study, we decided to focus on pre-service teachers due to the pressures that they have to perform on the mathematics portion of the Praxis II teacher-licensing exam. We also wanted to be able to improve the mathematics abilities and confidence of future teachers as well as broaden their perceptions about possible instructional interventions for improving mathematics learning. Research shows that elementary education teacher majors report the highest average scores on mathematics anxiety tests amongst all college majors (Hembree, 1990). Often, students within the elementary education major have neither sufficient number of mathematics experiences through coursework nor do they get appropriate interventions to help change their attitudes and perceptions, which further complicates and perpetuates the issue (Ma, 1999).

The goal of this book chapter is two-fold. First, we report a research study that examined the effects of recreational video gaming on various cognitive abilities that are crucial for mathematics learning, focusing primarily on spatial abilities, which are central to successful geometry and general mathematics problem-solving learning. This chapter reviews the importance of spatial abilities in Science, Technology, Engineering, and Mathematics (STEM) education, especially among pre-service elementary teachers. It suggests that spatial abilities deserve more attention as a major predictor of achievements in science, technology, engineering, and mathematical fields. Second, we present a challenge for future research focusing on the relationships among spatial skills and creativity in teacher candidates, and in turn how to develop the creativity. The chapter also discusses the idea of creative approaches to teaching with video gaming.

12.2 Spatial Skills

12.2.1 Spatial Skills and Mathematics

Spatial skills, or visuospatial ability, refers to skills in representing, transforming, generating, and recalling symbolic, nonlinguistic information (Halpern & LaMay, 2000; Linn & Petersen, 1985). Mathematics skills, coupled with spatial skills, are critical to success in the areas of STEM including professions, such as architecture, medicine, chemistry, and engineering (Ceci & Williams, 2007). Cohen and Hegarty (2007) attributed such disadvantages to the difficulties of understanding and using the intensely-spatial information. More emphasis was put on the effect of spatial abilities on achievement in mathematics, and specifically geometry (Holzinger & Swineford, 1946). Spatial abilities have been successfully used to predict STEM achievement with mathematics and verbal skills controlled: lower spatial abilities are likely to result in lower STEM performances (Lubinski & Benbow, 2006; Shea, Lubinski, & Benbow, 2001). Visual perception and spatial ability are connected to mathematical thinking, and important for mathematical problem solving (Hegarty & Waller, 2005). For students to be successful in 3D geometry thinking, strong predictors are spatial visualization, spatial orientation, and spatial relations factors (Pittalis & Christou, 2010).

The interest in research on spatial abilities is attributed to its impact on mathematics performance. In a meta-analysis of 75 studies, Friedman (1995) found a substantial relationship between mathematical and spatial abilities, ranging from 0.3 to 0.45. A more recent meta-analysis by Mix and Cheng (2011) further confirmed the positive correlation between these two constructs across ages and tasks. The results indicated that the students with higher spatial skills tend to achieve better in mathematics. Empirical evidence has shown that geometry education can improve students' spatial abilities. For instance, high school students that received Descriptive Geometry instruction outperformed their peers from a control group on spatial ability tasks (Gittler & Gluck, 1998). Further demonstrating a connection between geometry reasoning and spatial abilities, Pittalis and Cristou (2010) revealed the relationships a student performance predictor in the four types of reasoning in 3D geometry. Their study leads to an assumption that students can improve on 3D geometry tasks with spatial reasoning practice.

12.2.2 Spatial Skills and Gender Differences

Numerous researches have consistently shown that males have better spatial abilities than females. The sex differences in spatial ability were apparently observed in infants even at 5 months old (Moore & Johnson, 2008). In a study with over one thousand high school students, Fennema and Sherman (1976b) found firm empirical evidence on sex-related differences in spatial abilities, and spatial visualization. Both of the spatial entities had a stronger sex-related difference than mathematical ability itself. In terms of spatial orientation, males were found to significantly outperform females on spatial perception and prediction in elementary school (Maxwell, Croake, & Biddle, 1975). In addition, gender differences were revealed in a study on spatial skills with 274 undergraduate participants (Casey, Nuttall, Pezaris, & Benbow, 1995): male students scored higher than female students for mental rotation and math aptitude, especially in the high-ability samples. The same pattern was reported in an earlier study conducted by Burnett, Lane, and Dratt (1979). In both studies, after controlling for the effects of spatial ability, the gender difference in math ability became insignificant, thus providing a clear indication about the role of spatial ability in the gender difference in mathematical performance. Linn and Petersen (1985) concluded in the meta-analysis on sex differences in spatial ability that sex differences favoring males existed in spatial perception and mental rotation but not in spatial visualization. Another meta-analysis by Tracy (1987) reached similar conclusions among children but suggested future research to use children's sex role orientations instead of biological sex for more reasonable assessment when analyzing sex-related differences.

What contributed to the gender differences in spatial abilities? Tracy (1987) attributed the differences to children's toy preference. Her meta-analysis revealed overwhelming evidence in support of sex-related toy preference and playing behavior in children: boys tend to play with toys that are spatial in nature (i.e., vehicles), while girls' toys usually do not encourage manipulation and movement through space. Thus, boys' spatial abilities are promoted by their choice of spatial since childhood.

Prior experiences in sports can also contribute to enhanced spatial ability. Researchers have found that athletes perform better than non-athletes on mental rotation tests (Ozel, Larue, & Molinaro, 2004). In many cultures, as part of growing up, males experience greater amounts of physical activities that require spatial reasoning skills involving hand-eye coordination (i.e., playing football, climbing trees, etc.). Those activities are believed to promote men's development of spatial cognition, which finally leads to gender difference in spatial abilities (Bjorklund & Brown, 1998).

Socio-cultural factors are also regarded as an important cause of gender differences in spatial abilities (Fennema & Sherman, 1976b). Sex stereotyping of mathematics and spatial abilities resulted in less confidence and less involvement of girls than boys in math and spatial activities, which increased the gender difference in math and spatial performance. However, if females were given proper instruction and practice, the gender differences could be diminished and even removed (Maxwell et al., 1975).

In the elementary education major, few of the students have been athletes, and more have been females. With this current state of composition, some of the natural benefits that life experiences bring to spatial reasoning escape many prospective and current teachers. With the gender gap in visuospatial reasoning and even the intentional methods to lower the gap (Newcombe, 2007; Terlecki, Newcombe, & Little, 2008), the impact remains unclear as to whether the intervention practices are effective. The concern is that there is a gap itself and elementary teachers are predominantly female, teaching generation upon generation of students, when they themselves may very well fall prey to the downside of the gap. Thus, the importance of developing spatial skills increased among elementary education majors.

12.2.3 Spatial Abilities and Math Anxiety

It has been an accepted belief that students with higher math anxiety tend to perform more poorly in mathematics than their low-math-anxious peers. High math anxiety also leads to less participation in STEM disciplines, and lower possibility of selecting STEM-related careers (Casey et al., 1995). The impact of pre-service elementary teachers' (PSET) math anxiety might go beyond their own achievement and interests. According to Gresham (2008), pre-service teachers with high math anxiety have been shown to develop lower math teaching self-efficacy and negative teaching attitudes, which might negatively affect their students' achievement. Recent empirical evidence has showed that female teachers' math anxiety inhibited female students' math development (Beilock, Gunderson, Ramirez, & Levine, 2010).

Mathematics anxiety is also negatively correlated with spatial abilities (Casey et al., 1995). In one of our previous studies (Novak & Tassell, 2017) we explored whether spatial ability was a mediator of the relationships between math problem solving in geometry, word problem, and non-word problem content areas and math anxiety. Spatial ability negatively correlated with student mathematics anxiety and was found to be a significant positive predictor of performance in geometry and word problem-solving after controlling for the effects of math anxiety. The results were consistent with earlier literature, supporting positive correlations between spatial abilities and geometry, and complex word problem solving (Brown & Wheatley, 1997; van Garderen, 2006).

Spatial abilities were also successfully used to partly explain why females experienced higher math anxiety than males (Maloney et al., 2012). The data indicated that spatial processing ability mediated the relationship between sex and math anxiety among college students. One possible explanation for the mediating role of spatial abilities is that math anxiety is preceded by poor spatial processing abilities. Maloney et al. argued that both spatial skills and gender differences in mental rotation emerge early in development (Moore & Johnson, 2008), and tend to remain stable after early childhood (Mortensen et al., 2003). Therefore, poor spatial ability indirectly results in high math anxiety through low math achievement (Maloney et al., 2012). Therefore, it is plausible to assume that the enhancement of spatial abilities might reduce math anxiety and improve math competence.

12.2.4 Spatial Training to Improve Spatial Skills

Given the predicting power of spatial abilities in STEM achievement, and the relationship between spatial ability and math anxiety, spatial training seems to be an extremely valuable tool in fighting against math underperformance and math anxiety. Although the literature on the effects of spatial training on performance in STEM disciplines is extremely scarce (Uttal, Meadow, Tipton, Hand, Alden, & Warren, 2013), several studies have demonstrated that training spatial skills can be transferred to generalized improvement in mathematics tasks that were not directly trained (Cheng & Mix, 2014; Mix & Cheng, 2011).

Spatial training has recently received high interest by researchers. There has been promising evidence showing positive effects of spatial training on improving spatial skills [see Uttal's et al., (2013) detailed meta-analysis on the effects of spatial training]. Uttal and colleagues classified training programs into three categories: video games, instructional course, and psychology laboratory training. Each training program was found to be effective in increasing spatial skills and spatial training effects were stable, persisting, and transferrable (to novel tasks). Specifically, spatial training of playing action video games has been found to successfully reduce gender difference in spatial abilities in very short time of training (Feng, Spence, & Pratt, 2007).

12.3 Rationale for Instruction that Is Creative and "Outside" of the School Day

In most schools and situations, spatial skills are taught through the typical mathematics curricula. However, a more focused approach through video game training for strengthening spatial skills can be inserted into a student's life to enhance and potentially improve spatial skills. Whether the mission is to teach and improve spatial skills in children or to enhance skills in pre-service teachers, employing fun and engaging activities that can be implemented both out of and in school. Such video game play can be a method considered for a creative angle of instruction. If a teacher is considering how to improve spatial reasoning skills, creative instruction is an option that has been researched. Pittalis and Cristou (2010) note that students might gain better improvement in their visuospatial skills if they were encouraged to explore 3D geometry activities outside of school.

According to Gunter et al. (2008), we are seeing a big rush to incorporate educational content in video games and the hope is that the players are motivated to play and in turn learn the content that is embedded in the game. The US National Council of Teachers of Mathematics (NCTM, 2000) association published a document, *Principles and Standards for School Mathematics*, where technology was held up as essential for teaching and learning due to its efficient and effective way to help students explore mathematics concepts in a deeper and different way. However, concerns creep into the discussion regarding technology use for instruction as teachers may not have the pedagogical understanding for how to implement the tool effectively in the classroom (Hoyles et al., 2004). The effective way to intertwine technology requires meaningful connection to the instructional objectives (Lawrenz et al., 2006). Therefore, teachers need to plan for this complexity and ensure a tight alignment with the instructional objectives to achieve the optimal engagement and student achievement (Clark-Wilson et al., 2014; Dewey et al., 2009). The connection to the instructional objectives is essential to avoiding conflicting opinions about game-based instruction (Bragg, 2007). The biggest challenge for teachers is finding the right mix of technology and instruction to have the optimal engagement, motivation, and interest for the students (de Freitas, 2006).

The use of video games for instruction, especially playing at home, could be a cultural and creative pedagogical shift. If teachers and parents would consider allowing carefully selected video game play in their instructional practices, it may help students increase their spatial reasoning and working memory skills. The cycle of out-dated instructional practices can be impacted by breathing a new life of technology by using video game play strategically. Instructional practices with freedom for students to work at their own pace with choice can provide students with opportunities to think critically and can allow for explorative learning (Lin, 2011).

The classroom environment should be cultivated to facilitate learning. When teachers incorporate creative instructional techniques like encouragement, or using video games, could they be fulfilling Freudenthal's (1971) suggestion that students be able to experience mathematics? Freudenthal emphasized that it was important that mathematics be taught in a manner that students invent themselves—in other words, designing their own learning pathway. Video gaming, in the classroom or at home, on their own terms, can be this "personal pathway" to improving spatial skills—a "creative" approach to allowing students to control their own learning.

12.4 Video Game Impact on Spatial Reasoning

In terms of the impact of video game play on spatial reasoning, different games and amount of video game training time have been considered as influential in research over the years (Cherney, 2008). In one school setting, "The Factory" or "Stellar 7" were used with fifth, seventh, and ninth graders to study mental rotation skills with two 45-min training sessions over six weeks (McClurg & Chaillé, 1987).

The findings showed that both genders in computer games treatment group outperformed the control group. In a similar study, fifth graders played "Marble Madness" (involving spatial skills) and "Conjecture" (no spatial skills), for three 45-min sessions (Subrahmanyam & Greenfield, 1994). Students who had the lowest pretest spatial skills made the greatest gains, pre- to post-test. Another study involved "Tetris" with college students, targeting mental rotation speed and spatial visualization (Okagaki & Frensch, 1994). The researchers found that after six hours of training, both genders improved in both areas. More recently, research has shown that playing action video games for only 10 h can eliminate gender differences in spatial abilities with females benefitting more than males (Feng et al., 2007). Cherney (2008) conducted a study to determine the amount of time and delivery methods that might impact mental rotation skills. The research findings suggest that only four hours in computer game training was needed to improve mental rotation skills. Both genders improved, but females improved with a significantly larger margin.

Playing specific video games, such as puzzles and action video games, has been proven through research studies to help both females and males in their spatial reasoning skills (Dorval & Pepin, 1986; McClurg & Chaillé, 1987; Subrahmanyam & Greenfield, 1994). However, such games are "naturally" played most often by males, thus females who usually have lower spatial abilities than males further develop a gender gap by not playing video games that have a potential of improving their spatial ability (Cherney & London, 2006; Terleckie & Newcombe, 2005). To illustrate the impact of playing recreational video games on cognitive abilities that are important for mathematics learning, we describe below a study that we conducted with pre-service teachers.

12.5 Two Experimental Studies on Video Gaming and Mathematics Problem-Solving: Methodology and Theoretical Framework

In examining the impact video gaming could have on students, we chose to look at the mathematics proficiency angle. Gaming is a part of our culture. What if a focused treatment of gaming could serve as an intervention for pre-service teachers to improve in their approach to and confidence with mathematics? What can be gained from an unexpected outcome? In this study, we were interested in learning specifically about pre-service teacher knowledge and impact of the gaming on their mathematics achievement and spatial skills. We examined the cognitive abilities and math perceptions that affect their math proficiency, focusing primarily on the role of spatial reasoning in students' mathematics performance and academic achievement. The pre-service teachers in this study experienced a video game intervention, working with the popular game, *Angry Birds*, and first-person shooter action video game (AVG), *Unreal Tournament*.

Action video games and first personal shooter games (FPS), in particular, have an advantage in improving low-level functions such as selective spatial attention (Feng et al., 2007), spatial perceptual resolution (Green & Bavalier, 2007), and contrast sensitivity (Li et al., 2009). FPS games produce improvements in the key cognitive functions: sensory, perceptual, and spatial (Spence & Feng, 2010). The brain can be altered after play, more frequently in a positive way (Ferguson, 2007). The attentional field size is increased with improvements from FPS play enduring for a prolonged amount of time (Feng et al., 2007; Green & Bavelier, 2006c; Spence et al., 2009). These games also have impact on the complex mental rotation skills (Feng et al., 2007). Sensory, perceptual, and cognitive skills have the spatial reasoning skills as a foundation (Spence & Feng, 2010). They are the building blocks. Therefore, the exploration of video game play impact on spatial reasoning is a field necessary of further study. The implications for this gaming research provide an argument for a "fresh" approach to new instructional strategies that help students increase their visuospatial skills (Spence & Feng, 2010). New ways to train the brain are surfacing.

12.5.1 Experiment 1

To collect baseline information about education-majors' mathematics proficiency, spatial abilities, and attitudes toward mathematics, education-majors without prior AVG experience ($n = 30$) completed a series of tests that assessed their working memory performance (Inquisit 4 Web OSPAN software (Turner & Engle, 1989), mental rotation test (MRT) (Vandenberg & Kuse, 1978), mathematics performance, and mathematics anxiety and confidence in learning mathematics (Fennema & Sherman, 1976a). Our mathematics performance test mirrored the complexity and format of the Praxis II exam, a licensure exam in mathematics for pre-service elementary teachers. The mathematics performance test covered topics taught in elementary school in three distinct areas: geometry, word, and non-word problems.

To examine individual differences between students with high and low spatial abilities, we divided participants into high and low spatial abilities groups using the median of the MRT variable (*Median* = 8.00). We found significant differences ($p < .05$) between individuals with low and high spatial abilities on the following measures, favoring students with high MRT scores: Confidence in learning mathematics, ACT scores in mathematics, ACT scores in science, ACT composite scores, as well as geometry, word, and non-word math problems performance (see Figs. 12.1 and 12.2). In addition, students with lower spatial abilities had significantly ($p < .05$) higher math anxiety. Not surprisingly, when looking at the ACT scores of the low MRT and high MRT skills individuals, those with the high skills did better, especially on mathematics, science, and the composite score. This connects to the research regarding spatial skills enhancing performance in standardized assessments.

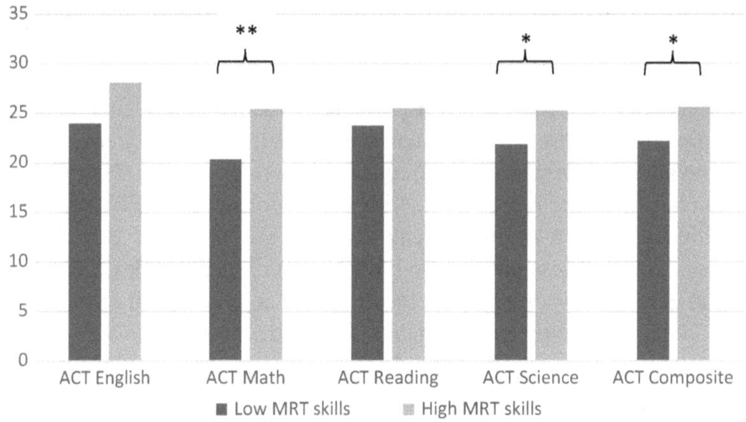

Fig. 12.1 Means of ACT scores for students with low and high MRT skills. *Note* $*\rho < .05$, $**\rho < .01$, $***\rho < .001$. ACT math, $F(3,24) = 11.71$, $p = .002$; ACT science, $F(3,24) = 6.30$, $p = .019$; ACT composite, $F(3,24) = 5.23$, $p = .027$

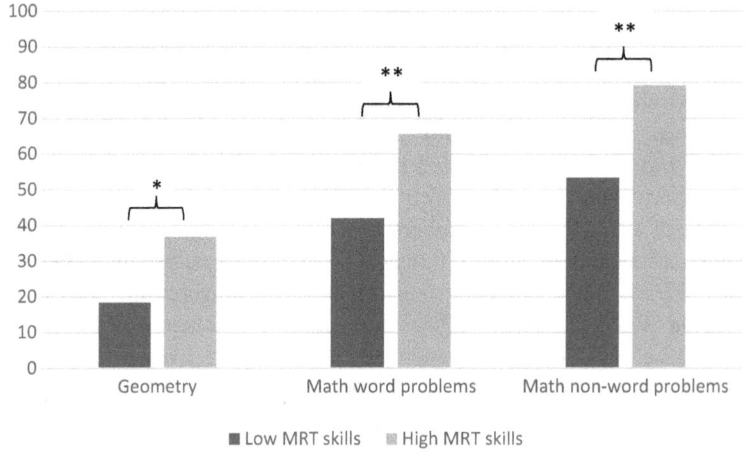

Fig. 12.2 Means of mathematics proficiency for students with low and high MRT skills. *Note* $*\rho < .05$. $**\rho < .01$. $***\rho < .001$. Geometry, $F(3, 26) = 5.66$, $p = .025$; word, $F(3, 26) = 11.91$, $p = .002$; non-word math problems performance, $F(3, 26) = 10.93$, $p = .003$

12.5.2 Experiment 2

In a follow-up study, we randomly assigned the students who participated in the previous experiment to two intervention groups (Novak & Tassell, 2015). One group (AVG; $n = 14$) was assigned to play an action video first-person shooter game, *Unreal Tournament 2004*. This game was successfully utilized in previous research to increase student attentional capabilities (Green & Bavalier, 2007).

The other group (non-AVG; $n = 16$) played the low-stress game, *Angry Birds*. In order to select a non-AVG game, we consulted with a video game specialist to ensure that our non-AVG game does not include AVG qualities that are believed to foster attentional capabilities (Cohen, Green, & Bavelier, 2007).

Each participant played the assigned game for 10 h in a computer lab supervised by the project co-investigators. All participants completed their 10-h video-game practice within the three weeks window. Both groups started playing the assigned games on the lowest possible level and were encouraged to constantly make progress attempting playing a game every time on a higher level. We recorded participant's game scores after each time they played a game in the lab. On average after 10 h of video gaming, the AVG group increased their video gaming skills by 318% and the non-AVG group improved their video game score by 546%. After completing the 10-h video-game practice, all students took the same assessment battery as in the first experiment.

After 10 hours of video game play, both AVG and non-AVG groups significantly improved ($p < .05$) on the following scales from pre- to post-test: working memory performance, mental rotation skills, and geometry scores. However, individuals with higher MRT abilities scored significantly higher than individuals with lower MRT abilities in each intervention group on the following post-test measures: working memory performance, confidence, geometry, and non-word problem solving. In addition, students with higher MRT skills reported significantly lower math anxiety post-test levels than students with lower MRT skills (see Table 12.1). Moreover, there was a significant interaction effect ($p < .05$) between the video game played and MRT abilities of a person on a geometry test, indicating that students with high and low MRT abilities were affected differently by the video game they played. Specifically, low and high MRT individuals from the AVG group demonstrated on average similar geometry post-test performance; but high MRT individuals performed significantly better on the geometry test than their low MRT peers from the non-AVG group (see Fig. 12.3).

Table 12.1 Analysis of variance for video gamers' performance (N = 30)

Source	df	F	p
Between subjects (post-test: high MRT vs. low MRT)			
Working memory	3	10.01	.004
Confidence in learning mathematics	1	9.61	.004
Geometry problem-solving	3	4.31	.048
Non-word problem-solving	3	6.52	.017
Math anxiety	1	14.22	.001
Within subjects (pre/post-test gains)			
Working memory	1	4.94	<.05
Spatial skills	1	38.38	<.0001
Geometry problem-solving	1	4.89	<.05

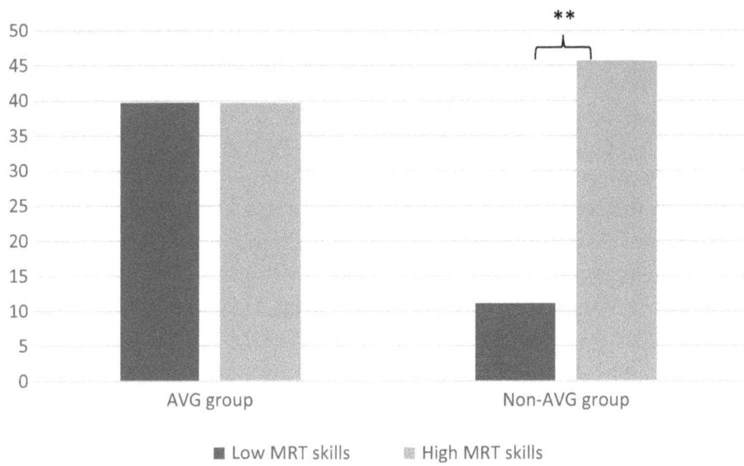

Fig. 12.3 Differences in means between students with high and low MRT skills by intervention group on geometry post-test problem solving. *Note* *$\rho < .05$. **$\rho < .01$. ***$\rho < .001$

The results provide preliminary evidence that both AVG and non-AVG participants improved their working memory performance, mental rotation skills, and geometry performance similarly from pre- to post-test. However, neither AVG nor non-AVG interventions did eliminate initial pretest spatial ability differences observed before the video gaming interventions. Both AVG and non-AVG interventions were effective for improving working memory performance and spatial skills in low and high MRT participants. In terms of performance outcomes, the only difference revealed between the AVG and non-AVG interventions was their effectiveness in improving participants' geometry problem solving. While the AVG game was equally effective in improving geometry performance in low and high MRT individuals, the non-AVG game was more effective for individuals with high MRT skills than for those with low MRT skills. These findings suggest that low MRT individuals can benefit more from playing the AVG game than from the non-AVG game. Nevertheless, it is important to note that a sample size in this experiment was relatively small and therefore the findings should be treated with caution.

Dedicated players of the video game would also assert that there are connections to the real world in the physics of how to angle the birds in Angry Birds to catapult for their correct destination. Gamers would also argue that the video game is "just plain fun" and would see a similar connection to how other strategies can enliven a classroom lesson. A typical comment made by a participant in this gaming study after playing the video game was the following statement: "This was such a mental break from my class today. This was really fun!"

In summary, video game play has the potential to open up the mathematical minds for students. Unfortunately, video game play has been seen by parents and the school community, oftentimes, as a "waste of time" or causing distractions when trying to get children to focus on what is really important with school—learning the core content area skills. However, we assert from the literature review and our research in video game play, as related to mathematics, that the potential for expanding mathematical spatial skills with video games is possible.

Moreover, the Angry Birds recreational video game has been used in other studies to teach mathematics content directly and connect children's recreational video gaming experiences to illustrate specific mathematical concepts. Russo et al. (2013) investigated how the Angry Birds game could be used in mathematics education. Findings suggested that the game should be utilized along with intentional planning and systemic implementation connected to mathematics concepts within the context of the game to gain the best impact. Russo's research team analyzed 54 *YouTube* instructional videos of "Angry Birds" as the data set. The team watched the videos and was asked to (1) figure out if there were noticeable difference in how participants in formal versus informal settings communicated their understanding of the mathematics concepts; (2) compare the video resources looking for common, effective and appropriate incorporation of the technology for instruction; and (3) evaluate the videos for whether they would use the video for their own personal classroom instruction.

In the Russo et al. (2013) study, the videos were analyzed for mathematical content with 'parabola' and 'angle' topping the frequency. The results suggest that teachers realize the potential for Angry Birds as an instructional tool to explore quadratic functions. Russo et al. (2013) assert that an argument could be made for the game as a natural set up for such concepts. The effective videos were categorized into two themes: (1) 'relevant connections' and (2) 'multiple representations.' In regard to 'relative connections,' Watson and Fang (2012) point out that game-based instruction gives teachers an opportunity to engage students in contextualized learning in an authentic way through solving the challenges of the game. Russo's study found that the most effective videos where making explicit connections to the mathematical ideas and the game context. For 'multiple representations,' Russo et al. (2013) found that the most exemplary uses of Angry Birds involved clear incorporation and connections to the mathematical concepts of the game into instruction.

12.6 Future Implications

12.6.1 Potential Impact of Video Gaming as an Instructional Intervention for Spatial Skill Development and Relationship to Creativity

Could video-gaming be used as a creative instructional approach to increase student skills in mathematics problem-solving? We assert this question needs to be weighed heavily as a creative instructional strategy, for school and home. The future research in this area could be expanded by looking at relationships between schools and homes working together with strategic gaming plans to examine the impact on spatial reasoning and problem solving. With this research, the potential benefits are promising: improved relations between school/home, excitement-infused curriculum planning, extended model of the classroom to maximize time by utilizing evening and weekend hours, personalized and preferential learning for students.

Teacher education programs could transform. Both pre-service and teachers can benefit from coursework and professional development programs if steps are made to incorporate gaming training into the instructional practices as a stop-gap measure in creating better prepared teachers for working with children in our schools. These instructional enhancements could be applied in a myriad of ways, especially through the elementary math content courses, and/or through the mathematics methods courses in the preparation program.

The video game industry has been under scrutiny over the past years with issues connected to school violence, disengaged learners, and cited as a waste of time. With the findings from this spatial reasoning study, positive outcomes for AVG play can be shared. With this connection to the spatial reasoning, then to problem solving and standardized testing, video gaming can experience a "face lift" and intentionally design a path to position themselves in the education arena. This study, and others, are showing that video game play does not take extensive amounts of time for the resulting impact. With ten hours devoted to fun for many, a new culture of stronger spatial reasoning may emerge. Parents can use this as a guideline for the amount of time that is productive and yet not consuming. This could sway parents towards videogame use, particularly for improving mathematics performance. The global and cultural impact on society is a factor. Spence and Feng (2010) state that the role of video games could revolutionize how spatial skills are taught and help reduce gender differences in children's' learning of spatial skills. The impact of this change would be societal and global. New instructional methods based on video games could also help people maintain and even improve the spatial skills through the aging process. Spence and Feng's research (2010) has implications that video games are not just for children, but can have an impact far-reaching into an older age group.

According to Wai, Lubinski, and Benbow's (2009) research, spatial ability is a predictor of achievement in STEM, with mathematics and verbal skills variables held steady. Uttal et al. (2013) confirmed this finding after meta-analyzing 217 related studies. This important finding supports the urgency of effective education in STEM disciplines. Studies that provided indirect learning via games were found equally effective as those that directly involved spatial tasks practice, producing positive durable improvements in spatial skills across all training methods, according to a recent meta-analysis of training studies aiming at improving spatial skills (Uttal et al., 2013). In addition, given the flexibility of spatial skills and the fact that both genders respond equally well to training, implementing spatial-focused interventions can help even adult students. These types of programs can be particularly critical for increasing students' preservation rate in STEM majors that writhe from high dropout rates (Price, 2010).

In Russo et al. (2013) study, we gain ideas for next steps for how to incorporate action video games effectively into instruction. Although the games in our study were played outside of the classroom instruction, could there be an intentional connection made between home and the classroom? Perhaps this could be a version of the flipped classroom where students are actually assigned to play a certain segment of a video game, and come in the next day to class ready for the teacher to make the connections in mathematics instruction. The use of "relevant connections" and "multiple representations" (Russo et al., 2013) could be the foundation for instructional goals for teachers for incorporating action-video games.

Although the study conducted for this chapter did not focus on an angle of creativity, the outcomes and related literature of recent studies add superb ideas for future contributing research to enhance the work begun with this research. As was stated with the 2011 study from Michigan State University, the link between video-game play and creativity was in a positive correlation (Moore, 2011). We would like now to look at the video games through the lens of creative impact. We would also like to continue with surveying elementary-age parent families, gather test data, including creativity test data, and find out the different forms of technology used at home, including game play.

12.6.2 Future Research Directions: Creativity, Spatial Skills, and Video Gaming

Researchers have been examining the link between spatial skills and creativity. In a 2013 study by Harrison Kell, David Lubinski, Camilla Benbow, and James Steiger published in *Psychological Science, they have* made an even stronger connection between early spatial talent and creativity in adult life. The study, showed that spatial skill had an increment of prediction for creativity over and above math and verbal skills (assessed at age 13) when looking at scholarly publications and patents —even those in STEM (Kell et al., 2013).

In a research update from Vanderbilt, Wetzel (2013) shares work by Lubinski and colleagues about how exceptional spatial ability at age 13 predicts creative and scholarly achievements more than 30 years later, according to results from a Vanderbilt University longitudinal study. Despite longstanding speculation that spatial ability may play an important role in supporting creative thinking and innovation, there are very few systems in place to track skill in spatial reasoning (Wetzel, 2013).

A positive link between video-game play and creativity has been surfacing in recent years. In a study of five hundred 12-year-old children, the findings point to a relationship between amount of video game play and creativity in writing stories and drawing pictures (Moore, 2011). Interestingly, other technologies, like cell phones, Internet, and computer use other than video games, did not have a relationship to increasing creativity. The aforementioned study, which is out of

Michigan State University (MSU) (Moore, 2011), is coined as the first study to demonstrate the relationship between creativity and video or computer game play. These findings are particularly relevant in light of the data published by Entertainment Software Association: 72% of United States households incorporate video/computer game play as part of their everyday life (Jackson, Witt, Games, Fitzgerald, von Eye, & Zhao, 2011).

With the findings of (MSU, 2011) linking creativity to video/computer game play, game designers may be motivated to identify aspects of the crafted games that enhance creative capacity in the players. The MSU researcher, Jackson, believes that once the creativity aspect identification occurs, games can maximize the effects of creativity while retaining the entertainment value, blurring the line between education and entertainment (MSU, 2011).

In another study out of MSU's Children and Technology Project, a group of researchers surveyed 491 middle-school students to find out how often they used different forms of technology compared then to their scores on the Torrance Test of Creativity-Figural. The study found that boys favored violent and sports games and played more often while girls favored interaction with others games and played less often. However, regardless of type of game played, or gender/race of the student, there was a positive relationship between video game play and increased creativity (Jackson et al., 2011).

12.7 Conclusion

We often notice the creative actors, dancers, artists, musicians, and writers in our world. However, we do not tend to notice a creative engineer or mathematician. Why is STEM not in the spotlight when it comes to being considered as a "creative" career? Research now shows a positive correlation between early spatial skills and later creativity, yet in our world we still lack an appreciation and respect for the highly creative aspects of science, technology, engineering, and mathematics (Wai, 2013).

Research evidence has revealed the critical role of spatial skills and reasoning in performance on standardized assessments and achievement in STEM areas. Several creative approaches that have been found effective in enhancing spatial reasoning skill are expected to enhance generalized performance in STEM disciplines. Some of these mentioned previously are video games such as Tetris, and more recently, Unreal Tournament and Angry Birds. With the knowledge that spatial skills are less developed in females than males, the research agenda for exploring how to help female pre-service elementary education teachers improve their spatial skills is imperative. The reality is that elementary teachers are predominantly females. Coupling the finding that pre-service teachers with lower spatial abilities have significantly higher math anxiety, it is time to try means to break the cycle for students to begin learning and reaching their potential in spatial reasoning. Teachers with teaching confidence and understanding of the importance of spatial reasoning instruction and interventions are more likely to benefit students. With a direct and personal experience with an appropriate video game, the changes could gradually begin.

References

Beilock, S. L., Gunderson, E. A., Ramirez, G., & Levine, S. C. (2010). Female teachers' math anxiety affects girls' math achievement. *Proceedings of the National Academy of Sciences of the United States of America, 107*(5), 1860–1863.

Bjorklund, D. F., & Brown, R. D. (1998). Physical play and cognitive development: Integrating activity, cognition, and education. *Child Development, 69*(3), 604–606.

Boot, W. R., Kramer, A. F., Simons, D. J., Fabiani, M., & Gratton, G. (2008). The effects of video game playing on attention, memory, and executive control. *Acta Psychologica, 129*(3), 387–398.

Bragg, L. (2007). Students' confilicting attitudes towards games as a vehicle for learning matheamtics: A methodological dilemma. *Mathematics Education Research Journal, 19*(1), 29–44.

Brown, D. L., & Wheatley, G. H. (1997). Components of imagery and mathematical understanding. *Focus on Learning Problems in Mathematics, 19,* 45–70.

Burnett, S. A., Lane, D. M., & Dratt, L. M. (1979). Spatial visualization and sex differences in quantitative ability. *Intelligence, 3*(4), 345–354.

Casey, M. B., Nuttall, R., Pezaris, E., & Benbow, C. P. (1995). The influence of spatial ability on gender differences in mathematics college entrance test scores across diverse samples. *Developmental Psychology, 31*(4), 697–705.

Ceci, S. J., & Williams, W. M. (2007). *Why aren't more women in science. Top researchers debate the evidence*. Washington, D.C.: American Psychological Association.

Cheng, Y. L., & Mix, K. S. (2014). Spatial training improves children's mathematics ability. *Journal of Cognition and Development, 15*(1), 2–11.

Cherney, I. D. (2008). Mom, let me play more computer games: They improve my mental rotation skills. *Sex Roles, 59*(11–12), 776–786.

Cherney, I. D., & London, K. (2006). Gender-linked differences in the toys, television shows, computer games, and outdoor activities of 5-to 13-year-old children. *Sex Roles, 54*(9–10), 717–726.

Clark-Wilson, A., Robutti, O., & Sinclari, N. (Eds.). (2014). *The mathematics teacher in the digital era: An international perspective on technology-focused professional development.* Berlin, Germany: Springer.

Cohen, C. A., & Hegarty, M. (2007). Sources of difficulty in imagining cross sections of 3D objects. In *Proceedings of the Twenty-Ninth Annual Conference of the Cognitive Science Society* (pp. 179–184). Austin, TX: Cognitive Science Society.

Cohen, J. E., Green, C. S., & Bavelier, D. (2007). Training visual attention with video games: Not all genres are created equal. In H. O'Neil & R. Perez (Eds.), *Computer games and team and individual learning* (pp. 205–227).

de Freitas, S. (2006). *Learning in immersive worlds: A review of game-based learning*. Retrieved from http://www.webarchive.org.uk/wayback/archive/20140615100504/http://www.jisc.ac.uk/media/documents/programmes/elearninginnovation/gamingreport_v3.pdf.

Dewey, B., Singletary, T., & Kinzel, M. (2009). Graphing calculator use in algebra teaching. *School Science and Mathematics, 109*(7), 383–393.

Dorval, M., & Pepin, M. (1986). Effect of playing a video game on a measure of spatial visualization. *Perceptual and Motor Skills, 62*(1), 159–162.

Feng, J., Spence, I., & Pratt, J. (2007). Playing an action video game reduces gender differences in spatial cognition. *Psychological Science, 18*(10), 850–855.

Fennema, E., & Sherman, J. A. (1976a). Fennema-Sherman mathematics attitudes scales: Instruments designed to measure attitudes toward the learning of mathematics by females and males. *Journal for Research in Mathematics Education, 7*(5), 324–326.

Fennema, E., & Sherman, J. (1976b). *Sex-related differences in mathematics learning: Myths, realities and related factors*. Paper presented at the American Association for the Advancement of Science, Boston.

Ferguson, C. J. (2007). The good, the bad and the ugly: A meta-analytic review of positive and negative effects of violent video games. *Psychiatric Quarterly, 78*(4), 309–316.

Freudenthal, H. (1971). Geometry between the devil and the deep sea. *Educational Studies in Mathematics, 3,* 413–435.

Friedman, I. A. (1995). Measuring school principal-experienced burnout. *Educational and Psychological Measurement, 55*(4), 641–651.

Gittler, G., & Glück, J. (1998). Differential transfer of learning: Effects of instruction in descriptive geometry on spatial test performance. *Journal of Geometry and Graphics, 2*(1), 71–84.

Green, C. S., & Bavelier, D. (2006). The cognitive neuroscience of video games. In L. Humphreys & P. Messaris (Eds.), *Digital media: Transformations in human communication* (pp. 211–223). New York, NY: Peter Lang.

Green, C. S., & Bavelier, D. (2007). Action-video-game experience alters the spatial resolution of vision. *Psychological Science, 18*(1), 88–94.

Green, C. S., & Bavelier, D. (2012). Learning, attentional control, and action video games. *Current Biology, 22*(6), R197–R206.

Gresham, G. (2008). Mathematics anxiety and mathematics teacher efficacy in elementary pre-service teachers. *Teaching Education, 19*(3), 171–184.

Gunter, G., Kenny, R., & Vick, E. (2008). Taking educational games seriously: Using the RETAIN model to design endogenous fantasy into standalone educational games. *Educational Technology Research and Development, 56*(5/6), 511–537.

Halpern, D. F., & LaMay, M. L. (2000). The smarter sex: A critical review of sex differences in intelligence. *Educational Psychology Review, 12*(2), 229–246.

Hegarty, M., & Waller, D. (2005). Individual differences in spatial abilities. In P. Shah & A. Miyake (Eds.), *The Cambridge Handbook of Visuospatial Thinking* (pp. 121–169). Cambridge, UK: Cambridge University Press.

Hembree, R. (1990). The nature, effects, and relief of mathematics anxiety. *Journal for Research in Mathematics Education, 21*(1), 33–46.

Holzinger, K. J., & Swineford, F. (1946). The relation of two bi-factors to achievement in geometry and other subjects. *Journal of Educational Psychology, 37*(5), 257.

Hoyles, C., Noss, R., & Kent, P. (2004). On the integration of digital technologies into matheamtics classrooms. *International Journal of Computers for Mathematical Learning, 9*(3), 309–326.

Jackson, L. A., Witt, E. A., Games, A. I., Fitzgerald, H. E., von Eye, A., & Zhao, Y. (2011). Information technology use and creativity: Findings from the children and technology. *Computers in Human Behavior*, *28*(2), 370–376. Retrieved from http://www.sciencedirect.com/science/article/pii/S0747563211002147.

Kell, H. J., Lubinski, D., Benbow, C. P., & Steiger, J. H. (2013). Creativity and technical innovation: Spatial ability's unique role. *Psychological Science*, *24*(9), 1831–1836. Retrieved from https://my.vanderbilt.edu/smpy/files/2013/01/Kell-et-al.-2013b1.pdf.

Lawrenz, F., Gravely, A., & Ooms, A. (2006). Perceived helfulness and amount of use of technology in science and matheatics classes at different grade levels. *School Science and Mathematics, 106*(3), 133–139.

Li, R., Polat, U., Makous, W., & Bavelier, D. (2009). Enhancing the contrast sensitivity function through action video game training. *Nature Neuroscience, 12*(5), 549.

Lin, S. F. (2011). Effect of opponent type on moral emotions and responses to video game play. *Cyberpsychology, Behavior, and Social Networking, 14*(11), 695–698.

Linn, M. C., & Petersen, A. C. (1985). Emergence and characterization of sex differences in spatial ability: A meta-analysis. *Child Development, 56*(6), 1479–1498.

Lubinski, D., & Benbow, C. P. (2006). Study of mathematically precocious youth after 35 years: Uncovering antecedents for the development of math-science expertise. *Perspectives on Psychological Science, 1,* 316–345.
Ma, L. (1999). *Knowing and teaching elementary mathematics: Teachers' understanding of fundamental mathematics in China and the United States*. Mahwah, NJ: Erlbaum.
Maloney, E. A., Waechter, S., Risko, E. F., & Fugelsang, J. A. (2012). Reducing the sex difference in math anxiety: The role of spatial processing ability. *Learning and Individual Differences, 22*(3), 380–384.
Maxwell, J. W., Croake, J. W., & Biddle, A. P. (1975). Sex differences in the comprehension of spatial orientation. *The Journal of Psychology, 91*(1), 127–131.
McClurg, P. A., & Chaillé, C. (1987). Computer games: Environments for developing spatial cognition? *Journal of Educational Computing Research, 3*(1), 95–111.
Michigan State University. (2011, November 9). Video game playing tied to creativity, research shows. *ScienceDaily*. Retrieved February 18, 2017 from www.sciencedaily.com/releases/2011/11/111102125355.htm.
Mix, K. S., & Cheng, Y. L. (2011). The relation between space and math: Developmental and educational implications. *Advances in Child Development and Behavior, 42,* 197–243.
Moore, D. S., & Johnson, S. P. (2008). Mental rotation in human infants a sex difference. *Psychological Science, 19*(11), 1063–1066.
Moore, E. A. (2011, November 2). New study links video gaming to creativity. [Web log post] *Cnet*. Retrieved from https://www.cnet.com/news/new-study-links-video-gaming-to-creativity/.
Mortensen, E. L., Andresen, J., Kruuse, E., Sanders, S. A., & Reinisch, J. M. (2003). IQ stability: The relation between child and young adult intelligence test scores in low-birthweight samples. *Scandinavian Journal of Psychology, 44*(4), 395–398.
NCTM. (2000). *Principles and standards for school mathematics*. Reston, VA: National Council of Teachers of Mathematics.
Newcombe, N. S. (2007). *Taking science seriously: Straight thinking about spatial sex differences*. Washington, D.C.: American Psychological Association.
Novak, E., & Tassell, J. L. (2015). Using video game play to improve education-majors' mathematical performance: An experimental study. *Computers in Human Behavior, 53,* 124–130. https://doi.org/10.1016/j.chb.2015.07.001.
Novak, E., & Tassell, J. L. (2017). Studying preservice teacher math anxiety and mathematics performance in geometry, word, and non-word problem solving. *Learning and Individual Differences, 54,* 20–29.
Okagaki, L., & Frensch, P. A. (1994). Effects of video game playing on measures of spatial performance: Gender effects in late adolescence. *Journal of Applied developmental Psychology, 15*(1), 33–58.
Ozel, S., Larue, J., & Molinaro, C. (2004). Relation between sport and spatial imagery: Comparison of three groups of participants. *The Journal of Psychology, 138*(1), 49–64.
Pittalis, M., & Christou, C. (2010). Types of reasoning in 3D geometry thinking and their relation with spatial ability. *Educational Studies in Mathematics, 75*(2), 191–212.
Price, J. (2010). The effect of instructor race and gender on student persistence in STEM fields. *Economics of Education Review, 29*(6), 901–910.
Russo, D. M., Diletti, J., Strzelec, J., Reeb, C., Schillace, J., Martin, A., et al. (2013). A study of how angry birds has been used in mathematics education. *Digital Experiences in Mathematics Education, 1*(2), 107–132.
Shea, D. L., Lubinski, D., & Benbow, C. P. (2001). Importance of assessing spatial ability in intellectually talented young adolescents: A 20-year longitudinal study. *Journal of Educational Psychology, 93*(3), 604–614.
Spence, I., & Feng, J. (2010). Video games and spatial cognition. *Review of General Psychology, 14*(2), 92.
Spence, I., Yu, J. J., Feng, J., & Marshman, J. (2009). Women match men when learning a spatial skill. *Journal of Experimental Psychology: Learning, Memory, and Cognition, 35*(4), 1097.

Subrahmanyam, K., & Greenfield, P. M. (1994). Effect of video game practice on spatial skills in girls and boys. *Journal of Applied Developmental Psychology, 15*(1), 13–32.

Terlecki, M. S., & Newcombe, N. S. (2005). How important is the digital divide? The relation of computer and videogame usage to gender differences in mental rotation ability. *Sex Roles, 53*(5), 433–441.

Terlecki, M. S., Newcombe, N. S., & Little, M. (2008). Durable and generalized effects of spatial experience on mental rotation: Gender differences in growth patterns. *Applied Cognitive Psychology, 22*(7), 996–1013.

Tracy, D. M. (1987). Toys, spatial ability, and science and mathematics achievement: Are they related? *Sex Roles, 17*(3–4), 115–138.

Turner, M. L., & Engle, R. W. (1989). Is working memory capacity task dependent? *Journal of Memory and Language, 28*(2), 127–154. https://doi.org/10.1016/0749-596X(89)90040-5.

Uttal, D. H., Meadow, N. G., Tipton, E., Hand, L. L., Alden, A., Warren, C., et al. (2013). The malleability of spatial skills: A meta-analysis of training studies. *Psychological Bulletin, 139*(2), 352–402.

Vandenberg, S. G., & Kuse, A. R. (1978). Mental rotations, a group test of three-dimensional spatial visualization. *Perceptual and Motor Skills, 47*(2), 599–604.

van Garderen, D. (2006). Spatial visualization, visual imagery, and mathematical problem solving of students with varying abilities. *Journal of Learning Disabilities, 39*(6), 496–506.

Wai, J. (2013). Why we need to value students' spatial creativity. *MindShift*. Retrieved from https://ww2.kqed.org/mindshift/2013/07/31/why-we-need-to-value-spatial-creativity/.

Wai, J., Lubinski, D., & Benbow, C. P. (2009). Spatial ability for STEM domains: Aligning over 50 years of cumulative psychological knowledge solidifies its importance. *Journal of Educational Psychology, 101*(4), 817–835. Retrieved from https://www.psychologytoday.com/files/attachments/56143/spatial-ability-stem-domains.pdf.

Watson, W. R., & Fang, J. (2012). PBL as a framework for implementing video games in the classroom. *International Journal of Game-Based Learning (IJGBL), 2*(1), 77–89.

Wetzel, J. (2013). Early spatial reasoning predicts later creativity and innovation, especially in STEM fields. *Research News @ Vanderbilt.* Retrieved from https://news.vanderbilt.edu/2013/07/15/early-spatial-reasoning-predicts-later-creativity-and-innovation/.

Chapter 13
Prototype Problem Solving Activities Increasing Creative Learning Opportunities Using Computer Modeling and 3D Printing

Antonia Szymanski

Abstract This chapter explores the use of Prototype Problem-Solving Activities and 3D printing (PPSA) as a curricular tool to develop mathematical understanding, creativity, and technological literacy. Prototype Problem Solving Activities (PPSA) are teaching and learning activities that have been designed for students to create artifacts that demonstrate their understanding and to find unique solutions to authentic problems. They represent an outgrowth of the maker movement and attempt to involve students in authentic problem-solving exploration. The thesis of this chapter is that by using PPSA as a teaching strategy teachers can (1) provide students with opportunities to develop mathematical and creative thinking, (2) encourage students who may not perceive themselves as talented in mathematics by providing new ways in which to demonstrate mathematical thinking, and (3) use authentic problems and interdisciplinary approaches to problem solving that simulates real-life behavior by practitioners in the STEM fields. PPSA emphasizes communication and problem solving which are two principles that are stressed in education and by business leaders as being critical for life-long success. A description is provided of the creative processes that are nurtured through the use of PPSA, as well as the instructional design principles, and specific connections to technological literacy that moves students beyond being mere consumers of information to generating ideas and reflecting on thinking. The use of authentic problems requiring a generation of prototype products allows learners to self-assess and reflect on their understanding. The process of using PPSA allows the students to develop higher order thinking skills of analysis and synthesis in their mathematical understanding.

Keywords Problem-based learning · 3D printing · 3D modeling
Creativity · Prototypes

A. Szymanski (✉)
Western Kentucky University, Bowling Green, KY, USA
e-mail: antonia.szymanski@wku.edu

© Springer International Publishing AG, part of Springer Nature 2018
V. Freiman and J. L. Tassell (eds.), *Creativity and Technology in Mathematics Education*, Mathematics Education in the Digital Era 10,
https://doi.org/10.1007/978-3-319-72381-5_13

13.1 Introduction

As 3D printers become more affordable, hobbyists and educators are now able to acquire them for use. These devices, which allow users to create products by using modeling software and plastic printing material, offer an opportunity to increase mathematical understanding by allowing students to physically create previously abstract components. The ability to engage in the engineering process of conceptualization, design, production, evaluation, and re-design fosters creativity in students and provides an opportunity to develop a deeper understanding of mathematical concepts than can be afforded with 2D representations.

The purpose of this chapter is to investigate the relationship of 3D printing with Prototype Problem-Solving Activities (PPSA) to develop creativity in mathematics. Accordingly, the following research topics will be examined: How is creativity in mathematics defined and developed? How can 3D printers be used to improve mathematical understanding and creativity? What is the role of Prototype Problem Solving Activities in mathematical understanding and creativity? These three questions will be explored through examining current literature on the topics. Using Prototype Problem-Solving Activities as a teaching tool allows a holistic combination of creativity in all domains while allowing students to work together to develop solutions to real problems thus developing true 21st Century skills of communication, collaboration, and higher order thinking along with technical literacy and is the focus of this chapter.

The chapter begins with a conceptualization of creativity and its role in mathematics, it is followed by a discussion of the role of 3D printing in mathematics education. After a brief description of makerspaces as the foundation for PPSA, the more practical aspect of the chapter focuses on Prototype Problem-solving Activities as a teaching tool and the design principles required to create appropriate educational models. The final aspect of the chapter discusses the use of 3D printers in the classroom, how they can be used to create models, and results of current research.

13.2 How Is Creativity in Mathematics Defined and Developed?

13.2.1 Conceptual Definition of Creativity

Numerous definitions of creativity, specifically mathematical creativity, make explicating instructional outcomes difficult. Thus, even after decades of researching creativity, thinking, and learning, very little has changed in educational strategies for enhancing creativity. It is important to understand the creative process because having frameworks for analyzing the ways in which learners create knowledge can provide important information for teachers regarding the ways in which students can work together while constructing new knowledge.

Sawyer (2012) noted that there are two main lenses by which to define creativity; individualist and sociocultural. "Individualist definition: Creativity is a new mental combination expressed in the world" (Sawyer, 2012, p. 7). The individualist creativity focuses on one single person who is experiencing creativity in thought or action. Snow's theory of individual aptitude fits well with this idea because aptitude is comprised of a person's ability, motivation, and affect (1992). Aptitude can explain why some people may seem to engage in creative activities more than others. Every time an individual has a new thought or does something in a new way, it does not necessarily result in an observable product therefore many people refer to this as "little c" creativity (Sawyer, 2012). "Little c" creativity occurs throughout daily life when people solve routine problems. Although many people may encounter similar problems, if the individual uses a new approach then he/she is displaying "little c" creativity. The second definition is the sociocultural. This type of creativity, also call "big C" creativity, focuses on an external product or performance that is evaluated by others. Society determines if the product is novel, useful, or valuable and thus awards the designation of being creative (Sawyer, 2012). It is important for teachers to provide opportunities for students to engage in both types of creativity. Students need to be taught to recognize and use "little c" creativity to adapt to new challenges and also to collaborate and develop "big C" creativity to solve complex societal problems. These modes of thinking are critical in a society that places increasing demands on the creativity of its citizens. Thus, when the goal is to develop mathematical creativity educators must be deliberate regarding the type of creativity they are targeting.

13.2.2 Operational Definition of Creativity

After reviewing the literature on creativity, Plucker, Beghetto, and Dow (2004) derived a working definition. "Creativity is the interaction among aptitude, process, and environment by which an individual or group produces a perceptible product that is both novel and useful as defined within a social context" (p. 90). Figure 13.1 depicts the conceptual model of creativity that was created by combining components of several robust theories of creativity and development for this chapter. Expanding on the definition by Plucker et al. (2004) it is important to understand the relationship between the components of aptitude and the components of creativity. The components of aptitude are ability, affective characteristics such as openness and curiosity, and motivation (Snow, 1992). The environment and process may influence both motivation and expression of ability. Restrictive processes that provide little room for individual expression or divergent thinking may reduce student motivation due to the lack of autonomy that is offered. Such an environment may also decrease external manifestations of ability due to student disinterest or preference for performing the task in another way. Similarly, environments that do

Fig. 13.1 The Szymanski conceptual model of creativity

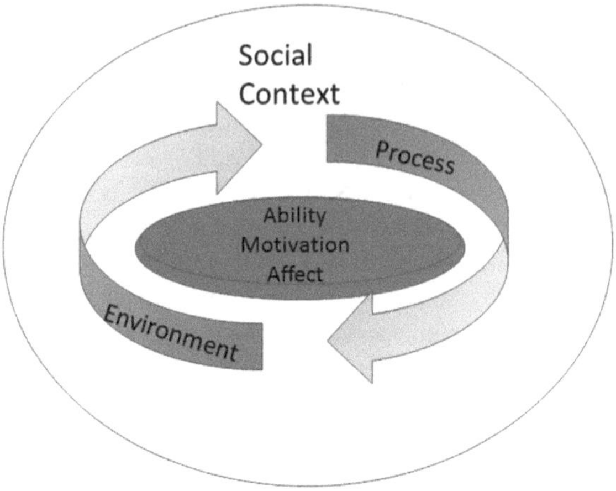

not value individual contributions and imaginative problem-solving may lead to reduced motivation and achievement as a means of students to express their frustrations. It is also important to keep in mind that the situated nature of the social context by definition implies that something that is considered creative in one social context may not be creative in another. Much like Bronfenbrenner's Ecological Model of Development, the development of creativity can be viewed as a system of interaction (Bronfenbrenner, 1986). Figure 13.1 depicts the dynamic relationship between each of these components. The outer areas of social context, environment, and process reflect what Sawyer describes as the "sociocultural" aspect of creativity. The inner oval represents aptitude and is comprised of ability, motivation, and affect and reflects "individual" aspects of creativity.

The interaction between aptitude and environment has important considerations for education. Classrooms can be designed to support and develop creativity thus enhancing the creative ability inherent in every student. Too often traditional, teacher-centered classrooms leave little room for creativity (Mann, 2006). Focusing on basic skill attainment has pushed teachers to emphasize memorizing algorithms rather than building on mathematical understanding and developing creativity by engaging in open-ended problem solving (Mann, 2006). A negative by-product of this type of educational experience is the level of discomfort shown by individuals when they are confronted with problems that do not have "right" answers.

13.2.3 The Development of Mathematical Creativity

Most elementary mathematics teaching focuses on low-level thinking skills, such as memorizing formulas (Lesh & Doerr, 2003). Using textbook based problems and worksheets rarely offer the opportunity to move beyond basic applications of formulas It is necessary for students to have the basic concepts of mathematics content, however memorizing formulas is not sufficient to prepare students to creatively solve problems in the ever-changing, real world (Chamberlin & Moon, 2005; Mann, 2006). Mann (2006) clearly states "Procedural skills without the necessary higher order mathematical thinking skills, however, are of limited use in our society" (p. 244). The demands of a world that increasingly relies on technology require global citizens who are capable of engaging in mathematical thinking not just solving equations that require basic skills.

Mann (2006) noted that encouraging mathematical creativity in all students is essential to the understanding of content and enhancing the enjoyment of the learning process. By allowing creative thinkers to flourish while learning the basic foundations inside of the classroom, future leaders will have practice in solving the problems that do not yet exist. The creation of authentic situations that require new and engaging ways to support higher level thinking allows students to expand their mathematical knowledge with more complex activities.

13.2.4 Mathematical Giftedness and Creativity

Recognizing creativity in mathematics often requires exceptional discernment on the part of teachers. Students who are able to calculate quickly or solve straightforward problems may be designated as gifted and provided additional mathematical challenges. However, these students may lack mathematical creativity and may require additional opportunities and instruction to develop this skill. Oftentimes students who perform well on standardized mathematics tests may be exceptional at computation and following algorithms; however, they may lack the understanding of creatively applying mathematical concepts to real world problems (Sriraman, 2005).

It may be by focusing on high test scores to determine eligibility for specialized programming we are targeting high achievers but ignoring the creative people who may be the true innovators. This is a problem that happens when we create artificial separations of domains. Sriraman (2005) supported this finding by noting "The dearth of specific definitions of mathematical creativity in the mathematics and mathematics education literature necessitates that we move away from the specific domain of mathematics to the general literature on creativity in order to construct an appropriate definition" (2005, p. 23). In order to enhance the development of creativity, learning opportunities must be designed with that goal in mind. Technology provides new tools for teachers to use to stimulate creative thinking.

13.3 How Can 3D Printers Be Used to Improve Mathematical Understanding and Creativity?

Numerous opportunities are developing for using technology to enhance creativity and learning as it becomes more accessible for consumers. Computer aided design tools are used in every industry from manufacturing, entertainment, and agriculture. Technology increases creativity by allowing users to create and modify designs quickly and easily without physically creating a product. Thus ideas can be trialed in a low-risk environment. 3D printing offers an extension to computer aided design by allowing users to create low cost prototypes. It also enhances creativity by allowing users to share designs easily. Designing and creating prototypes provides teachers new teaching tools to bring mathematics to life. The technology and the ability to be creative may increase student engagement in mathematics.

13.3.1 3D Designing as a Tool for Teaching

The advent of computer aided design technology has ushered in a new age in visualizing objects. Computer-Aided Design (CAD) allows objects to be created that are not limited to physical realities such as gravity and a single light source. CAD objects also allow users to view from multiple angles simultaneously. This allows the creator to determine how the object will function in the real world much easier than by having to imagine a 2D image such as that drawn on paper in 3D space (Kűcukozer, 2013).

The ability to view objects in 3D with multiple perspectives reduces the cognitive load on users who are able to visualize an onscreen image rather than having to use their imagination to rotate the shape and think about the ways in which it will interact with other objects. While this aspect is especially helpful when thinking about geometric problems and shapes, virtually any object can be represented using CAD. The three dimensions can be modeled as three related variables to determine how changes in one aspect may impact the other two. Allowing students to physically see the objects rather than trying to imagine abstract concepts has been found to improve understanding and retention (Kűcukozer, 2013; Matthews & Geist, 2002).

The use of 3D CAD models as a teaching tool has been well documented (Kűcukozer, 2013). CAD modeling has typically been used in education to teach math, science, and engineering concepts. Researchers have found that integrating 3D models aids students' understanding and has been shown to foster lasting cognitive change (Kűcukozer, 2013). Participants reported that the images they saw immediately disclosed any misconceptions and the pictures "stayed with them" over time. Thus, providing another means to communicate the concept rather than textual or even 2D images seems to cement the idea and fosters changes in understanding that last several years.

13.3.2 3D Designing in the Classroom

CAD modeling can also be used as learning tool directly by students. There are several free, or low-cost software programs that students can use to create 3D models. "These 3D modeling applications such as *Form Z* by autodessys, *Roboworks* by Newtonium, *Infini-D* by MetaCreations Corporation, *ModelMagic 3D* by ImageWare development, *Merlin VR* by digital immersion, and many other packages allow children to build 3 dimensional shapes on the computer and manipulate them in a simulated 3 dimensional space" (Matthews & Geist, 2002, p. 332). Most of these programs are relatively easy for students to learn how to manipulate and allow them the freedom to create their own models. Some great beginning programs for creative artistic students are *Autodesk Mesh* mixer and *Blender.* Other free, easy to use beginning software for engineering students are *Free CAD* and *Autodesk mesh editing and Animation suite.*

Creating their own objects and watching them interact on the screen provides students with a sense of ownership of the work and aids understanding. Students must think abstractly and mentally deconstruct physical objects in order to construct them in the virtual world. This process causes learners to think deeply about the relationships of size, shape, and material. CAD modeling is also a powerful teaching tool because the cost of mistakes is non-existent. Objects created in the virtual world can be modified, destroyed, and re-created with a few keystrokes thus providing freedom for creativity and testing ideas in a safe environment. The models give the students immediate feedback regarding their ideas and allow them to self-assess their thinking and make modifications. This metacognitive activity provides feelings of competence and independence as students no longer rely on the teacher for the solutions to the problem. Figure 13.2 depicts the mental processes of creating using 3D modeling (Matthews & Geist).

The act of creating computer objects and sharing their designs with one another has been found to empower students who previously may not have been engaged

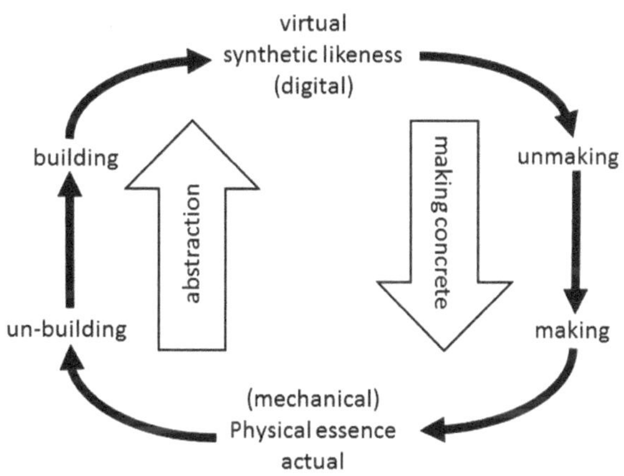

Fig. 13.2 Model of digital and physical integration. From technological applications to support children's development of spatial awareness. Information Technology in Childhood Education Annual. 2002. Reprinted with permission

with mathematics. In addition to gaining a greater conceptual understanding of the mathematical concepts, the students also develop logical thinking capabilities as they program the software to create the objects. Computer aided design modeling uses the recursive system of imagine, design, test, and modify to create models to solve problems (Matthews & Geist, 2002). This process exactly mimics the processes used by mathematicians, scientists, and engineers in their daily work, thus, providing students with a much more authentic learning environment than that typically found in classrooms where students attempt to imitate the methods used by the teacher to solve problems that have only one correct solution

13.3.3 3D Printing as a Means of Enhanced Creating

Rapid Prototyping or 3D printing is an additive process where a machine adds material layer by layer to create a form. 3D printing is referred to as additive manufacturing because the product is created by adding several layers together. Traditional product manufacturing is a subtractive process where products are created from materials by sanding, cutting, or melting away unwanted materials. The process of building a product by adding materials together eliminates the waste that is created from subtractive manufacturing because there is no unwanted pieces of the product. The advent of a process that allows users to create products as needed onsite also reduces waste due to transportation expenses. Instead of waiting for parts to ship users can simply create the parts that they need at a much lower cost. Just as the advent of computers allowed the sharing of information around the world, 3D printing will allow the sharing of products. This technology is at the inception of home and school use as machines are becoming less expensive.

Creating a product using 3D printing requires the use of computer design software that is able to take 2D renderings and create the internal structures necessary to build the final 3D product. When the product has been designed using the software it becomes a universally readable file similar to a pdf text file. This information is sent to a 3D printer and the machine begins to lay down thin layers of filament typically comprised of ABS plastic and gradually builds up the product. This process can take several hours depending on the size of the product and the type of filament being used. Although the process can take several hours or even overnight, the 3D printer does not require any monitoring or user interaction once it begins the printing process. One way to reduce the time necessary to print the product is to determine ways to deconstruct the model into component parts that could be adhered to one another after printing instead of printing out a solid object. An example could be a tetrahedron where the shapes are printed out and then glued together rather than printing out the solid object. Another example could be a bridge where the supports and road are created to interlock rather than printing out the solid bridge. The variety of designs and the ability for students to reflect their individual creativity is an advantage of 3D printing as a teaching tool that engages students in the learning process. Engagement with creating items is becoming very popular in the United States and Europe as seen by the development of Makerspaces.

13.4 Makerspaces, Creativity, and Learning

13.4.1 Definition of Makerspaces

Makerspaces, a series of informal places where people of varying ages gather to create using digital and physical resources, have been surfacing throughout the United States and are beginning to emerge in Europe. These sites exist as a place for individuals to create, tinker, and experiment. They use 3D printing and other high tech equipment along with traditional items such as wood and electronics to encourage people to explore their creativity. Makerspaces can be terrific entry points into STEM learning by sparking curiosity and supporting creativity. As Martinez and Stager note, "Using technology to make, repair, or customize things we need brings engineering, design, and computer science to the masses." (Martinez & Stager, 2013a, p. 11). Makerspaces offer a way to democratize access to technology and increase experimentation, tinkering, and imagining "what if".

The growth in popularity of makerspaces has spread through social media, publicly posted videos, magazines dedicated to making, and Maker Faires where people exhibit their creations. These spaces are found in universities, museums, libraries, schools, community centers, church basements, and areas leased for the specific purpose of creating a makerspace (Sheridan et al., 2014). The physical location, and even the technology present, do not define the makerspace. That is dependent on the participants and their interaction with the resources and one another. Makerspaces combine technology, resources, mentors, beginners, participants of all ages, and materials (Sheridan et al., 2014). The people involved are all working to create a physical or virtual answer to a question; what would happen if, how could I, is it possible? Participants demonstrate the physical embodiment of constructionism (Parpet & Harel, 1991). This theory posits that the creation of a physical object, the process of creating itself, enhances understanding of the concept. Indeed, the interaction between those in the process of creating offers numerous opportunities to expand conceptual understanding.

Makerspaces may be compared to studio art labs where students use materials to create art yet it is in the creation process where students are able to apply techniques to demonstrate their understanding of concepts (Sheridan et al., 2014). The act of creating, *the process*, is much more important than *the product*. It is the is commitment to voluntarily explore ideas and create that unify makerspaces despite differences in participants, materials and resources used, technology, and physical location (Peppler & Bender, 2013). As participants test out their ideas and improve designs, they increase their understanding of techniques and what might be possible with the resources.

13.4.2 Learning in a Makerspace

Learning can be seen in multiple ways in makerspaces. Both formal and informal learning happens. Formal learning can take place in the form of workshops, courses, and mentoring. Informal learning occurs when participants observe others' work, chat to offer suggestions for improvement, or interact with materials. In makerspaces learning happens as individuals become more involved in the community; however, the focus is on creating and designing the space to maximize participant interaction with one another and the resources (Sheridan et al., 2014). Learning is a by-product of satisfying individual curiosity and knowledge is expected to be shared throughout the community.

The purpose of learning in makerspaces is quite different from that in traditional classrooms. In makerspaces learning happens across disciplines. In traditional education teaching occurs in isolated disciplines such as math, art, and science with little discussion on how these areas may overlap in real-world problem solving. Another difference is "just in time learning" (Novak, Patterson, Garvin, & Christian, 1999, p. 11). Just in time learning occurs when students learn concepts and skills as they are needed to complete a task rather than learning in abstract without concrete application. Traditional classrooms may teach the concepts and skills without students' understanding their relevance or relationships. Makers learn skills and concepts to satisfy an immediate need.

The participants in makerspaces show strong evidence for the Self-Determination Theory of necessary components of intrinsic motivation, namely: perceived competence, autonomy, and relatedness (Deci & Ryan, 2000). The varying levels of expertise existing within the makerspace communities offer beginner participants the tools to explore and mentoring needed to make the project thus fulfilling the component of perceived competence. Autonomy is evidenced in that each participant is free to decide whether or not to work or continue to work on a project, when to work, and the amount of involvement in the group. Relatedness, feeling part of the community of practice, is a strong driving force in makerspaces. Observing others working, engaging in informal critique and problem-solving, and physically sharing space and expertise forms strong bonds among group members. These relationships encourage the development and sharing of new skills and products within the group. They also increase intrinsic motivation to persist when difficulties arise.

The purpose of products in makerspaces differ greatly from the purpose of products in traditional classrooms. In formal education, products serve as evidence to display the level of mastery or skill obtained. They are typically evaluated and graded. Participants engage in creating products in a makerspace due to curiosity and intrinsic motivation. When the curiosity is fulfilled or another topic is more compelling, the project may be abandoned. Similarly, the final product may actually fail yet that failure may be the launching pad for a new idea. Makerspaces do not make demands on participants to complete projects or provide evidence of learning. It is assumed that learning happens during the creation activities. As projects are deeply personal and are engaged in voluntarily, there are no requirements for assessment documentation. However, some participants do work on a single project for years and keep detailed notes on their attempts, failures, and progress (Sheridan et al., 2014).

13.4.3 Makerspaces and Education

Noting the interest in makerspaces and seeing their connection to STEAM education has many people wondering how to incorporate making into the educational process. Peppler and Bender observe, "The maker movement is an innovative way to reimagine education." (2013, p. 23). Martinez and Stager stated, "The maker movement may represent our best hope for reigniting progressive education." (2013, p. 11). Intrinsic motivation to learn and make has been found at every age level presented with the opportunity to participate in a makerspace. Thus lifelong learning is demonstrated in the maker community. Makers "identify their own challenges and solve new problems. Making provides ample opportunities to deeply understand difficult concepts (Makermedia, 2013, p. 3). This willingness to persist and learn with no external grade or reward demonstrates engagement that is enviable in education. Martinez and Stager point to Leonardo de Vinci as the ultimate maker and role model for the maker community (2013).

A necessary shift in the conceptualization of teaching and learning is necessary to bring making into the classroom. Numerous researchers have noted that focusing on superficial subject matter information such as vocabulary or reading about abstract theories actually diminishes student engagement and misses opportunities for building knowledge and skills (Clapp & Jimenez, 2016; Martinez & Stager, 2013a, b). Rather than viewing the elementary and middle school grades as building blocks to prepare students for high school and college level material, students in these grades should be viewed as beginning scientists, mathematicians and engineers early on (Martinez & Stager, 2013a, b). Instead of a teacher-centered classroom with the teacher lecturing on abstract concepts, and students responding to low level comprehension worksheets, classrooms need to be transformed into student-centered places of exploration. By giving students opportunities to experiment and play, teachers send the message that students have good ideas and there is more than one way to solve a problem (Martinez & Stager, 2013a, b). Increasing students' perceptions of competence and autonomy will also increase their intrinsic motivation toward learning the subject matter.

As noted by Martinez and Stager, "School, especially in science and math classes, typically only honors one type of learning and problem-solving approach, the traditional analytical step-by-step model. Often more non-linear, more collaborative, or more artistic problem-solving skills are often dismissed…" (2013, p. 37). This approach may severely restrict engagement of visual-spatial learners who may not think in linear patterns (Silverman, 2002). It also diminishes those who may be interested in math and science but approach it from a curious, non-linear way.

When schools focus on only one *correct* approach and one *right* answer it limits thinking and development. Children invent naturally. Given blocks or boxes they imagine, build, experiment, and revise. Nurturing these natural creative inclinations in an engineering context would help develop concrete understanding of math and science concepts (Martinz & Stager, 2013a, b). Using 3D printing, and other technology along with the ideas behind makerspaces as places to explore and solve problems, teachers can transform education from the bottom up starting within their classrooms and with their teaching practice. As teachers learn to create ill-structured problems with multiple entry points and solutions, the activities will support students' critical thinking and development of deep understanding of mathematical concepts.

13.5 What Is the Role of Prototype Problem-Solving Activities in Mathematical Understanding and Creativity?

Ill-structured, messy problems reflect real-world experiences of mathematicians and engineers. However, students are unfamiliar with these types of problems in the study of mathematics. Teachers need to create scaffolded activities to introduce learners to this new environment where the problems could have many possible solutions and perspectives. Such activities encourage creativity and critical thinking because there are no single correct answers. Creating activities that require students to produce a prototype further enhances student creativity in design decisions. Prototype Problem Solving Activities also allow students to self-assess and receive feedback from how well the physical prototype solves the problem.

13.5.1 Prototype Problem Solving Activities as a Tool for Learning

The idea of using concrete objects to represent mathematical ideas dates back to the early 1900s. Glas's historical account of the mathematician Felix Klein noted that Klein was one of the first people to blur the line between applied and theoretical mathematics (Glas, 2002). Klein's ideas regarding transforming abstract mathematical ideas into concrete representations are still relevant today as we endeavor to develop creativity in mathematical students. Klein supported the use of physical representations in mathematics because they represented different perspectives of the content. "Conceptual development depends on imagination, the ability to recognize patterns to discern relevant connections, and to model things from new points of view" (Glas, 2002, p. 100). Physical representations allow insights from other domains to be used with mathematics thus serving as a means of creating new understanding. Developing creativity that enables people to see math in their everyday lives and to make intuitive guesses that are able to be tested rather than learn mathematics as some distinct separate group of facts that exist on their own is critical to education and society. Creating concrete objects represents the highest levels of thinking in Bloom's taxonomy as people synthesize information from multiple areas of knowledge of the discipline.

13.5.2 Pedagogy for Effective Prototype Problem Solving Activities

Prototype Problem Solving Activities (PPSA) demand new ways of behaving for both students and teachers. Lesh and Doerr (2003) explain how teachers must adopt constructivist pedagogy to teach effectively through the use of physically creating objects to represent abstract ideas. PPSA build on this idea by asking teachers to support students as they identify the problem and create physical products. Students collaborate and typically work in groups of three or four and must document their thinking process as they create various prototypes to test their solutions. Teachers are seen as guides or coaches who ask key questions rather than providing answers for students. At the beginning of a PPSA the teacher does not provide instruction in the mathematics concepts. Instead the students are encouraged to invent their own methods and create physical prototypes to test the solution to see if they solved the problem. Students are required to discuss and document their ideas and the revisions to their prototypes as they undergo multiple iterations as a result of testing. Unlike most factual-based mathematics assignments, there is no singular correct answer in PPSA the correct prototype is the one that solves the problem and best meets the "client's" needs. "They are different from other problem-solving tasks because *the process is the product* and creativity is a major emphasis" (Chamberlin & Moon, 2005, p. 43 emphasis added) in solving PPSA. This aspect of using PPSA in the classroom is often the one requiring the most adjustment on the part of students. Especially those who are used to an assignment being graded upon completion.

Chamberlin and Moon (2005) detail six design principles that are required to create meaningful learning opportunities for students using hands-on problem solving. These design principles are construction, the reality principle, the construct documentation principle, the construct share-ability, the construct reusability principle, and the effective prototype principle (Chamberlin & Moon, 2005). The construction principle implies that the solution must be represented by a physical object which is a system of distinct elements and the relationships and interactions among the elements. For example, asking students to create a bridge that is a 1/100th scale that can hold the weight equivalent to 1/100th of 100 cars passing simultaneously and span 1/100th of a space measuring 4000 feet. The distinct elements would be the size and type of bridge, the weight of the cars, and the span.

The relationships and interactions could be the physical design (truss, suspension, or some other form of bridge), the width of the bridge to handle the weight of the cars, the relationship of width to span, and the consideration of whether the cars were moving or in a traffic jam. The students would design and physically print out the bridges with a 3D printer then test to see if they fulfilled the specifications and were aesthetically appealing. Figure 13.3 illustrates this process. The low cost of 3D printing and rapid results possible by printing allows students to create multiple iterations thus mistakes become part of the learning experience. The reality principle suggests that the problem could realistically happen in the life of the student. It is important to make the problem realistic because it encourages students to imagine the problem and engage meaningfully in finding a solution. When students can realistically imagine a problem, they are more familiar with various aspects of the solution and can bring unique insight. For example, the bridge could be described as a means to relieve traffic congestion. Students can virtually visit different types of bridges in multiple locations to gain a better understanding of the relationships between the structure, its purpose, and its place in the environment. The construct documentation principle may be the most important aspect in the PPSA. This principle provides many ways for students to show their thought processes as they work on the problem. The documentation principle allows students to self-assess and document the modifications of their prototype which provides evidence of higher-order thinking skills and develops metacognition. This aspect also helps

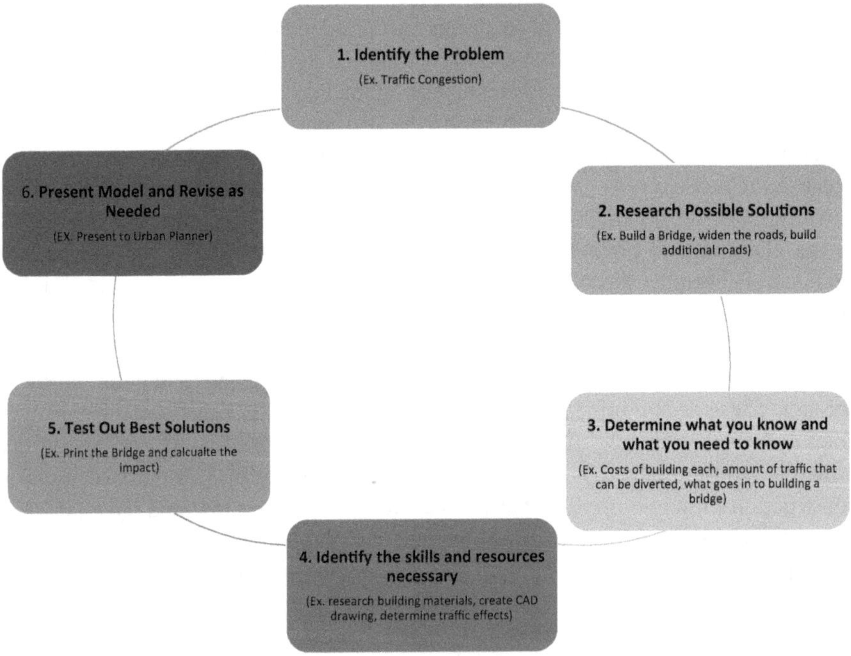

Fig. 13.3 The prototype problem solving activities model

teachers to focus on the thinking process rather than the end product. Indeed, the evidence produced by the construct documentation principle provides better examples of student understanding than could be obtained through simple worksheets and chapter tests. The final three principles, construct share ability, construct reusability, and the effective prototype principles are related. These principles apply to the final solution and indicate that the prototype should be able to be used in other similar situations and that the model should be able to be used and understood by others. In the bridge example students could present their bridges to a planning community, urban planner, or architect. Figure 13.3 illustrates the cycle that is involved in Prototype Problem Solving Activities for students.

PPSAs that are designed using the six principles produce curricular characteristics that increase creativity in the solution. Some of the characteristics include interdisciplinary thinking and communication. As students collaborate to understand the problem presented they bring in knowledge from other content areas and exchange ideas regarding the important aspects of the problem and some possible solutions. One of the hallmarks of creativity is the ability to develop a good question or identify a quality problem (Sawyer, 2012). PPSAs train students in this ability through metacognitive coaching. As the teacher asks guiding questions in response to students' inquiries, the students learn to question one another to clarify thinking and find their own answers. Kim and Kim (2010) note that one of the benefits of these activities is the ability for students to experience self-directed learning as they identify the problems and the necessary information to solve them. The curricular characteristic of documenting the thinking process is also developed through the construct documentation principle. It is important for students to document their thinking and to be able to look back and see how their thought process developed over time. The ability to see students' thinking also allows teachers an opportunity to refine the curriculum or make adjustments in their own teaching as a result of this information.

PPSAs are specifically applicable for teaching mathematics, technology, engineering, and increasing creativity because they rely on the iterative process of designing, testing, and documenting success or failure of the prototype which increases mathematical understanding and creativity as students continue to work to overcome deficiencies in their product. Hobson, Trundel, and Sackes (2010) demonstrated that by using computer simulation models children were able to learn much more sophisticated scientific content because the visualization provided support. Physically creating and viewing the objects allowed the students to create cognitive structures that supported their understanding. Coxbill, Chamberlin, and Weatherford (2013) documented improvement in third graders mathematical understanding and creativity by implementing such activities. It took students several interactions with the new pedagogical process to become comfortable in new ways of learning but once they became accustomed they were eager to demonstrate their understanding. Jaakkola and Nurmi (2008) conducted pairwise comparisons between students who used hands-on learning activities to those who attended traditional learning. They found that the students who used simulations and engaged in hands-on learning significantly higher learning outcomes.

Thus, engaging students in simulations and prototype creation learning may provide increased academic achievement and understanding.

13.5.3 Creating PPSA

Combining PPSA activities with computer modeling and 3D printing (PPSAC3D) offers unprecedented opportunity for students to learn in authentic environments and apply creativity to understand complex concepts. Currently most children interact with technology as consumers of animations and pre-created programs (Eisenberg, 2013). The use of 3D printing allows children to become fabricators or "makers" and become more self-directed in their interaction. This process requires students to become thoughtful in their choices. PPSAC3D allows educators to participate a process that "demands concentration, encourages creativity, and rewards expertise" (Eisenberg, 2013, p. 131). Thus, raising the level of thinking and increasing student engagement by allowing students to take ownership over their learning and apply their knowledge in an authentic environment (Kwan, Park, & Park, 2014).

A large advantage of using 3D printers in the classroom is that students are able to print out physical models quickly and at low cost. The ability to rapidly print out a prototype part and test it in the solution provides immediate real-world feedback to students. It also opens the door to creativity as challenges arise and innovative ways of thinking about the problem emerge from being able to physically see and manipulate parts of the model. Being able to imagine, design, and create physical products frees students from the limitations of only using parts that are easily available in their environment. Students who do not live close to home-improvement stores or art supply retailers can print out parts to help construct inventions and problem-solving models. Testing out smaller versions of physical products can also avoid the costs of producing full-sized models that may not work as intended. Reducing the monetary cost of failure by producing inexpensive plastic pieces frees students from the anxiety of wasting their budgets. For example, Kroll and Artzi (2011) estimated that using the 3D printed parts to create airplanes to be tested in a wind-tunnel for their engineering students cost between five and ten times less than creating the same model out of metal. This allowed the designers to create many more variants of the model than they would have if they typically had to endure the cost in time and money of creating metal models. Thus each variant improved the model design and these changes may not have existed if the students were limited to only one or two versions due to monetary constraints. Students also benefitted by being able to produce physical models to answer some of the questions related to aircraft engineering rather than having to rely on theoretical calculations.

Some schools are choosing to house and fund 3D printers through their library as a way of increasing the access and sharing resources. Similar to computer labs in schools, sharing a 3D printer allows multiple users to participate with a technology that would not be affordable for each individual classroom. It also creates opportunities for inter-disciplinary collaboration when users meet up at the printer. In a report on the use of a 3D printer at the University of Nevada, Reno, the top printer users were mechanical engineering, art, biochemistry, biomedical, electrical engineering, and anthropology (Cosgrove, 2014). As soon as the 3D printer was available for use it has been running at full capacity and often has a two week waitlist of projects to be printed thus reflecting the previously unmet need on campus.

The use of a 3D printer follows the cycle stated by Doorley, Witthoft, and Kelly (2012) "imagination begets fabrication, fabrication begets imagination" (p. 80). Holding a tangible object, viewing it from multiple angles, and imagining its interaction with other objects provides fuel for the imagination. It allows students to visualize how the object might be modified to fulfill other functions. Cosgrove (2014) states:

> Just as a document printer lets users create a tangible product of their creative writing, enabling further refinement and collaborative input as it is physically marked up and shared with others, ready access to 3D printer technology could enable learners and researchers to quickly produce real-world versions of otherwise intangible digital object.

Although it is in the beginning stages, 3D printing is projected to follow a similar developmental path as the personal computer. Thus it is not fanciful to imagine a time in the near future when students will have 3D printers in their home for personal use. Taking advantage of the recent cost reductions in 3D printing allows institutions to invest in technologies that will be relevant for many years to come.

13.5.4 Research on 3D Printing in the Classroom

As 3D printing is a new technology just beginning to become affordable for use in schools, research on its use is scarce. A comprehensive search of educational research databases yielded only nine empirical studies. Due to the limited employment of 3D printers in classroom learning, all of the studies were qualitative with using observational data. Nonetheless, the findings offer promising opportunities to increase student engagement and enthusiasm for STEM related content.

Student enthusiasm and engagement was demonstrated by observing student behaviors as they were working on their 3D projects. Researchers noted that students were willing to give up personal time to work on projects (Chu, Quek, Bhangaonkar, Ging, & Sridharamurthy, 2015; Kostakis, Niaros, & Giotitsas, 2015; Schelly, Anazlone, & Chia, 2015). Schelly et al., (2015) even noted the unintended consequence of a student attempting to break into school after hours to gain more access to the 3D printers. This willingness to spend time outside of class to work on homework is rarely reported. The teachers in the Kostakis et al., (2015) study reported greater engagement during class time and fewer discipline or disruption issues. Those teachers also noted an increase in parental involvement as students were able to bring their physical projects home and explain the challenges they were facing in the design process. Chu et al., (2015) noted that students in their study exhibited fun and excitement during the workshop but also expressed some tension and frustration when confronting new ways of learning and processing information.

Fonseca, Valls, Redondo, and Villagrassa reported on the effects of using virtual modeling on architecture students (2016). Students in this study used augmented reality and 3D software to design and present ideas for urban development. Students and the public expressed positive feedback regarding the use of technology. Because the software was easy to use, students reported improved participation and motivation which translated into better academic performance. The results of this study indicated that the student participants were able to demonstrate "significant improvements in spatial, research, and interaction skills" (Fonseca et al., 2016, p. 516) as a result of working with technology to create physical and virtual representations of their ideas.

Two studies focused specifically on the importance of 3D in mathematics (Kaufman & Schmalsteig, 2003; Kwan et al., 2014). Kaufman and Schmalsteig focused on 3D modeling to teach geometry and used technology to create and project 3D models of various shapes which were previously confined to line drawings. The students responded enthusiastically and moved about the classroom to explore the shapes from various viewpoints. "It was clear they were all proud of what they had built" (Kaufman & Schmalsteig, 2003, p. 343). The students offered multiple applications for creating 3D representations such as interactive conic sections, vector analyses, and building 3D worlds from 2D views (Kaufman & Schmalsteig, 2003). Thus, using 3D modeling not only increase their understanding but served as a launch pad for further exploration. The findings from Kwan et al., (2014) support the idea of authentic learning as a result of using 3D printers in mathematical tasks. For example, asking students to produce 3D objects that are the results of mathematical formulas allows them to physically interact with mathematics and deepen their understanding. This study found that students showed improvement in their cognitive, social, and behavior development as a result of using 3D printers to create tangible objects to represent theoretical concepts. Students engaged in authentic learning were able to connect new understandings that excited their learning (Kwan et al., 2014).

Flores and Springer (2013) provided a summary of their detailed evaluation of middle-school students participating in makerspaces. They noted that the process of iteration and self-directed learning had a positive impact on student achievement and their ability to learn new material. "Students learn that every failure teaches you something new and should be embraced as an opportunity for learning, growth, and improvement" (Flores & Springer, 2013, p. 2). This approach is a stark contrast to most educational experiences where failure is considered final.

Two final studies documented an effort to expose teachers to the potential of 3D printers in K-12 education (Kűcukozer, 2013; Schelly et al., 2015). These teacher workshops were created to provide teachers with pedagogical experiences as well as an opportunity to engage in the creative process as students. Teachers reported feeling proud of their accomplishments and that the process facilitated learning content. The teachers in the Schelly et al. (2015) study expressed the belief that working with 3D printers would empower their students. They identified the opportunity to engage in cross-curriculum collaboration. The teachers also noted how students could improve their understanding of conceptual or abstract problems by creating tangible, physical products. They also saw 3D printing as an opportunity to engage typically non-involved student and to provide challenge for gifted and creative students. One teacher wrote, "The printer has almost unlimited potential to make our departmental teaching units become more hands-on and student oriented" (Schelly et al., 2015, p. 234). This budding awareness represents the potential to revolutionize the educational environments to encourage more critical thinking and creativity in learning.

13.6 Conclusion

Understanding creativity to be both "little c" regarding solutions to everyday problems and "big C" resulting in products that are valued by society, teachers and researchers can begin to focus on providing opportunities to practice and develop both aspects. Allowing students the time and space to engage in creative learning may permit students to be identified who might otherwise be overlooked as gifted or in need of additional challenge. As teachers learn to discern the difference between those students who are able to apply algorithms quickly and accurately and those students who are able to identify and solve problems creatively they are better able to support the development of creativity in students.

Designing and implementing PPSAs in classrooms provide an opportunity for teachers to support the development of creativity and problem solving skills in mathematics. Jaakkola and Nurmi (2008) showed that students who engaged in simulations and hands-on learning activities had higher learning outcome measures than those who received traditional education. Hobson et al. (2010) showed how creating models allowed students to build cognitive supports that facilitated understanding complex scientific content that was thought to be beyond them. Creating Prototype Problem-Solving Activities according to established principles moves students into higher levels of thinking. Instead of focusing on remembering facts or understanding how to apply algorithms, students must apply what they have learned, evaluate their solutions and create alternative solutions. This process is closer to how professional mathematicians and scientists work and has been found to increase student engagement.

Digital fabrication using 3D printing allows students to take the idea of PPSAs into the 21st century. PPSAC3D provides students with tangible evidence of their ideas and allows them to physically engage with abstract ideas. This additional aspect of being able to create objects that can be manipulated, tested, and shared with others adds an element of "big C" creativity to learning. As physical objects are viewed, tested, and commented on by others, the creator gains a better understanding of the content. 3D printing is inexpensive and provides rapid results therefore students are able to test and fail multiple times learning something new with each iteration. Reducing the impact of failing increases students' self-confidence in their creativity and problem solving abilities. The process of understanding the problem, creating a solution, physically producing the product, and receiving feedback is much similar to how professional engineers, mathematicians, and scientists work than the traditional ways of learning these subjects. Allowing students the time and space to explore the real-world experience of those in the STEM fields may spark an interest that would carry over into advanced studies.

Professional mathematicians move beyond simply understanding content and employ the skills of decision-making, being able to generalize and infer principles and engage in recursive thinking. Therefore, if the goal is to support the development of mathematics creativity, it is important that these skills are taught and students are offered opportunities to practice them. Sriraman (2005) found that mathematics creativity generally develops along a stage model of "preparation, incubation, illumination, and verification" (p. 24). Creative mathematicians indicated that time available to work on the problem, freedom of movement, and the interest in contributing solutions to real problems supported their achievements. If the aim is to develop skills in students that enhance their ability to be creative in mathematics and STEM fields, it is important that they be nurtured in environments that support the process. PPSA is such an environment.

As a new technology, published research that exists regarding the use of 3D models in educational settings is scarce. The purpose of this chapter is to provide a catalyst for readers to imagine the possibilities that could be created as learning opportunities both inside and outside of the classroom as students are invited to become problem-solvers.

References

Bronfenbrenner, U. (1986). Ecology of the family as a context for human development: Research perspectives. *Developmental Psychology, 22*(6), 723–742.

Chamberlin, S., & Moon, S. (2005). Model-eliciting activities as a tool to develop and identify creatively gifted mathematicians. *The Journal of Secondary Gifted Education, 17*(1), 37–47.

Chu, S. L., Quek, F., Bhangaonkar, S., Ging, A. B., & Sridharamurthy, K. (2015). Making the maker: A means-to-an-ends approach to nurturing the maker mindset in elementary-aged children. *International Journal of Child-Computer Interaction, 5,* 11–19.

Clapp, E. P., & Jimenez, R. L. (2016). Implementing STEAM in maker-centered learning. *Psychology of Aesthetics, Creativity, and the Arts, 10*(4), 481–491. http://doi.org/10.1037/aca0000066.

Cosgrove, P. (2014). Making it real: 3D printing as a library service. *Educause Review*. Retrieved from http://er.educause.edu/articles/2014/10/making-it-real-3d-printing-as-a-library-service.

Coxbill, E., Chamberlin, S., & Weatherford, J. (2013). Using model-eliciting activities as a way to identify and develop mathematically creative students. *Journal for the Education of the Gifted, 36*(2), 176–197.

Deci, E. L., & Ryan, R. M. (2000). The 'what' and 'why' of goal pursuits: Human needs and the self-determination of behavior. *Psychological Inquiry, 11*(4), 227–268.

Doorley, S., Witthoft, S., & Kelly, D. (2012). *Make space: How to set the stage for creative collaboration*. Hoboken, NJ: Wiley.

Eisenberg, M. (2013). 3D printing for children: What to build next? *International Journal of Child-Computer Interaction, 1,* 7–13.

Flores, C., & Springer, T. (2013). Authentic learning and assessment in the self-directed environment of a middle school maker space. In *Proceedings from the Interaction Children and Design Conference*, New York, NY.

Fonseca, D., Valls, F., Redondo, E., & Villagrassa, S. (2016). Informal interactions in 3D education: Citizenship participation and assessment of virtual urban proposals. *Computer in Human Behavior, 55,* 504–518.

Glas, E. (2002). Klein's model of mathematical creativity. *Science & Education, 11,* 95–104.

Hobson, S. M., Trundle, K. C., & Sackes, M. (2010). Using a planetarium software program to promote conceptual change with young children. *Journal of Science Education Technology, 19,* 165–176.

Jaakkola, T., & Nurmi, S. (2008). Fostering elementary school students' understanding of simple electricity by combining simulation and laboratory activities. *Journal of Computer Assisted Learning, 24,* 271–283.

Kaufmann, H., & Schmalsteig, D. (2003). Mathematics and geometry education with collaborative augmented reality. *Computers and Graphics, 27,* 339–345.

Kim, S. H., & Kim, S. (2010). The effects of mathematical modeling on creative production ability and self-directed learning attitude. *Asia Pacific Educational Review, 11,* 109–120.

Kostakis, V., Niaros, V., & Giotitsas, C. (2015). Open source 3D printing as a means of learning: An educational experiment in two high schools in Greece. *Telematics and Informatics, 32,* 118–128.

Kroll, E., & Artzi, D. (2011). Enhancing aerospace engineering students' learning with 3-D printing wind-tunnel models. *Rapid Prototyping Journal, 17*(5), 393–402.

Kűcukozer, H. (2013). Designing a powerful learning environment to promote durable conceptual change. *Computers & Education, 68,* 482–494.

Kwan, O. N., Park, J. H., & Park, J. S. (2014). Providing students authentic learning experiences through 3D printing technology. In *Learning experiences to promote mathematics learning yearbook 2014*. Singapore, Singapore: World Scientific Publishing Company.

Lesh, R., & Doerr, H. (2003). *Beyond constructivism: Models and modeling perspectives on mathematical problem solving, learning, and teaching*. Mahawah, NJ: Lawrence Erlbaum Associates.

Makermedia. (2013). *Makerspace playbook school edition*. Retrieved from https://makered.org/wp-content/uploads/2014/09/Makerspace-Playbook-Feb-2013.pdf.

Mann, E. (2006). Creativity: The essence of mathematics. *Journal for the Education of the Gifted, 30*(2), 236–260.

Martinez, S. L., & Stager, G. S. (2013a, December). Invent to learn: Makers in the classroom. *Educational Digest*, 11–15.

Martinez, S. L., & Stager, G. S. (2013b). *Invent to learn: Making, tinkering, and engineering in the classroom*. Torrance, CA: Constructing Modern Knowledge Press.

Matthews, D., & Geist, E. (2002). Technological applications to support children's development of spatial awareness. *Information Technology in Childhood Education Annual*, 321–336.

Novak, G. M., Patterson, E. T., Gavrin, A. D., & Christian, W. (1999). *Just-in-time teaching: Blending active learning with web technology*. Upper Saddle River, NJ: Prentice-Hall Inc.

Parpet, S., & Harel, I. (1991). Situating constructionism. In I. Harel & S. Papert (Eds.), *Constructionism*. Norwood, NJ: Ablex Publishing Corporation.

Peppler, K., & Bender, S. (2013, November). Maker movement spreads innovation one project at a time. *Kaplan*, 22–27.

Plucker, J. A., Beghetto, R. A., & Dow, G. T. (2004). Why isn't creativity more important to educational psychologists? Potentials, pitfalls, and future directions in creativity research. *Educational Psychologist, 39*(2), 83–96.

Sawyer, K. (2012). *Explaining creativity: The science of human innovation* (2nd ed.). New York, NY: Oxford University Press.

Schelly, C., Anazlone, G., Wijnen, B., & Pearce, J. (2015). Open-source 3D printing technologies for education: Bringing additive manufacturing to the classroom. *Journal of Visual Languages and Computing, 28,* 226–237.

Sheridan, K. M., Halverson, E. R., Litts, B. K., Brahm, L., Jacobs-Priebe, L., & Owens, T. (2014). Learning in the making: A comparative case study of three makerspaces. *Harvard Educational Review, 84*(4), 505–531.

Silverman, L. K. (2002). *Upside-down brilliance: The visual-spatial learner*. Denver, CO: DeLeon Publishing.

Snow, R. E. (1992). Aptitude theory: Yesterday, today, and tomorrow. *Education Psychologist, 27*, 5–32.

Sriraman, B. (2005). Are giftedness and creativity synonyms in math education? *Journal of Secondary Gifted Education, 17*(1), 20–26.

Part V
Creativity and Advanced Mathematics

Chapter 14
Can a Kite Be a Triangle? Aesthetics and Creative Discourse in an Interactive Geometric Environment

Hope Gerson and Paul Woo Dong Yu

Abstract In this chapter we will use the lens of aesthetics (Sinclair in Mathematics and beauty: aesthetic approaches to teaching children. Teachers College Press, New York, NY, 2006) to explore mathematical creativity in an interactive geometric environment from three different perspectives: inquiry, teaching, and mathematical resolution. We will be illustrating the mathematical creativity with an episode where academically talented middle school students are working with Shape Makers (Battista in Shapemakers. Key Curriculum Press, Emeryville, CA, 2003) in Geometer's Sketchpad. At one point in the lesson, a student makes a triangle looking shape with the Kite Maker Tool and asks, "Can a triangle be a kite?" We see creativity reflected in three ways: in the generation of ideas, in the teacher's instructional choices, and in the resolution of the mathematical discussion by the students. The creative and aesthetic qualities of open inquiry, the Geometers' Sketchpad, and teacher moves created a setting where students and the teacher made aesthetically motivational, generative, and evaluative choices to build understanding of geometric properties of kites and triangles as well as the limitations of sets of geometric properties in classifying geometric shapes.

Keywords Interactive geometry · Aesthetics · Student discourse

14.1 Introduction and Background

When someone says, "that was creative," they are often referring to artistic qualities, imagination, or innovative ideas. We easily recognize disciplines like art, music, and design as creative but the term, like many others, is not well-defined in our everyday usage. When we access our image of creativity, we find components

H. Gerson (✉) · P. W. D. Yu
Grand Valley State University, Allendale, MI, USA
e-mail: gersonh@gvsu.edu

P. W. D. Yu
e-mail: yupaul@gvsu.edu

© Springer International Publishing AG, part of Springer Nature 2018
V. Freiman and J. L. Tassell (eds.), *Creativity and Technology in Mathematics Education*, Mathematics Education in the Digital Era 10,
https://doi.org/10.1007/978-3-319-72381-5_14

such as artistry, the act of creating, the element of surprise, and the clarity that comes from changing perspectives. Mathematicians often describe their work as creative. They use aesthetic terms such as "elegant" and "beautiful" when describing their work. They engage in the creative work of problem posing, invention, pattern noticing, and problem solving. However, unless you are a mathematician or a teacher, it might take a while to think of an example of mathematical creativity. Yet there are examples from mathematics that are almost universally accepted as creative. Also, if you read research on student thinking, you see similar examples of creativity as students make sense of relationships among numbers, shapes, and ideas.

How do we recognize creativity in a school mathematics setting? In some instances, mathematical creativity is clear, as in the story of Gauss who as a child found a way to add the numbers from 1 to 100. While his classmates added the numbers one at a time, Gauss split the group in half, 1–50 and 51–100. Then he paired the numbers into groups of 101 (Fig. 14.1). Fifty groups of 101 were much easier to count. And, some say, in less than a minute Gauss had the answer—an innovative, efficient, and some would say, beautiful solution (Tent, 2006). In other cases, it is harder to recognize. For instance, in the proof of the Four-Color Theorem, the assertion that any map can be colored using four colors without any adjacent regions sharing the same color, Appel and Hacken (1977) created a cumbersome proof by cases, assisted by a computer. A mathematician might have trouble finding beauty or elegance in such a proof. Others criticized the use of a computer as if it offended their sense of form or fairness. Yet, a new proof was created where none existed before.

Gauss' mathematical creativity is interesting because on the surface, it looks like number crunching (a decidedly uncreative activity). The creativity is not in the calculation, but in the organization of the numbers. It is not necessarily beautiful to look at (Fig. 14.1), but his solution was elegant and unexpected.

The proof of the Four Color Theorem is also interesting. The application of the Four Color Theorem itself certainly has aesthetic elements (Fig. 14.2). The proof was created by two mathematicians and a computer (Wilson, 2002). Some would call that innovative. But many would have trouble applying the word creative to the resulting proof.

The question of what mathematical creativity might look like, especially in a school setting, needs further development. Researchers have identified some of the key ways to identify creativity in school mathematics. Sriraman, Yaftian, and Lee (2011) suggest that in school mathematics, one can recognize creativity as the

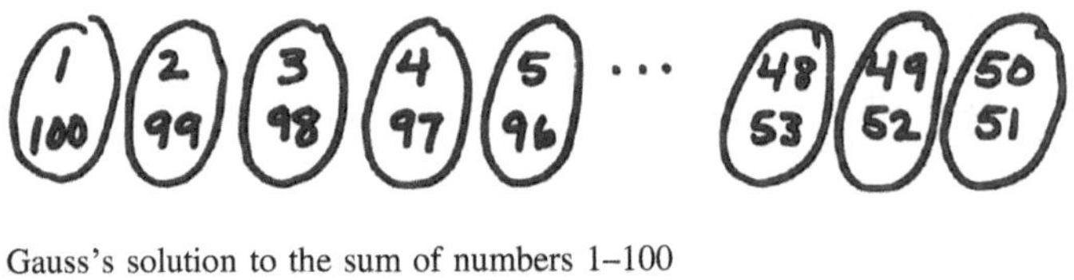

Fig. 14.1 Gauss's solution to the sum of numbers 1–100

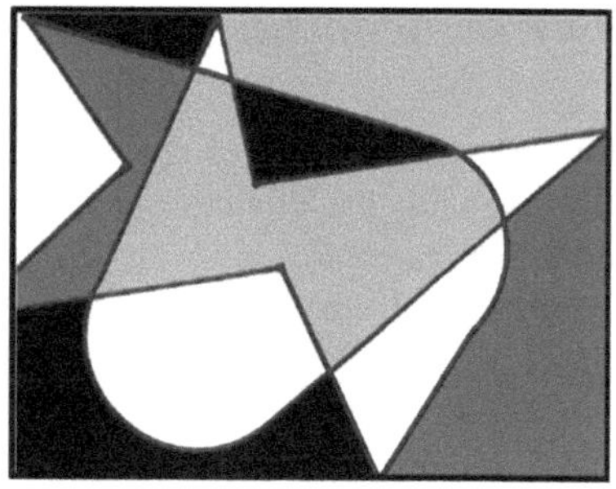

Fig. 14.2 Geometric design illustrating the Four Color Theorem

process that results in surprising solutions or a new way of looking at a problem. In addition, they recognize that creativity is an act of making choices that lead to new outcomes (for the students). Leikin and Lev (2007) suggest that student creativity is found in the originality, fluency, and flexibility expressed in problem solving. Liljedahl (2013) also suggests that one can identify student creativity through 'aha' moments that are often vocalized. Sinclair, Freitas, and Ferrera (2013) offer further help in identifying creativity as an active process that results in inventive moments that one can identify through students' gestures, written work, and work with technology.

Common to all of these definitions of creativity are surprise, perspective, novelty, originality, fluency, flexibility, moments of clarity, and invention, and the nature and role of aesthetics. There is more to creativity than the aesthetic that one normally associates with beautiful art. Here we conceptualize the aesthetic to be defined more broadly also to include the art of mathematical problem solving and sense making. Aesthetic sensibilities such as simplicity, visual appeal, connectedness, surprise, symmetry, and order may be invoked whenever one makes a mathematical choice (Sinclair, 2006). We suggest that viewing the choices students make through an aesthetic rather than a cognitive lens may allow us to more fully recognize and understand students' mathematical choices because our social constructs of creativity are based in the aesthetic world. We see aesthetics like an emotive gestalt—instead of an 'aha!' it's a "wow!" In other words, we are paying attention to an aesthetic that may lead to shifts in one's cognition.

Additionally, these characterizations of creativity all focus on an active process of creating mathematical ideas. In particular, creativity is enabled by choices students make in the process of sense-making and problem-solving. We suggest, therefore, that in order to recognize creativity in action, we need to view students' and the teacher's choices. We are focusing on creativity as expressed through aesthetic choices the students and the teacher make during problem-solving and sense-making activities.

14.1.1 Aesthetics as a Theoretical Framework

The characterizations of creativity, mentioned in the previous section, also suggest that viewing the choices students make through an aesthetic lens rather than a cognitive one may allow us to more fully recognize and understand students' mathematical creativity because our constructs of creativity are based in the aesthetic world. Sinclair (2006, 2008) provides us with such a framework through which we can study students' creativity. Sinclair suggests three different roles that the aesthetic plays in sense-making and problem-solving in school mathematics: motivational, generative, and evaluative. Each of the roles engages a set of aesthetic sensibilities, such as beauty, pleasure, or simplicity. Furthermore, each may be characterized by the types of actions, expressions, and choices that students make.

Sinclair notes that aesthetics play a motivational role when a student or teacher makes a choice that leads them to select a problem, pose a problem, or understand a mathematical idea. In the motivational case, these choices are made in an effort to lead to simplicity, visual appeal, mystery, relations between different modes of thinking, or a desire to connect different ideas. The most recognizable example, perhaps, is associated to when students have a cognitive 'aha!' moment. The pleasure and excitement of that moment of clarity has a motivational outcome making students want to explore more. One can recognize the motivational role when students are highly engaged, animated, make large gestures, or vocalize excitement.

Aesthetics play a generative role when the student or teacher pursues a path of mathematical inquiry because they expect it to reveal some insight or fact. For instance, students may see a pattern after they classify different shapes. This might lead them to pursue a general solution. A choice that invokes a generative aesthetic involves organizing or noticing pleasing properties such as symmetry, order, simplicity, liberating form, exactness, or fit. The choice generates a new, pleasing perspective on the mathematics that leads to a path of inquiry.

Finally, the aesthetic plays an evaluative role when students or the teacher are evaluating which of a set of options is best for choosing when to end the inquiry. Students using an evaluative aesthetic might see one solution strategy as more beautiful, simple, or efficient than another. For example, a student may prefer a geometric proof of the Pythagorean Theorem over an algebraic one because it is more visually appealing. The evaluation might be based upon beauty, elegance, simplicity, perfection, or simply because it is easier to understand or matches their thinking. Additionally, a student might feel a sense of closure as a question is answered or a path of inquiry is exhausted. Table 14.1 summarizes Sinclair's aesthetic framework.

There are other theoretical and professional considerations in the exploration of open-response environments that induce creative and aesthetic qualities to mathematics instruction, in particular, the consideration of affective elements of mathematics learning. In 2001, the U.S. National Research Council (NRC, 2001) identified five strands of mathematical proficiency: *procedural fluency, conceptual*

Table 14.1 The three forms of aesthetics in mathematics education

Role	When do we see it?	Aesthetic sensibilities	Possible indicators
Motivational: the student/teacher pursues a mathematical inquiry because it is interesting	In problem selection, problem posing, and throughout the inquiry process	Apparent simplicity, visual appeal, connectedness, mystery, relations between different modes of thinking, desire to connect different ideas	Pursue a question different from the teacher or peers, Have an 'aha!' moment, Raise voice, big gestures, show their work to someone else, Become highly engaged with an exploration
Generative: the student/teacher pursues a path of mathematical inquiry because they expect it to reveal some insight or fact	When representing the ideas, searching for patterns, and choosing the path of inquiry	Symmetry, order, simplicity, liberating form, exactness, fit, pleasure	Express pleasure especially tied to the sensibilities, Simplify a representation, Organize the information in a new way, Finds a new pattern
Evaluative: the student/teacher evaluates the worth of strategies, representations and solutions	When evaluating best strategies, representations, solutions, etc.	Beauty, elegance, simplicity, illuminating, perfection, easy to understand, matches with their thinking, closure	Express aesthetic reason why something is better than another. Evaluates the line of inquiry as being finished

understanding, adaptive reasoning, strategic competence, and productive disposition. Sinclair's framework helps to identify those elements, within the 5 strands, that are emotive, and exist in the affective domain. For example, consider constructs like 'out of the box' thinking in *strategic competence*, or students' self-efficacy in understanding and appreciating mathematical inquiry as described in having a *productive disposition.* Also, conceptualizing the U.S. Common Core Standards for Mathematical Practice (2010) along the aesthetics and creative domain may provide insight into the affective and emotive elements of the Mathematical Practices beyond just the cognitive elements. While the analysis of the data will not use specific elements of the Five Strands of Mathematical Proficiency and Standards for Mathematical Practice (2010), they are mentioned here as professional context in which aesthetics connect to mathematical instruction, and will be discussed further at the end of this chapter. Finally, we view creativity as an aesthetic construct that is predicated on student and teacher choices. Therefore, our analysis and discussion will focus predominantly on the aesthetic sensibilities and the roles they are playing as the students and teachers make mathematical choices.

14.1.2 Aesthetics and Instructional Technology

One application of instructional technology is as a platform for building procedural fluency (NRC, 2001) or skill-based transmission (Murphy et al., 2014; Baron, 2010; Niederhauser & Stoddart, 2001). Students may work on self-grading computer-based assignments, or use game-based contexts to solve mathematics problems to advance or to earn time to play a videogame that pops up after they have answered a set of questions. As a metaphor, we consider these forms of instructional technology like a painting by number activity. While there may be aesthetic elements, the design of the activity is directional and prescriptive. In these settings students are not encouraged to explore and make mathematical choices that are necessary for creativity (Kwon, Park, & Park, 2006).

In contrast, dynamic mathematical environments such as Geometer's Sketchpad, GeoGebra, Wofram Alpha and Desmos elicit mathematical construction, creation, and exploration. As students interact with geometric figures and graphs, they naturally make and test conjectures, choose what to build, what to pursue, what to notice and pay attention to, based on their own imaginations, aesthetic sensibilities, whims, and mathematical thinking (Johnson-Gentile, Clements, & Battista, 1994; Lavy & Shriki, 2010; Ng & Sinclair, 2015; Patsiomitou, 2008).

Students interacting with dynamic mathematical environments connect a virtual mathematical world with embodied action that allow them to engender abstract mathematics (Burbules, 2006; Sinclair et al., 2012). The aesthetic sensibilities are activated not only by the representations on the screen, which may have visual appeal, but in the actions of the students as they interact with those representations. Students act upon the representations to create symmetry, form, structure, collinearity, visual appeal, orientation, balance and order and they embody those actions. Thus, dynamic mathematical environments have the potential to both elicit aesthetic sensibilities through interaction with visual representations and affording students an environment where they are free to make choices. The interactions among choice and aesthetic sensibilities make dynamic mathematical environments rich for the study of mathematical creativity. Students' interactions with dynamic mathematical environments also affect the way they talk about mathematical ideas. Sinclair and Moss (2012) found that student discourse first focuses on visual cues without regard to properties and then cycles between visual and more formal mathematical discourse.

In particular, we chose the microworld Shape-Makers (Battista, 2001, 2003). In this world, students manipulate quadrilaterals and make and test conjectures about their properties. Students are able to click and drag on quadrilaterals and generate a large example space that allows them to actively build an image of each family of quadrilaterals (Jackiw & Sinclair, 2009). The instructional goal was to help students build geometric sophistication by moving from idiosyncratic spatial structuring to property-based mathematical definitions (Battista, 2001). The environment is highly structured, but the students are free to interact with the Shape-Makers on their own terms. Students choose the cases they will explore and how they will move the shape-makers about. They are free to notice properties, make and test conjectures, and explore as they see fit. The visual-mathematical structure of Shape-Makers also allows students to manipulate and attune to aesthetic properties. Students build a shared experience of working with the Shape-Makers, but the open nature of their exploration allows them to build diverse perspectives as they make different choices and then communicate about their experiences. In this chapter, we investigate the question: What roles does the creative aesthetic play and what aesthetic sensitivities are enacted by the students and teacher as they determine whether a triangle can be a kite? Additionally, we examined the question: What patterns emerge in the roles played by the aesthetic over the course of the class discussion?

14.1.3 Setting and Methodology

The data for this qualitative case study was taken from a twelve-minute videotaped excerpt of a class discussion during a Shape Makers (Battista, 2003) curriculum activity called "How are they the same?" Shape Makers is an inquiry-based unit consisting of interactive quadrilateral shapes, called Shape Makers, made with Geometer's Sketchpad (Jackiw, 1991). The class in which the video was taken was a special section of an accelerated high school geometry class (N = 15) taught to a select group of academically talented middle school students ranging mostly from grades 6 through 8.

In this particular activity, the students working individually had to list as many invariant properties of any shapes made with a particular Shape Maker as they clicked and dragged the vertices around creating many different examples of each shape. For example, commonly listed properties of any shapes made with the Kite Maker were: two pairs of congruent sides, at least one pair of congruent angles, and at least one line of symmetry. After a time of individual exploration, the students shared their observations to create a class list of properties for each shape. Each shape and its associated properties were written on large paper, and put at the front of the classroom for all the students to see. The questions and lesson procedures were designed by the teachers in order to lead to a set of student-generated and teacher-guided results and conclusions. In the case of this lesson, the goal was to develop a comprehensive list of properties for each of the seven quadrilaterals. We selected this particular class period because the students and teacher creatively chose to explore an unplanned topic.

In the analysis of data, two perspectives were considered. The first perspective was that of a traditional researcher, and the second was the researcher-as-teacher (Ball, 2000). The first perspective provided an objective view of the classroom activities and discourse. This provided an analytical point-of-view that had no assumptions about the classroom episode. The second perspective, the researcher-as-teacher, provided an 'insider's' view of both the classroom dynamics and motivations of the instructional decisions that are not evident in the videotaped data. The data was independently coded by both researchers using Sinclair (2006) aesthetic framework. After independent coding, the researchers discussed and refined the aesthetic codes. On a second qualitative pass of the data, the data was recoded based upon the refined coding scheme. Finally, axial coding was used to detect patterns in the data. A final element to the methodological framework was writing as a method of inquiry (Richardson & St. Pierre, 2005). This is reflected in both the tool of writing during the analytical process, as well as the use of narrative prose in storytelling form with embedded comments and analysis in the following sections. The intent is to provide word-pictures that describe to and invite the reader to consider elements of aesthetic sensibilities as enacted in the lesson.

14.2 Aesthetics and Inquiry: Students' Observations and Teacher Moves

In this section, the 12-min video is broken down into six narrative case-stories. The stories are chronological and connected. However, the choice of chunking the narrative into the six sections was to provide specific analysis and reflection of the aesthetic elements of the lesson based on the students' ideas and associated teacher's moves. The aesthetic framework is particularly useful as this discursive deviation from the original lesson was not guided by a predetermined lesson plan. Rather the teacher's response to the students' comments were based on the aesthetic elements perceived by the teacher and students. In all six narrative case-stories, pseudonyms are used for the students' names.

14.2.1 Generation of Ideas

Prior to the discussion, the students engaged with the Shape Makers in the computer lab. A number of students observed the Kite Maker in a degenerate form making a figure that looked like a triangle (Fig. 14.3), prompting one student, Hank, to ask, "It is possible that the kite can be a triangle?... Is a triangle a valid kite?... Because you can do more stuff with the computer that maybe you can't do with [pencil and paper]."

Initially, the teacher did not understand the issue, and tried to table the questions raised by the students. This degenerate case of a Kite Maker, in the shape of a triangle (Fig. 14.3), was not in the original lesson plan. However, the number of students that seemed interested in the issue prompted the teacher to pursue a discussion. It appears that the students were drawn to the aesthetic sensibilities from the visual appeal of the familiar looking triangle that was made with the less familiar Kite Maker. This triangle-kite figure became important to the students as they explored what is and is not a kite. What follows is a rich consideration of the definitions of angle, vertex, side, triangle and kite. The Kite Maker Tool activated student's creativity as they generated shapes, made and tested conjectures, and considered properties. The generative nature of the software led to both generative and motivational aesthetic choices that propelled the class discussion and led to students examining mathematical properties such as vertex and side, in a more rigorous way than they had prior to the discussion (Table 14.2).

Hank's question "Is a triangle a valid kite?" is aesthetically generative in that it seeks to organize the students' notion of triangle and kite in a new way. The question is also aesthetically motivational as Hank pursued a different question than the teacher proposed. Additionally, Hank's suggestion that "you can do more stuff with the computer," suggests both that the computer was a generative tool and also that it may create objects that would not be possible with pencil and paper. While paper and pencil representations also enact aesthetic sensibilities, as noted by Hank, the use of the interactive geometric kite-maker provides a unique, and widely recognized phenomenon by the students, thrusting the students and teacher into the following discussion.

Kaden begins a mathematical argument, "We're talking about quadrilaterals here, all these shapes [gestures towards the board] are quadrilaterals, but a triangle…is not one of these."

Kristie, speaking to the teacher, began, "On the first day you said—told us the Shape Maker Rule was 'that any shape made by a shape maker is that shape…and so any shape made with a Kite Maker is a kite shape…so wouldn't that make, that because you made the triangle with the kite [maker], that means that the triangle is a type of kite?"

Nick added, "There are only 3 sides [and 4 line segments]."

Fig. 14.3 The triangle-kite

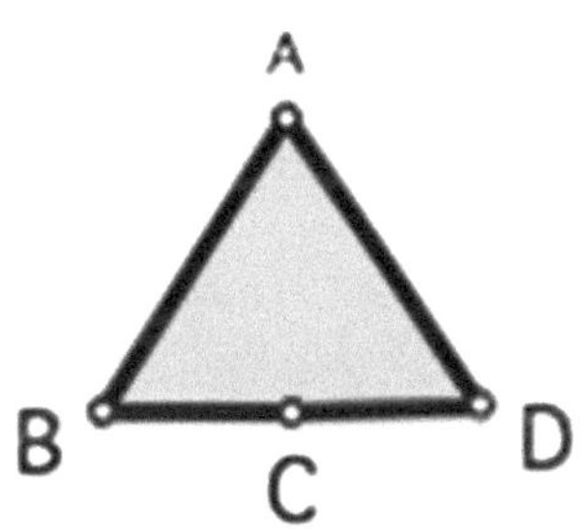

Table 14.2 Analysis of the opening question

Statement or event	Aesthetic roles	Aesthetic sensibilities
"Is a triangle a valid kite?"	Motivational	Visual appeal, mystery
	Generative	Order—organization
"Because you can do more stuff with the computer that maybe you can't do with [pencil and paper]"	Generative	Simplicity and order through example generation

Violet then said, "When we were messing around [with the seven quadrilateral makers] on the first day, all the shapes… ah, the only one that ever said that this is not whatever shape maker it is, was one time, the Quadrilateral Maker (Fig. 14.4), and that was when you made the hourglass shape."

Nick added, "This also proves Sheldon's point, correctly because there are three sides that are 60° and one of them is 180°. Three equal angles."

Nick continued to engage the issue through an examination of the triangle-kite's properties. Aesthetics played a motivational and generative role in Nick's thinking as he continued to pose ideas in an almost reactive manner characterized by elements of brainstorming, connection making, and thinking aloud. Kristi and Violet apply the evaluative aesthetic in an appeal to the authority of the teacher and the software, but Nick's generative redirection to angle properties propels the class into further exploration of the figure (Table 14.3).

The students used the drawing of the figure on the board as a visual point of focus, and began to pay attention to the four angles. This re-organization of their thinking played a generative role in their discussion as the students continued to pursue this mathematical inquiry, expecting it to reveal some insight or fact. This was a turning point in the class discussion as it turned their focus away from the software tool and towards the examination of the properties of the static figure drawn on the board.

14.2.2 *Teacher's Instructional Choices*

In this section, we will look at the teacher's instructional choices in response to the students' aesthetic sensibilities. Previously, when Hank asked, "It is possible that

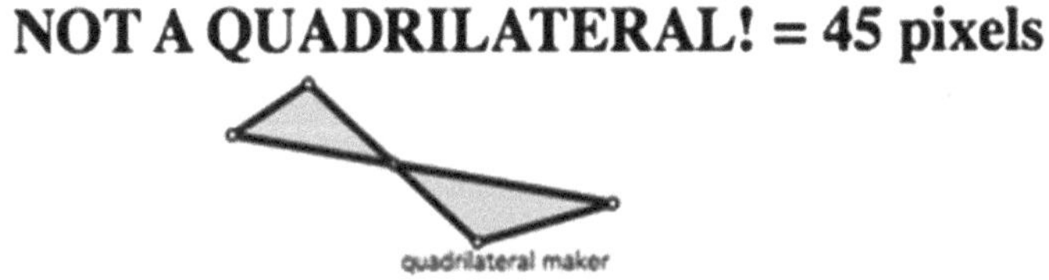

Fig. 14.4 The quadrilateral maker in a degenerate form

Table 14.3 Analysis of engagement in the question

Statement or event	Aesthetic roles	Aesthetic sensibilities
"All of these shapes are quadrilaterals, but a triangle…is not one of these"	Motivational	Connection making Relations between different modes of thinking
	Generative	Generating examples
"On the first day you… told us the Shape Maker Rule was that any shape made by a shape maker is that shape"	Evaluative	Appeal to authority of the teacher
"the only one that ever said that this is not whatever shape maker it is, was one time, the Quadrilateral Maker (Fig. 14.4), and that was when you made the hourglass shape"	Evaluative	Appeal to authority of the software
Nick's attention to angle properties	Generative	Reorganization of thinking around angle properties Continue to pursue the question

the kite can be a triangle?" initially, the teacher did not understand the issue. The teacher tried to table the questions raised by the students. This degenerate case of a Kite Maker, in the shape of a triangle (Fig. 14.3), was not in the original lesson plan. However, the number of students that seemed interested in the issue, as indicated by Kaden, Kristie, Nick, and Violet's dialogue in the previous section, prompted the teacher to pursue a discussion of the issue. Seeking clarification, the teacher asked, "You shift [the Kite Maker] into a triangle, you're asking is that thing really a kite?" When the teacher changed his mind and decided to pursue Hank's question, we see the generative aesthetic and motivational aesthetic extended. The teacher pursued the question even though it was outside the scope of the lesson because he recognized the motivational role the question was playing for the students. While unsure of the trajectory of the lesson, the teacher hoped the discussion would reveal insight into the exploration of the quadrilateral properties for the students. He reorganized the lesson to accommodate his own and his students' aesthetic sensibilities.

The teacher then referred to a sketch of the triangular shape made by the Kite Maker, "So the issue is, looking at this, I've got figure ABCD, where C is on the same [segment from B to D]." see Fig. 14.3. The students used the drawing of the figure on the board as a visual point of focus, and began to pay attention to the four angles. This re-organization of their thinking played a generative role in their discussion as the students continued to pursue this mathematical inquiry, expecting it to reveal some insight or fact. This was a turning point in the class discussion as it turned their focus away from the software tool and towards the examination of the properties of the static figure drawn on the board.

Matt, pointing to the shape ABCD drawn on the board, added, "But triangles' angles have to add up to 180, but if we have three 60s and a 180, that goes over the limit …so we can't have that it's a triangle" (Table 14.4).

Nick continued, "I see what [Matt's] going at…technically 4 angles isn't totally correct because it is possible, so it could be a vertice and not an angle."

Amber asked, while pointing to the teacher's drawing of ABCD on the board, "Does a triangle have only three vertices? Because that shape has 4 vertices, doesn't it?" Gesturing toward the bottom half of ABCD (Fig. 14.4), along line B to C to D, Amber clarified her idea, "Because the line where B and C, and C and D… converge at C, so C is where two lines come together." Amber's question is aesthetically generative as she tried to make a distinction between categorizing ABCD as triangle, as having three vertices, and a quadrilateral that has four vertices.

The teacher said, "So the issue is this. If I cover up this point C". Then covering point C with his hand, "How many vertices [in the figure] do we have?"

"Three," answered the class in unison.

"Three, but when I do that," continued the teacher as he removed his hand revealing point C, "How many vertices do we have?"

"Four," said the class in unison.

In this section, Nick's creative redefinition of vertex helped propel the discussion forward. It is an example of the generative aesthetic that helped him and other students imagine relations between points and vertices. Nick's statement was also motivational in that it prompted him and other students to pursue further the interconnectedness of the properties and their definitions. It invited students to creatively separate properties from each other and consider other new ideas such as whether a vertex can degenerate into a point on a polygon.

The teacher responded to an important question posed by Amber that is aesthetically generative as she tried to make a distinction between categorizing ABCD as triangle, as having three vertices, and a quadrilateral that has four vertices. The teacher's decision to draw the class into considering Amber's question was his response to what appeared to be a 'really good idea' to help the group see ABCD as

Table 14.4 Analysis of teacher moves

Statement or event	Aesthetic roles	Aesthetic sensibilities
"I see what [Matt's] going at…technically 4 angles isn't totally correct because it is possible, so it could be a vertice and not an angle"	Generative	Relations between different modes of thinking
	Motivational	Interconnectedness
"Does a triangle have only three vertices? Because that shape has 4 vertices, doesn't it?"	Generative	Reorganizing thinking around a different property
	Motivational	Gesture
"Four"	Motivational	Choral response

a quadrilateral. Furthermore, Amber's gesture and the class' choral answer, "four" both convey a motivational aesthetic that keeps the mathematical discourse moving forward. The students were smiling throughout this portion of the video, clearly enjoying the discussion, highly engaged, raising their voices and making gestures, reflective of multiple motivational aesthetic responses (Table 14.5).

Speaking to the entire class, the teacher returned to a previously stated idea, "Here's an interesting thing though…." Then turning to Amber, he continued, "Amber, could you go, I think you are going down a trail of thought. You said, okay, a triangle has to have three vertices. And you are saying, well, doesn't this have four. So, what is the point of your statement?"

"If it has more vertices, then it can't be a triangle," Amber replied.
The teacher asked, "So, it would have to be a what?"
"A kite, er, a quadrilateral," concluded Amber.

The teacher continued the discussion of figure ABCD. Using Amber's conclusion he said, "Let's go with Amber's point. She says look, how many vertices does it have? Four. Right? How many line segments is it made of? Four. But here is the question, is that thing [ABCD] a kite?" The question momentarily hung.

In the redirection to Amber's previously stated idea, that ABCD has four vertices, the teacher displayed a shift from a generative aesthetic to an evaluative aesthetic. Implicit in the teacher's tone and focus on Amber's statement is a value judgement about the worth, or potential worth of the student's idea in classroom discourse. In the previous section, the teacher pursued Amber's question as a 'good idea' in a generative manner, hoping or expecting her question to reveal some insight to the class. However, in this moment, the teacher explicitly validated Amber's line of thinking as he wanted the students to understand that ABCD must be a quadrilateral. Also, the teacher's choice to ask the question is "ABCD a kite?" after advocating for the position that C was a vertex was perhaps surprising to the students. They were highly engaged and expecting finality, and then the teacher asked a generative question again, inviting the students to reconsider (Table 14.6).

Table 14.5 Analysis of Amber's question and ensuing teacher choice

Statement or event	Aesthetic roles	Aesthetic sensibilities
"Does a triangle have only three vertices? Because that shape has 4 vertices, doesn't it?" Gesturing towards point C	Generative	Searching or attempting to confirm a conjecture to substantiate an idea
"So the issue is this. If I cover up this point C." Then covering point C with his hand, "How many vertices [in the figure] do we have?"	Generative	Choosing a path of inquiry
Teacher's interactions leading to class choral response, "Three…Four…."	Motivational	High engagement with the exploration

Table 14.6 Analysis of teacher's response to Amber's path of inquiry

Statement or event	Aesthetic roles	Aesthetic sensibilities
"Amber…I think you are going down a [correct] trail of thought. You said, a triangle has to have three vertices. And you are saying, well, doesn't this have four. So, what is the point of your statement?"	Evaluative	Resolving the line of inquiry
"If it has more vertices, then it can't be a triangle," Amber replied.	Evaluative	Resolving the question
Speaking to the entire class, the teacher returned to a previously stated idea, "Here's an interesting thing though," and "Let's go with Amber's point…"	Generative	Choosing a path of inquiry

The teacher continued, "But let's pretend… and go with Amber's point. She says 'Look, how many vertices does this thing have?' Four, right? How many line segments is it made of? Four. How many sides…clearly we see four line segments, right? But here's the question. Is that thing a kite?" Referring back to the list of properties of a kite the class generated through their explorations with the Kite Maker, the teacher continued by applying each of the listed properties to figure ABCD. Pointing to the figure the teacher asked the class, "Now, here is what we have. Does this, this shape ABCD, does it ascribe to everything that we've talked about so far? Does it have two pairs of equal sides?"

In unison the class responded, "Yes." The teacher then had students specify which sides were shape ABCD were congruent.

He then continued, "Does [ABCD] have at least one line of symmetry?"

A few students, pointing out the obvious, said, "Yes!"

"Does it [shape ABCD] have at least one pair of equal angles?"

"Yes," a few students answered.

"Ah… Is it made up of 4 line segments?" asked the teacher.

"Yes," said the class.

"…And 4 angles?" the teacher continued.

"Yes," said the class.

At this point the teacher slowed down the pace of the class discussion, "Is it made up of 4 angles?" he asked in a questioning tone.

The entire class responded emphatically, "Yes!"

Clarifying, Matt added, "C is one angle."

Picking up the cadence, "What type of angle is C?" asked the teacher.

"Straight," was the simultaneous response by a number of students throughout the class.

"[Angle C is] a straight angle, isn't it?" The teacher wanted to make sure that the class understood that while collinear with B and D, point C formed the vertex of a straight angle.

In this section, we see the teacher engage the students' generative aesthetic sensibilities of fit and pleasure as he led them to examine whether the triangle-kite

Table 14.7 Analysis of teacher examining each property

Statement or event	Aesthetic roles	Aesthetic sensibilities
"But let's pretend… and go with Amber's point. She says 'Look, how many vertices does this thing have?' Four, right? How many line segments is it made of? Four. How many sides…clearly we see four line segments, right? But here's the question. Is that thing a kite?"	Generative	Reopening the resolved question
"Now, here is what we have. Does this, this shape ABCD, does it ascribe to everything that we've talked about so far? Does it have two pairs of equal sides?"	Generative	Fit and pleasure

had the properties of a kite that they generated earlier in the lesson. The teacher's generative question invited students to reconsider the properties of a kite and apply them to the ambiguous triangle-kite figure (Table 14.7).

14.2.3 Resolution of the Mathematical Discussion

Nick, still unsure, but fishing for answers posed another idea to the group, "Can I make an argument here? I think I may be bending the rules here a little too much, but if we did put sides instead of line segments… I mean technically, C could just be a point there that would still have four sides…"

The teacher asks, "Well, how you do YOU define side?"

Nick pondered aloud, "I'm thinking about my definition… probably a line on an object…but it is still possible that C could only just a point there, and so most of what we already said would still be correct.… I'm just thinking for a second… is it possible that it doesn't…that a kite doesn't have to have four line segments, and have three and it would still be a kite, like in this case for example?"

Even though Nick had begun to disengage with the question earlier, he has been pulled back in. Nick and the teacher both asked generative questions leading to a discussion about what is a side. Nick was viewing point C as just a point on the segment BD rather than a vertex. Now he appears to be considering a new set of definitions as evidenced by his statement, "I think I may be bending the rules here a little too much" and his consideration that a kite could have three line segments. He is considering that both conceptions (the triangle with a degenerate vertex and the kite with vertex C) might be valid. The aesthetic is playing motivational, generative and evaluative roles as he is pursuing and imagining new ideas, reorganizing his thinking, and choosing one conception over another (Table 14.8).

The Teacher redirects the students back to their page of definitions and asks students about the definition of quadrilateral. The class discusses lines of symmetry and then Nick reasserts himself.

Table 14.8 Analysis of Nick's new argument

Statement or event	Aesthetic roles	Aesthetic sensibilities
"Can I make an argument here?"	Motivational	Pursue a new idea
"I think I may be bending the rules here a little too much, but if we did put sides instead of line segments… I mean technically, C could just be a point there that would still have four sides…"	Motivational	Bending the rules
	Generative	Imagining a new definitions of side, vertex and kite
"is it possible that it doesn't…that a kite doesn't have to have four line segments, and have three and it would still be a kite, like in this case for example?"	Motivational	Following a new pattern of thinking
	Generational	Imagining possibilities Reorganizing thinking
	Evaluative	Choosing between two conceptions, the point C as a vertex, and as a point on the object

Table 14.9 Analysis of Nick's attempt to settle the argument

Statement or event	Aesthetic roles	Aesthetic sensibilities
"Can I make one more thing here?" Nick tried to bring clarity to the issue, "I think an easy way to settle this argument now and still be correct, a kite [maker] can make a shape that looks like a triangle, but by definition is not a triangle…."	Motivational	Redirect the conversation
	Evaluative	Clarify End the argument

"Can I make one more thing here?" Nick tried to bring clarity to the issue, "I think an easy way to settle this argument now and still be correct, a kite [maker] can make a shape that looks like a triangle, but by definition is not a triangle…."

Nick's assertion had a motivational role in redirecting the conversation. Additionally it showed that he returned to a more absolute perspective. He no longer viewed the vertex C as both a point on the side BD and a potential vertex as evidenced by his assertion that it "looks like a triangle, but by definition is not a triangle." The aesthetic is playing an evaluative role as Nick invoked both clarity and finality in his assertion (Table 14.9).

However, the rest of the class had not yet reached finality, and continued the discussion.

Matt had a different way to look at it, "We know it's not a triangle, because we proved it has too many angles."

From the back of the room Travis asserted, "I can prove that a kite is not a legitimate triangle."

The teacher stepped into moderate the class conversation. In particular, the teacher chose to highlight Nick's assertion. "OK, we are going to go to you in a minute, Travis, when we make sense of what [Nick] said…" Turning towards Nick the teacher said, "Say that again."

Nick repeated, "A kite [maker] can make a shape that looks like a triangle, although but by definition is not a triangle…"

The teacher repeated his statement to make sure the class understood the point, "Ok, a kite [maker] can make a shape that looks like a triangle, but by definition is not a triangle…"

Violet asked, "What is the definition of a triangle then?"

The class had still not come to a consensus about whether or not the triangle-kite was a kite. Nick's assertion was important to the discussion because he articulated the benefit of paying attention to properties. Aesthetically, the teacher made an evaluative choice to pursue Nick's response rather than Travis' proof. This lead to Violet's generative question "What's the definition of a triangle then?" The students were still highly engaged in answering the original question pointing to the motivational role of the aesthetic (Table 14.10).

The teacher clarified, "Three vertices and three sides…even though we haven't formally discussed this in class, all of us can agree that a triangle has three vertices and three sides."

Travis then stated his point he was trying to make earlier, "A kite cannot be a legitimate triangle because properties of a triangle and properties of a quadrilateral conflict…because the insides of the angles of a quadrilateral, when added equal 360, but it will only equal 180 with a triangle."

"Let me ask you this Travis, this shape right here ABCD," asked the teacher pointing to the figure on the board, "do you think that shape right here is a kite?"

"Yes, because C is 180°," answered Travis.

"OK, so ABCD is a kite? What do you guys think? Chris?"

The class responded in unison, "Yes."

Playing devil's advocate, the teacher added, "OK, so is ABCD a triangle?"

This time the class responded, "No."

Violet added, "Not by definition."

The teacher continued, "Not by definition, but it…"

"…looks like one [a triangle]," said the class finishing his statement.

As the discussion neared its conclusion, the students were raising their voices, emphatically speaking in unison and sharing multiple perspectives on the question

Table 14.10 Analysis of Nick's assertion

Statement or event	Aesthetic roles	Aesthetic sensibilities
"Ok, a kite [maker] can make a shape that looks like a triangle, but by definition is not a triangle…"	Motivational	Redirect the conversation
	Evaluative	Chose Nick's assertion over his classmates
"What's the definition of a triangle then?"	Motivational	Students highly engaged, smiling, etc.
	Generative	States a new question, redirects inquiry

Table 14.11 Analysis of Travis' assertion

Statement or event	Aesthetic roles	Aesthetic sensibilities
"A kite cannot be a legitimate triangle because properties of a triangle and properties of a quadrilateral conflict…because the insides of the angles of a quadrilateral, when added equal 360, but it will only equal 180 with a triangle"	Evaluative	Chose a final answer to the question
"Let me ask you this Travis, this shape right here ABCD. Do you think that shape right here is a kite?"	Generative	Invites Travis to make an evaluative judgement
"Yes, because C is 180°,"	Evaluative	Brings argument to an end
"OK, so is ABCD a triangle?"	Generative	Invites the class to make an evaluative judgement
"No"	Motivational	Students highly engaged, speaking in unison
	Evaluative	The class is united in their evaluation

indicating motivational aesthetic, and the desire to use the properties to answer Hank's question. The teacher then summarized the whole discussion by pointing out that given the current list of properties for a kite, shape ABCD was a kite, in the form of a triangle. This conclusion may be inconsistent with one's visual intuition, and definitions of polygons that do not allow for three adjacent collinear vertices. However, many textbook definitions of quadrilaterals and polygons omit qualifying statements like, no three adjacent collinear vertices. In the absence of such qualifying statements, and given the mathematical context, the class's conclusion was indeed mathematically valid. Realizing this, and to conclude the class's discussion, the teacher stated, "This was an interesting discussion because it helps us think through these [listed] properties in what I would consider a very unique case of a kite that is really cool for discussion within the context of a mathematics discussion. But, within the context of the real world, is moot or irrelevant, because it looks like a triangle, so let's just call it a triangle" (Table 14.11).

14.3 Discussion

The nature and role of aesthetics in mathematics classroom discourse is multi-faceted, as illustrated in this relatively short collection of stories. As such, there are many different analytical directions one could take in summarizing the elements of the preceding case stories. In this chapter two are considered. First is a discussion of the overarching role of aesthetics in mediating the classroom discourse. Second is a discussion of the role of aesthetics in the sense-making of one particular student, Nick, during the classroom episode.

14.3.1 Patterns and a Hypothetical Trajectory of Aesthetics in Classroom Discourse

Returning to the aesthetic framework, as we analyzed the data we categorized the episode looking for the three roles of the aesthetic: motivational, generative, and evaluative and the aesthetic sensibility that was invoked. So, for instance, Hank's original question: "Is a triangle a valid kite?" was coded as a motivational (new question), evaluative (kite or triangle), and generative (can be viewed as a quadrilateral or a triangle). After the first pass of coding was complete, we looked for patterns in distribution of the codes.

The generative code appeared most frequently. This was surprising to us because we had originally suspected that the motivational aesthetic would occur with the highest frequency as the students engaged in the discussion. There are both cognitive and emotive reasons why the generative aesthetic was so common. First, the dynamic aspect of the Shape Makers activity is essentially generative because the students generate hundreds of examples from which they can examine a multitude of examples and counter-examples (Johnson-Gentile et al., 1994; Jackiw & Sinclair, 2009; Lavy & Shriki, 2010; Patsiomitou, 2008). The generative aesthetic was invoked in the discussion whenever students changed perspectives or organized their thinking in a new way. The flexible nature of the computer environment allowed students to change perspectives at will dragging different points, segments and figures paying attention to different properties such as side length, angle measure, orientation, and symmetry (Battista, 2001; Ng & Sinclair, 2015). That flexibility in perspective engendered in the tool carried over into the discussion that took place without the tool. The discussion started with the generative question, "Is a triangle a valid kite?" and continued with generative questions and statements especially in the first section of discussion.

The coding also revealed a pattern of motivational questions set apart from each other by a series of generative ideas. For instance, the episode began with Hank's motivational question whether a kite could be a triangle. Hank's question was followed by a series of generative actions and ideas: the teacher decided to pursue it, a student reorganized it as a figure with three sides, but four line segments, other students changed the focus to angle measure, number of angles, and the definition of triangle. So the initial question, set off a series of generative ideas—new perspectives from which to examine the triangle-kite ending with another motivational question when Amber asked, "Does a triangle have only three vertices? Because that shape has 4 vertices doesn't it?" Amber's motivational question set off another series of generative ideas which led to a moment of high engagement (motivational aesthetic) when all the students were talking at once and gesturing to one another.

This pattern continues several more times each time with a motivational question followed by a series of generative ideas. The motivation was necessary for students to engage, but the mathematics that they were building required new ideas and new ways of thinking about the task. The students posed new questions and new ideas in order to make sense of the interconnection of the various properties (Battista, 2001).

Another interesting pattern we saw in the data was that the generative organization aesthetic often occurred with or shortly after a generative question indicating that questions and reorganizing the information in a new way went hand-in-hand. For example, Amber asked a generative question that indicated that she was organizing her sense-making around the concept of vertices. "Does a triangle have only three vertices? Because that shape has 4 vertices, doesn't it?" Shortly after, Matt suggested that the angles of a triangle "have to add up to 180" which reorganized the thinking around interior angle sums. And shortly later, Nick proposed his first relativistic organization of the two figures, stating that C could be a vertex but not an angle. This is a moment were Matt takes a creative leap based on generative reorganization.

While one might expect the motivational aesthetic to be important at the beginning of the inquiry, we did not find that to be case. Instead the motivational aesthetic was much more active towards the middle and end of the episode seeming to build up over time. "We first see the motivational aesthetic when Nick initially proposes that C could be a vertex but not an angle and the students start responding emphatically, in chorus to the teacher's questions." In this episode the motivational aesthetic followed the pattern of a good story with the motivation at its peak during the climax and then continuing through to the resolution.

The discourse that was encouraged both by students' exploration of the Shape Makers and the teacher's questions during the ensuing discussion led the students to examine the properties deeply. Generative roles of the aesthetic played off one another with questions prompting a reorganization of the thinking that in turn generated more questions. Generative questions resulted in a deep consideration of properties that invoked the aesthetic sensibilities of fit, order, and clarity. In turn the pleasure tied to those sensibilities in addition to the visual appeal the ongoing development of connectedness between ideas and properties and the desire to solve the problem played a motivational role of increasing engagement.

14.3.2 The Five Strands

Paying attention to the roles, sensibilities, and indicators of the aesthetic allowed us to understand the students' mathematical sense-making and the teacher's choices. It was a productive way to identify creativity within the students' and teacher's choices. It gave us a new appreciation for the roles aesthetic sensibilities play in mathematical discourse and sense-making in a dynamic mathematical environment. Returning to our earlier discussion of the five strands, we found examples of conceptual understanding, adaptive reasoning, and productive disposition in our analysis of the aesthetics in this episode. For example, in this episode, students worked to develop conceptual understanding of the properties of vertex, side, and angle. Nick showed adaptive reasoning by moving between absolute and relativistic definitions of kite. And the student's persistence and engagement with the exploration indicates productive disposition. We suggest that each of these strands has both cognitive and aesthetic properties.

Additionally, we see aesthetic examples in this episode of persistence, reasoning, argument, precision and structure. For example, we see Nick's persistence as he engages in the exploration, loses interest and then re-engages. Multiple students reason about the definitions of the properties. They take a position and argue then choose whether to accept each other's arguments. They move from a position of low precision to high precision because they recognize the value of communicating precisely. And finally, they reason explicitly about the structure of both triangles and kites to answer Hank's initial generative question.

We believe that it would be illuminating to view the Five Strands of Mathematical Proficiency (National Research Council, 2001) and Common Core State Standards for Mathematical Practices (National Governors Association Center for Best Practices, & Council of Chief State School Officers, (2010) through an aesthetic lens to develop a broader understanding of the students' interactions with these constructs.

14.4 Implications and Further Directions

In this chapter, we set out to study creativity within a dynamic geometry environment of Battista's Shape Makers microworld. As creativity is often described in aesthetic terms, we used Sinclair (2006, 2008) aesthetic framework rather than a cognitive framework to identify creativity within the class discussion. Several attributes of the technology made the microworld a creativity-rich environment. As has been studied before the open, exploratory environment necessitates students to pose questions, explore ideas, and attune to different properties (Johnson-Gentile et al., 1994; Lavy & Shriki, 2010; Ng & Sinclair, 2015; Patsiomitou, 2008). Additionally, we found the generative nature of the environment to be particularly rich in allowing students to activate the generative aesthetic to change their focus, look for patterns, and reorganize their thinking in different ways. The generative aesthetic was driving the creativity forward allowing students to view the triangle-kite from many different perspectives. We believe that other dynamic mathematics environments such as GeoGebra, Desmos, and Wolfram Alpha would be similarly rife for student creativity and would provide further opportunities to explore the aesthetic motivations and sensibilities of mathematical creativity.

Furthermore, our analysis covered a class discussion that took place after students had explored mathematics in a dynamic mathematical environment. We are curious about how the aesthetic roles might differ if a similar study were done while students were interacting with the technology. Since the actions that students take on the mathematical representations, as they interact with the representations on the screen have an active, embodied component (Burbules, 2006; Sinclair & Moss, 2012), would the motivational role of the aesthetic be more pronounced through students use of gesture and discussion of movement? Would the generative aesthetic play a similar role during the investigation that it played afterwards in the class discussion?

Finally, the lesson excerpt presented in this chapter is a non-traditional discussion on the properties of a kite. Careful analysis of the case stories shows various elements of the aesthetic nature in the students' responses: student interest, genuine curiosity, brainstorming, and motivation. A broader element to be considered is an observer's response to the lesson. On one hand one can respond that the lesson was irrelevant to the original lesson plan and a waste of time. On the other is a response that considers the lesson to be artful in the engaged and fluid conversation between all the various students that participate in the discussion. In agreeing with either opinion, or somewhere in between, that in itself represents certain aesthetic sensibilities based on the values of the observer that reside in the affective domain.

References

Appel, K., & Haken, W. (1977). Every planar map is four colorable. Part I: Discharging. *Illinois Journal of Mathematics, 21*(3), 429–490.

Ball, D. (2000). Working on the inside: Using one's own practice as a site for studying teaching and learning. In A. E. Kelly & R. A. Lesh (Eds.), *Handbook of research design in mathematics and science education* (pp. 365–402). Mahwah, NJ: Lawrence Erlbaum Associates. https://eric.ed.gov/id=ED478318.

Baron, L. M. (2010). Helping teachers generate better homework: MAA makes time for WeBWorK. *MAA Focus, 30*(5), 18–19.

Battista, M. (2001). Shape makers. *Computers in the Schools, 17*(1–2), 105–120. https://doi.org/10.1300/jo25v17n01_09.

Battista, M. (2003). *Shapemakers*. Emeryville, CA: Key Curriculum Press.

Burbules, N. C. (2006) Rethinking the virtual. In J. Weiss et al. (Eds.), *The International handbook of virtual learning environment* (pp. 37–58). Dorderecht, The Netherlands: Springer.

Jackiw, N. (1991). *Geometer's sketchpad* [computer software]. Emeryville, CA: Key Curriculum Press.

Jackiw, N., & Sinclair, N. (2009). Sounds and pictures: Dynamism and dualism in dynamic geometry. *ZDM Mathematics Education, 41*(4), 413–426.

Johnson-Gentile, K., Clements, D., & Battista, M. (1994). Effects of computer and non-computer environments on students' conceptualizations of geometric motions. *Journal of Educational Computing Research, 11*(2), 121–140.

Kwon, O. N., Park, J. S., & Park, J. H. (2006). Cultivating divergent thinking in mathematics through an open-ended approach. *Asia Pacific Education Review, 7,* 51–61.

Lavy, I., & Shriki, A. (2010). Engaging in problem posing activities in a dynamic geometry setting and the development of prospective teachers' mathematical knowledge. *The Journal of Mathematical Behavior, 29,* 11–24.

Leikin, R., & Lev, M. (2007). Multiple solution tasks as a magnifying glass for observation of mathematical creativity. In *Proceedings of the 31st International Conference for the Psychology of Mathematics Education* (Vol. 3, pp. 161–168). Seoul, South Korea: The Korea Society of Educational Studies in Mathematics.

Liljedahl, P. (2013). Illumination: An affective experience? *ZDM Mathematics Education, 45*(2), 253–265.

Murphy, R., Gallagher, L., Krumm, A., Mislevy, J., & Hafter, A. (2014). *Research on the use of Khan Academy in schools*. Menlo Park, CA: SRI Education.
National Governors Association Center for Best Practices, & Council of Chief State School Officers. (2010). *Common core state standards for mathematics: Standards for Mathematical Practice*. Washington, DC: Authors.
National Research Council. (2001). In J. Kilpatrick, J. Swafford, & B. Findell (Eds.), *Adding it up: Helping children learn mathematics*. Washington, DC: National Academy Press.
Ng, O., & Sinclair, N. (2015). Young children reasoning about symmetry in a dynamic geometry environment. *ZDM Mathematics Education, 47,* 421–434.
Niederahauser, D. S., & Stoddart, T. (2001). Teacher's instructional practices and uses of educational software. *Teaching and Teacher Education, 17*(1), 15–31.
Patsiomitou, S. (2008). The development of students geometrical thinking through transformational processes and interaction techniques in a dynamic geometry environment. *Issues in Informing Science and Information Technology, 5,* 353–393.
Richardson, L., & St. Pierre, E. A. (2005). Writing: A method of inquiry. In N. K. Denzin & Y. S. Lincoln (Eds.), *The Sage handbook of qualitative research* (pp. 959–978). Thousand Island, CA: Sage Publications.
Sinclair, N. (2006). *Mathematics and beauty: Aesthetic approaches to teaching children*. New York, NY: Teachers College Press.
Sinclair, N. (2008). Notes on the aesthetic dimension of mathematics education. In *Proceedings of the 100th International Commission of Mathematics Instruction Symposium* (https://www.unige.ch/math/EnsMath/Rome2008/WG5/Papers/SINCL.pdf). Rome, Italy: The International Commission of Mathematics Instruction.
Sinclair, N., De Freitas, E., & Ferrera, F. (2013). Virtual encounters: The murky and furtive world of mathematical inventiveness. *ZDM Mathematics Education, 45*(2), 239–252.
Sinclair, N., & Moss, J. (2012). The more it changes, the more it becomes the same: The development of the routine of shape identification in dynamic geometry environment. *International Journal of Educational Research, 51,* 28–44.
Sriraman, B., Yaftian, N., & Lee, K. H. (2011). Mathematical creativity and mathematics education. In *The elements of creativity and giftedness in mathematics* (pp. 119–130). Rotterdam, The Netherlands: Sense Publishers.
Tent, M. B. W. (2006). *The prince of mathematics: Carl Friedrich Gauss*. Wellesley, MA: A K Peters.
Wilson, R. J. (2002). *Four colors suffice: How the map problem was solved*. Princeton, NJ: Princeton University Press.

Chapter 15
Technology and the Development of Mathematical Creativity in Advanced School Mathematics

Sergei Abramovich

Abstract The availability of sophisticated computer programs capable of complex symbolic computations has created challenges for mathematics educators working with mathematically motivated students. Whereas technology may be praised for enabling educators to bridge the gap between the past—when only some students were able to do mathematics, and the present—when an average student is able to enjoy finding an answer to a difficult problem using a computer, it can also put a barrier in the way of developing students' creative mathematical skills. This dichotomy between positive and negative affordances of technology in the teaching of mathematics calls for the development of new curriculum enabling the outcome of problem solving not to be dependent on students' ability to simply enter correctly all data into a computer. Towards this end, the chapter proposes a way of modifying traditional problems from advanced mathematics curriculum to be both technology-immune and technology-enabled in the sense that whereas software can facilitate problem solving, its direct application is not sufficient for finding an answer.

Keywords Affordances of technology · Teacher education · TITE problems
Problem reformulation · Einstellung effect

15.1 Introduction

What is creativity? Educators, in general, see creativity as "one of the essential 21st century skills … vital to individual and organizational success" (Beghetto, Kaufman, & Baer, 2015, p. 1). Whereas creativity has always been considered an important factor (not necessarily just a skill) for any kind of success, the reference to the 21st century has major didactic implications for its development in the context of mathematics education. The modern day availability of sophisticated

S. Abramovich (✉)
SUNY Potsdam, Potsdam, USA
e-mail: abramovs@potsdam.edu

© Springer International Publishing AG, part of Springer Nature 2018

V. Freiman and J. L. Tassell (eds.), *Creativity and Technology in Mathematics Education*, Mathematics Education in the Digital Era 10,
https://doi.org/10.1007/978-3-319-72381-5_15

computer programs capable of intricate symbolic computations has made many traditional problems from the advanced school mathematics curriculum kind of outdated as they can be solved by software almost at the push of a button. Whereas technological advances of the digital era have opened new research opportunities for professional mathematicians (Arnold, 2015; Borwein & Bailey, 2004; Epstein, Levy, & de la Llave, 1992), the advent of powerful computational tools into the modern mathematics classroom has created additional challenges for mathematics educators (Abramovich, 2014a). It has been almost three decades since Schoenfeld (1988) noted that the task of integrating technology at the college level of teaching mathematics is much more difficult in comparison to the elementary and secondary levels: "It's one thing to build a drill-and-practice program or a computer-based tutor for work on the quadratic equation. It's quite another to do the same for max-min problems!" (p. 1). This comment may be taken to mean that software tools available at that time were not sophisticated enough to enable their use in the teaching of advanced topics in mathematics without significant recourse to computer programming … unless the very programming (alternatively, coding) has been one of the foci of instruction, like in the case of LOGO (Feurzeig & Lukas, 1972), robotics (Bers, Ponte, Juelich, Vieram, & Schenker, 2002) or, more recently, digital fabrication (Dittert & Krannich, 2013) and digital making (Hughes, Gadanidis, & Yiu, 2016). In the early 1980s, outside the context of using LOGO at the elementary level, the pedagogic expectations and the didactic emphasis of technology integration were mostly on drill and practice in arithmetic and basic shape construction in geometry. Maddux (1984) called such procedural perspective on teaching with computers Type I application of technology, advocating, instead, for Type II applications, "which constitute new and *better* ways of teaching" (p. 38, italics in the original). The concept of Type I versus Type II technology application has become a powerful theoretical shield against often persistent skepticism regarding the worth and purpose of using computers in the schools (Maddux & Johnson, 2005). It underscored the difference between instructivist and constructivist learning environments which measure learning, respectively, by tests and by one's ability to analyse, guess, and invent as a way of wrestling with big ideas (Brooks & Brooks, 1999). Just as the boring and impractical uses of mathematics can motivate the development of its more effective concepts (e.g., repeated addition motivates multiplication, which, in turn, motivates the use of logarithms), one's mundane experience with Type I application of technology can motivate the development of creative ideas for educational computing, thus bringing about technology applications of Type II.

Nowadays, the use of technology at the elementary level varies significantly. Whereas in a regular mathematics classroom, by the author's observations of local schools, technology use is often limited to taking advantage of frolic effects of multiple computer programs designed for "creative" (or, rather, entertaining) activities of young children (like KidPix Studio Deluxe) and enhanced by interactive white boards, digital experiences of young children may also include the use of coding as a tool of digital making (Hughes et al., 2016) or the use of electronic spreadsheets integrated with the images of familiar modern technologies in solving and posing grade appropriate real life problems (Abramovich, Easton, & Hayes, 2014). At a higher level, such tools as computer algebra systems, dynamic geometry software, and spreadsheets are commonly available for the teaching and learning of mathematics, requiring almost no need for computer programming. In particular, these tools "favor development of such important qualities … as initiative, invention and creativity" (Freiman, Kadijevich, Kuntz, Pozdnyakov, & Stedøy, 2009, p. 107). Much, however, depends on what we understand by integrating technology in the teaching of mathematics, assuming that it reaches the Type II level enabling conceptual understanding of and engagement with the subject matter.

The goal of this chapter is to argue that the availability of mathematically sophisticated computer programs may not only help to bridge the gap between creative abilities of average and above average students (teacher candidates included), but put a barrier in the way of fostering mathematical creativity. Indeed, nowadays, the outcome of problem solving may simply become a function of one's procedural mastery to enter correctly all data into a computer or even to push the right button on the keyboard. With this in mind, the chapter, by reflecting on the author's work with teacher candidates (referred to below as teachers) of the United States and Canada[1], underscores the importance of developing their skills in using technology in the context of *academic* work. Just as their future students, despite being 'digital natives' (Prensky, 2001), the teachers do need thorough assistance and proper training to make their own learning experience, reminiscent (to a certain extent) of mathematics research, the most effective agency of teaching, something that would affect the students' learning to be creative. Conference Board of the Mathematical Sciences (2012), an umbrella organization consisting of 16 professional societies in the United States concerned (in particular) with the mathematical preparation of schoolteachers, found that those teachers "who have engaged in a research-like experience for a sustained period of time frequently report that it greatly affects what they teach, how they teach, what they deem important, and even their ability to make sense of standard mathematics courses" (p. 65).

In this context, the chapter discusses positive and negative affordances (Gibson, 1977) of the modern technology tools associated with mathematics teacher education and by virtue of technology-enabled problem posing proposes a way of modifying traditional problems used to foster mathematical creativity so that to make them immune from applying purely technological know-hows typical for 'digital natives'. The selection of traditional problems is from the books by authors as diverse as Pólya (1954)—a notable mathematician and mathematics educator (Sect. 15.7), Tchekoff (1970)—one of the greatest short story writers of all time

[1]The university where the author works is located in the United States in close proximity to Canada, and many of the author's students are Canadians pursuing their master's degrees in education.

(Sect. 15.8), and Sivashinsky (1968)—an influential educator of in-service teachers and expert in afterschool activities for mathematically motivated students (Sect. 15.9).

15.2 Analysing the Use of Technology by 'Digital Natives' in Academic Contexts

Does effective Type II application enable problem solving or just make it "easy" to solve problems by outsourcing mathematical thinking to a computer? It appears that new challenges for teaching mathematics with technology at the advanced level of problem solving are due to the availability of sophisticated computer programs capable of complex symbolic computations. Some programs of that type were designed to enable solving problems in mathematics research that are not solvable otherwise even at the highest level of professionalism in mathematics (e.g., Leonov & Kuznetzov, 2013). Also, there exist computer programs specifically designed to solve school mathematics problems of different levels of complexity providing users with detailed solutions both in written and oral forms (e.g., Universal Math Solver (UMS), http://www.universalmathsolver.com). Furthermore, when powerful computer programs become available free on-line (e.g., *Wolfram Alpha*—a computational knowledge engine capable of accepting a natural language input without requiring any computer programming), it is difficult to pretend that they don't exist when dealing with students "who are digital natives comfortable with the use of technologies" (Ministry of Education Singapore, 2012, p. 2) and whose "sources of knowledge are significantly influenced by current technology" (Advisory Committee on Mathematics Education, 2011, p. 13) because they "spent their entire lives surrounded by and using computers, video games, digital music players, video cams, cell phones, and all the other toys and tools of the digital age" (Prensky, 2001, p. 1).

Classroom observations and educational research suggest that being a member of the generation of digital natives does not necessarily imply that one is capable of appropriately using technology for mathematical learning. Consequently, knowing how to appropriately teach (and learn) mathematics with technology is crucial for all academic contexts. In general, this has been confirmed by a number of studies. For one, Prensky (2001), who coined the term 'digital natives', noted (without going into details) that in the context of mathematics "the debate must no longer be about *whether* to use calculators and computers—they are a part of the Digital Natives' world—but rather *how* to use them" (p. 5, italics in the original). Bers et al. (2002), advocating for the appropriate use of computers by elementary teachers argued that "one of the most dangerous traps of using technology in the classroom is to turn the computer into a TV set for children to sit an watch CD-ROMS or use video games that do not invite creativity" (p. 134), that is, to promote the Type I application of technology. On the contrary, "by creating learning contexts for young children that incorporate powerful machines, such as robotics … [teachers skilled in Type II applications help students to become] active learners who are producers of new knowledge rather than consumers of what already exists" (Bers, 2010, pp. 243–244). Needless to say, new knowledge (occurring by serendipity or not) results from creative thinking.

By the same token, in the context of the Type II technology application to support the secondary level mathematical investigations, 'digital natives' may not necessarily be prepared to take into account a possibility of "the incoherent representations of mathematical objects … [leading] to incorrect conclusions or solutions" (Freiman et al., 2009, p. 128). At the tertiary level, the study by Kennedy, Judd, Churchward, Gray, and Krause (2008) of the first year students' use of technology at the University of Melbourne, questioning "the cultural and environmental assumptions underpinning the construct of the Digital Natives" (p. 108), dismissed the notion "that being a member of the 'Net Generation' is synonymous with knowing how to employ technology-based tools strategically to optimise learning experiences" (p. 118) and concluded that "it is difficult to expect students to have the expertise to judge how to best use emerging technologies for educational purposes" (p. 119). Research by Kirkwood and Price (2005) at the Open University in England produced a similar conclusion suggesting "that the medium itself is not the most important factor in any educational programme—what really matters is how it is *creatively* exploited and constructively aligned" (p. 272, italics added).

Likewise, in the United States it was found that "freshman students … report lower skill levels in course-related technologies … and can make technology work but cannot place these technologies in the service of (academic) work" (Kvavik & Caruso, 2005, p. 7). Similarly, more recent studies in the United States indicated that whereas "[college] students are generally tech inclined, they do not necessarily use technology to the full extent in supporting their academic endeavors" (Dahlstrom & Bichsel, 2014, p. 35). The above findings suggest that just being comfortable and savvy with different technologies is not enough for their appropriate use in support of mathematical creativity. Furthermore, it is worth noting that the United States government recommended that educationalal technology research addresses "a growing need for new instructional materials … that are aligned with higher standards and provide much richer learning experience and more vibrant sources of information" (President's Council of Advisors on Science and Technology, 2010, pp. 80–81). In particular, one of the implications of this recommendation in the context of advanced school mathematics is the need for new teaching ideas and curriculum materials conducive to the development of creative thinking despite or perhaps because of the easy to use mathematically sophisticated computational tools.

15.3 Duality in the Affordances of Technology

From the first glance, digital tools may be praised for enabling educators to bridge the gap between the past—when only some students were able to do mathematics, and the present—when an average student, without significant preparation in either mathematics or technology, is able to enjoy finding an answer to a difficult (non-standard) problem using a computer. In doing so, educators take advantage of "the possibility of developing creativity in people who don't display much of it" (Luchins, 1960, p. 138). For example, in the absence of technology one may not be able to construct a graph of the system of inequalities $y^{0.5} > 2x^2$, $y < 3$, but once it is constructed by a tool capable of graphing relations from two-variable equations and inequalities, e.g., by the *Graphing Calculator* (Avitzur, 2011), one can (or be prompted to) realize that the graph could be used to digitally fabricate (Dittert & Krannich, 2013) a vase and to (creatively) control its shape by changing the coefficient in x^2. At the same time, one can argue that technology, when not used appropriately, can also put a barrier in the way of developing creative skills in the students of mathematics, teachers included.

In terms of the theory of affordances (Gibson, 1977) frequently used by educators when talking about technology (Angeli & Valanides, 2009; Kieran & Drijvers, 2006; Lingefjard, 2012), it appears that the more (commonly recognized) positive affordances a tool offers, the smaller is the number of (mostly hidden) negative affordances that users of the tool are aware of. Unlike "[t]he affordances of danger … [which] are usually perceivable directly, without an excessive amount of learning" (Gibson, 1977, p. 82), negative affordances of a computational tool are not always visible. In mathematics education, positive affordances of technology include the possibility of teaching traditional topics earlier than usual (Kaput, 1992), the increase of the number of students comprehending abstractness of mathematics (Noss & Hoyles, 1996), the emergence of new pieces of learnable mathematics (Kaput, Noss, & Hoyles, 2008), and even the discovery (not to be confused with re-discovery) of new mathematical knowledge (Abramovich & Leonov, 2011). Negative affordances that are visible include the reduction of computational skills, frequent emphasis on drill and practice, and the ease of constructing graphs of functions without understanding their behavior. It comes at no surprise that, following the ideas of Common Core State Standards (National Governors Association Center for Best Practices, Council of Chief State School Officers, 2010), the major educational document in the United States at the time of writing this chapter, the use of technology encourages one "to visualize the results of varying assumptions, explore consequences, and compare predictions with data" (Beghetto et al., 2015, p. 90).

One hidden negative affordance aspect of using technology that calls for creative thinking deals with notational discrepancy of human-computer interaction when semantic codes of the user and the software (or different software tools) do not coincide (Abramovich & Cho, 2013; Kadijevich, 2002; Peschek & Schnieder,

2001). Simpy put, mathematical notation may vary from tool to tool and because of this nuisance a user might become a hostage of a particular coding protocol. For example, the notation INT is understood by Excel as the greatest integer function, is ignored by the *Graphing Calculator* (Avitzur, 2011), and is treated by *Wolfram Alpha* as integral (antiderivative). So, typing "INT($2x - 1$) = 2" into the input box of *Wolfram Alpha* instructs the program to integrate $2x - 1$. This yields the equation $x^2 - x = 2$ with the roots $x = 2$ and $x = -1$ which, obviously don't satisfy the original equation as $\text{INT}(2 \cdot 2 - 1) = 3$ and $\text{INT}(2 \cdot (-1) - 1) = -3$. At the same time, typing "floor($2x - 1$) = 2" yields the equation $\lfloor 2x \rfloor = 3$, thereby, still leaving it to the user to find x. Likewise, the input "floor($2x - 1$) = 2" with a space separating the word "floor" from the expression "($2x - 1$)" yields the equation $2x - 1 = 2$ with the single solution $x = 3/2$, something that completely ignores the presence of the command "floor" in the input. However, if the user is able to appreciate the meaning of the lack of expected response produced by the software and enter an augmented request "solve the equation floor($2x - 1$) = 2", thus making "a switch from one system of semiotic text awareness to another at some internal structure level" (Lotman, 1988, p. 43), *Wolfram Alpha* yields the correct solution in the form of the inequality $3/2 \leq x < 2$. In that way, in the context of *Wolfram Alpha*, the values $x = 2$ and $x = -1$ when selected as solutions to the equation $\text{INT}(2x - 1) = 2$ are irrelevant, leaving the answer in the form $\lfloor 2x \rfloor = 3$ does not demonstrate understanding, and the solution $x = 3/2$ is incomplete.

This dichotomy of human-computer interaction requires an acute awareness of conceptual meaning of the procedures involved and their expected outcomes. In turn, this requires a combination of conceptual understanding and procedural knowledge as pillars of mathematical problem solving without which creative thinking in the context of notation used is kind of fluky. One of the signs of creativity, in the milieu of a problem-solving uncertainty, is the ability to "check the result" (Pólya, 1957, p. 59). In the author's own pre-college learning experience, verification of the correctness of a solution/answer to a problem was a mandatory conclusion for any mathematical task. This verification can be done either through the plug in strategy (for a numeric answer) or by recourse to a special case (whatever the nature of uncertainty). In the case of the greatest integer function which returns the largest integer not exceeding a given number, one can enter into the input box of *Wolfram Alpha* the command "INT(1/2)" and expect the program to return zero. Otherwise, the basic conceptual understanding of the notation used would suggest it has a different meaning for the program which, thereby, delivered an incorrect result. Likewise, even the direct verification of an answer through the plug in strategy, while being purely procedural, may bring about a counter-example to one's conceptual assumptions, something that does entail creativity.

15.4 The Interplay Between Positive and Negative Affordances

Through a brief glance at the last section's references related to affordances, one may note that studies focusing on positive affordances of technology preceded those revealing its negative affordances. Such timing difference between the discussion of the positive and the negative is not surprising for a new tool is almost always introduced first from a positive perspective, and the longer it is in use, the more ill-advised applications emerge calling for improvement. For example, the first appearance of dice, made of clay, dates back to the 3rd millennium BC (David, 1970) and people who played the games of chance were originally not aware of the negative affordances of their tool of entertainment. But in the several millenniums span they not only learned about negative affordances of unfair dice but better still, learned how to make a fair die. A similar interplay between positive and negative affordances can be observed in the case of mathematical ideas. Quite often a statement that something is true (a positive affordance) is made (e.g., inductively, without proof), yet later the statement is found to be false and thereby its negative affordance is revealed.

For example, in 1919, Pólya conjectured that for any natural number n the number of positive integers smaller than or equal to n with an odd number of prime factors is at least the same as those with an even number of factors (counting repeated factors, so that $12 = 2 \cdot 2 \cdot 3$ is considered having an odd number of prime factors). One can use *Wolfram Alpha* to illustrate the case $n = 15$ (Fig. 15.1): among 14 integers only six have an even number of prime factors. In 1962 a counterexample to Pólya's conjecture was found experimentally: $n = 906{,}180{,}359$ (Stark, 1987, p. 7). More specifically, if all natural numbers smaller than or equal to 906,180,359 are put in two sets, one with an odd number of prime factors and another with an even number of prime factors, then the cardinality of the latter set is greater than that of the former set by one. That is, as the negative affordance of Pólya's conjecture was revealed through a counter-example, a positive affordance of the counter-example had enriched the area of number theory for which the

Input:

Table [Ω(n), {n, 2, 15}]

Ω(n) gives the number of prime factors counting multiplicities in n

Result:

n	2	3	4	5	6	7	8	9	10	11	12	13	14	15
$\Omega(n)$	1	1	2	1	2	1	3	2	2	1	3	1	2	2

Fig. 15.1 Illustrating Pólya's conjecture for $n = 15$ using *Wolfram Alpha*

conjecture was important. Pea (1993) describes affordance as "the perceived and actual properties of a thing [e.g., Pólya's conjecture], primarily those functional properties that determine how the thing could possibly be used" (p. 51). The fallacy of the conjecture provides both negative and positive affordances for number theorists. In a mathematics classroom, a creative teacher can blur the distinction between positive and negative affordances of an incorrect answer offered by a student, when using their mistake as a thinking device and a generator of new meaning for the whole class (Mehan, 1979); in other words, the teacher "does not simply declare students' answers to be "right" or "wrong" but encourages students to evaluate the validity of their solutions for themselves or to try multiple solution paths" (Baumert et al., 2010, p. 145).

Another number-theoretical example of a plausible, but eventually refuted, conjecture is due to Euler who in 1788 suggested that for $n \geq 3$ it is not possible to represent a perfect n-th power through the sum of fewer than n like powers. The importance of this conjecture is that, if true, it would have implied the correctness of the statement (known as Fermat's Last Theorem) that for $n \geq 3$ it is not possible to represent the n-th power of a natural number as a sum of two like powers. In the case of Euler's conjecture, it took almost two centuries to find a counter-example (in 1966, via computer search): $144^5 = 27^5 + 84^5 + 110^5 + 133^5$ (Stark, 1987, p. 146).

As for refuting both Pólya's and Euler's conjectures the use of computers was critical, one can see the important role of the digital tools in enabling mathematical creativity—an artistry, which is essential for the advancement of mathematical ideas by negating those ideas that seemed to cast a positive affordance. The above two examples support the notion (discussed in Sect. 15.2) that being 'digital natives' may be necessary but not sufficient for teachers' academic success in a technology-enhanced mathematics classroom. Even using *Wolfram Alpha* for illustrating Pólya's conjecture (Fig. 15.1) requires skills that teachers do not necessarily possess. And to find a counter-example to inductively developed conjecture is far beyond the skills used for its illustration. By being introduced to such examples, teachers can appreciate the value of mathematical creativity necessary for defying such conjectures in the age of computers. In turn, teachers would need to foster mathematical creativity and computer skills of their students in the context of learning mathematics in the digital world.

Furthermore, the noticed interplay between the positive and the negative can also be interpreted in terms of Einstellung effect (Luchins, 1942)—a tendency to use previously learned workable strategy in situations that either can be resolved more efficiently or to which the strategy is not applicable at all. For example, whereas dividing both sides of the inequality $5x > 10$ by 5 yields an equivalent inequality $x > 2$, dividing both sides of the inequality $\sin x > 2\sin x \cos x$ by $\sin x$ yields simultaneous reduction and extension of its solution set, quite a complex outcome of this ill-advised transformation. That is, a strategy proved to be successful in arithmetic (a positive affordance of cancelling out a common factor), becomes fallacious in algebra (a negative affordance of canceling out a common factor).

Einstellung effect in problem solving is sometimes referred to as the rigidity of thinking—a hindrance to creativity, which, on the contrary, requires cognitive flexibility.

As was mentioned above, negative affordances of a computational tool are often hidden and, therefore, may simply be overlooked by mathematics educators in the context of technology integration into the subject matter. In addition to the issue of notational inconsistency across different computer programs, ready-made spreadsheet-based computational environments may require significant programming revision as any new version of MS Office is released. Likewise, *Wolfram Alpha* is a powerful program with a fluid knowledge base, which can be modified/updated at any time. Also, the availability of mathematical problem-solving software (e.g., UMS mentioned above) reduces opportunities for a meaningful homework assessment by a teacher as it could be completed with the help of software. The old argument banning calculators from the mathematics classroom can now be given a new emphasis at a higher level including algebra, trigonometry, calculus, and discrete mathematics.

15.5 Conceptualizing the Appropriate Use of Computers in Mathematics Education

It appears that the appropriate use of computers in mathematics education can be conceptualized as a process that maximizes positive and minimizes negative (both explicit and implicit) affordances of technology. This process calls for a change in the curriculum. In particular, many problems traditionally used in the preparation of students for mathematical Olympiads or used for the development of mathematical creativity in a regular classroom may become somewhat outdated. For example, the task of finding the last digit of the number 777^{777} found in several mathematical Olympiad problem-solving books of the 20th century can be outsourced to *Wolfram Alpha* or *Maple*: each program displays the above number as an integer with the last digit seven. (If required, knowing the answer can prompt explaining it formally, which, after all, is not a bad thing. But still, an opportunity for creative thought is kind of diminished by computing). Thus, the use of technology in support of the emergence of conceptual knowledge in mathematics education is not a simple matter for it depends on a number of factors such as problem type, tool used, population of problem solvers involved, pedagogical goal, and instructional expectation.

Understanding negative affordances of technology makes it possible to minimize their effect by modifying technology-enabled instruction and, in doing so, to bring about new positive affordances that *appropriate* use of technology provides. One way to foster mathematical creativity in the digital era is to furnish students with problems that, on the one hand, are immune from the straightforward computer-based symbolic computations as a problem-solving method and, on the other hand, motivate and enable conceptual development and insightful inference through the use of technology. In the sections that follow, such problems will be referred to as TITE (technology-immune/technology-enabled) problems (Abramovich, 2014b). In particular, one goal of a TITE problem-solving mathematics curriculum is to develop the appreciation of the mutual importance of conceptual knowledge and algorithmic skills—two major components of mathematical creativity and insight.

15.6 Symbolic Computations and TITE Problems

The growth in computational capabilities of mathematics software tools, allowing for an automatic answer to a multistep problem, blurs the distinction between the Type I and Type II technology applications as introduced originally by Maddux (1984). Mathematical activities that until recently belonged to the latter type may become less and less cognitively complex. Such reduction in complexity of mathematical problem solving enables the corresponding epistemic game—"the set of rules and strategies that guide inquiry" (Collins & Ferguson, 1993, p. 25) within a particular representational structure—to be reduced essentially to a simple push of a button. Consequently, the type of technology application may depend on what kind of technology is used to support the game. In order to continue securing educational benefits from the distinction between the two types, Type II application of technology can be advanced to a higher level where one deals with TITE problems. Such problems cannot be automatically solved by software, yet the role of technology in dealing with them is critical. The appropriate use of technology in the context of TITE curriculum can make symbolic computations more cognitively demanding despite or perhaps because of automatic problem-solving capability of the modern day software tools. An example of such computations will be given in Sect. 15.9 when the result of solving Eq. (15.3) by *Wolfram Alpha* is used to find a range for one of the parameters of the equation using paper and pencil alone. Such use of symbolic computations in the context of TITE problems provides a springboard into a follow-up mathematical work requiring conceptual understanding of the symbols involved.

While this perspective on the revision of mathematics curriculum is not limited to a specific educational context, using TITE problems may be a way of developing creative thinking of mathematically motivated students in the digital era. With this population of students in mind, two directions of using technology in the context of school mathematics can be suggested: the use of technology by prospective teachers of mathematically motivated students and the use of technology by the students themselves. In the former case, the teachers can learn posing new problems by modifying/expanding already existing difficult (non-traditional) problems through the use of technology. They need to be able to recognize whether a problem is technology immune and if not, to know which tool to use in order to turn the problem into a TITE one. By the same token, the students can learn exploring TITE problems created either by their teachers or by themselves. Such explorations are consistent with the contemporary approach to mathematical research when (both old and new) problems become solvable only due to the capability of computers to carry out symbolic computations not possible otherwise (e.g., Leonov & Kuznetsov, 2013).

15.7 Using Well-Known Mathematical Tasks to Pose TITE Problems

Whereas mathematicians have been using problem posing for educational purposes quite a long time [e.g., see *Mathematical Questions with Their Solutions* published in (Hodgson, 1870)], educators, in general, consider problem posing as an educational philosophy. It started with "the *liberty of the pupil*" (Montessori, 1965, p. 28, italics in the original) being declared as the fundamental pedagogical principle that "must tend to help the children to advance upon this road of independence" (ibid, p, 94). This tendency towards independence in Montessori classrooms has been an encouraging factor in students' posing their own problems [e.g., Lillard (1996, p. 147)], thereby developing their creative thinking. From the standpoint of psychology, "creativity often is manifested in the ability to formulate or create problems or in the ability properly to reformulate problems" (Luchins, 1960, p. 132).

With this in mind, consider the task of guessing "the rule according to which the successive terms of the following sequence are chosen: 11, 31, 41, 61, 71, 101, 131, …" (Pólya, 1954, p. 8). This task can be outsourced to *Wolfram Alpha*, which, in turn, refers to the On-line Encyclopedia of Integer Sequences (OEIS®, http://oeis.org) where one can find out that the sequence represents prime numbers of the form $10n + 1$. The OEIS® also offers several other interpretations of the sequence. In terms of positive affordances, this may encourage mathematically motivated students to find yet another rule not included into the OEIS®. In terms of negative affordances, less energetic students may just either pick up the first interpretation or, even worse, select an interpretation they don't understand or which is totally irrelevant. This raises an issue of how one can use properly what is available when technology affords an easy access to large quantity of information (Conole & Dyke, 2004). From the TITE perspective, the Pólya's task, which in the digital age can be outsourced to technology, may be didactically enriched by asking a technology-immune question: What is the smallest number of divisions one needs to decide whether 131 is a prime or not? In other words, while recourse to the OEIS® can inform a problem solver that 131 is a prime number with the $10 \cdot 13 + 1$ representation, the didactical enrichment of the (otherwise technology-enabled) task is in requesting explanation of why 131 is a prime number that is based on the most efficient (in terms of the number of divisions) algorithm. One can use technology in making those divisions; yet technology, in the absence of mathematical reasoning, does not give an answer about the smallest number of divisions. The students can also be introduced to the on-line sieve of Eratosthenes (Meyer, 2016) and even do all divisions using this tool. In doing so, one has to realize that after 131 survived divisibility by 11 (as well as by 2, 3, 5, and 7), no division is necessary and 131 is a prime number. One can also be asked to use spreadsheet conditional formatting to eliminate (or highlight) all primes of the form $10n + 1$ for n not greater than a certain integer (Sugden, Baker, & Abramovich, 2015).

A more difficult task from the same book (Pólya, 1954, p. 31) is to find the value of the infinite product $\prod_{i=2}^{\infty}(1 - \frac{1}{i^2})$. Entering "find the product $(1 - \frac{1}{i^2})$ from $i = 2$ to infinity" into the input box of *Wolfram Alpha* yields 1/2 and the value of the partial product $\prod_{i=2}^{n}(1 - \frac{1}{i^2})$ in the fractional form $\frac{n+1}{2n}$ (Fig. 15.2). The latter assumes the values $\frac{3}{4}, \frac{4}{6}, \frac{5}{8}, \frac{6}{10}, \cdots$ and so on. One can be asked: Why is every second fraction in this sequence reducible to the simplest form? In the case or reducibility, is the number 2 always the greatest common factor? Technology cannot answer such (conceptual) questions. That is, seeking the values of the infinite and partial products is a request for information, something that can be outsourced to *Wolfram Alpha*. But the questions seeking explanation of numeric properties of an algebraic expression that represents the sequence of the partial products turns Pólya's rather complex task into a relatively simple TITE problem. Through solving this problem, one can recognize the duality of the fraction $\frac{n+1}{2n}$ (or any other algebraic expression for that matter)—first using it as a process and then seeing it as a concept (Tall et al., 2001).

Another TITE problem might be to develop identities between the infinite product and other infinite sums or products converging to 1/2. For example, one can be asked to carry out the following explorations:

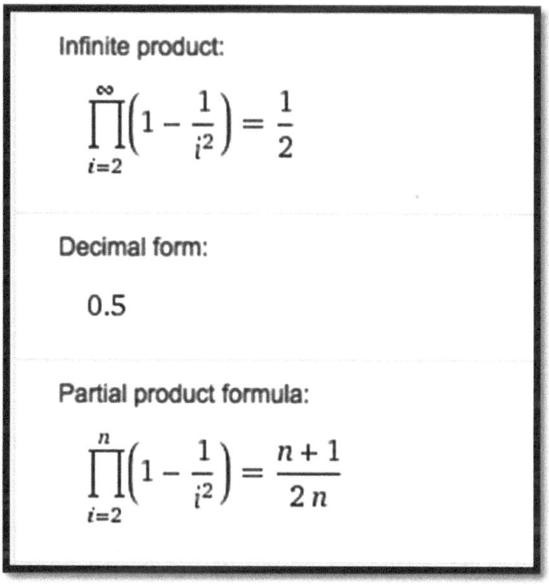

Fig. 15.2 Using *Wolfram Alpha* in solving a problem from a classic book

using *Wolfram Alpha* find $\prod_{i=2}^{\infty}(1-\frac{1}{i^2})$;
prove the identity $\prod_{i=2}^{\infty}(1-\frac{1}{i^2})=\sum_{i=2}^{\infty}\frac{1}{2^i}$;
prove that $x_n=\prod_{i=2}^{n}(1-\frac{1}{i^2})$ and $y_n=\sum_{i=2}^{n}\frac{1}{2^i}$ are, respectively, monotonically decreasing and increasing sequences; and
using a spreadsheet find N such that for all $n > N$ the inequality $|x_n - y_n| < \varepsilon$ holds true for $\varepsilon = 0.01$ and $\varepsilon = 0.001$.

Note that through the suggested use of a spreadsheet one can conclude that y_n converges to 1/2 faster than x_n (Fig. 15.3). Why is it so? Even if one uses *Wolfram Alpha* to establish the relation $\sum_{i=2}^{n}\frac{1}{2^i}=\frac{1}{2}-\frac{1}{2^n}$, one still has some explaining to do regarding the difference between the infinitesimal sequences $\frac{1}{n}$ and $\frac{1}{2^n}$. In that way, a TITE problem consists, following the idea of Isaacs (1930), of two types of questions: an informational type question that can be outsourced to technology and an explanatory type question that requires one to connect procedural and conceptual knowledge.

A similar task from Pólya's (1954, p. 31) book is to find the value of the infinite product $\prod_{i=3}^{\infty}(1-\frac{4}{i^2})$. Slightly modifying the last expression by substituting $i + 1$ for i and entering "find the product $(1-\frac{4}{(i+1)^2})$ from $i = 2$ to infinity" into the input

	A	B	C	D	E	F
1	*i*			Product		Sum
2	2		0.75	0.75	0.25	0.25
3	3		0.888889	0.666667	0.125	0.375
4	4		0.9375	0.625	0.0625	0.4375
5	5		0.96	0.6	0.03125	0.46875
6	6		0.972222	0.583333	0.015625	0.484375
7	7		0.979592	0.571429	0.007813	0.492188
8	8		0.984375	0.5625	0.003906	0.496094
9	9		0.987654	0.555556	0.001953	0.498047
10	10		0.99	0.55	0.000977	0.499023
11	11		0.991736	0.545455	0.000488	0.499512
12	12		0.993056	0.541667	0.000244	0.499756
13	13		0.994083	0.538462	0.000122	0.499878
14	14		0.994898	0.535714	6.1E-05	0.499939
15	15		0.995556	0.533333	3.05E-05	0.499969
16	16		0.996094	0.53125	1.53E-05	0.499985
17	17		0.99654	0.529412	7.63E-06	0.499992
18	18		0.996914	0.527778	3.81E-06	0.499996
19	19		0.99723	0.526316	1.91E-06	0.499998
20	20		0.9975	0.525	9.54E-07	0.499999
21	21		0.997732	0.52381	4.77E-07	0.5

Fig. 15.3 Modeling partial products and sums within a spreadsheet

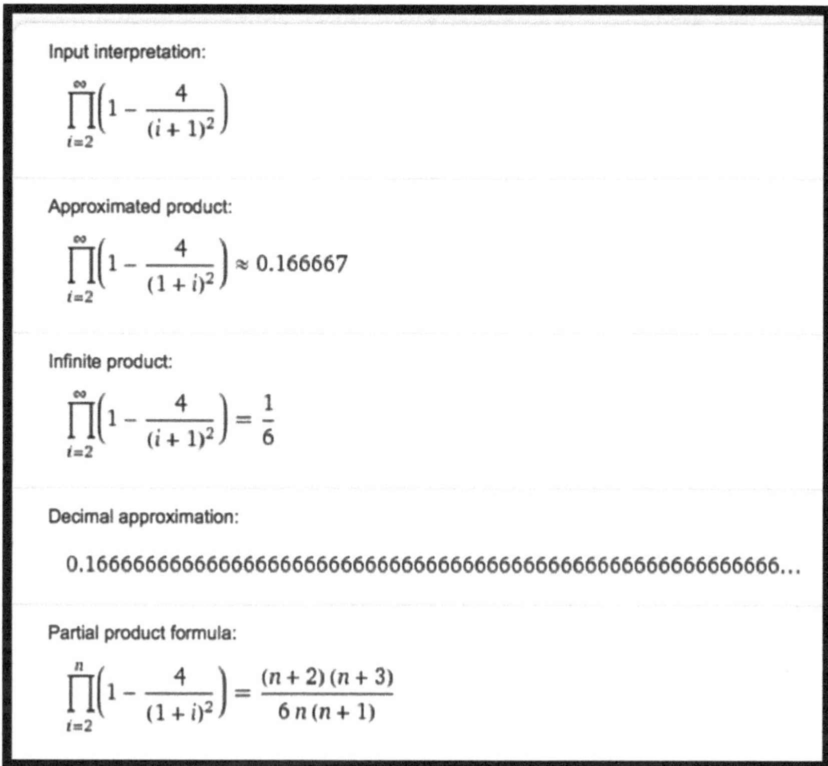

Fig. 15.4 Another problem about infinite product

box of *Wolfram Alpha* yields 1/6 (preceded by its approximation) and the value of a partial product in the fractional form $\frac{(n+2)(n+3)}{6n(n+1)}$ (Fig. 15.4). Note that the suggested modification of the form of the product is an example of how mathematical creativity can be used to correct some negative affordances of technology when information it generates is not well suited for the goal of instruction—to formulate a coherent TITE problem. Entering the product without modification would yield a fraction not defined for $n = 1$ thus complicating some issues which can be avoided by using technology in a creative way. That is, mathematical creativity is a factor in turning a negative affordance of technology into a positive affordance enabling a smooth transition to the technology-immune phase of problem solving.

One can be asked to investigate the fraction $\frac{(n+2)(n+3)}{n(n+1)}$, an expression defined for $n = 1, 2, 3, \ldots$. Through this investigation, it can be connected to triangular numbers $t_n = \frac{n(n+1)}{2}$ and thus can be replaced by $\frac{t_{n+2}}{t_n}$. The first question to be formulated here is: What is the greatest common divisor (*GCD*) of two triangular numbers separated by another triangular number? A simple spreadsheet

Fig. 15.5 The sequence $GCD(t_n,\ t_{n+2})$ forms a cycle of period three

C1 | fx =GCD(B1,B3)

	A	B	C	D
1	1	1	1	
2	2	3	1	
3	3	6	3	
4	4	10	1	
5	5	15	1	
6	6	21	3	
7	7	28	1	
8	8	36	1	
9	9	45	3	
10	10	55	1	
11	11	66	1	
12	12	78	3	

investigation (Fig. 15.5) gives an answer to this question requesting information: the sequence $GCD(t_n, t_{n+2})$ forms the cycle (1, 1, 3) where the number 3 appears each time when n is a multiple of three. Now, one can be asked to explain this phenomenon conceptually. To this end, one can show that the difference $t_{n+2} - t_n$ is equal to $2n + 3$, so that when n is a multiple of three, the difference is not only divisible by three but $GCD(t_n, t_{n+2}) = 3$; otherwise, $GCD(t_n, t_{n+2}) = 1$. That is, the inquiry into the behavior of the sequence of greatest common divisors is a TITE problem: the property of the sequence forming the cycle (1, 1, 3) is technology motivated and its explanation requires formal demonstration which, if necessary, may be enhanced by technology.

Similarly, using a spreadsheet, one can try to explore the behavior of the sequence $GCD(t_n, t_{n+1})$ only to discover the absence of an interesting pattern. Likewise, no simple pattern stems from the greatest common divisors of other pairs of triangular numbers. Was the connection of the infinite product to triangular numbers accidental? To answer this question, one can be asked to explore another problem from Pólya's (1954, p. 32) book and to find the infinite product $\prod_{i=0}^{\infty}(1 - \frac{16}{(2i+1)^2})$ using *Wolfram Alpha*. In response, the program generates the value of a partial product in the form $\frac{(2n+3)(2n+5)}{4n^2-1}$. This time, using a spreadsheet, one can see that the sequence $GCD(4n^2 - 1,\ (2n+3)(2n+5))$ for n = 3, 4, 5, … forms the cycle (1, 1, 3) already observed in the previous example. This suggests at least three things. First, the cyclic behavior of the sequence of the greatest common divisors of the elements of the partial products is perhaps due to the products themselves and the observed connection to triangular numbers was indeed accidental. Second, formulating a coherent TITE problem requires creativity in a sense that such a problem should allow for a reasonable extension. Third, creativity can be a motivating factor for the formal study of a mathematical content that was used to develop TITE problems. In particular, one can become motivated to study infinite products in a way it was designed in the classic book by Pólya (1954) written in the pre-digital era.

15.8 Developing Creativity Through Posing Arithmetical Word Problems

In the presence of technology, mathematical creativity can be developed through posing arithmetical word problems that bear a TITE flavor. Although arithmetic is commonly described as a window to algebra, arithmetical problem solving and posing become part of advanced school mathematics when conceptual understanding is expected to supplant algebraic routines provided by the use of variables. Freire (2003) critical education theory emphasized that "problem-posing education … corresponds to the historical nature of humankind … for whom looking at the past must only be a means of understanding more clearly what and who they are so that they can more wisely build the future" (p. 84). By looking at the past, one can recall elementary school mathematics curriculum when students were expected not only to "check the result" (Polya, 1957, p. 59), as described at the conclusion of Sect. 15.3, but to solve word problems without using algebra. Instead, their solution process was rooted in asking conceptual questions to be answered in a purely numeric form.

One such problem can be found in the mathematics curriculum of the 19th century as described by Tchekoff (1970, p. 70) in a story *Tutor*: "If a merchant buys 138 yards of cloth, some of which is black and some blue, for 540 roubles [*sic*], how many yards of each did he buy if the blue cloth cost 5 roubles [*sic*] a yard and the black cloth 3?" To solve this problem without using algebra (something that the tutor could not do), one can begin with "guessing" any additive partition of 138 in two positive integers, e.g., 138 = 100 + 38, and then proceed to calculating the payment that would have been made under this guess. In doing so, the linear combination of the prices and the meters, $3 \cdot 100 + 5 \cdot 38 = 490$, has to be subtracted from the actual payment, 540, to get 50. The next consideration is that the difference between the actual and guessed payments has to be a multiple of the difference in prices for a yard of each type of cloth, $5 - 3 = 2$. Therefore, $50 \div 2 = 25$ is an error made in the guessed partition of 138. This makes it possible to offset the original guess through subtraction and addition: $100 - 25 = 75$, $38 + 25 = 63$. That is, the merchant bought 75 yards of the black cloth and 63 yards of the blue cloth.

How can one pose problems of that type? How can one create similar didactic materials in order to support the notion that "often a child left to himself will go back to the same puzzle he solved yesterday, simply for the pleasure of getting it right" (Mayer, 1965, p. xxxii)? Creativity and pleasure of doing something go hand-in-hand and similar problems can be used to enhance both qualities in the context of mathematical development. Even if a strategy of solving a problem is known, a cursory change of data does not lead to a solvable problem. For example, replacing 138 by 139, partitioning 139 = 100 + 39, and subtracting 540 – ($3 \cdot 100 + 5 \cdot 39$) yields 45, a number not divisible by the difference in prices, 5 − 3. Likewise, replacing 540 by 541 without changing the rest of the data yields 51, another number not divisible by two. The *Tutor* problem data can be presented through a conceptual bond (Fig. 15.6) in which the apex holds the whole pyramidal structure in a sense that the money spent on 138 yards of cloth with the given costs for a meter may vary from the smallest sum, $416 = 3 \cdot 137 + 5 \cdot 1$, to the largest sum, $688 = 3 \cdot 1 + 5 \cdot 137$. In connection with the reference to the smallest and the largest sums, one can recall Euler's explanation of the general significance of this issue: "since the fabric of the world is the most perfect and was established by the wisest Creator, nothing happens in this world in which some reason of maximum or minimum would not come to light" (Pólya, 1954, p. 121). Indeed, problems seeking maxima and minima have been "of interest not only in mathematics but also in everyday life where people often deal with questions of the best or worst and the least or most of a certain quality or quantity of behavior and of features of the social and physical world" (Luchins & Luchins, 1970, p. 301).

One can note that because the difference between the costs for a meter of each type of cloth is two, the sum of money spent should be a multiple of two. In addition, 138 is an even number and therefore, its additive partitions in two parts are either both even or odd. Furthermore, both prices for a meter are odd numbers and therefore, a linear combination of the moneys spent on each type of cloth is always an even number. So, one has to find the maximum and minimum values of the linear combination $3x + 5y$ for $x + y = 138$. The search can be reduced to exploring the function of one variable, say, $f(x) = 5 \cdot 138 - 2x$ when $1 \leq x \leq 137$. Thus, $f_{max} = f(1) = 688$ and $f_{min} = f(137) = 416$. That is why, any even sum of money in the range [416, 688] works for the apex 138 provided the two bottom vertices don't change. In fact, the bottom vertices may be changed to another pair of odd (or even) numbers, but this will affect the range for the vertex at the top. Likewise, any even number of yards of cloth varying in the range [110, 178] can be bought for the given prices and total sum (Fig. 15.7). Indeed, it follows from the equation $5x + 3y = 540$ that $y = \frac{5}{3}(108 - x)$ and therefore, in order to make the difference $108 - x$ divisible by three, the largest $x = 105$ yields the smallest $y = 5$ whence $x + y = 110$ and the smallest $x = 3$ yields the largest $y = 175$ whence $x + y = 178$. This result can be confirmed by using a spreadsheet, which can be also used to generate a variety of problems similar to the one described in the above story (Tchekoff, 1970). That is, formal reasoning can be enhanced by the use of technology that provides a TITE problem-posing environment informed by understanding the role of conceptual bonds, like those shown in Figs. 15.6 and 15.7, in guiding mathematical creativity.

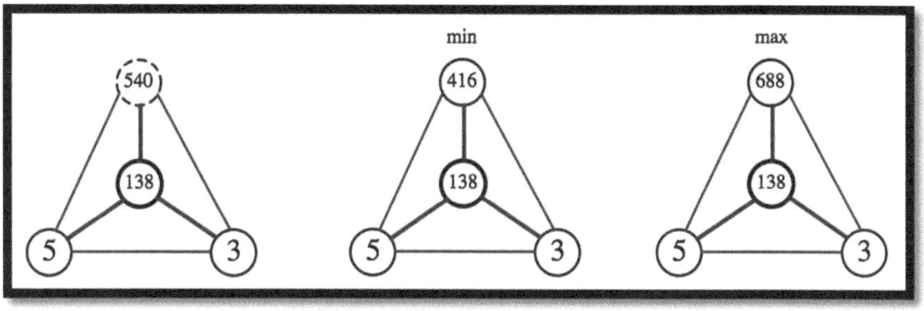

Fig. 15.6 The apex (yards bought) holds the problem structure

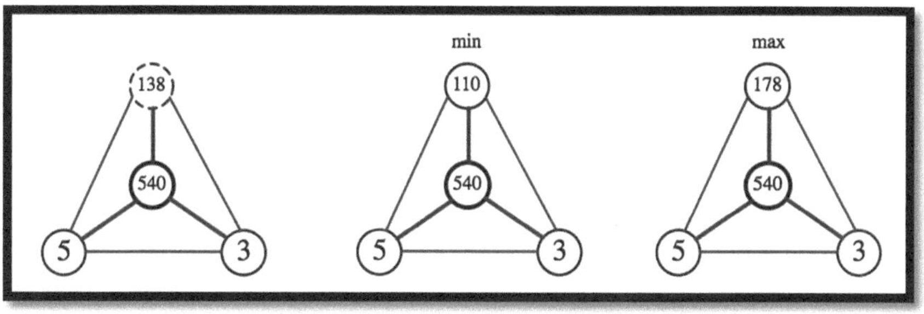

Fig. 15.7 The apex (money paid) holds the problem structure

15.9 Equations with Parameters as TITE Problems

The advent of computing technology in the mathematics classroom has made it possible to use computational experiment as the modern day signature pedagogy (Gurung, Chick, & Haynie, 2009; Shulman, 2005) of mathematics, something that draws on the power of computers to carry out sophisticated numeric/symbolic computations and geometric/graphic constructions. The availability of these tools and new teaching methods they bring about, "create an opportunity for reexamining the fundamental signatures we have so long taken for granted" (Shulman, 2005, p. 59). One such signature to be reexamined in the digital era is the use of non-standard, extra-curriculum problems as means of fostering creative thinking in secondary school mathematics. As an illustration, consider the equation (Sivashinsky, 1968)

$$(x - 2.5)^4 + (x - 1.5)^4 = 1 \qquad (15.1)$$

the real roots of which are to be found—a problem used in the pre-digital era in afterschool activities of secondary students with special interest in mathematics. Suppose a modern student uses *Wolfram Alpha* to solve Eq. (15.1). The program responds to the quest by showing two real roots, 2.5 and 1.5. Several questions about Eq. (15.1) can be asked by a teacher at that point, such as:

What property of Eq. (15.1) makes it having two real roots?
Does a change in the numeric value of the positive right-hand side of Eq. (15.1) affect the number of real roots?
How can one decide when Eq. (15.1) ceases having real roots; that is, what property of its left-hand side is responsible for having two real roots?
How can one, proceeding from Eq. (15.1), develop a family of similar equations with two real roots that depend on a parameter?

In connection with the last question, a teacher may come up with an equation

$$(x - a)^4 + (x - a + 1)^4 = 1 \tag{15.2}$$

with parameter a and, using *Wolfram Alpha*, determine that it has two real roots, $x = a$ and $x = a - 1$, for all real values of a. Note that a pedagogy of introducing a parameter into an equation, which, like Eq. (15.1), is solvable due to its friendly form, followed by exploring conditions in terms of the parameter under which the so generalized equation is solvable analytically, was described by Mason (2000) as turning a doing into an undoing. Undoing is essentially a reflection on what was done, something that seeks to understand why the strategy worked and whether it can be applied to similar problems. Such problems may comprise a family of problems depending on a parameter. A typical question to be asked about a problem with parameter is: For which values of the parameter does the problem have a solution? Problems depending on parameter may not be immediately solvable by software. Alternatively, when software offers some kind of response to a query involving parameter rather than a complete solution, the response has to be interpreted, which leads to a TITE problem.

Once again, the notion of a TITE problem can be interpreted in terms of two types of questions: those seeking information or explanation as such and questions requesting specific types of explanation (Isaacs, 1930). Here, one can distinguish between questions that don't presuppose reflection and questions that motivate reflective inquiry into a quantitative result. The latter type of questions, often stemming from the thorough analysis of computer-generated data, may be considered more intelligent. Often, this intelligence requires just a slight modification of questions that request explanation of the information obtained. Problems asking for explanation are reflections on problems requesting information. In the context of mathematics, problems that are seeking explanation of a certain phenomenon may indeed be more challenging than those through which a certain phenomenon formulated in quantitative terms can be revealed.

For example, solving the equation $|x| = 1$ seeks information about the roots of the equation. The roots found through a routine algorithm do not explain why the equation is solvable unless the algorithm is conceptualized. Knowing that not any equation has real roots, one can seek the explanation of this specific phenomenon by asking the question: for what values of parameter a does the equation

$|x| = a$ have real roots? In order to answer this question conceptually, one has to articulate the definition of the absolute value of a real number. Indeed, a purely procedural perspective on solving the equation $|x| = 1$ may result in replacing it by $x = \pm 1$ prompting the same "algorithm" in the case $|x| = -1$ leading to $x = \pm(-1)$, that is, presenting $x = \pm 1$ as the solution to both equations. Put another way, in the absence of conceptual understanding, procedural skills may result in Einstellung effect (Luchins, 1942). However, knowing that the absolute value of a real number may not be negative, immediately rejects the second case and, consequently, suggests that in the general case $a \geq 0$.

Returning to Eq. (15.2), note that it can be used by a teacher in order to offer to students its multiple versions for different values of parameter a assuming that *Wolfram Alpha* would not be available for finding a solution. In other words, technology here is used for problem posing and not for problem solving. Students can be asked to compare their answers with the goal to conjecture that Eq. (15.2) has always two real solutions (mentioned above), which can then be found in the general form. The next step could be to introduce students to *Wolfram Alpha* asking them to explore the equations such as $(x-a)^4 + (x-a+2)^4 = 1$ and $(x-a)^4 + (x-a+1.1)^4 = 1$. In doing so, one can see that whereas the former equation does not have real roots, the latter equation does have such roots. This computational discovery motivates finding the range for parameter b in the equation

$$(x-a)^4 + (x-a+b)^4 = 1 \qquad (15.3)$$

that provides two real roots.

Using *Wolfram Alpha* (Fig. 15.8), two candidates for real roots of Eq. (15.3), $x = 0.5\,(2a \pm \sqrt{2\sqrt{2}\sqrt{b^4+1} - 3b^2} - b)$, can be selected. In turn, the inequality $2\sqrt{2}\sqrt{b^4+1} - 3b^2 \geq 0$ has to be solved yielding $|b| \leq \sqrt{48} \approx 1.6818$. Note that the described scenario regarding Eq. (15.3) is an example of a TITE problem. Indeed, whereas the very values of x have to be selected by a student from the list of solutions generated by *Wolfram Alpha* and the program does not automatically provide the b-range, its contribution to finding those values is critical.

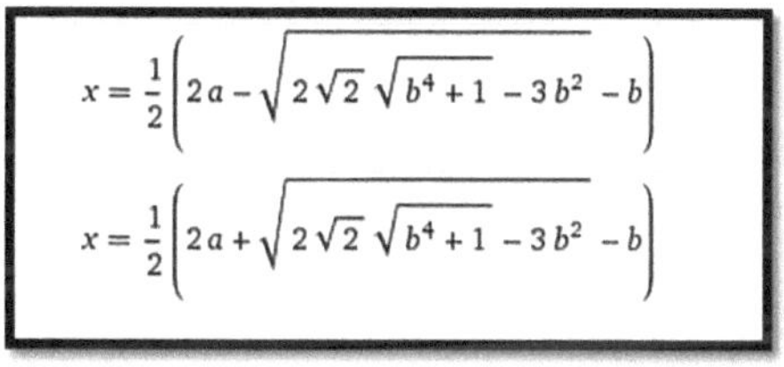

Fig. 15.8 Solving Eq. (15.3) for real x using *Wolfram Alpha*

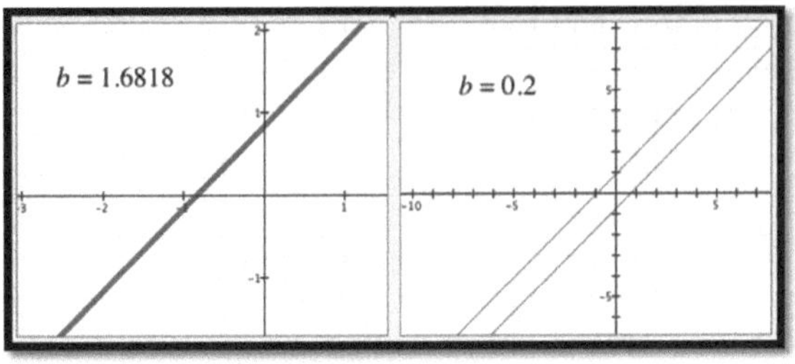

Fig. 15.9 Graphical exploration of Eq. (15.3)

15.10 From Computer Graphing to Data Analysis

As an alternative to *Wolfram Alpha* used in the previous section starting from Eq. (15.1) and ending up with Eq. (15.3) by gradually adding parameters, by setting $a = y$, Eq. (15.3) can be graphed in the plane (x, y) of the *Graphing Calculator* (Avitzur, 2011) for different values of parameter b. In doing so, one can observe (Fig. 15.9) that when $|b| < 1.6818$ the locus of Eq. (15.3) consists of two parallel lines that merge into a single one when $|b| = 1.6818$ and disappear when $|b| > 1.6818$. Computational experiments of that kind using new digital tool provide an alternative verification of the b-range in Eq. (15.3) obtained through a combination of analytical and technological approaches. Furthermore, by changing the exponent in the left-hand side of Eq. (15.2), one can construct graphs of multiple equations using the new tool and, by analyzing computer-generated data, discover that all equations of the form $(x-a)^n + (x-a+1)^n = 1,\ n \in N$ have either one or two real roots in common. This raises new questions requesting explanation. How can one explain this property of the last equation in mathematical terms? Is this property, nowadays easily revealed through the use of technology, at the very core of including Eq. (15.1) into afterschool mathematics curriculum of the past for reasons known to professional mathematicians only? Could one's creative thinking be put to work to comprehend these reasons? Such conceptual inquiry into a result generated by technology is another source from where TITE problems can be developed.

In that way, teaching practices commonly known in the United States as "model with mathematics, use appropriate tools strategically … [including] a spreadsheet, a computer algebra system, … look and make use of structure … that mathematics educators at all levels should seek to develop in their students" Common Core State Standards (National Governors Association Center for Best Practices, Council of Chief State School Officers, 2010, pp. 6–8) are appropriate for fostering creative thinking in students, including those with a special interest in mathematics. By learning to use these technology-enabled mathematical practices, teachers can develop research-like experience in advanced school mathematics and its 21st century signature pedagogy. In particular, this experience includes the development of skills in posing problems, especially TITE ones, and the appreciation of using multiple software tools in experimenting with mathematical ideas in a creative way. As was already mentioned in the introduction, the findings of the Conference Board of the Mathematical Sciences (2012) support the author's assumption that teachers' creative experiences in posing TITE problems, appropriately using commonly available tools, provide strong foundation for imparting such experiences to their future students.

15.11 Concluding Remarks

In the pre-digital era, the development of mathematical skills needed for creativity, invention, and productive thinking had been studied by both mathematicians and educational psychologists (e.g., Hadamar, 1996; Krutetskii, 1976; Lakatos, 1976; Luchins, 1960; Pólya, 1965; Stern & Stern, 1971; Wertheimer, 1959). For one, Luchins (1960) noted that because sometimes "value-laden conceptions of creativity depend on social norms and may differ for different cultures and for different people in the same culture … [it is important to] indicate certain invariants common to some or all of the conceptions of creativity … in the form of relationships that assume different concrete forms in different specific contexts" (p. 131). Nowadays, technology, the use of which may vary across the spectrum of educational contexts, can be considered as such an invariant due to its ubiquitous status within the manifold of contexts, especially in the context of mathematics education. From this perspective, one can say that computer technology is an agency for fostering mathematical creativity in a classroom setting and beyond, including recreation, individual investigations, preparation for mathematical Olympiads, and online mathematical communities of practice (Freiman et al., 2009).

A strong focus on the use of technology in mathematics education has brought to light the need for new pedagogical approaches to work with mathematically motivated students in the presence of computers equipped with programs capable of sophisticated symbolic computations and graphic constructions. Likewise, in a regular classroom, computers should not be considered as tools making problem solving just "easy"; rather, they have to be used for supporting the development of creativity and insight. This chapter proposed an approach of revisiting the traditional advanced mathematics curriculum to include a new type of problems, TITE ones, the solution of which is both dependent on the power of technology and requires the use of creative thinking. The idea of using TITE problems in mathematics teacher education can be seen as an interaction between their TI and TE components in much the same way as the concept of TPCK (technological pedagogical content knowledge) provides teachers with "understanding of the interaction of the knowledge of technology and the knowledge of their subject area" (Niess, 2005, p. 520). In terms of the theory of affordances (Gibson, 1977), a TITE problem affords using technology in a way that minimizes its possible negative impact on the development of creativity.

As was shown in this chapter using books geared towards very different readers, posing TITE problems can be accomplished through the reformulation of traditional problems with the focus on questions that seek explanation of computer-generated data and require human intellect to deal with. The process of reformulation of a mathematical problem in the digital era can be seen not only as a reflection on a solved problem but also as an attempt to replace its procedural orientation, often being trivial in the presence of technology, by conceptual orientation derived from information provided by computing. As shown elsewhere (Abramovich, 2015), this kind of problem reformulation enables new ways of linking procedural and conceptual knowledge towards the development of TITE problems to be used in work with mathematically motivated students.

Therefore, the role of technology in the appropriate reformulation of a problem continues being critical for it is through computing that one can understand the underlying relationships structuring numeric data used to formulate the problem. Such inquiry into the problem's structure supports the development of productive thinking and creative ideas in agreement with highly mathematically oriented Gestalt theory (Ellis, 1938). One of the major tenets of Gestalt asserts that individual parts of a given whole are determined by the inner structure of that whole. This epistemological position that guides creative problem solving in mathematics was demonstrated in the chapter through the notion of conceptual bond (the inner structure of a problem) when a cursory change of the data (the parts of problem) may not necessarily lead to a solvable problem, as "a whole [e.g., a problem] is meaningful when concrete mutual dependency obtains among its parts" (Werthiemer, 1938, p. 16). Mathematical creativity can be developed through analyzing the relationships among the elements of a problem's conceptual bond and how technology can be used to facilitate this analysis. Through learning to pose TITE problems prospective mathematics teachers develop research-like experiences that, in turn, by the teachers' own admission (Conference Board of the Mathematical Sciences, 2012, p. 65), are helpful in intellectually demanding work with the 21st century school students. By the same token, through the proposed approach the students can significantly advance their ability to do mathematics and genuinely enrich their interest in the subject matter.

To conclude, several other directions in the area of developing creativity in the digital era can be suggested. Given this chapter's focus on revisiting known problems from number theory (prime numbers, sequences, series), arithmetic (word problems), and algebra (higher order polynomial equations) in the context of duality of computational affordances and expanding/modifying the problems under the TITE umbrella, another conceptually rich area of advanced secondary school mathematics curriculum is worth mentioning. It deals with transcendental functions, including circular, exponential, and logarithmic functions. In Sect. 15.4 of this chapter, a trigonometric inequality was briefly mentioned in connection with the notion of Einstellung effect in problem solving when the cancellation of a variable factor common to both sides of the inequality resulted in a complex transformation of its solution set. Such problems cannot be easily formulated unless positive affordances of technology have been realized. Seeing the use of technology in posing problems of a given type as an important aspect of fostering creative thinking, one could develop a mechanism conducive to formulating a variety of solvable trigonometric, exponential, and logarithmic equations and inequalities having a pre-determined conceptual structure responsible for a certain outcome of Einstellung effect. This can lead to new explorations dealing with error analysis which is critical in search for explanation of why a specific problem-solving strategy is incorrect. Teachers' ability to provide that kind of explanation is an indicative trait of their creativity. By the same token, as the request for explanation and fostering creativity in the context of technology-supported problem solving go hand-in-hand, creative teachers are major custodians of the unfolding creative potential of their students.

References

Abramovich, S. (2014a). *Computational experiment approach to advanced secondary mathematics curriculum*. Dordrecht, The Netherlands: Springer.

Abramovich, S. (2014b). Revisiting mathematical problem solving and posing in the digital era: Toward pedagogically sound uses of modern technology. *International Journal of Mathematical Education in Science and Technology, 45*(7), 1034–1052.

Abramovich, S. (2015). Mathematical problem posing as a link between algorithmic thinking and conceptual knowledge. *The Teaching of Mathematics, 18*(2), 45–60.

Abramovich, S., & Cho, E. K. (2013). Technology and the creation of challenging problems. *Mathematics Competitions, 26*(2), 10–20.

Abramovich, S., Easton, J., & Hayes, V. O. (2014). Integrated spreadsheets as learning environments for young children. *Spreadsheets in Education, 7*(2), 3.

Abramovich, S., & Leonov, G. A. (2011). A journey to a mathematical frontier with multiple computer tools. *Technology, Knowledge and Learning, 16*(1), 87–96.

Advisory Committee on Mathematics Education. (2011). *Mathematical needs: The mathematical needs of learners*. London, UK: The Royal Society.

Angeli, C., & Valanides, N. (2009). Epistemological and methodological issues for the conceptualization, development and assessment of ICT-TPCK: Advances in technological pedagogical content knowledge (TPCK). *Computers & Education, 52*(1), 154–168.

Arnold, V. I. (2015). *Experimental mathematics*. Providence, RI: The American Mathematical Society.

Avitzur, R. (2011). *Graphing Calculator* (Version 4.0). Berkeley, CA: Pacific Tech.

Baumert, J., Kunter, M., Blum, W., Brunner, M., Voss, T., Jordan, A., et al. (2010). Teachers' mathematical knowledge, cognitive activation in the classroom, and student progress. *American Educational Research Journal, 47*(1), 133–180.

Beghetto, R. A., Kaufman, J. C., & Baer, J. (2015). *Teaching for creativity in the common core classroom*. New York, NY: Teachers College Press.

Bers, M. U. (2010). When robots tell a story about culture … and children tell a story about learning. In N. Yelland (Ed.), *Contemporary perspective on early childhood education* (pp. 227–247). Maidenhead, UK: Open University Press.

Bers, M. U., Ponte, I., Juelich, K., Viera, A., & Schenker, J. (2002). Teachers as designers: Integrating robotics in early childhood education. *Information Technology in Childhood Education Annual, 2002*(1), 123–145.

Borwein, J., & Bailey, D. (2004). *Mathematics by experiment: Plausible reasoning in the 21st century*. Natick, MA: AK Peters.

Brooks, J. G., & Brooks, M. G. (1999). *In search of understanding: The case for constructivist classrooms*. Alexandria, VA: Association for Supervision and Curriculum Development.

Collins, A., & Ferguson, W. (1993). Epistemic forms and epistemic games: Structures and strategies to guide inquiry. *Educational Psychology, 28*(1), 25–42.
Conference Board of the Mathematical Sciences. (2012). *The mathematical education of teachers II*. Washington, DC: The Mathematics Association of America.
Conole, G., & Dyke, M. (2004). What are the affordances of information and communication technologies? *ALT-J. Research in Learning Technology, 12*(2), 113–124.
Dahlstrom, E., & Bichsel, J. (2014). *ECAR Study of Undergraduate Students and Information Technology, 2014*. Research report. Louisville, CO: ECAR, October 2014. Retrieved from http://www.educause.edu/ecar.
David, F. N. (1970). Dicing and gaming (a note on the history of probability). In E. S. Pearson & M. G. Kendall (Eds.), *Studies in the history of statistics and probability* (pp. 1–17). London, UK: Griffin.
Dittert, N., & Krannich, D. (2013). Digital fabrication in educational contexts—Ideas for a constructionist workshop setting. In J. Walter-Herrmann & C. Büching (Eds.), *FabLab: Of machines, makers and inventors* (pp. 173–180). Bielefeld, Germany: Transcript Verlag.
Ellis, W. D. (Ed.). (1938). *A source book of Gestalt psychology*. New York, NY: Harcourt, Brace.
Epstein, D., Levy, S., & de la Llave, R. (1992). About this journal. *Experimental Mathematics, 1* (1), 1–3.
Freiman, V., Kadijevich, D., Kuntz, G., Pozdnyakov, S., & Stedøy, I. (2009). Technological environments beyond the classroom. In E. J. Barbeau & P. J. Taylor (Eds.), *Challenging mathematics in and beyond the classroom* (pp. 97–131). Dordrecht, The Netherlands: Springer.
Freire, P. (2003). *Pedagogy of the oppressed* (with an introduction by D. Macedo). New York, NY: Continuum.
Feurzeig, W., & Lukas, G. (1972). LOGO–A programming language for teaching mathematics. *Educational Technology, 12*(3), 39–46.
Gibson, J. J. (1977). The theory of affordances. In R. Shaw & J. Bransford (Eds.), *Perceiving, acting and knowing: Toward an ecological psychology* (pp. 67–82). Hillsdale, NJ: Lawrence Erlbaum.
Gurung, R. A. R., Chick, N. L., & Haynie, A. (2009). *Exploring signature pedagogies: Approaches to teaching disciplinary habits of mind*. Sterling, VA: Stylus.
Hadamard, J. (1996). *The Mathematician's mind: The psychology of invention in the mathematical field*. Princeton, NJ: Princeton University Press.
Hodgson, C. F. (1870). *Educational times*. London, UK: CF Hodgson & Son.
Hughes, J., Gadanidis, G., & Yiu, C. (2016). Digital making in elementary mathematics education? In *Digital experiences in mathematics education* (pp. 1–15). New York, NY: Springer, May 27. Retrieved from http://link.springer.com/article/10.1007/s40751-016-0020-x.
Isaacs, N. (1930). Children's why questions. In S. Isaacs (Ed.), *Intellectual growth in young children* (pp. 291–349). London, UK: Routledge & Kegan Paul.
Kadijevich, D. (2002). Towards a CAS promoting links between procedural and conceptual mathematical knowledge. *The International Journal of Computer Algebra in Mathematics Education, 9*(1), 69–74.
Kaput, J. J. (1992). Technology and mathematics education. In D. A. Grouws (Ed.), *Handbook of research on mathematics teaching and learning* (pp. 515–556). New York, NY: Macmillan.
Kaput, J. J., Noss, R., & Hoyles, C. (2008). Developing new notations for a learnable mathematics in the computational era. In L. D. English (Ed.), *Handbook of international research on mathematics education* (pp. 693–715). New York, NY: Lawrence Erlbaum.
Kennedy, G. E., Judd, T. S., Churchward, A., Gray, K., & Krause, K. L. (2008). First year students' experiences with technology: Are they really digital natives? *Australasian Journal of Educational Technology, 24*(1), 108–122.
Kieran, C., & Drijvers, P. (2006). The co-emergence of machine techniques, paper-and-pencil techniques, and theoretical reflection: A study of CAS use in secondary school algebra. *International Journal of Computers for Mathematical Learning, 11*(2), 205–263.

Kirkwood, A., & Price, L. (2005). Learners and learning in the twenty-first century: What do we know about students' attitudes towards and experiences of information and communication technologies that will help is design courses? *Studies in Higher Education, 30*(3), 257–274.

Krutetskii, V. A. (1976). *The psychology of mathematical abilities in school children*. Chicago, IL: University of Chicago Press.

Kvavik, R.B., & Caruso, J.B. (2005). *ECAR study of students and information technology, 2005: Convenience, connection, control, and learning* (Vol. 6). Boulder, CO: EDUCAUSE. Retrieved from http://www.educause.edu/ecar.

Lakatos, I. (1976). *Proofs and refutations: The logic of mathematical discovery*. Cambridge, UK: Cambridge University Press.

Leonov, G. A., & Kuznetsov, N. V. (2013). Hidden attractors in dynamical systems. From hidden oscillations in Hilbert-Kolmogorov, Aizerman, and Kalman problems to hidden chaotic attractor in Chua circuits. *International Journal of Bifurcation and Chaos, 23*, 1330002 (2013) [69 pages]. https://doi.org/10.1142/s0218127413300024.

Lillard, P. P. (1996). *Montessori today: A comprehensive approach to education from birth to adulthood*. New York, NY: Shaken Books.

Lingefjärd, T. (2012). Mathematics teaching and learning in a technology rich world. In S. Abramovich (Ed.), *Computers in education* (Vol. 2, pp. 171–191). New York, NY: Nova Science Publishers.

Lotman, Y. M. (1988). Text within a text. *Soviet Psychology, 24*(3), 32–51.

Luchins, A. S. (1942). Mechanization in problem solving: The effect of Einstellung. In *Psychological Monographs, 54*(6, whole No. 248). Evanston, IL: The American Psychological Association.

Luchins, A. S. (1960). On some aspects of the creativity problem in thinking. *Annals of the New York Academy of Sciences, 91*, 128–140.

Luchins, A. S., & Luchins, E. H. (1970). *Wertheimer's seminars revisited: Problem solving and thinking* (Vol. I). Albany, NY: Faculty-Student Association, SUNY at Albany.

Maddux, C. D. (1984). Educational microcomputing: the need for research. *Computers in the Schools, 1*(1), 35–41.

Maddux, C. D., & Johnson, D. L. (2005). Type II applications of technology in education: New and better ways of teaching and learning. *Computers in the Schools, 22*(1/2), 1–5.

Mason, J. (2000). Asking mathematical questions mathematically. *International Journal of Mathematical Education in Science & Technology, 31*(1), 97–111.

Mayer, M. (1965). Introduction. In M. Montessori (Ed.), *The Montessori method* (pp. xxiii–xli). Cambridge, MA, Robert Bentley.

Mehan, H. (1979). *Learning lessons*. Cambridge, MA: Harvard University Press.

Meyer, H. B. (2016). *Eratosthenes' sieve*. Retrieved from http://www.hbmeyer.de/eratosiv.htm.

Ministry of Education Singapore. (2012). *N(T)-level mathematics teaching & learning syllabus*. Curriculum Planning & Development Division. Retrieved from http://www.moe.gov.sg/education/syllabuses/sciences/files/normal-technical-evel-maths-2013.pdf.

Montessori, M. (1965). *The Montessori method*. Cambridge, MA: Robert Bentley.

National Governors Association Center for Best Practices, Council of Chief State School Officers. (2010) *Common core state standards (Mathematics)*. Washington, D.C.: National GovernorsAssociation Center for Best Practices, Council of Chief State School Officers. Retrieved from http://www.corestandards.org/Math/.

Niess, M. L. (2005). Preparing teachers to teach science and mathematics with technology: Developing a technology pedagogical content knowledge. *Teaching and Teacher Education, 21*(5), 509–523.

Noss, R., & Hoyles, C. (1996). *Windows on mathematical meanings: Learning cultures and computers*. Dordrecht, The Netherlands: Kluwer.

Pea, R. D. (1993). Distribted intelligence and designs for education. In G. Salomon (Ed.), *Distributed cognitions—Psychological and educational considerations* (pp. 47–87). New York, NY: Cambridge University Press.

Peschek, W., & Schneider, E. (2001). How to identify basic knowledge and basic skills? Features of modern general education in mathematics. *The International Journal of Computer Algebra in Mathematics Education, 8*(1), 7–22.

Pólya, G. (1954). *Induction and analogy in mathematics* (Vol. 1). Princeton, NJ: Princeton University Press.

Pólya, G. (1957). *How to solve it.* New York, NY: Anchor Books.

Pólya, G. (1965). *Mathematical discovery: On understanding, learning, and teaching problem solving*. New York, NY: Wiley.

Prensky, M. (2001). Digital natives, digital immigrants. *On the Horizon, 9*(5), 1–6.

President's Council of Advisors on Science and Technology. (2010). *Prepare and inspire: K–12 education in science, technology, engineering and math (STEM) for America's future.* Washington, D.C.: Author. Retrieved from https://www.whitehouse.gov/sites/default/files/microsites/ostp/pcast-stemed-report.pdf.

Schoenfeld, A. H. (1988). Users of computers in mathematics instruction. In D. A. Smith, G. Porter, L. Leinbach, & R. Wenger (Eds.), *Computers and mathematics: The use of computers in undergraduate instruction* (Vol. 9, pp. 1–11). Washington, D.C.: The Mathematical Association of America.

Shulman, L. S. (2005). Signature pedagogies in the professions. *Daedalus, 134*(3), 52–59.

Sivashinsky, I. H. (1968). *Zadachi po matematike dlja vneklasnyh zanjatii (Mathematical problems for after school activities).* Moscow, Russia: Prosveschenie. (In Russian).

Stark, H. M. (1987). *An Introduction to number theory.* Cambridge, MA: MIT Press.

Stern, C., & Stern, M. B. (1971). *Children discover arithmetic: An introduction to structural arithmetic.* New York, NY: Harper & Row.

Sugden, S., Baker, J. E., & Abramovich, S. (2015). Conditional formatting revisited: A companion for teachers and others. *Spreadsheets in Education, 8*(3), 2.

Tall, D., Gray, E., Ali, M. B., Crowley, L., DeMarois, P., McGowen, M., et al. (2001). Symbols and the bifurcation between procedural and conceptual thinking. *Canadian Journal of Science, Mathematics and Technology Education, 1*(1), 81–104.

Tchekoff, A. (1970). *Russian silhouettes* (Translated from Russian by M. Fell). Freeport, NY: Books for Libraries Press.

Wertheimer, M. (1959). *Productive thinking.* New York, NY: Harper & Row.

Wertheimer, M. (1938). Gestalt theory. In W. D. Ellis (Ed.), *A source book of Gestalt psychology* (pp. 1–11). New York, NY: Harcourt Brace.

Chapter 16
"Integrating" Creativity and Technology Through Interpolation

Bharath Sriraman and Daniel Lande

Abstract The digital age of the 21st century is ubiquitous with easy access to information. Students of mathematics find at their fingertips (literally) immense resources such as Wolfram Math and other digital repositories where anything can be looked up in a few clicks. The purpose of this chapter is to convey to the reader that Mathematics as a discipline offers examples of how hand calculations using first principles can result in deep insights that present students with the opportunities of learning and understanding. By first principles we are referring to fundamental definitions of mathematical concepts that enable one to derive results (e.g., definition of a derivative; definition of a Taylor series etc.). We also highlight the value of integrating (pun intended) technology to understand functions that were obtained via mathematical interpolation by the likes of John Wallis (1616–1703), Lord Brouncker (1620–1684), Johann Lambert (1728–1777) and Edward Wright (1558–1615). The interpolation techniques used by these eminent mathematicians reveals their creativity in deriving representations for functions without the aid of modern technology. Their techniques are contrasted with modern graphing techniques for the same functions.

Keywords Circular functions · Π · John wallis · Quadratures (areas) Conformal maps · Secant function · Integration · History of calculus History of infinite series · Interpolation · Mathematical creativity

B. Sriraman (✉)
Department of Mathematical Sciences, University of Montana, Missoula, MT, USA
e-mail: SriramanB@mso.umt.edu

D. Lande
Sentinel High School, Missoula, MT, USA

© Springer International Publishing AG, part of Springer Nature 2018
V. Freiman and J. L. Tassell (eds.), *Creativity and Technology in Mathematics Education*, Mathematics Education in the Digital Era 10,
https://doi.org/10.1007/978-3-319-72381-5_16

16.1 Introduction

The title of this chapter in a book that addresses the relationship between technology and creativity in mathematics education is a pun. By integrating we mean the mathematical technique of finding anti-derivatives, and by interpolation we literally mean mathematical interpolation to find missing information. Finding a piece of knowledge or a mathematical fact is very different from actually being able to remember it or deriving it from first principles. Yet "knowing" among students is increasingly becoming associated with or even synonymous with "looking it up" as opposed to understanding first principles to be able to derive a result or understanding ways in which results are arrived at. To this end the examples presented in this chapter are deliberately chosen from the history of series and continued fractions to illustrate the value of interpolation techniques as a forgotten art of hand computation. The history of infinite series played a significant role in how functions were dealt with before modern day integration techniques were established. For instance, manipulating the series representation of circular functions led to insights about their anti-derivatives; similarly, the series for the logarithm function allowed better facility for calculation purposes. This is further explored in an ensuing section of the chapter. Ironically any modern Calculus textbook in the U.S typically presents series after techniques of integration when the former actually led to the latter. While deduction is emphasized in textbooks for the sake of presenting the subject matter in a logical manner, the advent of the digital age with readily accessible results from Wolfram Math or other mathematical repositories presents the danger of a student conceiving of mathematics as a platonic and deductive subject. The examples presented will hopefully convey the inductive aspect of the mathematician's craft.

Interpolation which literally means, "inserting between other things" can be viewed as the original "cut and paste" but in mathematical parlance one that required intuitive and systematic thought as opposed to present day pastiche. Can we present hand computations requiring interpolation and *per modum inductionis* (Wallis, 2004) as a contrast to the "digital habit" of invoking a function on Mathematica or looking up the end result in Wolfram Math? The expression "digital habit" is used colloquially here to refer to observations from the authors in their classrooms where students rely on hand-held devices to access information. For instance when students are asked for the series representation of a well-known function, they rely on retrieving the representation from Wolfram Math. As a contrast to the way things are today the examples presented in this chapter reveal both a forgotten art of mathematical creativity and extol the virtues of hand computation as a necessary complement to the "looking it up" and "cutting and pasting" habits of many present day students. The definition of creativity adapted here are those from Paul Torrance and Alex Osborn. To paraphrase these two individuals who furthered the study of creativity, Torrance (1974) referred to creativity as being able to sense difficulties or gaps or missing elements or something askew when confronted with information. Osborn (1953) on the other hand suggested creativity was the process of finding a solution by first finding a mess and then finding data to explain the mess which in turn leads to defining the real problem and coming up with the ideas for a solution. The examples we present are mathematically messy and involved numerical data, but as will be evident the solutions found by the mathematicians in the past are nothing short of ingenious.

16.2 Three Examples

16.2.1 π

Many learners of mathematics at the undergraduate level may simply think of π as a button on a calculator (or calculator app). The number π is often used in formulas for the area and circumference of circles without any real thought to where this number comes from or what it means, unless initiated by the teacher. Some early high school lessons attempt to enlighten students on its origin by having them measure the circumference and diameter of many different sized circles in order to derive π by using the equation: $\frac{C}{d} = \pi$. This exercise often passes them by without any increase in understanding the intricacies and ingenuity involved in deriving this extraordinary constant. A study of past methods of derivation often results to a greater understanding and appreciation for the creativity involved in these calculations given the tools available to the mathematicians of their time.

16.2.1.1 Derivation of π from Tables

John Wallis's *Arithmetica Infinitorum* contained tables, which required creative interpolation to derive formulas we take for granted today. Many of these tables have served as fodder for mathematicians of today who attempt to explain the ingenuity involved in their construction. Stedall (2000) describes how problems in number theory began to capture the attention of early 17th century mathematicians in England, preceding the age of Newton. In particular focus is brought on the work of John Wallis and Lord William Brouncker and their collaboration on the problem of calculating quadratures (areas) of circles. It should be noted that at this time period the Calculus of Newton was yet to be invented, and most of the work requiring integrals was accomplished by using infinite series or continued fractions. The infinite product formula for π that is attributed to John Wallis is often featured in undergraduate mathematics textbooks, more often in an honors section of the course. Honors courses in U.S universities typically offer higher-level or more academically challenging assignments. The formula in question is presented as $\frac{2}{\pi} = \left(\frac{1 \cdot 3}{2 \cdot 2}\right)\left(\frac{3 \cdot 5}{4 \cdot 4}\right)\left(\frac{5 \cdot 7}{6 \cdot 6}\right) \cdots$

To derive the formula for $\frac{2}{\pi}$ as the quotient of infinite products, students are told to work with the integral $I_n = \int_0^{\frac{\pi}{2}} \sin^n(x)dx$ without any explanation as to how it pertains to the formula in the first place. The problem at hand concerned the quadrature of the circle, which Stedall (2000) poetically described as "Squaring the circle". To tackle the problem, Wallis began with tables that gave values of $\frac{1}{\int_0^1 \left(1-x^{\frac{1}{Q}}\right)^P dx}$ for convenient values of P and Q that resulted in integers as the answer (see Table 16.1). The goal was to interpolate the value of $\int_0^1 \sqrt{1-x^2}dx$ by generating tables for $\int_0^1 (1-x^2)^0 dx$ and $\int_0^1 (1-x^2)dx$, but taking the reciprocals facilitated ease of computation.

Stedall (2000) provides details of the process that Wallis went through as well as his correspondence with Lord Brouncker during his investigation of this table. In modern terms, using integration by parts and a change of variable, the integral can be transformed easily to $\frac{1}{Q\int_0^1 (1-x)^P x^{Q-1} dx}$ which can be then be evaluated to generate the table for values of P and Q. However, *not knowing* the binomial theorem called for a creative leap that resulted in the astonishing formula from the tables, namely $\frac{2}{\pi} = \left(\frac{1\cdot 3}{2\cdot 2}\right)\left(\frac{3\cdot 5}{4\cdot 4}\right)\left(\frac{5\cdot 7}{6\cdot 6}\right)\ldots$

The actual value calculated by Wallis was $\frac{4}{\pi} = \left(\frac{3}{2}\right)\left(\frac{3\cdot 5}{4\cdot 4}\right)\left(\frac{5\cdot 7}{6\cdot 6}\right)\ldots$ which meant the quadrature was the reciprocal, namely $\frac{\pi}{4}$. The mathematical constraints of the time period called for ingenuity in methods. Haught and Stokes (2017) argue that domains are defined by constraints, which in turn lead to specific goal setting. For them creativity is mastering the basic constraints and the competency to achieve the goals despite the constraints. In the same vein, having access to the table produced by Wallis led Lord Brouncker to an entirely different interpretation to solve the quadrature problem. Lord Brouncker's work on the same tables led to a continued fraction for π, namely $\frac{4}{\pi} = 1 + \frac{1^2}{2+}\frac{3^2}{2+}\frac{5^2}{2+}\ldots$

Again, for lack of a better descriptor, this is another astonishing formula for π, which is often found in books without any reference to the context of the problem. In modern terms, this interpolation can be understood by looking at the integral $I_n = \int_0^1 \frac{x^n}{1+x^2}dx$ and setting up a recursive formula for I_n to arrive at $\frac{4}{\pi}$. Graphing the

Table 16.1 From arithmetica infinitorum proposition 169 (Stedall, 2000, p. 299)

P	Q								
	0	$\frac{1}{2}$	1	$\frac{3}{2}$	2	$\frac{5}{2}$	3	$\frac{7}{2}$	4
0	1		1		1		1		1
$\frac{1}{2}$			?						
1	1		2		3		4		5
$\frac{3}{2}$									
2	1		3		6		10		15
$\frac{5}{2}$									
3	1		4		10		20		35
$\frac{7}{2}$									
4	1		5		15		35		70

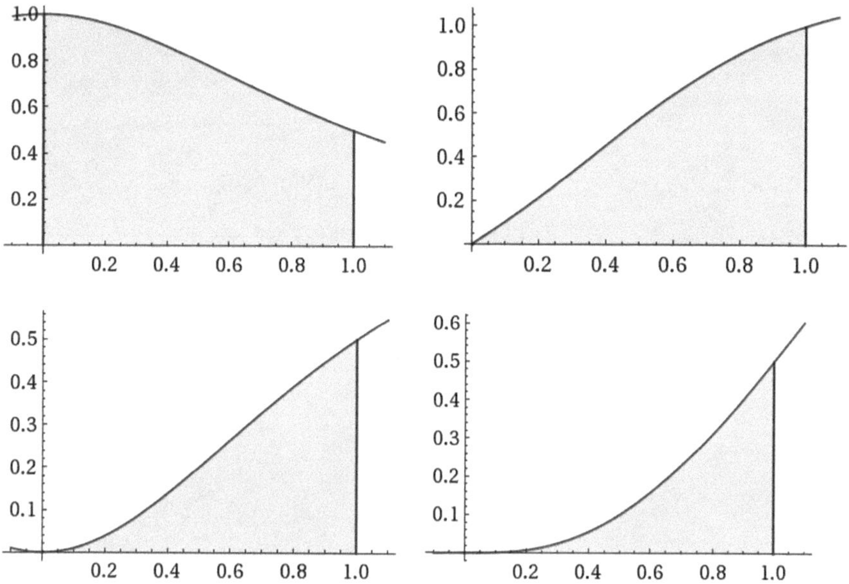

Fig. 16.1 Graphs of $I_n = \int_0^1 \frac{x^n}{1+x^2}dx$ for I_0, I_1, I_2, and I_3

integral for various values of n is the advantage that technology offers today's student (see Fig. 16.1). The graphs can be used to generate a discussion on how the value of $\frac{4}{\pi}$ relates to the integral.

Similarly, as mentioned earlier, Wallis's infinite product is evaluated by starting with the integral $I_n = \int_0^{\frac{\pi}{2}} \sin^n(x)dx$, which involves a circular function, and evaluating $\lim\limits_{n \to \infty} \frac{I_{2n}}{I_{2n+1}}$ after arriving at the fact that $\frac{I_{2n-1}}{I_{2n+1}} = \frac{2n+1}{2n} \geq \frac{I_{2n}}{I_{2n}+1} \geq 1$. Again, there is a recursive formula involved and the necessity to integrate by parts; tools, which were not available to John Wallis! However, technology again helps us understand the nature of the integrals and the value that is obtained by the infinite product. Graphing the integral (see Fig. 16.2) is an advantage that was not available to the 17th century mathematician, but available to us today.

In both these examples, integrals involving circular functions need to be invoked by a 21st century student. However, the fact remains that interpolating tables played a significant part in arriving at the astonishing closed forms seen earlier. Dutka (1990) in his exposition of the history of the factorial function presents a different route available to us today since closed forms for the integral that Wallis had to tackle are now available to us. He suggested that Wallis's computations are understood better by integrating $(x - x^2)^n dx$ between the limits of 0 and 1. By knowing the closed form for this integral, namely

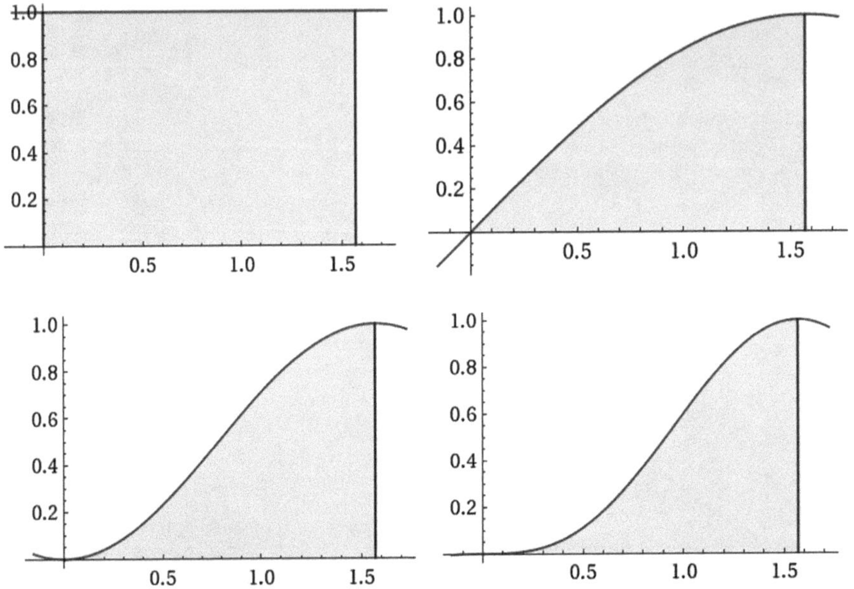

Fig. 16.2 Graphs of $\int_0^{\frac{\pi}{2}} \sin^n(x)dx$ for I_0, I_1, I_2, and I_3

$$\int_0^1 (x - x^2)^n dx = \frac{1}{2n+1}\frac{n!n!}{(2n)!} \text{ for } n = 0, 1, 2, 3, \ldots$$

we can calculate the following values for n = 0,1,2,3,...

For $n = 0$

$$\int_0^1 (x - x^2)^0 dx = 1$$

For $n = 1$

$$\int_0^1 (x - x^2)^1 dx = \frac{x^2}{2} - \frac{x^3}{3}\bigg|_0^1 = \frac{1}{2} - \frac{1}{3} = \frac{1}{6} = \frac{1}{2(1)+1}\frac{1!1!}{2!} = \frac{1}{3}\cdot\frac{1}{2} = \frac{1}{6}$$

For $n = 2$

$$\int_0^1 (x-x^2)^2 = \int_0^1 (x^2-2x^3+x^4)dx$$
$$= \left(\int_0^1 x^2 - 2\int_0^1 x^3 + \int_0^1 x^4\right) dx$$
$$= \frac{x^3}{3}\Big|_0^1 - 2\frac{x^4}{4}\Big|_0^1 + \frac{x^5}{5}\Big|_0^1$$
$$= \frac{1}{3} - \frac{2}{4} + \frac{1}{5}$$
$$= \frac{1}{30} = \frac{1}{2(2)+1}\frac{2!2!}{4!} = \frac{1}{5}\cdot\frac{4}{4\cdot 3\cdot 2\cdot 1} = \frac{1}{30}$$

We arrive at the sequence of values $1, \frac{1}{6}, \frac{1}{30}$, etc. but we want the value at $n = \frac{1}{2}$. Using trigonometric substitution we can show:

$\int_0^1 (x-x^2)^{\frac{1}{2}}dx = \frac{\sin^{-1}(2x-1)+(4x-2)\left(x-x^2\right)^{\frac{1}{2}}}{8} = \frac{\pi}{8}$; when evaluated between the limits of 0 and 1. So, we now have the sequence of values $1, \frac{\pi}{8}, \frac{1}{6}, \frac{1}{30}$, etc.

Knowing

$$\int_0^1 (x-x^2)^{\frac{1}{2}}dx = \frac{\pi}{8} = \frac{\left(\frac{1}{2}\right)!\left(\frac{1}{2}\right)!}{\left(2\cdot\frac{1}{2}\right)!}\cdot\frac{1}{2\left(\frac{1}{2}\right)+1}$$

We arrive at the astonishing fact that

$$\frac{\pi}{8} = \frac{\left(\frac{1}{2}\right)!\left(\frac{1}{2}\right)!}{2}$$

or

$$\frac{\pi}{4} = \left(\frac{1}{2}\right)!\left(\frac{1}{2}\right)!$$

which means:

$$\left(\frac{1}{2}\right)! = \frac{\sqrt{\pi}}{2}$$

While Wallis interpolated this value between 1 and $\frac{1}{6}$ by "unknowingly" constructing values for P and Q, which gave binomial coefficients, we have today at our disposal both the binomial theorem, methods of integration, and the ability to graph the said integral to arrive at value. However, hand computation leads to the surprising discovery that $\left(\frac{1}{2}\right)! = \frac{\sqrt{\pi}}{2}$ which is the value of the gamma function at $\frac{1}{2}$. The pedagogical point is that even today a student would be hard pressed to find a calculator that gives the value of $\left(\frac{1}{2}\right)!$, and in the event a numerical answer is obtained from a calculator, it conceals the fact that it is connected to π.

16.2.2 The Irrationality of π

Our second example involving old-fashioned calculations that can be illuminating to students is establishing irrationality of transcendental numbers like π. Establishing irrationality of π. is relegated to being a difficult problem even in undergraduate mathematics courses because it is assumed that students do not have the mathematical tools necessary to construct a proof. Ivan Niven (1947) started with $\int_0^{\pi} \frac{x^n(\pi-x)^a}{n!} \sin x\, dx$ to give a simple proof by contradiction that π. is irrational. However, irrationality was established much earlier by the polymath Johann Lambert in 1761 by starting with the tangent function, namely $= \frac{x - \frac{x^3}{3!} + \frac{x^5}{5!} - + \cdots}{1 - \frac{x^2}{2!} + \frac{x^4}{4!} - + \cdots}$, and using long division along with the Euclidean Algorithm to calculate a sequence of remainders that could produce a continued fraction for the tangent function.

In other words, if we write out the remainders at each stage of the long division, we get:

$$R_1 = \sin x - x\cos x = \frac{x^3}{3} - \frac{x^5}{2\cdot 3\cdot 5} + \cdots$$

$$R_2 = \left(3 - x^2\right)\sin x - 3x\cos x = \frac{x^5}{3\cdot 5} - \frac{x^7}{2\cdot 3\cdot 5\cdot 7} + \cdots$$

$$R_3 = \left(15 - 6x^2\right)\sin x - \left(15 - x^3\right)\cos x = \frac{x^7}{3\cdot 5\cdot 7} - \frac{x^9}{2\cdot 3\cdot 5\cdot 7\cdot 9} + \cdots$$

Each of these remainders "measures" the tangent function in terms of truncations of an infinite contained function. The representation constructed by Lambert (1768) was $\tan x = \cfrac{x}{1 - \cfrac{x^2}{3 - \cfrac{x^2}{5 - \cfrac{x^2}{\cdots}}}}$ and now using the result that an infinite continued fraction is irrational, we quickly arrive at a contradiction when we assume π is rational and substitute the value of $\frac{\pi}{4}$ for x in the expression. This method of establishing irrationality of π is quite different from Niven's proof which also relies on contradiction. The pedagogical point here is that Lambert's method requires one to perform long division of infinite series, which is rarely done in mathematics courses. Also, hand computation results in some astonishing results that interpolate intermediate values of functions by truncating the series representation of a function. Lambert also played an important role in the field of map projections and write extensively on the topics of conformal projections. Our third and final example shows a fascinating integral that relates to interpolation and conformal map making.

16.2.3 Integrating the Secant Function

An interesting application of interpolation involves integrating the secant function using the series representation of $\log(1+x)$. A practical application for this calculation resulted in the need to construct navigational charts. Deriving the integral of the secant function is often a difficult task for calculus students. It involves making a trigonometric substitution that seems contrived and logically convenient. However, understanding how this integral was approximated before there was even an understanding of calculus can be illuminating to students and give them a full appreciation for the creativity that can arise through hand calculations and approximations. Again, the notion of domain constraints leads to unexpected results when required to solve a pressing problem, which in this case was conformal maps needed for navigation in the 16th century.

16.2.3.1 Mercator Projection Map

In 1569, the Flemish cartographer Gerhardus Mercator created a map now known as the Mercator Projection Map. This map allowed navigation using lines of constant course also known as rhumb lines. These maps allow a navigator to draw a line between two points on a map, find a bearing, and then follow a compass reading to their destination. This is allowed through the scaling of the space between latitudes on Mercator projection maps. This scaling caused distortion in sizes of landmasses on the map but provided a great advance in navigational abilities. Mercator did not document his method of construction, but it is thought that it was constructed using a compass and straight edge (Carslaw, 1924).

The English mathematician Edward Wright mathematically derived Mercator's projection in 1599 by creating a table that provided the scale factor as a function of the latitude. This table allowed for the accurate construction of Mercator projection maps by converting latitudes into distances from the equator. Wright's table was constructed using approximate sums, what would now be known as Riemann sums. He constructed his table with an interval of one minute of arc, or $\frac{1^\circ}{60}$ for all latitudes to 75°. His table was later found to be an accurate table of the integral of secants. The creation of his table is even more astounding given the lack of understanding of logarithms or calculus that existed in his time.

16.2.4 Derivation of Wright's Table

Wright realized that in order to preserve angles on the Mercator projection, i.e., to keep conformality, the vertical and horizontal direction on the map needs to be stretched by the same factor. This allowed the meridians to be parallel and intersect the equator at a right angle. By simultaneously scaling in the vertical direction, the Mercator map can be constructed to allow the constant course navigation that so greatly aided sailors of the time, see Fig. 16.3.

Wright recognized that by choosing a common interval he could determine the position of a line of latitude on the Mercator projection by summing the distance of all of the lines separated by the previous intervals. The smaller the interval that is used, the more accurate is the approximation. In order to build his table, Wright used a trigonometric table of secants available during his time. For example, if we want to find the distance from the equator to the 30th line of latitude with an interval of 5°, we would use the following table of secants, see Table 16.2.

Taking the total of these secants multiplied by the interval results in $15.6163 * 5° = 78.0815°$. The 60th line of latitude should be placed on the Mercator projection map at 78.0815°.

16.2.4.1 Modern Derivation

Wright determined an approximation of the integral of the secant function using numerical summation with an interval of one minute ($1'$) or $\frac{1°}{16}$ for all lines of latitudes up to 75°. This results in $16 * 75 = 1200$ entries. These calculations would be very tedious without the use of calculus (or modern technology). Wright's method is not exact, but provides a very reasonable approximation. The method

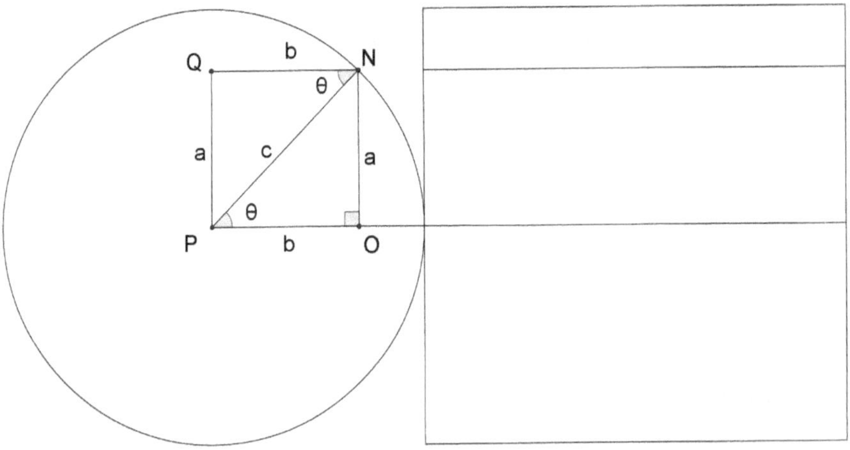

Fig. 16.3 $f(\theta) = \sec(\theta)$

Table 16.2 Table of secants approximated at an interval of 5°

Table of secants	
Secants 5°	1.0038
Secants 10°	1.0154
Secants 15°	1.0353
Secants 20°	1.0642
Secants 25°	1.1034
Secants 30°	1.1547
Secants 35°	1.2208
Secants 40°	1.3054
Secants 45°	1.4142
Secants 50°	1.5557
Secants 55°	1.7434
Secants 60°	2.0000
Total	15.6163

could be improved by looking at even smaller interval widths, though again this would require a very tedious number of calculations. The real improvement comes with the development of the logarithmic function and calculus. The following modern proof demonstrates the closed form of the integral of the secant.

$$\begin{aligned}
\int \sec\theta d\theta &= \int \frac{1}{\cos\theta} d\theta \\
&= \int \frac{\cos\theta}{\cos^2\theta} d\theta \\
&= \int \frac{\cos\theta}{1-\sin^2\theta} d\theta \\
&= \int \frac{\cos\theta}{(1-\sin\theta)(1+\sin\theta)} d\theta \\
&= \int \frac{1}{2}\left(\frac{\cos\theta}{1-\sin\theta} + \frac{\cos\theta}{1+\sin\theta}\right) d\theta \\
&= \frac{1}{2}\int \frac{\cos\theta}{1-\sin\theta} + \frac{\cos\theta}{1+\sin\theta} d\theta \\
&= \frac{1}{2}[-\ln|1-\sin\theta| + \ln|1+\sin\theta|] + c \\
&= \frac{1}{2}\ln\left|\frac{\sin\theta+1}{\sin\theta-1}\right| + c \\
&= \frac{1}{2}\ln\left|\frac{(1+\sin\theta)^2}{(\cos\theta)^2}\right| + c \\
&= \frac{1}{2}\ln\left|\frac{1+\sin\theta}{\cos\theta}\right| + c \\
&= \ln|\sec\theta + \tan\theta|.
\end{aligned}$$

This closed form of the integral of the secant function is assumes a radian measure. If θ is measured in degrees the result would be the following equation:

$$\int_0^\theta \sec\theta d\theta = \frac{180}{\pi}\ln|\sec\theta + \tan\theta|$$

Performing this calculation for 60° results in a value of $\frac{180}{\pi}\ln|\sec 60 + \tan 60| = 75.4651°$. Using Wright's method of approximation with a large interval of 5° results in a value that is only off by 2.6164°. Using an Excel spreadsheet to calculate Wright's approximation at an interval of $1'$ would result in a value equal to 75.4874°, a difference of only 0.0223°. An approximation of this accuracy likely approaches the limits of the measurement and mapmaking tools of Wright's time. A tool such Wolfram-Alpha can readily provide the value of the secant function, but insight into the meaning and origin of this calculation is lost. There is a beauty and elegance in Wright's approximation that students have to experience to appreciate. Another point to note is that many Calculus textbooks evaluate the integral by using the trick of multiplying and dividing the given function by $\sec(\theta) + \tan(\theta)$ to convert the integral $\int \sec\theta d\theta = \int \frac{1}{u} du$ where u = $\sec(\theta) + \tan(\theta)$.

Such a trick though resulting in the correct answer is devoid of any mathematical insight whatsoever and as far removed from Wright's interpolation as the secant of a right angle!

16.3 Concluding Remarks

The three examples provided in this chapter illustrate the complementary nature of hand calculations to the affordances provided by the digital age. As argued by the editors of this book, the connection between technology and creativity in mathematics education is still unexplored territory. If the past serves as a reminder, it is important to remember that graphing calculators and computer algebra systems did not really change the nature of college Calculus much other than the fact that instructors needed to actually think about the type of questions that could be asked of students if the hand held utility more or less did everything that was taught in a traditional course. Unlike the resistance to technology in the 1990's, present day learner's experience in any classroom is ubiquitous with the use of ICT. The advent of big data as the next frontier for computing to reveal patterns and trends relating to human behavior requires "interpolation" to fill in incomplete data sets no different from the gaps in information encountered by mathematicians in the past.

Learning environments that utilize the wealth of information provided by mathematical repositories in addition to computing tools like Mathematica and Wolfram Alpha provide a much richer experience than the classrooms of yore.

However, simply using multimodal software to show multiple representations does not necessarily mean that it results in any deeper understanding or insight unless the complementary and "beautiful" nature of hand calculations are also incorporated into the environment. Paper and pencil mathematics has long been the shibboleth of mainstream mathematics[1] and is unlikely to change even with the advent of the digital age. One may think of such orthodoxy as simply the motivation for teachers to show the ingenuity and beauty of hand calculations with historical examples such as those described in this chapter. These examples also illustrate the creative nature of mathematical interpolation to complement other modes of representation and serve as useful lessons to those that think "big data" is a recent product of the information age!

References

Carslaw, H. S. (1924). The story of Mercator's map. *The Mathematical Gazette, 12*(168), 1–7.

Dutka, J. (1990). The early history of the factorial function. *Archive for History of Exact Sciences, 41,* 225–249.

Haught, C., & Stokes, P. (2017). Constraints, competency and creativity in the class room. In R. Leikin & B. Sriraman (Eds.), *Creativity and giftedness: Interdisciplinary perspectives from mathematics and beyond* (pp. 105–116). Dodrecht, The Netherlands: Springer.

Lambert, J. H. (1768). Mémoire sur quelques propriétés remarquables des quantités transcendantes circulaires et logarithmiques. In L. Berggren, J. M. Borwein, & P. B. Borwein (Eds.), *Pi, a source book* (3rd ed., pp.129–140). New York, NY: Springer.

Niven, I. (1947). A simple proof π is irrational. *Bulletin of the American Mathematical Society, 53*(6), 509.

Osborn, A. (1953). *Applied imagination: Principles and procedures of creative problem solving.* New York, NY: Charles Scribner's Sons.

Stedall, J. A. (2000). Catching Proteus: The collaborations of Wallis and Brouncker I squaring the circle. *Notes and Records of the Royal Society of London, 54,* 293–316.

Torrance, E. P. (1974). *Norms-technical manual: Torrance tests of creative thinking.* Lexington, MA: Ginn and Company.

Wallis, J. (2004). *The arithmetic of infinitesimals* (J. A. Stedall, Trans). New York, NY: Springer (from Latin).

[1]Personal communication with Reuben Hersh, October 8, 2015.

Chapter 17
Ancient School Without Walls: Collective Creativity in the Mathematics Village

Elçin Emre-Akdoğan and Gönül Yazgan-Sağ

Abstract This study is designed as a qualitative research in order to examine (i) how the Mathematics Village promotes mathematical creativity and (ii) the transformation of Mathematics Village from a non-virtual environment to Social Media, which is a virtual environment. Our data collection tools include individual interviews with two mathematicians, who teach at the Mathematics Village as well as focus group interviews with seven high school, undergraduate, and graduate students and classroom observations. We have analyzed the collected data via content analysis. The findings of this study reveal that the Mathematics Village promotes mathematical creativity of students and enables mathematicians to activate their own creativity. From that perspective, having an educational setting that provides freedom can positively affect students' state of mind and creativity. Therefore, it is of importance to transfer basic characteristics of a non-virtual environment (Mathematics Village) into a virtual environment (Social Media), which brings people together with the aim of doing mathematics.

Keywords Mathematical creativity · Promoting creativity · High school students Undergraduate and graduate students · Mathematicians · Social media

17.1 Introduction

Social interaction is a characteristic of mathematical creativity (Sriraman, 2004). In addition to social interaction, the learning environment has the potential to activate creativity (Amabile, 1996; Csikszentmihalyi, 2000). Social media, which could serve as such a virtual environment, is defined as various networked tools or technologies to communicate, collaborate, and creatively express such as Twitter, Facebook, Blogs, YouTube, etc. (Dabbagh & Reo, 2011). Social media also makes it possible to have informal conversations and share ideas (Ito et al., 2009;

E. Emre-Akdoğan (✉) · G. Yazgan-Sağ
Ankara, Turkey
e-mail: elcinemre@gazi.edu.tr

© Springer International Publishing AG, part of Springer Nature 2018

V. Freiman and J. L. Tassell (eds.), *Creativity and Technology in Mathematics Education*, Mathematics Education in the Digital Era 10,
https://doi.org/10.1007/978-3-319-72381-5_17

McLoughlin & Lee, 2011); also enables learning to start in an informal setting and allows for independent thinking. Social media has an impact on individuals' learning both at a personal and a social level. Studies have presented that students' social media usage for learning mathematics enables them to gain significant perspectives (Baya'a & Daher, 2013). In the literature, there is a classification for social media usage consisting of three levels: Level (1) Personal information management, Level (2) Basic interaction or sharing, and Level (3) Social networking (Dabbagh & Reo, 2011). The personal information management level focuses on collecting information through social media and organizing them in order to increase productivity. Basic interaction or sharing level comprises of communication, social interaction, and collaboration via social media. For example, students could make a comment on a blog or create a collaborative workspace using a wiki (Dabbagh & Kitsantas, 2012). Social networking denotes social interaction at the highest level that contains aggregating and synthesizing information from level (1) and level (2) (Dabbagh & Reo, 2011).

In contrast to traditional learning contexts, learning environments with social media goes beyond the walls and does not need to occur real time. Thereby, in contrast to the structured technology-rich learning environments, the social media learning environments, which enable students to be creative, give students the freedom of thinking (Lu, Hao, & Jing, 2016). The existence of social media caused recent literature to reconceptualize creativity from a social perspective; researchers now address creativity from a collaborative stance, rather than from a solely individual perspective (Peppler, 2013). Furthermore, the virtual environment of social media represents a means of interaction and can play a significant role in proposing and spreading new ideas among people (Peppler & Solomou, 2011). Social media provides participants with equal opportunity to share ideas even when they are not in the same physical environments. Thus, the virtual environment of social media offers a democratic and moderate environment, which is a key aspect for creativity in both teachers and students (Levenson, 2011).

According to Cropley (2001), environments that facilitate the creation of novelty and play significant roles in the development of creativity share a set of properties, including "openness, positive attitude to novelty, acceptance of personal differentness, and willingness to reward divergence (p. 67)." In-group experts, colleagues, and/or friends can promote creativity in a social environment characterized by the creation of novelty (Cropley, 2001). Csikszentmihalyi (1988) and Amabile (1983, 1996) provides substantial insight to investigate creativity from a social perspective (Plucker, Beghetto, & Dow, 2004; Meusburger, 2009). Amabile (1988) defines creativity as "the production of novel and useful ideas by an individual or small group of individuals working together" (p. 126). Zaman, Ananda rajan, and Dai (2010) summarizes Amabile's (1996) definition of creativity as "interaction between individuals, environment, and socio-cultural context". Csikszentmihalyi (1996) states that creativity does not occur in "individuals' head", it is rather an interaction between the individual and the context. So, it can be concluded that environment plays a significant role in creativity, from this stance, Csikszentmihalyi (1988) proposes the very fundamental question regarding creativity: "where is creativity" and not "what is creativity" (Fleith, 2000). Csikszentmihalyi (1988) figures out the creativity as a system, which can be explained as follows:

> What we call creative is never the result of individual action alone; it is the product of three main shaping forces: a set of social institutions, or *field*, that selects from the variations produces by individuals those that are worth preserving; a stable cultural *domain* that will preserve and transmit the selected new ideas or forms to the following generations; and finally the *individual*, who brings about some change in the domain, a change the field, will consider to be creative…. Creativity is a phenomen that results from interaction between these three systems (Csikszentmihalyi, 1988, pp. 325–326).

Thus, we can not study creativity by solely focusing on individuals, since their work should not be separated from their environment (Csikszentmihalyi, 1988).

Evaluating creativity and social media together could give us important perspectives (Peppler, 2013), because as an environment social media has the potential to reveal creativity (Peppler & Solomou, 2011). Previous studies support the fact that the environment has an impact on students' ability to foster their own mathematical creativity (Levenson, 2011). The Vittra Schools which is an example of non-virtual environment in Sweden claim to promote students' creativity and curiosity through the school environments they create, based on their principle of "school without classroom" ("Personalize learning", 2017). In Vittra schools, students are expected to discover the approach that is most suitable for them, to develop an understanding of how they learn, to learn by experience, to develop confidence in themselves and their abilities, to improve their communication and interaction skills, and to obtain what they require in order to work or study on an international platform ("Personalize learning", 2017).

Within this context, we have dealt with the role of environment plays in mathematical creativity through focusing on the Mathematics Village, which has existed in a natural environment in Turkey for ten years. The Village aims to bring together both novices (students at high school, undergraduate, and graduate level) as well as experts (mathematicians and mathematics teachers) to do mathematics, to discuss on mathematics topics, to have lectures, and to talk about mathematics without having any obligations. The Village does not directly aim at constructing a creative context for doing mathematics. The nature of the Village possesses a moderate and democratic environment. More specifically, we have been planning to analyze the context of the Mathematics Village, where democratic environment and free thought is promised. Similarly, social media also enables students the freedom of thinking in a collaborative way (Lu et al., 2016; Peppler, 2013; Zaman et al., 2010). Teenagers and the twenty-something age group as students continually connected to the social media for gathering information, interaction or sharing, and networking (Dabbagh & Reo, 2011). As researchers, it is crucial to provide students with social media platforms as a learning environment which is an indispensable part of current generations' milieu and the reality of our current world. Thus, it is worth to transfer basic characteristics of non-virtual environment (Mathematics Village) that brings people together with the aim of doing mathematics on the virtual environment (Social Media). In this regard, our aim is to examine (i) how the Mathematics Village promotes mathematical creativity and (ii) the transfer of

Mathematics Village that is a non-virtual environment to Social Media, which is a virtual environment.

17.2 Methodology

17.2.1 Context of the Study

The Mathematics Village is a unique institution in Turkey that offers short but intense courses, and at which high school and college students are taught by professors and engage in seminar discussions (Alladi & Rino Nesin, 2015). It is located in Şirince, a small village of İzmir in the western part of Turkey that is surrounded by olive trees and is isolated from the downtown. In the Village, you can hear the sound of crickets day and night (Fig. 17.1).

In their study, in which they introduce the Mathematics Village, Alladi and Rino Nesin (2015) explained the duties of students at the Village, as follows: The Village mostly operates as a commune, since there are few paid personnel. High school and university students who arrive at the Village are divided into small groups of approximately the same size. For the next two weeks in the Village, the students in these groups do chores, which must be completed in order for the Village to function, such as peeling potatoes for the cook, taking out trash, and replenishing water coolers. Surprisingly, the majority of students genuinely believe they are contributing to Village life, and do not complain about their assigned chores. Moreover, these responsibilities provide them with a sense of ownership and community that endures for years.

Teaching is voluntary and the Village provides meals and accommodations. To ensure that the academic schedule matches the length of stay of the students, the courses are all only two weeks long.

The Village resembles an ancient historical school where, no TV or broadcast music is available but movies are sometimes displayed on the library's projection

Fig. 17.1 View from Nesin Mathematics Village (from http://nesinkoyleri.org/eng/)

Fig. 17.2 View of the library (from http://nesinkoyleri.org/eng/)

screen (Fig. 17.2). Students often play their musical instruments at night, such as guitars or the traditional Turkish musical instruments—the saz and kemençe—which provide a musical ambience. In order not to disturb those who are working, these mini concerts are usually given on Wednesday evenings, since everyone is given the day off on Thursdays, when students engage in extracurricular activities (Alladi & Rino Nesin, 2015).

17.2.2 Teaching and Learning in the Mathematics Village

At the Village, lessons are conducted in outdoor lecture theaters under the olive trees (Fig. 17.3). Ali Nesin, the founder of the Village, said: "One of the aims of the Mathematics Village is making the students feel the presence of an incredibly beautiful world besides mathematics by showing them the mathematics they had probably never seen before" ("Nesin Mathematics Village", 2017). It is assumed that the context of the village, which enables students to have freedom and to

Fig. 17.3 Lecture in an open amphitheater (from http://nesinkoyleri.org/eng/)

participate as they wish, will increase their curiosity about mathematics and promote their creativity.

Another key principle in the Village is the importance of ensuring the freedom of the students' ideas. Within that context, another Village norm is "mathematicians without borders" (Alladi & Rino Nesin, 2015), which is also viewed as one of the contextual features by which creativity is activated. For example, a mathematician who visited and stayed at the Mathematics Village for a while explained his experience as follows: "[…] my supervisor and I sat down to work on a problem that had been eluding us for a long time at 9 a.m. after a nice breakfast. Three hours later, we had the basis of a paper […]" (Alladi & Rino Nesin, 2015, p. 657). In addition to that, there are other articles, certain parts of which were completed in the Mathematics Village (i.e., Ayık, Ayık, Bugay, & Kelekci, 2013; Göral & Sertbaş, 2017). The researchers thanked the Mathematics Village for the support and warm hospitality they had experienced during their stay. These cases could set an example as to how the Village's environment encourages researchers to be more productive.

Voluntary lecturers, who are employed at national and international universities, teach the high school, undergraduate, and graduate level courses. The Village offers students the opportunity to meet mathematicians from universities around the world. Students are informed of the various programs through updates shared on the Village's social media pages and website. Programs last for two weeks and their schedule is organized according to lecturers' availability in the Village. Courses are generally in Turkish but some courses could be instructed in English based on the audience. For example, there are 14 courses organized for the students in 2016 undergraduate and graduate summer school from July 18th to September 25th ("Nesin Mathematics Village", 2017). The courses such as Inequalities, Calculus, and Basics of Analysis are open for high school students. Other courses such as Introduction to Group Theory, Introduction to Graph Theory,—Introduction to Social Choice Theory, Affine and Projective Geometry are instructed for beginners (i.e. 1st, 2nd and 3rd year undergraduate students), advanced undergraduate (i.e. 3rd and 4th year undergraduate students), and graduate students (i.e. master and Ph.D. students, researchers).

We have summarized a course named "Basics of Analysis" as a result of our observations that lasted for four hours at the Village. The content of the course contains axioms for the real numbers, Archimedean property, real numbers, natural numbers, whole numbers, rational numbers, induction, ordered filed, Cauchy sequences, non-standard numbers. The lecturer informed the audience that infinity would be discussed in the course which took approximately 2 h. At the beginning of the course, the lecturer stated "Hilbert Hotel" problem and associated it with the concepts of limit and infinity. After introducing the problem to the students by saying: "There is a hotel named Hilbert Hotel. The hotel has an infinite number of rooms, each room has a number such as 1, 2, 3, 4, … Then a bus with an infinite number of passenger stops by the hotel, the passengers can be numbered as 1, 2, 3, 4, … The hotel gives the first room to the first customer, the second room to the second customer and so on. What do we do if the next day another customer wants to stay at the hotel?" Then the lecturer changed the given numbers of hotel rooms and customers as: (i) the number of the rooms as (0,1) interval and the number of customers as (0,1) interval, (ii) the number of the rooms as (0,1) interval and the number of customers as [0,1] interval, (iii) the number of the rooms as (0,1) and the number of customers as $(3,\infty)$ interval, (iv) the number of the rooms as (0,1) and the number of customers as $(0,1) \times (0,1)$ open squares, (v) the number of the rooms as (0,1) and the number of customers as $\mathbb{R}$ real numbers. At the end of the course, he considered the number of the rooms as $\mathbb{N}$ and the number of customers as $\mathbb{R}$ real numbers, then he explained why it would not work in this situation and added that $\mathbb{N}$ and $\mathbb{R}$ has two different kinds of infinity. He also talked about George Cantor, who was the first person to introduce this approach. We observed that the lecturer took every idea of the students as a new way to solve the problem. He waited until a student proposed an idea to the solution, which took approximately 7 min for every situation. He also thought on and discussed every idea with other students and spent approximately 12 min to further develop each idea. Then he demonstrated how some of them could and/or could not be useful for solution. He did not push to get an answer or idea out of the students during the lesson. Instead, he patiently waited until the class came up with an idea. This observation has provided us with insight into the context of the courses instructed in the Mathematics Village. Alladi and Rino Nesin (2015) also characterize the instructional approaches used at the Mathematics Village by comparing them to the approaches applied in conventional Turkish schools, as follows:

> Ordinary high school education in Turkey is geared toward the university entrance examinations. As a result, the emphasis is on memorization, mindless competition, and end result rather than thought processes. The Mathematics Village aims to counteract this by giving high school students a glimpse of what university-level mathematics is. They learn to think for themselves, argue coherently, and spot logical fallacies. Most importantly, they see the process of solving a problem whose solution is not known beforehand. The university-level teaching is organized around themes if the lecturers' time permits, allowing not only for concentration of research interests but also preparing the ground for cooperation between colleagues in the same or related fields (p. 655).

As stated above, the instructional scope of the Mathematics Village coincides with its creative activities, required tasks, and the implications for the classroom, including thinking in new ways and making coherent arguments. For instance, one of the high school students we interviewed in the Village says that "the lecturer teaches the course by playing cards as if it is a game. Then we understood what we had to do." Below is one of the examples that is related to the probability concept mentioned by a student (Nesin, 2008).

> This is how the game goes: You are one of the players. There are three playing cards. One of these cards has blue sides, let's name this card as BB card. The other has two red sides, let's name this card as RR card. The last one has both blue and red sides, let's name this card as BR card. One of the players shuffles the cards without skunking, then randomly picks up one of the cards and puts it on the table. You only see the color of the upside, not the color of the downside of the card. You have to guess the color of the card's down side. If you guess the right color, that means you win. Suppose that the up side of the card is blue, it means that the card could not be the RR card. It is either the BB card or the BR card. Therefore the other side of the card would be either red or blue. The probability is 50

percent for both colors. The chance of winning is 50 percent. On the other side the card on the table will be BB or RR with a probability of 2/3. So, if you confuse the downside color with the upside color, your chances of winning is 2/3. Thus, if a card's upside is blue and you assume that its downside is also blue, your chances of winning is 2/3, not 1/2 (50%). Which of the calculations above is correct? Why is the other wrong?" (p. 141–142).

Another example mentioned by the students during the interviews is related to "sum of the angles in a triangle". In the interviews, a high school student explained one of their lectures' statements as: "The sum of the angles in a triangle does not always equal to 180°." The lecturer gave the students time to think and discuss on this statement in the course. Then, the students have found out that the statement was correct when the triangle is on spherical surface. The students consider this such courses as surprising and inspirational journeys for themselves.

17.2.3 Participants and Data Collection

This study has been designed as a qualitative research to examine mathematical creativity in the context of the Mathematics Village and the transformation of the environment of Mathematics Village from a non-virtual environment to the Social Media, which is a virtual environment. In addition, we explored whether the Mathematics Village, which can be considered to be an alternative learning environment, promotes the mathematical creativity of high school, undergraduate, and graduate students, as well as professional mathematicians. We attended one of the Mathematics Village summer schools as participant observers. The data collection tools included individual interviews with two mathematicians, Deniz and Derya (pseudonyms), who teach at the Mathematics Village. We conducted focus group interviews with four high school students—Alp, Cem, Aylin, and Nazlı (pseudonyms)—and three undergraduate and graduate students—Özge, Metin, and Emre (pseudonyms). We have also conducted classroom observations that lasted for 4 h

Fig. 17.4 Setting for interviews with students (from http://nesinkoyleri.org/eng/)

in the "Basics of Analysis" course. The focus group interviews with students were conducted in a natural setting at the Mathematics Village, as shown in Fig. 17.4.

We asked the students five open-ended questions, and asked the mathematicians four open-ended questions, as listed in Table 17.1. We aimed to explore reflections of the Mathematics Village's environment through the perspective of students and mathematicians. The interview questions led us to examine social interaction as a characteristic of mathematical creativity in the Village (Amabile, 1996; Csikszentmihalyi, 2000; Sriraman, 2004). We also asked the participants whether they used any digital technology (i.e. social media) in the Village. Focus group interviews with students lasted for approximately 45 min, and individual interviews with mathematicians lasted for approximately 40 min. Prompts were employed as necessary during the interviews to encourage the interviewees to share more about their ideas. All the focus and individual interviews were videotaped.

The data retrieved from the interviews were coded and categorized with respect to the experiences of the students and mathematicians (Patton, 2002). We analyzed the data by focusing on the environmental aspects of creativity from a social perspective (Amabile, 1996; Csikszentmihalyi, 2000) and by concentrating on the Village's properties of "openness, positive attitude to novelty, acceptance of personal differentness, and willingness to reward divergence" (Cropley, 2001, p. 67). Afterwards, we have discussed the transformation of the creativity culture, which emerged from the Mathematics Village, into the social media platform as a learning environment (Dabbagh & Reo, 2011).

Table 17.1 Interview questions

Interview questions for students	*Interview questions for mathematicians*
Questions on Mathematics Village 1. Could you compare the lectures you attend in the Mathematics Village with those you attend at your school/university? What are their differences and similarities? 2. What are your opinions about the Mathematics Village? Could you tell us about your experiences at the Mathematics Village? 3. Have there been any changes in your point of view regarding mathematics after attending lectures at the Mathematics Village? If so, could you explain them in detail?	*Questions on Mathematics Village* 1. Could you compare the lectures you give at the Mathematics Village with those you give at your school/university? What are their differences and similarities? 2. What are your opinions about the Mathematics Village? Could you tell us about your experiences in the Mathematics Village?
Questions on Digital Technology (Social Media) 1. How did you find out about the Mathematics Village? 2. Do you follow the Mathematics Village's social media accounts?	*Questions on Digital Technology (Social Media)* 1. How did you find out about the Mathematics Village? 2. Do you follow the Mathematics Village's social media accounts?

17.3 Findings

In this study, we have examined the experiences of the students and mathematicians within the context of the Mathematics Village, in order to explore how the Mathematics Village promotes mathematical creativity. For that purpose, we created two main categories: the experiences of the students and the experiences of the mathematicians. The sections below discuss these categories and their sub-categories, and Table 17.2 lists the various responses given by students and mathematicians.

17.3.1 *Experiences of Students in Mathematics Village*

17.3.1.1 Different Approaches on Instruction

One of the sub-categories that emerged from the student experiences category for the Mathematics Village is the different approaches to instruction. Since mathematical instruction in Turkish schools is often rule-based, the students defined the approaches they experienced at the Mathematics Village as *different.* For example, Bora stated: "education in here [mathematics village] is different from that of the school" and he further explained: "For example, they teach us about a mathematical rule in the school and explain it as this is it! However, here they clarify the reason behind the mathematical rule. That is really good and increases my curiosity on mathematics." Similarly, Cem explained how the instruction differs from his school experience as follows:

> We start an abstract object here and add on this abstract object for a while. And then, we realize that we have spent two hours proving this rule, however, the teacher just tell us to 'memorize this mathematical rule' in our school. It is incredible to learn where mathematical rules emerge from.

Cem also provided an example for his above statement: "For example; they did not introduce us to the summation symbol "n(n + 1)/2" in the beginning of the course, we saw it at the end of the course after conducting long algebraic

Table 17.2 Experiences of Students and Mathematicians at the Mathematics Village

Experiences of Students	*Experiences of Mathematicians*
Different approaches on instruction	Increasing productivity
Activating Curiosity	Activating Collective creativity
Having more time to think	Having no pressure
Being in a natural environment	
Realizing the nature of mathematics	
Living in the same habitat with mathematicians	

operations, and then lecturer applauds!" Moreover, Aylin shared that at school "they have difficulty with understanding mathematics" and she explains: "I do not know the reason behind the mathematical arguments." Aylin further explained her thoughts within the context of polynomials:

> In high school we learned polynoms but here [at Mathematics Village] it is really different, first lecturer starts to teach and you ask yourself 'what is it?' You try to understand and at the end of the course you realize they are polynoms. In high school we learned the concept of polynoms in a 40-minute course, but here we learned it an in two-hour course. It surprises us a lot because we are used to just having the formula and solving problems according to it.

Students state that in their school they learn formulas and mathematical statements at the beginning of their courses and then must solve problems based on these formulas. However, when they attend the Mathematics Village, students realize that they have difficulty understanding the meaning behind the formulas, since they were never given the chance to question the reason why mathematical formulas and statements were developed. In contrast to their school, they are not introduced to formulas or mathematical statements at the beginning of the course at the Mathematics Village. Learning the formulas and mathematical statements happens at the end of the course and being taught the rationale behind them is considered to provide students with a unique instructional approach, which can also excite them.

17.3.1.2 Activating Curiosity

A second sub-category in the category of experiences of Mathematics Village students is activation of curiosity. Students mentioned that being in the Mathematics Village and learning different perspectives helped to foster their interest in mathematics. For example, Aylin explained why the mathematics courses that she took at the Mathematics Village promoted her curiosity, as follows:

> Since elementary school we have been used to memorizing mathematical rules, they gave us rules and then we solved the problems by using those rules and applying them to the numbers given in the question. However, when I came here, my perspective has widened and I have started to think about why this rule is like that and why it has been discovered. This gives you curiosity. Following this curiosity you start to make an effort for mathematics even if you could not previously succeed at it.

Similarly, Bora mentioned that the different instructional approaches increased his "interest in mathematics a lot, which is really good." To summarize, students argued that the Mathematics Village's instructional approaches that differ so much from those they experience in their schools have helped them to be aware of the rationales behind mathematical formulas and rules, which has activated their curiosity.

17.3.1.3 Having More Time to Think

A third sub-category in the category of experiences of Mathematics Village students is having more time to think. Students indicated that having a different routine from the ones they have at school provides them with more time to think. For example, Cem explained how the environment of the Mathematics Village affected his thinking as follows:

> When you are here, you are in the world of this village, far away from the external world. Here, in this little world, you have more potential to think for example when something just falls down, you can discuss about it, like if you feel the wind […] you can get in the detail about the issue.

Cem perceived the Mathematics Village as an isolated environment, which helped him focus on specific topics that may or may not be related to mathematics. Nazlı shared her ideas by comparing school life to life at the Mathematics Village, as follows:

> We have more time to think here. In school, you have 40-minute lessons and 10-minute breaks, during which you try to relax; then go home and do your homework. However, our purpose of being here is to think, so we have more time to think.

According to Nazlı, thinking is the fundamental objective of being at the Village. She identified the routines of her life as a cycle of things that must be done and which do not involve deep thinking. Aylin explained how she spends her time at the Mathematics Village and how the environment affects her thinking, as follows:

> Here you can think more because of the environment; you can just sit and think. For example, I write essays when I am alone here. You can get into your inner world here. The city where I live is very crowded and it is not possible to have time for yourself. After school we get back home and do our homework, I do not like school because it makes me exhausted, which does not allow me to have time for myself. But here, it is not like that, I always have time to think about mathematics or life or different issues.

Comparing their routine here to their days in school, the students mentioned that at the Mathematics Village they have time to think, study, or focus on subjects that have attracted their interest. In addition, the students emphasized the positive impact of having time to think.

17.3.1.4 Being in a Natural Environment

A fourth sub-category in the category of experiences of Mathematics Village students is being in a natural environment. Students stated that living in an environment that differs from their home and school environments had a positive influence. For example, Nazlı explained how her school environment differs from that of the Mathematics Village: "We have neatly aligned traditional school desks in a concrete building and 40-min courses. Here [in the village] during the course a dog/cat can come and sit near us. After school when you come here you feel different." Aylin shared her ideas and feelings of when she was in Mathematics Village by saying:

> We grew up in the concrete jungle, the issues that we think here [Mathematics Village] is different from there [their daily routine]. Because the environment has a big influence on the amount of time we have to think. When you are in a different environment, you feel different; you force yourself to be creative here. When you are in a concrete jungle, you see the same things every day and it does not promote you to think of different issues.

Aylin defines in a striking way the environment she lives in as a "concrete jungle." This definition has led her to compare these two environments [Mathematics Village and her town] to determine whether they nurture her creativity or not. Cem explained "I am saying this literally, not joking at all, here while you are walking you can find a new formula that is what kind of an environment this place has." Özge expressed her ideas about living in the Mathematics Village, as follows: "Here you realize that you need to work harder and this motive you a lot. Even during lunch or dinner you talk about mathematics. You can ask each other question while having dinner. Here is the environment of concentration."

As described above, students discussed how their school environment differs from that of the Mathematics Village with respect to the physical features. Along with the physical features of the Mathematics Village, seeing other people studying mathematics also positively affected their creativity, productivity, and motivation.

17.3.1.5 Realizing the Nature of Mathematics

A fifth sub-category in the category of experiences of Mathematics Village students is realizing the nature of mathematics. Students stated that realizing how and why mathematics has developed has positively influenced them and promoted their curiosity. For example, Aylin mentioned that most people ask questions such as "What is the purpose of learning mathematics? Why do we learn it?" She asserts that, thanks to the Mathematics Village, she now has the answers to these questions, which are as follows:

> For example, when we have asked the lecturers [in the Village] why we learn mathematics, they have answered our question by explaining it in detail; moreover their instruction is also based on the roots of mathematics. Because of that mathematics now seems more useful to me.

When Aylin realized why she was being taught mathematics, her perspective on mathematics also changed: "Before visiting here, mathematics was just a course that I needed to pass for me, but now I view it differently. If there are so many people working in that field, there should be something interesting about mathematics." Furthermore, Bora made the following statement: "For example, we learn a formula or a new subject here, we realize that this subject has emerged from a need or a curiosity."

Being in the Mathematics Village has promoted students to think at the meta-level about what mathematics is and how it has emerged. Thinking in such a manner has also raised the awareness of students regarding the nature of mathematics. Moreover, it is significant that students have engaged in reflective thinking on the need for and the benefits of mathematics.

17.3.1.6 Living in the Same Habitat with Mathematicians

A sixth sub-category in the category of experiences of Mathematics Village students is living in the same habitat with mathematicians. Due to the structural design of the Mathematics Village, the mathematicians and students live in the same habitat, and students recognize the benefits of doing so. For example, Nazlı explained the advantage of being in the same environment with the lecturer in this way: "Here we have chance to ask questions whenever we want to. Lecturers also gave us the opportunity to ask them questions during the day." Bora discussed the positive effects of seeing the people who work in the field of mathematics, as follows:

> Everywhere you go here, there are people working on mathematics, you can ask any question you want. Everywhere there are mathematicians and mathematics teachers. There are lots of books about mathematics, these all have a positive effect in terms of learning and improving yourself in mathematics.

With respect to the opportunities presented to him by the Mathematics Village, Bora valued the freedom to ask questions anytime, being able to discuss mathematics with anyone who was around, and having access to all sorts of resources for mathematics. Cem explained his experience of living in the same environment with mathematicians, as follows:

> When I was in the mathematics village, an English professor was proving a theorem and it got everyone excited in the village. The Professor was explaining in English, even though I could not understand anything, it was still very exciting. In the library there was no noise except for the sound of chalk and the professor, it was really very fascinating.

Cem stated that witnessing a mathematician prove a theory was an exciting and fascinating process. Besides, Cem also mentioned that even though he was unable to understand the proof process, he was impressed by the environment he was in. Metin, one of the graduate students, argued the benefits of meeting scholars, who have different perspectives and who come from different countries, by saying:

> The main advantage of being here is having the opportunity to network with students and mathematicians coming from not only different parts of the country but also from different countries throughout the world. Secondly, you have the advantage of receiving seminar and courses from mathematicians, who are from other universities; this enables you to see different domains and perspectives.

The students mentioned that one of the advantages of being in the same habitat with mathematicians is they can easily ask questions and discuss mathematical issues whenever they wish. Also, getting to know students and mathematicians from different universities and schools was another advantage for them, especially in terms of the networking opportunities afforded by this community. Finally, students specified that watching a mathematician while he/she proved a theorem in the classroom was fascinating and very impressive.

17.3.2 Mathematicians' Experiences About Mathematics Village

17.3.2.1 Increasing Productivity

One sub-category in the category of experiences of the mathematicians at the Mathematics Village is increased productivity. For instance, Dr. Deniz explained how the Mathematics Village increases the productivity of mathematicians, as follows:

> Mathematicians who came here were able to write their articles thanks to the mathematics village. One of the remarks made by one of the researchers visiting the village states: "For a long time we were working on a problem with my advisor and while we were in mathematics village, we completed the draft of the article in just three hours". There are mathematicians who just came here for research and there have been 10 articles produced in the mathematics village.

When we asked Dr. Deniz why so many articles have been produced at the Village, she answered: "there are not responsibilities such as cooking, you just have your dinner and go back to work. There is a magnificent library, when you go inside you are impressed and you tell yourself here is the place for studying."

Dr. Deniz stressed the fact that scholars could easily focus on their research due to not having everyday life responsibilities while at the Mathematics Village. She explained how such an environment makes mathematicians more productive, mentioning that some mathematicians, while resident at the Village, have produced up to ten scientific papers.

17.3.2.2 Activating Collective Creativity

Another sub-category in the category of experiences of mathematicians at the Mathematics Village is activating collective creativity. Mathematicians value the connections between the virtual and non-virtual environments, and emphasized the importance of the collective creativity associated with an online platform (virtual). For instance, Dr. Deniz explained her views about online platforms, which activate collective creativity, as follows:

> There is a blog, through which many researchers can get together and prove an important theorem. Hence, the concept of creativity has been changed and network has gained more

significance. A researcher writes a blog, where people can get together and prove significant things. Therefore, the conception of creativity in mathematics has changed. The importance of network has been lately revealed in the theorem of twin prime connections. A researcher has started a blog, discussed proving a theorem, and written explanatory articles about it. Thanks to this method, the articles accumulate under one place, which results in collective creativity. In other words, it is not done by only one person, multiple people write small pieces and it is shaped in time. Thus its results are stronger than those of proven by only one person. This is a new type of creativity, and such environments can trigger it.

Mathematicians have indicated that creativity occurs collectively through the internet and, as such, researchers do not need to physically be in the same place, which accelerates the productivity process. Considering networking within the context of the Mathematics Village, Dr. Deniz stated:

We have evening seminars here. For instance, whenever I lecture a seminar, new questions about my own research raise in my head. That's because the researchers discuss their own subjects or the questions in their head in such a comfortable environment, which takes about an hour. Afterwards, we have the part where questions are asked and answered, which is also quite beneficial. People learn about the research subjects of scholars. Mathematicians get the chance to listen to other mathematicians' lectures since they want to learn from them and to develop a new perspective on their own researches. Thus, it is a fruitful place for creativity.

Within the context of the Mathematics Village, mathematicians highlighted the importance of collectively producing new ideas and describe how collective creativity is realized in online platforms. The productivity obtained from an online platform can also be attained in the seminars organized at the Mathematics Village.

17.3.2.3 Having No Pressure

A third sub-category in the category of the experiences of mathematicians at the Mathematics Village is having no pressure. Mathematicians stated that the absence of any pressure in their environment allows them to do their research more easily and promotes their creativity. For example, Dr. Deniz elaborated:

We are not under pressure here unlike the environment in the university. You have to produce something in university; there is a pressure coming from the administration. But you do mathematics only for yourself here, not because you have to do. When the pressure disappears, the environment becomes more relaxing/comforting.

Dr. Deniz stated that they have more responsibilities at the university, whereas they have no obligations while at the Mathematics Village. Being in such an environment provided her with more freedom to think. Dr. Derya also explained how researchers function when they are under no pressure, by providing a historical example:

Philosophy, mathematics, and art are luxury. Why were mathematics and philosophy discovered in ancient Greece, and not in South Africa or Zambia? Because people in Greece lived in a relaxing environment, weather is nice, economy is good and they were teasing,

> this is teasing. And now, mathematics village provides people with an environment to tease. They do not think about what to eat for dinner, washing clothes or doing the dishes. They just come here and do their research.

Dr. Derya explained the kinds of places where people have been more productive throughout the history and compared the features of these places with those of other places. He explained that the Village provides a kind of environment similar to those historical places such as ancient Greece, where mathematics was deeply discussed and studied.

17.3.3 *Transfer of the Context of Mathematics Village (Non-virtual Environment) to Social Media (Virtual Environment)*

In this section, we primarily discuss the role digital technologies plays in the Mathematics Village. Digital technology usage in Mathematics Village is not explicitly connected to creativity or mathematics. It is rather associated with socializing with other peers and gathering information via the websites of Mathematics Village and social media accounts. Then, we propose a social media platform that could be utilized as a learning environment with the creative characteristics, which we have investigated in the Mathematics Village. The creativity emerging within the Village is not related to the digital technology. Hence, it is worth to transfer the creative environment of the Village into a non-virtual environment in order to ensure the sustainability of creativity in such a unique environment.

17.3.3.1 The Role Digital Technologies Play in the Mathematics Village

Students use the Village's website and social media accounts to gather information about the Village and the courses available before coming to the Village. They also used the Village's website in order to register for the courses given at the Village. Furthermore, after leaving the Village, they used social media accounts to keep in touch and to get together with the other students they met at the Village. While at the Mathematics Village, students do not use the television, computers, or the Internet. Only undergraduate and graduate students use computers and the Internet for the purposes of research. Aylin, who is a high school student, made the following statement about technology usage at the Mathematics Village:

> We do not use computers. Undergraduate students can use computers; and if we have a research to do we can use their computers. We do not have television here, which makes me feel more relaxed. Before I came here, I was watching television quite often. After I came

here I realized that I do not need television. I started to allocate my own time and socialize more.

The students stated that they have mostly used technology to obtain information about the Village from its website and social media accounts. Bora explained this in the following way: "Before coming to the village I was so excited and I was constantly checking Village's website and reading all of the information about it, I have almost memorized everything. I was also checking their social media accounts." Nazlı stated: "I know that instruction of mathematics is different from those in other schools. When I check the website online, I realize that opportunities given in the village are extensive and different from those provided in other schools." Aylin said: "After taking a look at the Village's website, I got motivated and excited about going to the village."

The students also asserted that they visited the Village's website in order to register for courses given at the Mathematics Village. Nazlı claimed that the application process is very competitive, stating "When it is open for registration, you can see the courses immediately get full." Aylin constantly followed the application period online, and added "if registration is open, I should sign up immediately."

Emre stated that international students who have been at the Village "have information about the village through their lecturers or friends on social media." In addition, one of the international students was informed about the Mathematics Village through an online forum. Metin explained: "A student from Netherlands has asked a question in an online forum and another student has answered his question. Afterwards, they have kept in touch and the other student has provided the students from Netherlands with information about village." In addition to information about the students' use of technology, the founder of the Village stated that he uses social media for "making announcements regarding the mathematics village."

Özge reported that students who have attended the Village have social gatherings after they return home: "When we have met each other here [Mathematics Village], we have also become friends via our social media accounts. If we are ever in the same city, we meet in order to see each other again."

We conclude that the role the digital technology plays in the Mathematics Village is neither a part of teaching and learning mathematics nor for promoting creativity. Besides, the structure of digital technology usage in the Village does not bare creative features.

17.3.3.2 Transformation of the Creative Characteristics of the Mathematics Village (Non-virtual Environment) into Social Media (Virtual Environment)

In this section, we summarize the creative characteristics of the Mathematics Village. Then, we discuss how these features of Mathematics Village could be transformed into a Social Media platform that can be utilized as a learning environment.

We have explored experiences of the students and mathematicians within the context of the Mathematics Village, in order to find out how the Mathematics Village promotes mathematical creativity. According to the experiences of students and mathematicians, we identified that Mathematics Village provide participants with an environment that has the following characteristics: different approaches on instruction, activating curiosity, having more time to think, realizing the nature of mathematics, living in the same habitat with mathematicians, increasing productivity, activating collective creativity, having no pressure. Thus, we could argue that the Mathematics Village provides students and mathematicians with a free and democratic environment without having any obligations, which is a key aspect for creativity (Levenson, 2011). In a similar manner, social media enables students to think freely by collaborating with their other peers (Peppler, 2013). Hence, transforming a non-virtual environment, which bares creative aspects in its structure, into a virtual environment ensures continuity of the real environment in a digital way. Thus, transforming the environment of Mathematics Village into a social media platform is can be named as beneficial. Besides, we believe that utilizing a social media platform as a learning environment, which emerges from real environment, makes the social media platform, which is created due to inspirations from a real environment, more practical and realistic.

We propose a social media platform that can be used as a learning environment by structuring the classification of Dabbagh and Reo (2011) for social media usage. Due to its nature which promises free thought and a democratic environment, social media involves creative characteristics such as activating curiosity, having more time to think and applying no pressure without any obligations (Levenson, 2011). Thus, the basic structure of our social media platform consists of these creative characteristics. In Fig. 17.5, we propose a social media platform, which includes creative features for learning mathematics, with three classifications: (1) Information-gathering, (2) Interaction and sharing, (3) Social networking. Information-gathering level consists of realizing the nature of mathematics and different approaches on instruction, because these characteristics enable participants to collect and organize information through social media for their own performance. Interacting and sharing level comprises of increasing productivity through social interaction and collaboration as well as sharing, interacting, and collaborating with mathematicians. Social networking level involves activating collective creativity, since this level denotes social interaction at the highest level that contains aggregating and synthesizing information from level (1) to level (2).

A social media platform could contain instructional activities including creative characteristics such as different approaches on instruction and realizing the nature of mathematics. The course titled "Ancient Greek Mathematics" given by David Pierce in the Mathematics Village could be an example of an instructional activity as given below in the platform.

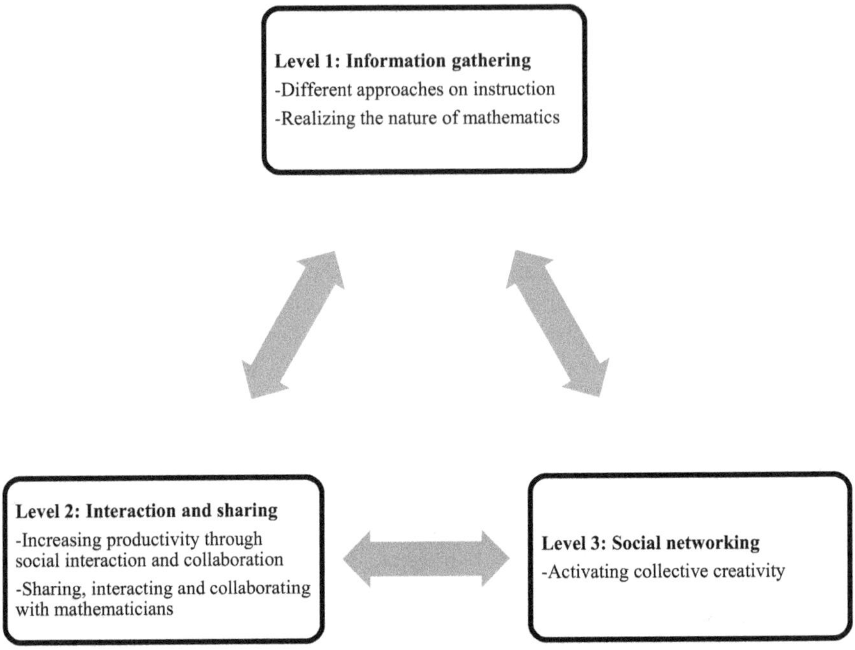

Fig. 17.5 The social media platform as a mathematically creative learning environment

> We shall investigate mathematics as it was done in the Mediterranean region (Alexandria, Syracuse, Anatolia) in ancient times. This way of doing mathematics has been largely forgotten since the development of symbolic algebra. Our point of view is not that mathematics has been improved by algebra, but simply that it has changed. We cannot understand this change unless we look at ancient mathematics in its own terms, as best we can. Possible topics of study include the following: (i) The theory of ratio and proportion expounded in Books V, VI, and VII of Euclid's Elements. (ii) Archimedes's quadrature of the parabola. (iii) Archimedes's proofs of the theorems about spheres now remembered in the formulas $V = 4\pi r^3/3$ and $A = 4\pi r^2$. (iv) The reason why Apollonius of Perga gave to conic sections the names parabola, hyperbola, and ellipse ("Nesin Mathematics Village", 2017).

The course aims at providing insight into how mathematics were done in ancient times. The course enables students to think mathematically without using any symbolic algebra. The students could appreciate what algebra brings for doing mathematics. This course could be viewed as an out-of-the-box one, which enables students to maintain their curiosity in order to promote their creativity. It also provides insight into the nature of mathematics.

Such a social media platform could also include instructional activities with different approaches on instruction, which involve enabling participants to think and question the nature of mathematical concepts through associating them everyday life examples. Here is an excerpt from the lecture notes named "How natural are natural numbers?" provided by Ali Nesin in the Mathematics Village.

> I want to understand 5. First, I have to know what is 5? I mean, I have to define 5: let's define 5 by using the fingers of one hand. Let's try to understand 5 through this definition. Which 5, that I have to understand […] How will I define bigger numbers by using such a definition? In a more general sense, how will I define the concept of a "number"? There is a difference between defining the numbers individually and defining the concept of a number. What will we do? We will differentiate the 'real world 5' and 'the mathematical world 5'. Mathematical 5 is not related or is only related a very small extent to the fingers of your hand. We will define a 5 that is brand new. […] The important issue here is to provide the properties of defined numbers and operations, not how we define numbers and operations. This is one of the most important characteristics of what makes mathematics mathematics. In a more accurate statement, this is what makes modern mathematics what it is ("Nesin Mathematics Village", 2017).

This discussion demonstrates the beauty and the nature of mathematics. The excerpt given above also is an example, which helps one to realize the mechanism of mathematics as a discipline. This indicates how a mathematician discusses the nature of a mathematical concept and how it is different from its everyday life usage.

Social media platform could enable participants to interact and share with their other peers and mathematicians. Besides, interactions and collaborations have the potential to activate curiosity of participants through discussions and sharing within the context of the instructional activities given in the platform. For example, participants could interact, share and collaborate with other participants by discussing about the division within the structure of the following instructional activity, which is a part of Alexandre Borovik's lecture notes used in the Mathematics Village:

> Dividing apples between people. I take the liberty to tell a story from my own life; I believe it is relevant for the principal theme of the paper. When, as a child, I was told by my teacher that I had to be careful with "named" numbers and not to add apples and people, I remember asking her why in that case we can divide apples by people: (1) 10 apples: 5 people = 2 apples. Even worse: when we distribute 10 apples giving 2 apples to a person, we have (2) 10 apples: 2 apples = 5 people Where do "people" on the right hand side of the equation come from? Why do "people" appear and not, say, "kids"? There were no "people" on the left hand side of the operation! How do numbers on the left hand side know the name of the number on the right hand side? ("Nesin Mathematics Village", 2017).

Thinking about the relation between apples and people could make students realize the meaning of the division. Social interaction and sharing this kind of problem could promote students to think creatively on not only the conceptual understanding of concepts, but also procedural understanding of them.

Such as social media platform could provide participants with an environment that has the potential to activate collective creativity. For instance, one of the mathematicians in the Village has highlights the potential of social media platforms to activate collective creativity, as follows:

> A researcher has started a blog, discussed proving a theorem, and written explanatory articles about it. Thanks to this method, the articles accumulate under one place, which results in collective creativity. In other words, it is not done by only one person, multiple people write small pieces and it is shaped in time. Thus its results are stronger than those of proven by only one person. This is a new type of creativity, and such environments can trigger it.

In this social media platform, participants could start a discussion about solving a problem or proving a theorem. Other participants could write their solutions under this discussion board, thereby all of the proposed solutions could be found under one section. The problems could be solved or theorems could be proved by not just one participant, but by all of the participants attending the discussions in a collective manner.

17.4 Discussion and Conclusion

In this study, we have examined how the Mathematics Village could promote mathematical creativity, as well as the transformation of the culture of mathematical creativity that emerged from the Mathematics Village (non-virtual environment) into Social Media (virtual environment). The results of this study contribute to the literature due to the fact that this study examines how providing a context, in which people with different levels of education and interest in mathematics could activate mathematical creativity, and how the creative characteristics of the Village could be transformed into a virtual environment.

We have found out that the environment of the Mathematics Village, which do not have any digital technology, promotes the creativity of the students and mathematicians. We have discussed the creative characteristics of Mathematics Village from a social perspective rather than a solely individual perspective (Amabile, 1996; Csikszentmihalyi, 2000). We found that what really matters in promoting mathematical creativity depends on the mathematical content itself, the teacher and his/her practices, and the teaching environment. To promote the creativity of its students, the Mathematics Village—a building complex situated in a natural environment—applies unique instructional approaches rather than those that are rule-based. The Mathematics Village activates the curiosity of its students about mathematics, which helps to increase their awareness of what there is to learn about mathematics and how best to learn it. We found that students feel much more confident and have more freedom in their thinking about mathematics, compared to the standard education provided in their schools. Pehkonen (1997) stated that the freedom to work on mathematics provides students with the possibility to identify and use their own methods of problem solving and to engage with mathematics.

While teachers in the disciplines of language arts, science, and social sciences often encourage their students to explore, question, interpret, and be creative in their studies, mathematics teachers have often relied on rule-based activities rather than working to activate student creativity (Mann, 2006). In contrast to rule-based instruction, which high school students receive in their schools, at the Mathematics Village, students are given instruction that activates their creativity and curiosity. With respect to mathematicians, in addition to the natural setting of the Mathematics Village, the absence of pressure from the university administration and being removed from the obligations and responsibilities of daily life made them more productive. At the same time, by living close by other colleagues, mathematicians can discuss and critique their research, learn from each other, and activate collective creativity.

The findings of this study align with those of the Vittra schools ("Personalize learning", 2017), in that the Village environment promotes mathematical creativity of students. In addition, the Mathematics Village enables mathematicians to activate their own creativity. From that perspective, having an educational setting that provides freedom can positively affect the states of mind and creativity levels of its students. Therefore, it is important to make available schools with educational settings in which the creativity of students is activated.

Digital technology usage (e.g. social media) could possess the potential to trigger creativity and enhance mathematics learning (Baya'a & Daher, 2013; Peppler, 2013). We strongly believe that additional adjustments to the effective and purposeful use of digital technology could enrich the contribution of the Mathematics Village to the people interested in mathematics. The architecture of the Mathematics Village resembles ancient school only uses online tools for introducing itself to the society (Mathematics Village's website and social media accounts: http://nesinkoyleri.org/eng/; https://www.facebook.com/matematik.koy; https://twitter.com/mat_koyu?lang=en). Not only can Turkish people attend the Mathematics Village, but so can those from around the world. Together, by welcoming national and international scholars and students to join the school, the Village provides opportunities to establish global networks. In order to ensure the sustainability of this global network, we have transformed the culture of mathematical creativity, which emerges from the Village, into a social media platform. We have proposed a social media platform that can be utilized as a learning environment, promises free thought and a democratic environment and has the features of activating curiosity, having more time to think and applying no pressure without any obligations. The social media platform has the following classifications: (1) Information-gathering, (2) Interaction and sharing, (3) Social networking which includes creative features for learning mathematics. Students have the opportunity to access to this social media platform, which consists of mathematically creative activities and enables interaction among students and mathematicians. The social media platform plays a key role in enabling participants to interact and share their ideas with one another even when they are not physically in the same place (Peppler & Solomou, 2011). Interacting and sharing information via social media could also increase the productivity of participants (Dabbagh & Reo, 2011). Besides, social media platform have the potential to bring mathematicians and students from all levels together to discuss specific topics about mathematics, to solve problems or to prove theorems, which pave the way for promoting collective creativity. Through collective creativity, participants of the social media platform produce new ideas or propose solutions on specific mathematical topics, which increase their productivity by providing interaction among participants throughout the world.

In this study, we only provide examples from the courses given in the Mathematics Village, which could serve as mathematically creative activities could be placed in the social media platform. Further studies could be focused on designing and determining the creative mathematical content of the social media platform for students from different levels. Besides, mathematically creative problems could be placed in the social media platform, so as to enable mathematicians and students to interact and share knowledge with one another.

Acknowledgements We would like to thank Dr. Ali Nesin and Dr. Özlem Beyarslan for their contributions to the study.

References

Alladi, K., & Rino Nesin, G. A. (2015). The Nesin Mathematics Village in Turkey. *Notices of the American Mathematical Society, 62*(6), 652–658. https://doi.org/10.1090/noti1258.

Amabile, T. M. (1983). *The social psychology of creativity*. New York, NY: Springer.

Amabile, T. M. (1988). A model of creativity and innovation in organizations. In B. M. Staw & L. L. Cummings (Eds.), *Research in organizational behavior* (Vol. 10, pp. 123–167). Greenwich, CT: JAI Press.

Amabile, T. M. (1996). *Creativity in context: Update to the social psychology of creativity*. New York, NY: Westview Press.

Ayık, G., Ayık, H., Bugay, L., & Kelekci, O. (2013). Generating sets of finite singular transformation semigroups. *Semigroup Forum, 86*(1), 59–66.

Baya'a, N., & Daher, W. (2013). Facebook as an educational environment for mathematics learning. In G. Mallia (Ed.), *The social classroom: Integrating social network use in education* (pp. 171–191). Hershey, PA: IGI Global.

Cropley, A. J. (2001). *Creativity in education and learning: A guide for teachers and educators*. London, UK: Kogan Page.

Csikszentmihalyi, M. (1988). Society, culture, and person: A systems view of creativity. In R. J. Sternberg (Ed.), *The nature of creativity* (pp. 325–339). Cambridge, MA: Cambridge University Press.

Csikszentmihalyi, M. (1996). *Creativity: Flow and the psychology of discovery and invention*. New York, NY: Harper Collins.

Csikszentmihalyi, M. (2000). Implications of a systems perspective for the study of creativity. In R. J. Sternberg (Ed.), *Handbook of creativity* (pp. 313–338). Cambridge, UK: Cambridge University Press.

Dabbagh, N., & Kitsantas, A. (2012). Personal learning environments, social media, and self-regulated learning: A natural formula for connecting formal and informal learning. *The Internet and Higher Education, 15*(1), 3–8. https://doi.org/10.1016/j.iheduc.2011.06.002.

Dabbagh, N., & Reo, R. (2011). Back to the future: Tracing the roots and learning affordances of social software. In M. J. W. Lee & C. McLoughlin (Eds.), *Web 2.0-based e-learning: Applying social informatics for tertiary teaching* (pp. 1–20). Hershey, PA: IGI Global.

Fleith, S. D. (2000). Teacher and student perceptions of creativity in the classroom environment. *Roeper Review, 22*(3), 148–153. https://doi.org/10.1080/02783190009554022.

Göral, H., & Sertbaş, D. C. (2017). Almost all hyperharmonic numbers are not integers. *Journal of Number Theory, 171*, 495–526.

Ito, M., Baumer, S., Bittanti, M., Boyd, D., Cody, R., Herr-Stephenson, B., et al. (2009). *Hanging out, messing around, and geeking out: Kids living and learning with new media*. Cambridge, MA: MIT press.

Levenson, E. (2011). Exploring collective mathematical creativity in elementary school. *Journal of Creative Behaviour, 45*(3), 215–234. https://doi.org/10.1002/j.2162-6057.2011.tb01428.x.

Lu, J., Hao, Q., & Jing, M. (2016). Consuming, sharing, and creating content: How young students use new social media in and outside school. *Computers in Human Behavior, 64,* 55–64. https://doi.org/10.1016/j.chb.2016.06.019.

Mann, E. L. (2006). Creativity: The essence of mathematics. *Journal for the Education of the Gifted, 30*(2), 236–262. https://doi.org/10.4219/jeg-2006-264.

McLoughlin, C., & Lee, M. J. W. (2011). Pedagogy 2.0: Critical challenges and responses to web 2.0 and social software in tertiary teaching. In M. J. W. Lee & C. McLoughlin (Eds.), *Web 2.0-based e-learning: Applying social informatics for tertiary teaching* (pp. 43–69). Hershey, PA: IGI Global.

Meusburger, P. (2009). Milieus of creativity: The roles of places, environments, and spatial contexts. In P. Meusburger, J. Funke, & E. Wunder (Eds.), *Milieus of creativity: An interdisciplinary approach to spatiality of creativity* (pp. 97–153). Dordrecht, The Netherlands: Springer.

Nesin, A. (2008). *Matematik ve develerle eşekler* [Mathematics with camels and donkeys] Istanbul, Turkey: Nesin Yayınevi.

Nesin Mathematics Village. (2017). Nesin Mathematics Village from http://nesinkoyleri.org/eng/. Accessed January 15, 2017.

Patton, M. Q. (2002). *Qualitative research and evaluation options.* Thousand Oaks, CA: Sage.

Pehkonen, E. (1997). The state-of-art in mathematical creativity. *International Journal on Mathematical Education, 29*(3), 63–67. https://doi.org/10.1007/s11858-997-0001-z.

Peppler, K. (2013). Social media and creativity. In D. Lemish (Ed.), *International handbook of children, adolescents, and media* (pp. 193–200). New York, NY: Routledge.

Peppler, K. A., & Solomou, M. (2011). Building creativity: Collaborative learning and creativity in social media environments. *On the Horizon, 19*(1), 13–23. https://doi.org/10.1108/10748121111107672.

Personalize Learning. (2017). Schools Without Classroom. Retrieved from http://www.personalizelearning.com/2014/06/schools-without-classrooms.html.

Plucker, J. A., Beghetto, R. A., & Dow, G. T. (2004). Why isn't creativity more important to educational psychologists? Potentials, pitfalls, and future directions in creativity research. *Educational Psychologist, 39*(2), 83–96.

Sriraman, B. (2004). The characteristics of mathematical creativity. *The Mathematics Educator, 14*(1), 19–34.

Zaman, M., Ananda rajan, M., & Dai, Q. (2010). Experiencing flow with instant messaging and its facilitating role on creative behaviors. *Computers in Human Behavior, 26*(5), 1009–1018.

Part VI
Learning from the Theories and Patterns of Students' Creativity

Chapter 18
APOS Theory: Use of Computer Programs to Foster Mental Constructions and Student's Creativity

Draga Vidakovic, Ed Dubinsky and Kirk Weller

> The body of mathematics is a model of creativity, and it also rests on a process of reflective abstraction.
>
> (Piaget, 1981b, p. 227).

Abstract According to Piaget, the root of all intellectual activity is reflective abstraction. In this context, mathematical creativity arises through students' abilities to make reflective abstractions. Considering that reflective abstraction is the main premise of APOS Theory, the theory provides a theoretical tool to guide the development of instruction that supports mathematical creativity. The letters that make up the acronym—A, P, O, S—represent the four basic mental structures—Action, Process, Object, Schema—that an individual constructs as he or she reflects on and reorganizes content in coming to understand a mathematical concept. Much of the instruction that involves the application of APOS Theory has been delivered using the ACE Teaching Cycle, a lab-oriented pedagogical approach that facilitates collaborative activity within a computer environment (programming and/or dynamic). The letters that make up the acronym—A, C, E—represent the three components of a pedagogical cycle—Activities, Classroom Discussion, Exercises—that facilitate reflection and collaboration. Numerous studies have demonstrated the efficacy of this approach when applied to the teaching and learning of a variety of mathematical topics at the elementary, secondary, and collegiate levels. We illustrate this with a description of instruction for the topics of cosets, infinite repeating decimals, and slope. To introduce these examples, we provide a brief overview of APOS theory with all its components in the context of learning the

D. Vidakovic (✉)
Georgia State University, Atlanta, USA
e-mail: dvidakovic@gsu.edu

E. Dubinsky
Miami, FL, USA
e-mail: edd@math.kent.edu

K. Weller
Ferris State University, Big Rapids, USA
e-mail: KirkWeller@ferris.edu

© Springer International Publishing AG, part of Springer Nature 2018

V. Freiman and J. L. Tassell (eds.), *Creativity and Technology in Mathematics Education*, Mathematics Education in the Digital Era 10,
https://doi.org/10.1007/978-3-319-72381-5_18

concept of function. Opportunities for development of mathematical creativity are emphasized throughout the entire chapter.

Keywords Creativity · Technology · APOS theory · ACE teaching cycle Writing and running computer programs

18.1 Introduction

Over the past two decades many researchers have focused attention on mathematical creativity in undergraduate mathematics teaching and learning (e.g., Silver, 1997; Savic et al., 2016; Tang et al., 2015; Zazkis & Holton, 2009; among others). The recent guidelines of the MAA's Committee on the Undergraduate Program in Mathematics state that successful mathematics teaching should help students develop critical and analytical skills along with 'creativity and excitement about mathematics' (Schumacher & Siegel, 2015, p. 9). However, there is sparse evidence in the literature about pedagogical strategies that support development of mathematical creativity in the classroom (Savic et al., 2017).

In this chapter, we illustrate how APOS, a well-established learning theory predominantly used in research in collegiate mathematics education, and the ACE teaching cycle, its complementary pedagogical method, foster development of mathematical creativity in a collaborative and technology enhanced classroom.

In Sect. 18.2, we offer a perspective on mathematical creativity and provide a general introduction to APOS Theory and its application in the design of instruction. In the remaining sections, we illustrate how these general ideas have been used in the development of instruction for specific concepts. In Sect. 18.3, we show how the theory and pedagogical approach are applied to the teaching and learning of three examples. In Sect. 18.3.1, we describe instruction on cosets (Dubinsky & Leron, 1994) where students write short computer programs. In Sect. 18.3.2, we show how the theory guided development of a computer application package which students used to deepen their understanding of infinite repeating decimals (Weller et al., 2009, 2011). In Sect. 18.3.3, we outline the details of a unit on slope using Geometer's Sketchpad (Jackiw, 2001). Throughout these sections, we explain how our approach supports students' mathematical creativity. We have designed the diagram in Fig. 18.1 to illustrate this organizational structure.

For each example—cosets (Sect. 18.3.1), infinite repeating decimals (Sect. 18.3.2), slope (Sect. 18.3.3)—we show how the general theory is applied to devise a *genetic decomposition* for the concept being taught. A genetic decomposition is a theoretical description that explains how a student may come to understand a concept. This includes a description of the mental structures that a student may construct along with the mental mechanisms that make those constructions possible. After we describe the genetic decomposition for each concept, we show how the theoretical description leads to the development of instruction. In each case, the instruction involves an application of technology. For cosets, students

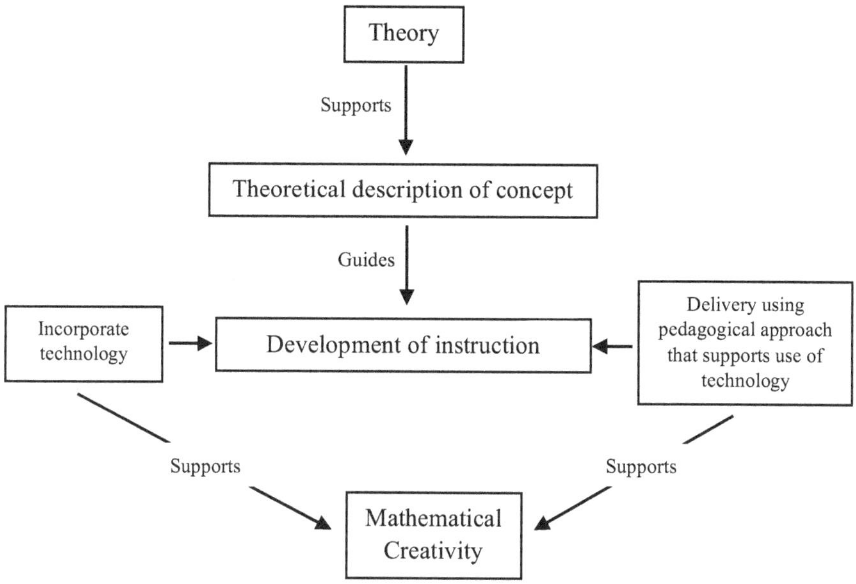

Fig. 18.1 Organizational structure involving the theory, application, and relation to creativity

were asked to write simple computer programs. This use of technology helped the students to think of cosets as mathematical objects that could be collected into sets to which a group structure could be applied. This supports students' creativity through the expansion of thinking, as students applied their knowledge of group theory in the construction of quotient groups. For infinite repeating decimals, the genetic decomposition guided the design of a pre-loaded computer package that was used to strengthen students' understanding of the relation between rational numbers and their decimal representations. This supported students' creativity through connection, as students developed a stronger grasp of rational number representations. In consideration of the concept of slope, we show how the principles of the theory were used to design an instructional unit involving Geometer's Sketchpad (GSP). The work with GSP supported development of students' creativity by helping them to conceive of the slope of a line as an invariant construct. Exploring the concept of slope with GSP allows groups of students to prove in different ways that the slopes of two parallel lines are the same no matter the positions of the lines. Sharing multiple perspectives and approaches and ways of exploration further students' conceptual knowledge that strengthens their creative thinking.

In what follows, we describe APOS Theory, a theoretical perspective based on Jean Piaget's theory of reflective abstraction. We explain how the theory guides the development of instruction to support students' mathematical creativity and their learning of mathematics (Mann, 2006).

18.2 The Theoretical Perspective: APOS Theory and Mathematical Creativity

This section is divided into five subsections in which we describe the general development of the theory and its application. In Sect. 18.2.1 we offer a perspective on mathematical creativity inspired by our review of related literature. In Sect. 18.2.2, we describe the elements of APOS Theory, a theory of learning that has proven to be effective in helping students learn mathematics in creative ways (Weller et al., 2003). In the next three subsections we show how the theory is applied to instruction. We present the idea of a genetic decomposition (Sect. 18.2.3), a description of the mental constructions students may make to learn a particular concept. In Sect. 18.2.3 we show how a genetic decomposition guides the development of instruction that typically involves use of technology. We conclude this general discussion in Sect. 18.2.3.4 with a description of the ACE Teaching Cycle, a pedagogical approach that accompanies instruction based on APOS Theory. Throughout this section, we use the concept of function as an example to illustrate these general ideas and to explain how instruction based on APOS Theory supports students' development of mathematical creativity. In Sect. 18.3, we consider specific examples of how the theory applies and how creativity is nurtured in the teaching and learning of cosets, repeating decimals, and slope.

18.2.1 A Perspective on Mathematical Creativity

Our literature review revealed numerous definitions and descriptions of mathematical creativity (Nadjafikhah, Yaftian, & Bakhshalizadeh, 2012). Considering the purpose of this chapter and our particular theory of learning, we settled on the following working definition:

> An individual's creativity in mathematics consists of the ability to observe patterns, to combine or to reorganize ideas, or to apply techniques or approaches in possibly new (novel) and useful ways when dealing with an unfamiliar situation. This type of activity often involves the creation of new objects, new insights into the relation among one or more existing objects, or reorganization of the structure among objects being studied (Haylock, 1987; Liljedahl & Sriraman, 2006; Nadjafikhah, Yaftian, & Bakhshalizadeh, 2012).

In the realm of the professional mathematician, this type of activity leads to original work that "extends the body of mathematical knowledge" (Liljedahl & Sriraman, 2006, p. 19). In K–16 school and university mathematics, this type of activity can lead to "unusual (novel) and/or insightful solution(s)" and the "formulation of new questions and/or possibilities that allow an old problem to be regarded from a new angle" (Liljedahl & Sriraman, 2006, p. 19).

In his description of the development of intelligence, Piaget (1952, 1981b) discussed creativity (in forms of innovation and invention of "new means", 1981b,

p. 331) and considered two of its aspects: the origin of creativity and the mechanism of creativity. Application of the mechanism of creativity, the focus of this chapter, is rooted in Piaget's notion of reflective abstraction. According to Piaget (1973), reflective abstraction consists of reflection and reorganization. The former involves awareness and contemplative thought, in the sense of reflecting content and operations from a lower level of cognitive thought to a higher level or stage. The latter, reorganization, refers to reconstruction of content and operations on a higher stage or level that leads to the operations themselves becoming content to which new operations can be applied. This general perspective on reflective abstraction forms the basis of APOS, a constructivist theory of mathematical learning that is based on the following premise about the nature of mathematical knowledge and how it is developed (Asiala et al., 1996):

> An individual's mathematical knowledge is her or his tendency to respond to perceived mathematical problem situations and their solutions by reflecting on them in a social context and constructing or reconstructing mathematical actions, processes and objects and organizing these into schemas to use in dealing with the situations (Asiala et al., 1996, p. 7).

Reflective abstraction connects the premise of APOS theory with mathematical creativity. Piaget himself stated that "… all actions, all acts of intellectual creativity, are processes of reflective abstraction" (Piaget, 1981b, p. 225). In using their reflective capacity to respond to problem solving situations, students construct or reconstruct certain mental structures—actions, processes, objects, and schemas. These mental structures, as we describe and exemplify in the paragraphs that follow, represent the cognitive building blocks that support mathematical creativity. A student's ability to construct these structures depends in part on motivation and curiosity—the student's tendency to explore and respond to a perceived mathematical problem situation—and instruction—the teacher's formulation of activities and experiences that lead students to reflect on problem situations in a social context.

18.2.2 A General Discussion of APOS Theory

APOS is an acronym that refers to the basic structures that individuals construct as they come to learn mathematical concepts. The letters that make up the acronym—A, P, O, S—represent the four basic mental structures—Action, Process, Object, Schema—that an individual constructs as he or she reflects on and reorganizes content in coming to understand a mathematical concept. These mental structures are developed by the mental mechanisms of Interiorization, Reversal, Coordination, Encapsulation, De-Encapsulation, and Thematization. The theory indicates how these mental mechanisms are linked and how they are applied in construction of the mental structures described by the theory (Arnon et al., 2014).

APOS Theory is rooted in the notion of mathematics as the study of mental objects (Dubinsky & McDonald, 2001). Similar to the natural sciences, this involves knowing and studying the properties of these objects. Like the physical world, the properties of mental objects are ascertained by acting on them. The difference between mathematics and the physical world lies in the nature of the actions that are applied to the objects as well as to the objects themselves. In the physical world, the objects are concrete and tangible; they are part of our sensory experience. The actions applied to them, such as measuring, weighing, and moving, are also part of that realm. By way of contrast, the objects in mathematics are abstract and, in terms of sensory experience, intangible; they are purely mental, as are the actions applied to them (Dubinsky, 1991). APOS Theory serves as a model to describe how mental mechanisms are activated in an individual's mind and how the activation of these mechanisms leads to the construction of mental structures that help students make sense of mathematical concepts (Dubinsky & McDonald, 2001).

According to APOS Theory, a concept is first conceived as an *action*, that is, an externally directed transformation applied to an existing mental object, or objects. In the case of functions, students first think of a mathematical function as a procedure for taking an element of one set as input and calculating another element of the same or different set as output according to explicit instructions. The sets and their elements are existing *objects* to which the action is applied (Asiala et al., 1996). At elementary levels, the objects are numbers and the assignment is externally directed, which, in the mind of many students, must be done by an algebraic formula. With such an understanding, an individual is able to determine the composition FoG of two functions F and G, provided each is represented by a single expression in a variable, say x, by replacing the x in the expression for F by the expression for G and calculating, if necessary, to simplify. Thus, if the expressions for F and G are represented by

$$F(x) = x^2 + 5$$
$$G(x) = x^3,$$

the action is external, carried out principally through direct substitution and simplification, as illustrated below:

$$x \xrightarrow{G} x^3 \xrightarrow{F} (x^3)^2 + 5.$$

However, if the situation is more complicated, for example, if one or more of the functions is defined in parts, as in,

$$T(x) = \begin{cases} -2x^2, & \text{if } x<0 \\ 5x+1, & \text{if } x \geq 0 \text{ and } x<3 \\ 3x^2+1, & \text{if } x \geq 3 \text{ and } x<10 \end{cases}$$

$$S(x) = \begin{cases} x, & \text{if } x<0 \\ -x, & \text{if } x \geq 0 \end{cases},$$

the learner needs to think how each part from T is linked with each part from S. This goes beyond substitution. In order to put the expressions together correctly, the individual needs to grasp the essence of how a function works. The individual must think of evaluating T for the given x, determining which branch of S the value $T(x)$ belongs to and calculating $S(T(x))$. This occurs as a learner reflects on the action and *interiorizes* that which is explicit into a more general *process*.

Mentally speaking, the newly developed process serves the same purpose as the action that underlies it. The difference is that the action is no longer externally directed but has been internalized (Asiala et al., 1996). For example, in the case of a function given by an algebraic formula, the learner, when thinking about a function as a process, grasps the general idea and can state as such: that a function defined by a formula involves applying that formula to each input value (an element of the domain) to obtain a unique output value (an element of the range). In this case, the learner does not need, nor does he or she refer to, a particular formula, nor is it necessary to explicitly calculate output values for specific input values (op. cit.).

The new mental structure of process enables the learner to engage in more creative mathematical thinking. When thinking about a function as a process, as opposed to an action, the learner no longer needs to think of, or to refer to, a particular formula. This type of ability becomes essential when thinking about situations where certain components of a function are not inherently obvious. For example, to calculate the derivative of a function F represented by an expression like

$$F(x) = \sin\sqrt{x},$$

which requires application of the chain rule, a student needs to realize that F is the composition of two functions, square root and sine (Clark et al., 1997). To see this, a learner needs to think of a function as a process.

For an instructor, the theory provides a descriptive framework that guides instruction. By knowing how a learner may come to understand a concept, an instructor has the potential to develop activities and experiences that lead to meaningful learning. For instruction based on APOS Theory, this has often involved use of a mathematical programming language. Having students write computer programs has been shown to facilitate interiorization (Breidenbach et al., 1992). In the case of the concept of function, a process conception enables a student to work with a wider class of functions than those represented by a single algebraic or trigonometric expression (op. cit.). This supports mathematical creativity by broadening the realm in which students can think about and work with functions. Additionally, the interactive nature of computer programming activities enables students to think more deeply about the concept of function. For example, when a group of students writes incorrect programming code, they need to think more deeply about the function concept in order to correct their errors. Discussions that occur in such situations involve considerations of multiple perspectives that students test on the computer. As a result of their efforts to write and to correct code, they re-evaluate their work and make revisions until they obtain correct answers and representations.

The use of a mathematical programming language has also proven to be useful in helping learners to think of a mental process as a mental *object*. This occurs when a learner tries to apply an action to a process. In order to apply an action successfully, the learner needs to make an *encapsulation*, that is, to reconceive the dynamism of the interiorized action as a static entity, or mental "thing," that can be acted on. For example, a learner who sees a function as a mental object can move beyond specific procedural steps to think in terms of unified entities that can be combined arithmetically or organized as elements of sets. In the case of the composition of two functions, say F and G, a learner with an object level of understanding of function may think of G as an object to which a function F is applied when determining the composition FoG (op. cit.).

Since mental objects arise by applying actions, the instructor plays an important role in designing activities to help students to construct and to apply actions that trigger encapsulation. When using a mathematical programming language, students write programs in which they treat functions as objects—for example, a function can be used as the input to another function. An object conception moves students to a higher plane of mathematical activity; this supports the learner's mathematical creativity. Indeed, an individual's ability to conceive of functions as objects is essential to understanding differential equations, whose solutions are also functions.

When a process is encapsulated, the underlying process is not lost. Rather, the individual, when necessary, can *de-encapsulate* the object back to its underlying process when the situation calls for it. This is exemplified by the chain rule. Taking the derivative of a composition of one or more functions is an action applied to the functions. To conceive of this action, the functions that make up the composition need to be encapsulated. In order for the individual functions to be recombined as a single function, the individual functions need to be *de-encapsulated* into two processes which are then *coordinated* into a single composition process. Once encapsulated, the composition can be differentiated (Arnon et al., 2014). Application of these mental mechanisms and the structures that result from them are illustrated in Fig. 18.2, which summarizes this discussion for two arbitrary functions f and g.

In this particular case, mathematical creativity is supported by having students work with multiple representations in a variety of contexts. For example, rather than having students think about the chain rule in a purely algebraic context, a more creative approach involves having students apply the chain rule to non-algebraic representations of functions. This type of activity can be facilitated by use of a mathematical programming language (Clark et al., 1997).

The A → P → O progression of the development of a concept such as the concept of function creates a system called a *schema*. According to APOS Theory, a schema is the collection of mental structures and mechanisms that make up a concept. Like a process, a schema can be *thematized* into a static structure that can itself be acted on. In order for thematization to take place, the schema needs to be

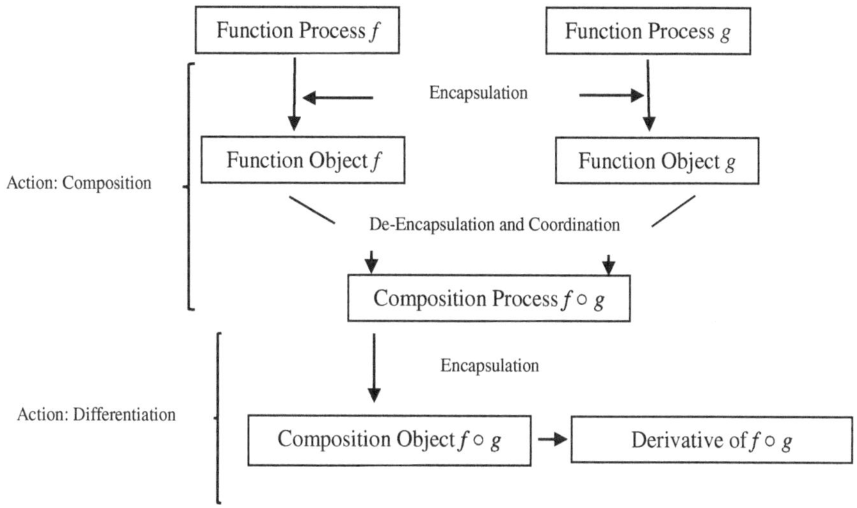

Fig. 18.2 Diagrammatic representation of chain rule applied to composition of functions

coherent. Coherence refers to the mechanism by which an individual ascertains whether the schema can be used to deal with a mathematical problem solving situation. For a mathematical function, an individual with a thematized function schema can determine whether a relation between two sets defines a function. In a more advanced context, an individual with a thematized set schema can create sets of functions to which actions can be applied. The schema structure, and the role of mathematical creativity in development of that structure, is illustrated in Fig. 18.3.

In addition to the mental structures and the mechanisms that link them, Fig. 18.3 illustrates the connection of the schema structure to mathematical creativity. Actions applied to an existing object, or objects, depend principally on the design of instruction, that is, on an instructor's ability to create activities and/or experiences that help a learner to re-organize, to expand, or to extend her or his thinking. The same idea applies to activities that foster encapsulation or that support thematization since both are triggered by actions and reflections on them. On the other hand, interiorization, encapsulation, and thematization depend more on internal activity, specifically, the students' ability to carry out internally that which is directed externally, to re-think something which is dynamic in terms of something that is static, and to begin to think about a concept as a coherent whole. These activities, on the part of both the instructor and student, support mathematical creative activity since they lead to the formation of new mental structures; this leads to new insights and new connections. We illustrate this further in Sect. 18.3 for the specific examples of cosets (Asiala et al., 1997b), infinite repeating decimals (Weller et al., 2009, 2011; Dubinsky et al., 2013), and slope (Reynolds & Fenton, 2011). For these examples, we show how the theory guides the development of the instruction,

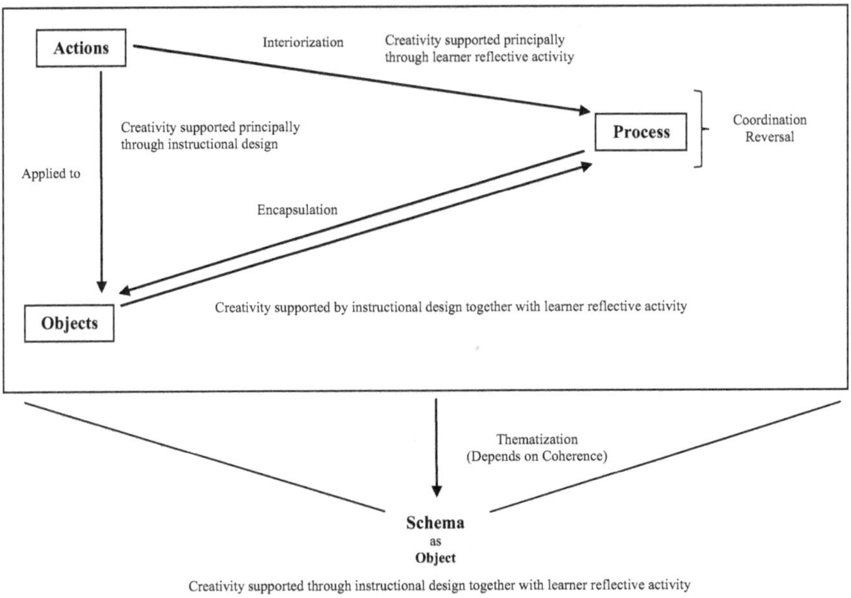

Fig. 18.3 Diagram of mental structures

which includes the use of technology, and explain how our theory-based approach to the instruction may support mathematical creativity.

18.2.3 Genetic Decompositions

In this subsection, we define the term genetic decomposition. We describe how the general theory is used to develop a genetic decomposition for a given concept, and explain how the decomposition and its implementation support the development of mathematical creativity.

18.2.3.1 Definition of a Genetic Decomposition

The basic idea of APOS Theory is to elucidate Piaget's notion of reflective abstraction in the context of learning mathematics by describing the development of thinking about mathematics in terms of a progression of the constructions of three modes of thinking about a mathematical concept called action, process, and object (Dubinsky, 1991). The progression is often complex and can involve subproges-sions, relations to other concepts whose understandings have been previously constructed by an individual, and the application to specific problems. A description of a possible progression for a particular concept is called a *genetic decomposition.*

A genetic decomposition is a description of the construction of mental structures and relationships among them that an individual might need to make in order to construct an understanding for a concept (Arnon et al., 2014).

A genetic decomposition for the function concept consists of a (not necessarily linear) progression from action to process to object. We give a diagrammatic representation of the genetic decomposition for the function concept in Fig. 18.4.

18.2.3.2 Genetic Decomposition as a Tool in the Design of Instruction

APOS Theory can be used as a tool to design instruction. The implementation of instruction provides an opportunity to gather data to determine whether students make the mental constructions called for by a genetic decomposition. Analysis of the data may lead to refinement of the genetic decomposition, which in turn may lead to revision of the instruction. This creates a cycle of continuous improvement; the goal is development of instruction that aligns with how students might construct their understanding of the concept in question. We give a diagrammatic representation of the cycle in Fig. 18.5.

While a *genetic decomposition* for a concept may be quite complex, every decomposition is rooted in the theory, which describes the mental structures learners may need to construct to understand that concept. As students work with the concept, certain mental mechanisms are activated. Each of these mechanisms represents an instance of reflective abstraction. Thus, a genetic decomposition provides a mental roadmap that guides instruction. On the basis of Piaget's perspective, that "… all acts of intellectual creativity are processes of reflective abstraction" (Piaget, 1981b, p. 225), it follows that instruction based on a genetic decomposition has the potential to support a learner's mathematical creativity. One of the instructional tools that is often used as part of APOS-inspired instruction is a mathematical programming language. The use of such a tool is discussed in the next subsection.

18.2.3.3 The Use of a Mathematical Programming Language

A *mathematical programming language* is a computer language that supports major mathematical structures using a syntax that is identical or close to standard mathematical notation. Among the mathematical programming languages available, our preference has been ISETL (Interactive SET Language), a language based on the language SETL (SET Language), which was developed by J. Schwartz (Dubinsky, 1995) to write the first successful ADA compiler.

In our use of a mathematical programming language, students analyze and write simple computer programs. By simple, we mean short (usually not more than about 10 lines), with a minimum of complex programming structures. The general way in which a mathematical programming language works in relation to APOS Theory is illustrated in Fig. 18.6.

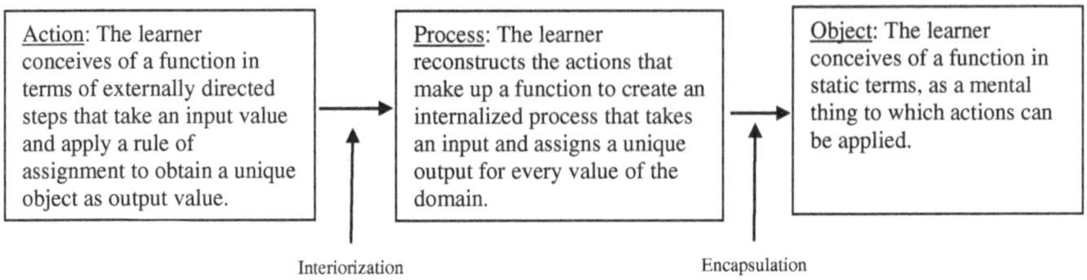

Fig. 18.4 Genetic decomposition for the concept of function

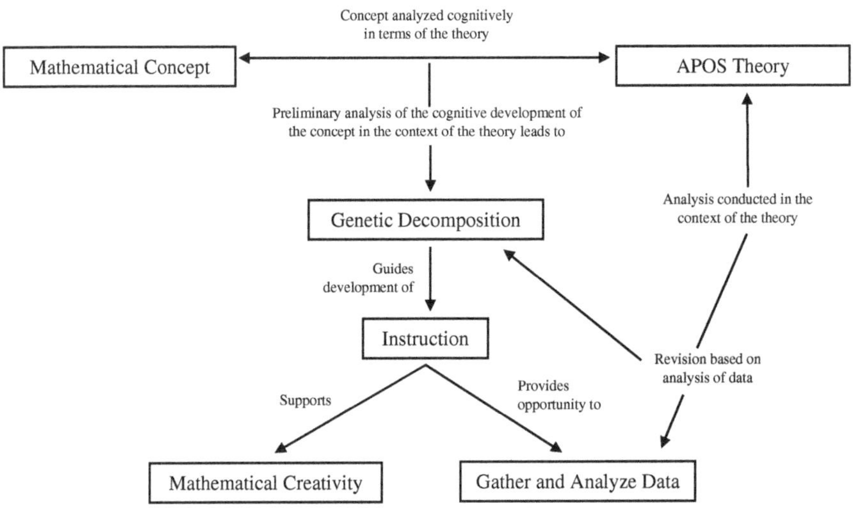

Fig. 18.5 Relation of theory, genetic decomposition, instruction, data analysis, opportunity for creativity

Typically, the construction of a new concept, which starts as an action applied to an existing mental object, involves having students carry out computational tasks with code that may be given to them. After writing and running their programs, they are often asked to explain what the program does and how. They may be asked to use the code to repeat procedures, to predict the results, or to make modifications. All of this is intended to foster reflective abstraction.

Reflection on an action leads to interiorization of the action into a mental process (Dubinsky, 1991). This is supported by having students replace code that carries out a specific calculation with a simple computer program that carries out the calculation for unspecified values, that is, the learner transforms a specific calculation into a general procedure. Our research strongly suggests that when students perform these activities they tend to move the externally driven operations or actions to internally driven mental operations or processes (Arnon et al., 2014). Or, to recall a

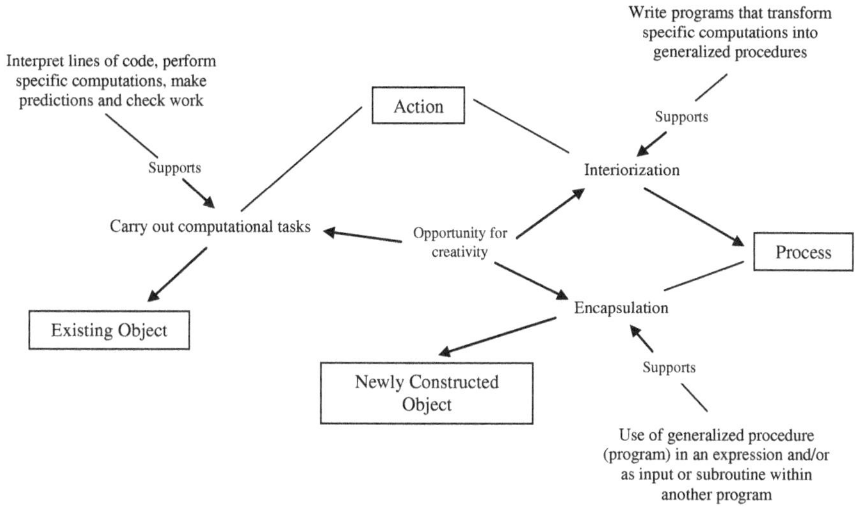

Fig. 18.6 Relationship between A → P → O sequence of mental structures and instructional activity as opportunities for creativity (Arnon et al., 2014)

phrase, not totally facetiously spoken in the early days of computer programming 60 years ago, "writing computer programs affects your mind."

As an action is applied to a process, the reflective abstraction of encapsulation may be performed to conceptualize the process as a cognitive object (op. cit.). Again, our research strongly suggests that encapsulation tends to occur as a student begins to treat the process as the input or output of a computer program, to use the process as a subroutine in a more elaborate program, and/or to operate on the process within a program. In each case, he or she thinks of and treats the process as a static entity rather than focusing exclusively on its dynamism.

As these mechanisms of reflective abstraction are applied (usually subconsciously), and lead to the formation of the new mental structures, opportunities for mathematical creativity arise as students obtain the mental tools to create higher level objects to which new actions can be constructed and applied (op. cit.). We illustrate these ideas in the case of functions.

For the concept of function, which begins as an action of assigning elements of one set to elements of another set, students might first be asked to predict the output returned by a `func` (subroutine) when given specific input values. A `func` is an ISETL command for a mathematical function. It accepts variables, returns output, and can include lines of code to assign a unique output for every input. The way in which a `func` is used is illustrated in the following lines of ISETL code represented by a `func f` in which students are asked to predict the outcome for specified input values, in this case −3, 2, 5, 8, and 11, and then to verify their predictions by running the code.

```
>    f:=func(x);
≫      if x<0 then return 2*x**2;
≫      elseif x>=0 and x<3 then return 3-x**3;
≫      elseif x>=3 and x<10 then return 3*x+1;
≫      else return "function not defined";
≫    end;
≫    end;
>

>

>    f(-3); f(2); f(5); f(8); f(11);
18;
-5;
16;
25;
"function not defined";
```

Unlike an algebraic formula in which students substitute values and apply operations to obtain a result, the programming feature encourages students to think about how elements of the domain are transformed to yield elements of the range. This type of activity is indicative of mathematical creativity: the analysis of simple computer programs helps students to grasp the essential idea of the concept of function in a broader context. To trigger interiorization, students may be asked to explain in general what the computer is doing when the `func f` is executed.

In devising a `func` that carries out the instructions specified in the exercise, students need to go beyond superficial features of its representation to think more critically about what a function is and how it behaves. According to our working definition (Sect. 18.2.1), this type of activity exemplifies mathematical creativity—the learner applies a technique (constructing a function) in a possibly new and useful way (using programming) when dealing with an unfamiliar situation (the concept of function).

In addition to interpretative activity, students might be asked to construct their own `funcs` using the mathematical programming language. For instance, for a problem situation, like the following:

> You have a rectangular piece of cardboard given by the dimensions 20 inches by 30 inches. You plan to cut squares out of each corner and then fold up the sides to create a box. How would you determine the volume of the box for different cuts that are made?

Students use the mathematical programming language to construct a `func`, which might take the following form:

```
>    V:=func(x);
≫      if 0 < x and x  <  20 then
≫      return (20-x)*(30-x)*x;
```

```
≫   end;
≫   end;
```

After constructing this func, students are asked to test their code for different input values.

In writing a func, students begin to generalize the assignment action of a function. In order to write code for a func, students move away from carrying out specific actions, such as determination of an output for a specified input, and think more in terms of a generalized procedure, or process. Our experience as teachers suggests that it is doubtful that this more general conception of a mathematical function would be as apparent, or even possible, without the use of a computer program such as ISETL. In this sense, programming supports the development of creativity as students are able to think of functions in a more general context.

As we have said, applying an action or process to an existing process triggers encapsulation. When working with functions, this can occur when a student is asked to write a function that accepts one or more funcs as input.

> Write a computer function D that accepts a function f and returns a computer function that gives, for any value of x in the domain of the function, the value of the difference quotient for a value of h.

In the code that follows, the func D enables ISETL to compute the difference quotient for any specified function f at any domain point x using a difference h (h here is a global variable although it does not have to be and can be included as a second parameter).

```
>   D:=func(f);
≫     return func(x,h);
≫       return (f(x+h)-f(x))/h;
≫     end;
≫   end;
```

For instance, if the function k is defined by $k(x) = x^2$, which would appear in ISETL as

```
>   k:=func(x);
≫     return x**2;
≫   end;
```

the call D(k)(1,0.01); returns the value of the difference quotient for k at $x = 1$ when the difference $h = 0.01$.

```
>  D(k)(1,0.01);
2.010;
```

In writing a `func` that accepts a function as input and returns a function as output, the learner moves from thinking of a function exclusively in dynamic terms to seeing a function as a "thing" that can itself be transformed.

Asking students to carry out these activities, and using programming to do so, involves creative instruction on the part of the teacher and creative thinking on the part of the learner. The creative aspect for the instructor is the use of programming. The creative aspect for the student is the possibility for new connections and new insights (Weller et al., 2003). The act of writing computer programs that involve the application of functions, operations on functions, and use of functions as subroutines within other functions helps a student to grasp the essence of what a function is and what it does. Development of this ability is essential for advanced thinking involving function concepts. For example, to form a function space and to equip it with a topology, one must first think of functions as mental objects. In order to see a relationship between two sets as defining a function, particularly in situations where it is not obvious, one must construct a coherent function schema. The ability to think of a concept structurally and then to apply it, without having to rely on external cues, is, according to our working definition (Sect. 18.2.1), at the heart of the ability to think creatively.

18.2.3.4 The ACE Teaching Cycle

One instructional sequence that often accompanies APOS Theory is the ACE Teaching Cycle (Arnon et al., 2014). The ACE Teaching Cycle consists of three components: (A) Activities; (C) Classroom Discussion; and (E) Exercises. For the Activities phase, students work cooperatively in a laboratory setting in which they use a mathematical programming language or dynamic software to complete tasks designed to help them to make the mental constructions suggested by the genetic decomposition. The focus of these tasks is more to promote reflective abstraction rather than to obtain correct answers. That is, the activities are designed to help students to make abstractions, not from the object itself but from their own actions on the objects (Piaget, 1981a). For the Classroom Discussion phase, students work on tasks that build on the lab activities completed in the Activities phase. As the instructor guides the discussion, he or she may provide definitions, offer explanations, and/or present an overview to tie together what the students have been thinking about and working on. For the Exercises phase, the instructor typically assigns fairly standard problems to reinforce the reflective activity that has taken place during the first two phases. In addition to supporting continued development of the mental constructions suggested by the genetic decomposition, the exercises may call for students to apply what they have learned and to consider the concept they are studying in relation to other mathematical ideas (Asiala et al., 1996). Every step in the instructional sequence provides an opportunity for the development of creative thinking. The computer activities help students to make the mental constructions called for by a genetic decomposition; this type of activity may help the learner to think of or to identify certain actions (from which properties of the object of study are derived), to compare them with similar actions (or analogies), or to modify and apply them to a new situation. In these activities students are often asked to observe and state in complete sentences their conjectures and to explain their thinking behind those conjectures. In this context, creativity is supported as students construct mental structures that lead to higher levels of thinking. Since the activities are usually done in small groups, this usually triggers discussions, negotiations, and clarification of the statements that are part of a group's response. The classroom discussion offers an opportunity for students to refine and to reflect on their thinking. Creativity is supported as students learn to revise their thinking in a social context. The exercises help students to solidify their fluency in working with the concepts they are learning. Creativity is supported as students are prepared to study new concepts.

While instruction based on APOS Theory lends itself to instruction involving the use of a mathematical programming language, its use is not required. It is simply the case that activities based on a mathematical programming language have been effective in helping students to learn mathematical concepts (see Weller et al., 2003; Arnon et al., 2014).

18.3 Use of APOS Theory and Technology in the Learning of a Mathematical Concept

18.3.1 Teaching and Learning of Cosets

Cosets play an important role in the study of abstract algebra. They form the basis for theoretical results in finite group theory, such as LaGrange's Theorem, and, when certain criteria are met, enable the creation of new group structures such as quotient groups.

Given a group G and subgroup H, the cosets of H in G partition G according to a particular equivalence relation. Specifically, if $a, b \in G$, then $a \sim b$ means that $ab^{-1} \in H$. If the subgroup H is normal,[1] one can define an operation on the set of cosets of H in G to yield a new group structure called a quotient group. From teaching experience and an APOS point of view, the main difficulty students have in working with cosets is to encapsulate the coset process into an object. One of the reasons for this difficulty is that the definition strongly suggests process to the exclusion of object. Therefore, they do not understand how cosets can be elements of a set or that a binary operation on a set of cosets is possible (Asiala et al., 1997b).

[1]A subgroup H of a group G is normal if $aH = Ha$ for every $a \in G$.

These difficulties, if not dealt with productively through instruction, can limit students' creativity and their potential to understand important concepts in mathematics beyond calculus.

One approach to address students' difficulties with cosets has been to design activities using a mathematical programming language. The activities and theory behind their development is the subject of this subsection (Dubinsky & Leron, 1994).

18.3.1.1 Genetic Decomposition of Cosets

As discussed in Sect. 18.2.3.1, a genetic decomposition is a description of the mental constructions that a student might need to make in order to come to understand a concept. A coset is formed as an action applied to an individual's group schema (Dubinsky et al., 1994). An action conception involves forming cosets in familiar situations that tend to be formulaic and explicit. An example[2] is the set of integers Z under addition with a subgroup bZ ($b \in Z$) where a student can construct (left) cosets[3] $a+bZ$ by carrying out specific computations, listing elements, and describing patterns. An action conception is insufficient in more complicated situations where cosets cannot be represented by formulas or simple recipes (Asiala et al., 1997b). An example would be the symmetric group S_n beyond familiar cases of small size such as S_3 or S_4.

A student who has developed a process conception can think of forming a left coset without the need to work with a specific group, subgroup, or group element. When thinking in terms of a process, the formation action becomes a generalized procedure: given a group (G, o), a subgroup H, and an element $g \in G$, a (left) coset is a set of the form $\{g\ o\ h : h \in H\}$. Thus, a process is indicated by a student's ability to internalize and generalize the action of coset formation (op. cit.).

In order to form sets of cosets, for example, to count them, or to define an operation that defines a new group structure (under the assumption of normality), the student needs to make a cognitive transition from the process of formation to an ability to think of a coset as an entity that can be acted on. For this transition to occur, the process of forming a coset needs to be encapsulated into a mental object. Through encapsulation, the dynamic process of forming is transformed mentally into an entity that one can imagine as already having been constructed. Figure 18.7 expresses this progression in diagrammatic form.

This cognitive progression guides the development of instruction which is described in the next subsection.

[2]For $b \in Z$, the subgroup bZ is of the form $\{bx : x \in Z\}$.

[3]For $a, b \in Z$, the left coset $a+bZ$ is a set of the form $\{a+bx : x \in Z\}$.

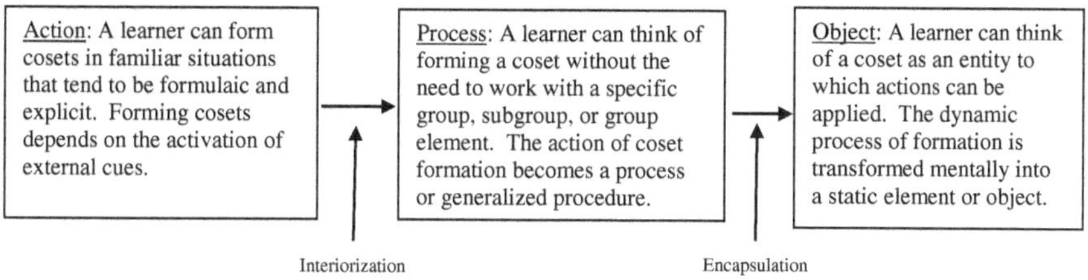

Fig. 18.7 Genetic decomposition of cosets

18.3.1.2 Instruction on Cosets with the Use of a Mathematical Programing Language

The genetic decomposition guided the development of instruction on cosets. According to Asiala et al. (1997b), the instruction consists of three parts, each of which corresponds to the progression of the development of the mental structures of action, process, and object. The treatment of cosets is part of an abstract algebra course (Dubinsky & Leron, 1994) that involves use of a mathematical programming language (ISETL) imbedded in collaborative lab activities that are delivered through the ACE Teaching Cycle (Dubinsky & McDonald, 2001).

Although not considered explicitly, the idea of cosets appears early in the course as students become familiar with the computer language (ISETL) in the context of learning about modular arithmetic. In one activity, students are asked to predict the result of code for different `funcs`. A `func` is an ISETL command for a mathematical function. It accepts variables, returns output, and can include lines of code to assign a unique output for every input. For the sets `Z12 = {0,1,2,3,4,5,6,7,8,9,10,11}` and `H = {0,3,6,9}` and the `func coset` defined as

```
>    coset := func(x);
>>      return {(x+h) mod 12 : h in H};
>>    end;
```

students predict the result of applying the `func`, for example, to determine `coset(2)`. They also consider whether alternative formulations, such as

```
>    modH := |x -> {(x+h) mod 12 : h in H}|;
```

yield the same results.

In another part of the activity, they determine the union of a set of sets such as

```
>   %union{coset(g) : g in Z12};
```

After they make predictions, students run the code to check their work. This type of activity supports development of an action conception. Although the term coset is not defined at this particular point in the course, students carry out the action that underlies its formation. The usual practice in a lecture course is for the instructor to present the definition first and then have students work with examples to make sense of it. In the APOS course, it is the reverse; students experience the definition before it is formally presented to them. This likely supports a better intuitive notion that has the potential to support more creative thinking (Asiala et al., 1997b).

As students learn about groups and subgroups, the concept of coset appears in several related lab activities. One such activity focuses on the concept of normality; students consider the conjugate gHg^{-1} of H by $g \in G$ for a group G and subgroup H by analyzing groups and subgroups encountered in previous activities. Students determine at least one example in which gHg^{-1} is a subgroup of G and at least one example in which it is not. They then repeat this activity for right cosets. On one hand, this activity introduces the notion of normality, a necessary condition for a quotient group. On the other hand, the activity supports development of a process conception. In order to determine whether a right coset is a subgroup, one must carry out the action involved in the formation of a coset. For some of the examples a student might consider, this will be difficult to do without the mental construction of the concept as a mental process.

At the end of students' introductory study of groups and subgroups, they make a major construction that provides additional support for the development of a process conception and sets the stage for encapsulation. In this activity, they write a generalized product PR that accepts a group and an operation as input and returns a func that represents the generalized product. A version of that program follows.

```
>   PR := func(G,o);
≫          return func(x,y);
≫             if x in G and y in G then
≫                return x .o y;
≫             elseif x in G and y subset G then
≫                return {x .o b : b in y};
≫             elseif x subset G and y in G then
≫                return {a .o y : a in x};
≫             elseif x subset G and y subset G then
≫                return {a .o b : a in x, b in y};
≫          end;
≫       end;
≫end;
>
>   oo : = PR(G,o);
```

The `func oo` is the `func` that a student can use after applying `PR` to a specific group, binary operation pair. This `func oo` accepts two inputs among the following possibilities: each input is an element of G; one input is a subset of G and the other is an element of G; both inputs are subsets of G. If each input is an element, the `func oo` returns the result of applying the binary operation of the group to those elements. If one input is a subset and the other is a fixed element, the `func oo` returns either the left coset or the right coset, depending on the order of the inputs. If the input pair is two subsets, the `func oo` returns the result of applying the binary operation to every pair of elements, one from the first subset and the other from the second. This latter application of the `oo` is called the coset product of two cosets (Asiala et al., 1997b).

Once they have constructed `PR` and applied `oo` to a particular set/operation pair, students are asked to use `oo` to test whether a subgroup of a given group is normal. They are also asked to think about various counting questions that lead to consideration of LaGrange's Theorem (op. cit.). One example of such an activity is the following:

> In this activity, you are going to use `PR` again for a group G with input pairs consisting of an element and a fixed subgroup H, and the task is to run through all elements x of G. Thus, your result will be a set of subsets of G. These are the sets $H\ .oo\ x, x \in G$. The questions we ask are: How many elements in each of the subsets? How many subsets are there? Which elements of G are in two or more of these subsets? Which elements are not in any of them? Calculate the set of subsets and answer the questions in each of the following situations.
>
> (a) $G = Z_{24}$, $H = \{0, 6, 12, 18\}$, the subgroup generated by 6.
> (b) $G = S_4$ and $H = \{(1), (12), (34), (12)(34)\}$, where each element of H is a permutation cycle or product of cycles.
> (c) $G = S_4$ and $H = \{(1), (12)(34), (13)(24), (14)(23)\}$ (op. cit.).

Students also use `oo` to build `funcs` that return cosets. An example of the code a student might write is given below (op. cit.).

```
>   lett_coset := func(x,H);
≫       return x .oo H;
≫   end;
```

If `G = {0,1,2,3,4,5,6,7,8,9,10,11}`, `H = {0,3,6}`, and `o` is addition mod 12, then the `func left_coset(4,{0,3,6})` returns the left coset `4 + {0,3,6}` when applied to the pair `(4,{0,3,6})`, where 4 is an element of `G` and `{0,3,6}` is one of its subgroups.

```
>   Z12:={0..11};
>
>   o := func(x,y);
≫       return (x+y) mod 12;
≫   end;
>
```

```
>  oo := PR(Z12,o);
>
>  left_coset(4,{0,3,6});
{10, 7, 4};
>
```

In order to write funcs such as PR, oo, left_coset, right_coset, as well as answering questions related to the application of these funcs, students need to think about the action of coset formation in terms of a generalized procedure, that is, the student must think of how to carry out the formation of a coset when presented with an arbitrary group. In short, to construct these funcs, the student has to move beyond specific formulas and recipes.

To construct an object conception of coset, encapsulation needs to occur. This is supported in part by having students use ISETL to construct the set GmodH of left cosets of H with elements in G, and then investigate the properties of those sets once they have constructed them. If the subset H is a normal subgroup, students use the func oo to define a binary operation on the set of left cosets, and then run is_group, a func they construct in their study of groups, to test whether the set/operation pair (GmodH, oo) forms a group in its own right. To work through these types of activities productively, students need to begin to see cosets as mental objects, that is, to conceive of cosets as elements of a set to which actions can be applied (op. cit.). In this context, encapsulation is likely to occur.

18.3.1.3 Fostering Creativity in Learning About Cosets

Mathematical creativity is impacted by how much mathematics a learner knows and what the learner can do with what he or she has learned. From an APOS perspective, this depends in part on a learner's ability to make the mental constructions called for by the genetic decomposition. In that context, the question, as it relates to the concept of cosets, is whether students who completed the instruction described in the previous sections (Sects. 18.3.1.1 and 18.3.1.2) actually made the mental constructions called for by the genetic decomposition illustrated in Fig. 18.7.

One way to answer this question is to determine how well students performed on problems involving cosets. In a study of students' thinking about cosets (Asiala et al., 1997b; Weller et al., 2003), the authors collected interview and written data from students who completed a course based on APOS Theory and the ACE Teaching Cycle. The study involved 31 students. All of the students completed written instruments (individual and group tests), and 24 of the 31 participated in individual interviews. For coset-related items that appeared on the individual written test, the students' performance indicated that they made substantial progress in making the mental constructions called for by the genetic decomposition:

- About half of the students (16/31) provided a correct proof that the kernel of a group homomorphism is a normal subgroup of the domain.
- About three-quarters of the students (24/31) computed the cosets S_3/A_3, where S_3 is the symmetric group on 3 letters, and A_3 is the alternating subgroup.
- Nearly all of the students (28/31) worked correctly with the two binary operations in the set of cosets $2Z/6Z$, where Z is the ring of integers under the usual binary operations, $2Z$ is the subring of even integers, and $6Z$ is the subring of multiples of 6.
- Nearly all of the students (28/31) were able to prove that the coset ring $2Z/6Z$ is isomorphic to the ring $3Z$ of multiples of 3.

Another aspect of mathematical creativity is a learner's ability "...to combine or reorganize ideas..." (Sect. 18.2.1). In the study of cosets, this was indicated by a learner's ability to go back and forth between a coset as an object and a coset as a set of elements. In the following excerpt, taken from Asiala et al. (1997b), Jocelyn shows evidence of this:

I: OK. That is good enough. Can you explain how you did the last one that you did?
Jocelyn: $K(12)$.[4]
I: $K(12)$ is right. You did $K(13)$ too, times $K(12)$, how did you get that answer?
Jocelyn: I picked representatives out of each coset. So out of $K(13)$ I picked the cycle (13) and out of $K(12)$ I picked the cycle (12) and multiplied them.
I: OK. Is that how a coset product is defined?
Jocelyn: It's uh, for subgroups one you can pick representatives and just multiply them and then your answer will be the coset that contains the product.
I: Right, but that is not the original definition of this.
Jocelyn: Why not?
I: Do you remember what the original definition is?
Jocelyn: Uh, I think we had to go through and multiply every single element in the first coset by every single element in the second coset.

In addition to being able to go back and forth between process and object conceptions, the researchers reported that most of the students were able to see cosets as mental objects (Asiala et al., 1997b). This has implications for mathematical creativity: the ability to see a coset as an object enables a learner to apply actions to her or his group schema to construct a higher level group structure that is necessary for work with quotient groups.

[4]K refers to the Klein 4-group.

Thus, the instructional unit supports creativity in three ways. The students appeared to learn more; this enabled the possibility for new insights. As they made the mental constructions called for by the genetic decomposition, students demonstrated an ability to move from one conception (process) to another (object); this offered the potential for increased flexibility of thought. In the construction of the mental structure of object, the students expanded their ability to work with group structures; this enhanced their capability for higher order thinking about the coset concept. According to our working definition (Sect. 18.2.1), "an individual's creativity in mathematics consists of the ability to observe patterns, to combine or to reorganize ideas, or to apply techniques or approaches in possibly new (novel) and useful ways when dealing with an unfamiliar situation." The unit on cosets shows evidence of a technology-based instructional approach that helped students to make substantial progress in this regard with respect to the learning of this concept.

18.3.2 Teaching and Learning of Repeating Decimals

18.3.2.1 Theory and Instruction Related to Infinite Repeating Decimals

Rational numbers are studied extensively at the elementary and middle school levels. An infinite number of rational numbers have infinite repeating decimal representations. For many students, and even for their teachers, infinite repeating decimals are a mystery. They often see them as nothing more than infinite strings of numbers that arise from performing long division on integers (Dubinsky et al., 2005a, b).

Repeating decimals are one of several mathematical conceptions that involve the paradoxical duality between potential and actual infinity. On one hand a repeating decimal can be thought of as an instance of potential infinity—a process of continually forming digits to express a rational number through long division. On the other hand, a repeating decimal is an instance of actual infinity—the representation of a number with fixed value (Dubinsky et al., 2005a, b).

Dubinsky et al. (2005a, b) studied the apparent tension between these seemingly contradictory notions in an APOS-based analysis of the historical development of the concept of mathematical infinity. The authors used these ideas to explain individuals' difficulties with the repeating decimal $0.\overline{9}$ and 1. Their analysis formed the basis of a preliminary genetic decomposition for infinite repeating decimals (Weller et al., 2009). The genetic decomposition was used to design an instructional unit on repeating decimals for pre-service elementary and middle school teachers. We illustrate the genetic decomposition in Fig. 18.8.

According to the analysis, a student begins by constructing certain actions on whole numbers. This involves reciting, either verbally or in writing, an initial sequence of digits, which may be seen as the beginning of a repeating decimal

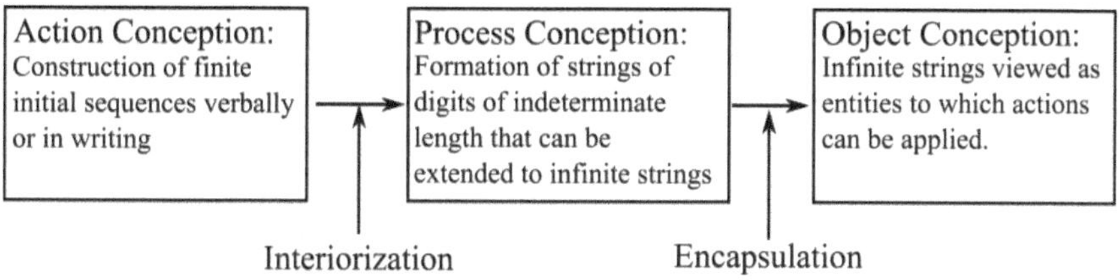

Fig. 18.8 Preliminary genetic decomposition of infinite repeating decimals

expansion. These actions are interiorized into a process of forming sequences of digits of indeterminate length to form an infinite string. As the student reflects on the process and begins to see an infinite string as an entity to which mental actions or processes can be applied, the process of forming an infinite string may be encapsulated into a mental object. The actions that may be applied to an infinite string include various arithmetic and comparison operations, determination of whether an infinite string satisfies certain relations or arithmetic equations, and the ability to see a repeating decimal as a number that equals a fraction or integer (op. cit.).

This genetic decomposition for repeating decimals served as a framework for development of the instructional unit, which consisted of three iterations of the ACE cycle. Each iteration of the cycle spanned two class days, one for computer activities and one for classroom discussions. Homework exercises were assigned at the end of each session and collected at the beginning of the next session (see Dubinsky et al., 2013; Weller et al., 2009, 2011).

For the instructional sequence, students performed calculations in ISETL using a pre-loaded decimal expansion package developed by the researchers (Dubinsky et al., 2013; Weller et al., 2009, 2011). Instead of writing programs, the students used the pre-defined package to carry out calculations to support the mental mechanisms of interiorization and encapsulation. Students used preloaded `funcs` to look at a single place or finite range of places of a repeating decimal. This type of activity supported interiorization by helping students to reflect on the action of writing out the terms of a decimal expansion. The students used the predefined `funcs` to perform arithmetic operations and comparisons on repeating decimals, as well as fraction to decimal and decimal to fraction conversions. These types of activities supported encapsulation by having students perform actions on repeating decimals. In addition to preloaded `funcs`, the decimal expansion package stored several examples of repeating decimals for use in different activities. For many of the activities, students performed calculations by hand, and then checked their results with the computer.

The purpose of the first iteration of the ACE Cycle was two-fold: to help students (1) to interiorize the action of listing digits to a mental process (in order to conceive of an infinite string of digits comprising a repeating decimal), and (2) to begin to see a repeating decimal as a mental object by agreeing on a notational scheme for its representation. Typically, when studying repeating decimals, students learn that certain fractions, when divided using long division, yield infinite expansions. They also learn that a particular type of notation (e.g., a bar over a series of digits) indicates a repeating decimal. To move beyond the vague notion of indefinite continuation suggested by these conventions, the instructors had students use a preloaded `func` called `View` to uncover the identity of eight unknown predefined expansions. The activities served as a guide to help the students to determine the identity of each expansion, to predict the value of the digit in any given position, and to generalize the determination of digits for arbitrary locations. This activity supported interiorization of the action of listing the first few digits of a decimal string into a mental process by which a meaningful description of an infinite decimal representation could arise. The instructors' approach constituted a mathematically creative way to guide students to expand their conceptions. Through reflective activity, students enhanced their grasp of infinite decimal expansions and increased their potential for broader, more creative thinking about the number system. Thus, the activities in this iteration of the instruction promoted students' creativity by helping them to abstract the mathematical idea of a repeating decimal into a more useful construct.

The second iteration of the cycle focused on encapsulation—to help students to transform infinite digit strings conceived as processes into mental objects to which actions could be applied. The third iteration emphasized development of the relation between an infinite digit string and its corresponding fraction or integer. The researchers considered development of this relation to be an important aspect of development of an individual's rational number schema. The construction and subsequent encapsulation of different rational number representations enable an individual to expand her or his schema, offer the potential to develop the coherence of the schema, and increase the likelihood for an individual to see that a rational number has the same standing, mentally speaking, as a whole number.

In the activities based on this idea, students worked with two preloaded `funcs` called `Frac2Dec` and `Dec2Frac`. The former returned the decimal expansion of a fractional representation (using the bar notation mentioned above), and the latter returned the fraction corresponding to a given decimal. Although the students could have made either conversion using standard arithmetic approaches, the `funcs` enabled the students to carry out arithmetic computations on repeating decimals, in support of encapsulation and encouraged the students to begin to connect different number representations, in support of schema development. The activities promoted mathematically creative behavior by helping the students to see that a number has multiple, equivalent, representations (Dubinsky et al., 2013; Weller et al., 2009, 2011).

18.3.2.2 Fostering Creativity in Learning About Repeating Decimals

Implementation of the unit offered the opportunity to gather data. The two studies (Weller et al., 2009, 2011) reported on results of a comparative analysis of the APOS-based instructional sequence with traditional instruction on repeating decimals. A third study (Dubinsky et al., 2013) analyzed students' thinking from the perspective of the genetic decomposition.

In the first study, Weller et al. (2009) surveyed students' views on repeating decimals prior to instruction. On the question of whether $0.\bar{9} = 1$, more than 70% of the students expressed their view that $0.\bar{9} \neq 1$. Nearly equal percentages in both the control and experimental groups treated repeating decimals as approximations and/or infinite processes of long division.

An important aspect of creativity is the development of new objects and new insights. Among students who received the APOS-based instruction, there were substantial gains in this regard. When compared with students in the control group, more than twice the percentage of students who received the APOS instruction expressed belief that $0.\bar{9} = 1$, and three times as many of the students in the APOS group expressed unequivocal belief that every repeating decimal has a fraction to which it corresponds.

The second comparative study (Weller et al., 2011), conducted several months after the instructional sequence, with a focus on the strength and stability of the students' beliefs, confirmed the gains in learning reported in the first study. The analysis revealed that students who received the APOS-based instruction developed stronger and more stable (over time) beliefs that a repeating decimal is a number; a repeating decimal has a fraction or integer to which it corresponds; a repeating decimal equals its corresponding fraction or integer; and $0.\bar{9} = 1$.

The third study (Dubinsky et al., 2013), based on interviews with the students, uncovered evidence of deep thinking, particularly among those who had received the APOS instruction. The following excerpt from Heidi exemplifies this:

> Heidi: Then that means that a repeating decimal—then it can have a—a fraction and a decimal that are exact, the exact same because of the, like you said, the theoretical stopping point, it makes that number useable and that way you can start comparing it to numbers with other stopping points.

Because consideration of repeating decimals is typically limited to carrying out the long division process, students often see repeating decimals exclusively from a dynamic point of view. The instruction helped Heidi to expand her view. This enhanced her mathematical creativity in the sense that she could view repeating decimals as "useable" numbers that could be compared and operated on.

These comparative studies, which showed conceptual gains for the students who experienced the APOS-based experimental instruction, relate to the development of mathematical creativity. The pre-loaded computational package helped students to deepen their connection among different representations of rational numbers. For decimals with infinite repeating representations, students began to see all rational numbers as having fixed value. As a result of this realization, students could carry out operations with infinite repeating decimals, something they could not do before the instruction, and to grasp the notion of an infinite repeating decimal as a number, a conception that previously eluded them. This supported creativity in three ways: a new insight—repeating decimals are not exclusively dynamic; a more robust conception—rational numbers have multiple, equivalent representations; and enhanced computational fluency—in terms of arithmetical operations, rational numbers behave like integers.

18.3.3 *Teaching and Learning of the Concept of Slope*

18.3.3.1 Theory and Instruction Related to the Concept of Slope

Basic geometric concepts are introduced in middle school. Based on the Common Core State Standards Initiative (National Governors Association Center for Best Practices, Council of Chief State School Officers, 2010), high school geometry provides students with opportunities to formalize their experience in geometry using more precise definitions and by developing more formal proofs. High school geometry experiences prepare students for more formal development of Euclidean and non-Euclidean geometries.

In our experience, most universities require future high school teachers to take at least one geometry course, most often a course focusing on Euclidean geometry. At Georgia State University (GSU), future high school teachers take a similar course 'College Geometry'.

The textbook used in teaching the GSU course is based on APOS Theory and use of GSP. The GSP has several features that support instruction. The most important feature is that it makes it easy for students to draw or construct geometric objects (points, lines, segments, rays, circles, etc.), drag them with respect to one another, and observe relationships among those objects. In the GSP there are two ways to create geometric objects: drawing with use of "free-hand' tools and construction using the construct menu. Constructions through use of the construct menu are based on Euclid's constructions and involve the use of properties of the particular objects being constructed. Such constructions allow for manipulation (dragging) of the objects while preserving their main properties. For example, if a right angle triangle *ABC* has been constructed with right angle at *C*, dragging vertices *A* or *B* preserves the right angle triangle. This would not be the case if the triangle were drawn with free-hand tools. Therefore, the GSP allows students to explore various mathematical ideas by creating dynamic constructions and manipulating them. The above mentioned textbook was based on the premise of using the GSP to help students develop various geometric concepts as suggested by APOS Theory and guided by the ACE instructional sequence.

We illustrate these ideas below with an example involving the construction of a line in analytic geometry. A preliminary genetic decomposition, illustrated in Fig. 18.9, was used to develop the unit in which students were asked to study the concept of slope.

During the Activity phase of the instructional cycle, students are asked to work in groups on activities designed to inspire thinking of the slope of a line as 'rise' over 'run' i.e., the ratio of change in y coordinates to change in x coordinates between any two points. Using the GSP, students are asked to plot two distinct

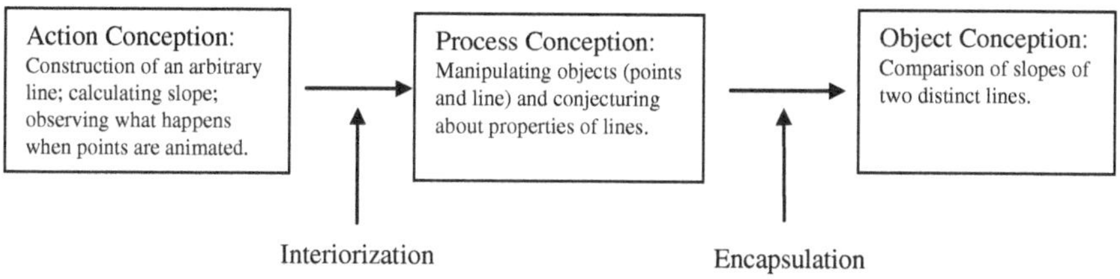

Fig. 18.9 Preliminary genetic decomposition of a line in analytic geometry

points *A* and *B*, construct a line through those points, and then 'explore' the slope as a particular property of the line. At the *action* level the students carry out specific steps such as to read the coordinates (x_A, y_A) and (x_B, y_B) of the given points *A* and *B* and to calculate the ratio $\frac{y_B - y_A}{x_B - x_A}$ (Fig. 18.10).

To trigger interiorization of the action of calculating the ratio for different pairs of points, students drag (or animate) one point, for example point *A*, along the line and observe the value of the ratio as point *A* is moved. Similarly, students carry out the same types of manipulations with point *B*. Then, the students explain what they see happening. Subsequently, they are instructed to move the line and explain why their observations remain valid. Finally, they are asked to formulate their observations as a statement or conjecture. Typically, when writing a conjecture, students first write them individually, discuss their individual statements with their group partners, and then 'negotiate' on a single 'group conjecture'. Students exhibit their creativity, not just in performing the constructions, but also in verbalizing their observations, i.e., by stating their conjectures clearly and concisely.

As stated earlier, applying an action to an existing process, for example, the action of comparing the ratios, fosters encapsulation of that process into an object. As illustrated in Fig. 18.11, students are asked to plot a point *C* outside the given line *AB* and then to construct (using the GSP menu) a line through *C* that is parallel to *AB*. Students are then asked to construct a point *D* on the new line, compute the slope of the line they have constructed, and compare it to the slope of line *AB*. To further foster encapsulation, the students are asked to transform (rotate, translate, drag) one of the lines and observe how the slope of the transformed line changes. In order to compare the slopes of lines *AB* and *CD*, students need to think of the concept of slope as a static entity instead of looking at it as a dynamic process. As they reflect on the actions they apply to the lines *AB* and *CD*, and notice that the relationship between the slopes remains invariant, they are able to generalize the relationship between the slopes of any two parallel lines. This is an act of abstraction that constitutes creative activity as students generate a general idea on the basis of their exploratory work with the GSP.

During the Classroom discussion, students work on tasks that build on these GSP activities. The instructor guides discussion about the ideas of slope that students explored in the activity phase. He or she provides a definition of the slope as the ratio of vertical to horizontal change between two points *A* and *B*. The instructor then introduces the notation for slope, $\frac{\text{vertical change}}{\text{horizontal change}} = \frac{\Delta y}{\Delta x}$. Next, the instructor

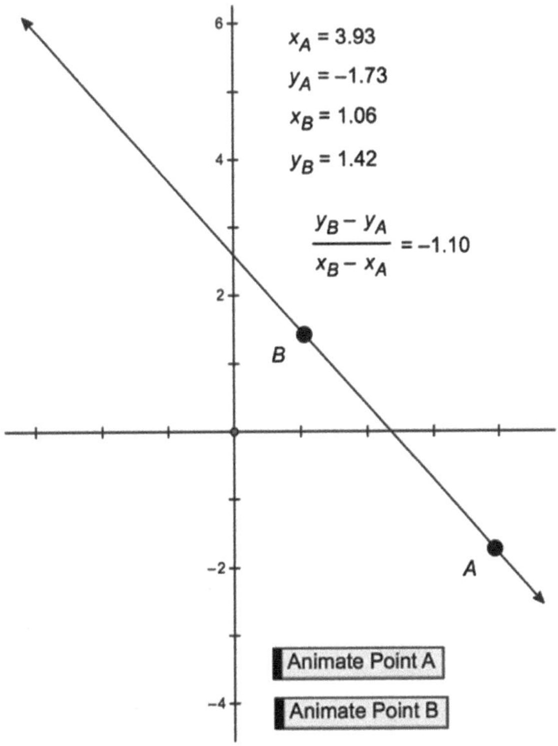

Fig. 18.10 Investigating the slope of a line

discusses with the students their conjectures and guides them to the statement of a general theorem about slope:

> For a non-vertical line, the slope is well defined. In other words, no matter which two points $A(x_A, y_A)$ and $B(x_B, y_B)$, are used to calculate the slope, the value $\frac{y_B - y_A}{x_B - x_A} = \frac{\Delta y}{\Delta x}$.
>
> (Reynolds & Fenton 2011, p. 119).

The instructor may then ask students to outline a proof of this theorem in the context of the following task (Fig. 18.12).

The basic idea in this proof is for students to show that the ratio of the legs Δy_i to Δx_i, respectively, remains the same, regardless of the position of the points on the line. One way to go about the proof is to prove that the triangles $A_1B_1C_1$ and $A_2B_2C_2$ are similar. Once similarity is established, it follows that the ratios of the corresponding sides of these two triangles are equal. With algebraic transformation of these two ratios, one can see that the ratios representing the slopes from A_1 to B_1 and from A_2 to B_2 are equal (i.e. that $\frac{B_1C_1}{A_1C_1} = \frac{B_2C_2}{A_2C_2}$).

This activity offers vast opportunities for creative thinking that may lead to a correct proof. For example, when exploring the similarity of the two right angle triangles constructed over the segments A_1B_1 and A_2B_2, some students may use the angle-angle (AA) theorem of similarity of two right triangles while others may use

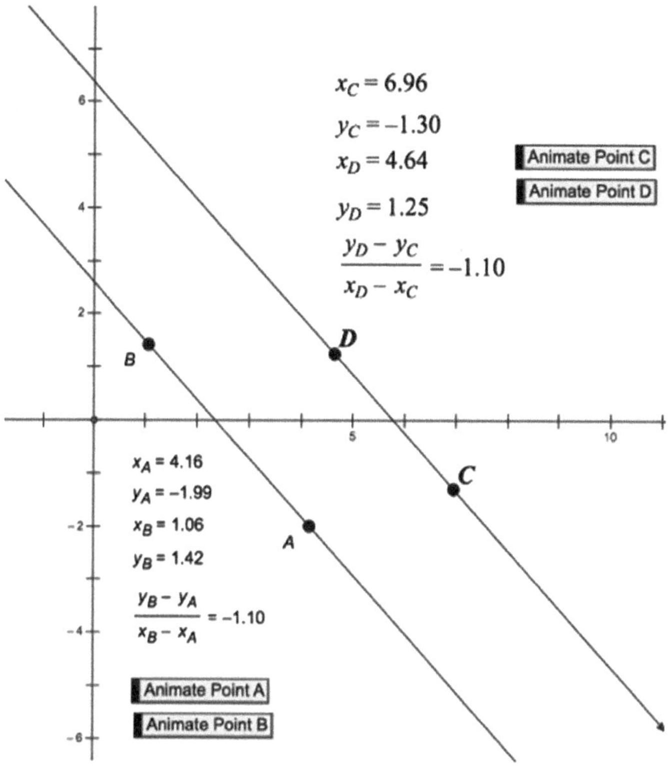

Fig. 18.11 Investigating the slopes of parallel lines

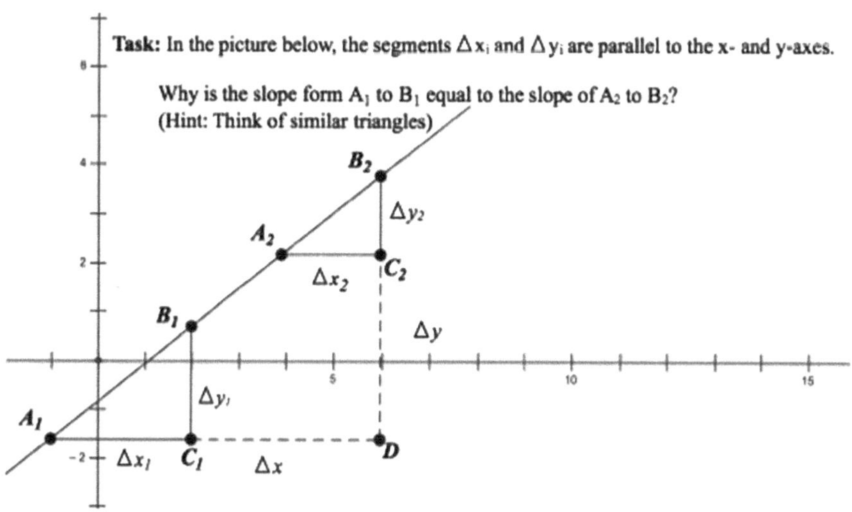

Fig. 18.12 Investigating slope as a well-defined property of a line

the side-side (SS) theorem. By the AA theorem, two right angle triangles are similar if their two corresponding angles are congruent. By the SS theorem, two right angle triangles are similar if their corresponding legs are proportional. Both theorems are familiar to students from their study of Euclidean geometry. The GSP allows for easy calculations and 'exploration' of either of these two similarity theorems that offer opportunities for creative thinking to lead to a correct proof. For example, in answering the given question, students may start by examining the hint and outlining a way of showing that the two triangles $A_1B_1C_1$ and $A_2B_2C_2$ are similar. Students using the AA theorem of similarity of triangles may use the GSP to 'measure' the angles at A_1 and A_2 to observe their congruence. This observation, together with the congruence of corresponding right angles C_1 and C_2, leads to the conclusion that the triangles are similar. On the other hand, students using the SS theorem may use the GSP to 'measure' the lengths of segments A_1C_1 and A_2C_2 as well as the lengths of segments B_1C_1 and B_2C_2. They then calculate the ratios of the corresponding sides $\frac{A_1C_1}{A_2C_2}$ and $\frac{B_1C_1}{B_2C_2}$, observe the equality of the ratios, and use this information to arrive at the same conclusion. However, these are not the only two ways students may outline this first step in the proof. Another way to prove the similarity is to identify the bases of each triangle as segments that are part of two parallel lines cut by a transversal, where the line in question acts as the transversal. Students can then observe the congruence of the resulting alternate angles to conclude similarity.

These approaches represent just a few choices the students may select to outline the proof that the slopes from A_1 to B_1 and from A_2 to B_2 are the same. The fact that students can consider a proof from different perspectives exemplifies how creative mathematical activity is supported through use of the GSP tool. The GSP allows students to explore the properties of triangles by measuring their parts (sides and angles), comparing them and determining if, for example, the sides of these triangles are proportional. Typically, different groups of students identify and use different properties and use a different approach in their proof, including the three proofs mentioned above. It is then the instructor's role to bring all of the different proofs to the students' attention and discussion. Multiple interactions that take place in such a classroom setting (among students during exploration with GSP, among students and instructor, and among all students in class) ultimately help students move to a higher level of conceptual understanding. This enables the students to express their thinking in new and creative ways (Mann, 2006; Savic et al., 2017).

In the Exercise phase of the instructional cycle, the instructor assigns homework problems that reinforce the reflective activities that have taken place during the first two phases. For example, an exercise may call for a student to write a detailed step-by-step proof of this theorem. This task requires the students to reflect on all special cases they considered in the GSP environment (every manipulation of an object such as a point or a line represents a set of special cases), to extend their reflections to a general case by means of relating their algebraic and geometric schemas, and to use those relationships to produce a formal proof. In doing so, students' creativity is manifested through the process of generalization and their efforts to write a complete, 'elegant' proof of the theorem.

18.3.3.2 Fostering Creativity When Learning the Concept of Slope

One aspect of creativity is a learner's ability to apply techniques or approaches in new or novel ways. The GSP tool supports this type of activity. By being able to manipulate the graphical representations of different linear functions, students are able to deepen their understanding of slope as an invariant property of a linear function. As a result of their explorations, the students are introduced to the meaning of being well-defined. In that context, they can engage in significant mathematical activity—stating a conjecture—followed by offering a proof in which they can apply the notion of similarity. The genetic decomposition provides a theoretical framework that guides this type of inquiry; it supports the students' creativity by enabling them to explore and to reflect on the concept of slope in a collaborative setting.

18.4 Conclusion

The goal of the research and curriculum development in which we have been engaged over the past 30 years has been to develop student creativity by enhancing their abilities to overcome difficulties reported in the literature. We have been guided in our work by a particular theory of learning, APOS Theory, based on Piaget's theory of reflective abstraction. APOS Theory points to specific reflective abstractions students are required to make. We use technology to help students to analyze and to make the mental constructions called for by theoretical analyses using the theory. This includes having students write simple programs using a mathematical programming language, working with pre-loaded computation packages, and using dynamic software. Creativity is supported in several different ways. When asked to write computer programs, students reformulate mathematical ideas by expressing them using code. When using a pre-loaded computational package, students apply their developing understandings to carry out various mathematical operations. When working with dynamic software, students connect analytical and graphical representations and articulate different insights through active experimentation. In general, we have implemented instruction based on these ideas within an instructional approach called the ACE Teaching Cycle, a framework that encourages exploration, collaboration, and discussion.

In this chapter, we have shown how the application of APOS Theory supports mathematically creative activity. At the school level, by which we mean K–16 mathematics education, creativity consists of students' ability to observe patterns, to combine or to reorganize ideas, or to apply techniques or approaches in possibly novel and useful ways when dealing with unfamiliar situations. This type of activity often involves the creation of new objects, new insights into the relation among one or more existing objects, or reorganization of the structure among the objects being studied. Mathematics instruction plays a central role in creating a learning environment that fosters this type of activity. The approach we have discussed in this chapter provides instructional support in two contexts, one cognitive and the other pedagogical.

APOS Theory supports the former by providing a theoretical framework to describe the types of mental mechanisms that lead to the formation of mental structures. The theory provides an opportunity for researchers and instructors to develop genetic decompositions for mathematical concepts. A genetic decomposition guides instruction by offering insight into how students learn. This enables an instructor to design activities that facilitate reflective abstraction. Reflective abstraction leads to the formation of new mental structures. As a result of the construction of these structures, students make deeper insights and engage in higher level reasoning, the essence of creative thinking. Mathematical programming languages, pre-loaded computer packages, and dynamic software have proven to be useful tools to support this type of theoretically-based instructional activity.

The ACE Teaching Cycle supports the latter; it provides a pedagogical environment that inspires collaborative, productive struggle in which students can use technology to explore their thinking and to nurture individual creativity and imagination to trigger construction of the mental mechanisms called for by the theory. As the examples in this chapter illustrate, the pedagogical approach offered by the ACE Cycle works in tandem with APOS theory to lead to reflective activity inherent in establishing opportunities for creative thinking.

For example, during the A-phase, students usually work in small-groups on computer activities to explore new ideas (mathematical concepts) by performing certain actions (for example, writing computer programs or constructing geometric figures), making observations, reflecting, discussing, and negotiating in their groups on what to write as their group's response for a given computer 'exploration'. This supports creativity, as students are encouraged to reflect on their thinking and construct new mental structures. During the C-phase, students engage in small-group and class discussions that focus on the mathematical formalization of ideas/concepts from the A-phase. Negotiated mathematical definitions, statements of theorems, proofs, and solutions to challenging problems exemplify the results of students' activities and engagement in this phase. The instructor's role in guiding students' negotiations and concept formalization is essential during class discussions (Arnon et al., 2014; Mann, 2006). This type of activity supports creative thinking, as these social interactions, both with other students and with the instructor, may lead to new individual insights. During the E-Phase, students are engaged in problem solving activities that may serve to reinforce ideas learned during the previous two phases and/or to extend or apply mathematical concepts to real-world problem situations. Opportunities for creative thinking may arise as students see how the content applies in different settings. Thus, the ACE Teaching Cycle provides opportunities for students to develop and improve their creative abilities and to build mathematical competence and confidence (Katz & Stupel, 2015; Weller et al., 2003).

Research reports published in the last three decades (see Weller et al., 2003 for a summary of the first two decades and Arnon et al., 2014 for more recent references) show the promise of our approach, which often involves the use of technology, and its support for creativity, which arises as students work with that technology. As students engage in mathematical thinking as a result of instruction based on APOS Theory, they create new objects, develop new insights among one or more existing objects, and reorganize structures among objects. According to the perspective articulated in Sect. 18.2.1, inspired by Haylock (1987), Liljedahl and Sriraman (2006), and Nadjafikhah, Yaftian, and Bakhshalizadeh (2012), this type of activity embodies the essence of creative thinking. As this work continues, we have no reason to believe it will not prove to be applicable to most, if not all topics in mathematics education. We also believe our approach can continue to support mathematical creativity by providing a means for the development of empirically-based instruction that aligns with how students learn.

References

Arnon, I., Cottrill, J., Dubinsky, E., Oktaç, A., Roa Fuentes, S., Trigueros, M., et al. (2014). *APOS theory: A framework for research and curriculum development in mathematics education*. New York, NY: Springer.

Asiala, M., Brown, A., DeVries, D., Dubinsky, E., Mathews, D., & Thomas, K. (1996). A framework for research and curriculum development in undergraduate mathematics education. In *Research in collegiate mathematics education II. CBMS issues in mathematics education* (Vol. 6, pp. 1–32). Providence, RI: American Mathematical Society.

Asiala, M., Cottrill, J., Dubinsky, E., & Schwingendorf, K. (1997a). The development of students' graphical understanding of the derivative. *Journal of Mathematical Behavior, 16*(4), 399–431.

Asiala, M., Dubinsky, E., Mathews, D., Morics, S., & Oktac, A. (1997b). Development of students' understanding of cosets, normality and quotient groups. *The Journal of Mathematical Behavior, 16*, 241–309.

Breidenbach, D., Dubinsky, D., Hawks, J., & Nichols, D. (1992). Development of the process conception of function. *Educational Studies in Mathematics, 23*, 247–285.

Clark, J., Cordero, F., Cottrill, J., Czarnocha, B., DeVries, D., St. John, D., et al. (1997). Constructing a schema: The case of the chain rule. *Journal of Mathematical Behavior, 16*(3), 345–364.

Dubinsky, E. (1991). Reflective abstraction in advanced mathematical thinking. In D. Tall (Ed.), *Advanced mathematical thinking* (pp. 95–103). Dordrecht, The Netherlands: Kluwer.

Dubinsky, E. (1995). ISETL: A programming language for learning mathematics. *Communications in Pure and Applied Mathematics, 48*, 1–25.

Dubinsky, E., & Leron, U. (1994). *Learning abstract algebra with ISETL*. New York, NY: Springer.

Dubinsky, E., & McDonald, M. (2001). APOS: A constructivist theory of learning in undergrad mathematics education. In D. Holton (Ed.), *The teaching and learning of mathematics at university level: An ICMI study* (pp. 273–280). Dordrecht, The Netherlands: Kluwer.

Dubinsky, E., Weller, K., McDonald, M., & Brown, A. (2005a). Some historical issues and paradoxes regarding the concept of infinity: An APOS analysis: Part I. *Educational Studies in Mathematics, 58*(3), 335–359.

Dubinsky, E., Weller, K., McDonald, M., & Brown, A. (2005b). Some historical issues and paradoxes regarding the concept of infinity: Part II. *Educational Studies in Mathematics, 60*(2), 253–266.

Dubinsky, E., Arnon, I., & Weller, K. (2013). Preservice teachers' understanding of the relation between a fraction or integer and its decimal expansion: The case of $0.\bar{9}$ and 1. *Canadian Journal of Science, Mathematics and Technology Education, 13*(3), 232–258.

Dubinsky, E., Dauterman, J., Leron, U., & Zaskis, R. (1994). On learning fundamental concepts of group theory. *Educational Studies in Mathematics, 27*(3), 267–305.

Haylock, D. W. (1987). A framework for assessing mathematical creativity in school children. *Educational Studies in Mathematics, 18*(1), 59–74.

ISETL: A Mathematical Programming Language. Retrieved from http://homepages.ohiodominican.edu/~cottrilj/datastore/.

Jackiw, N. (2001). *The geometer's sketchpad [software]*. Berkeley, CA: Key Curriculum Press.

Katz, S., & Stupel, M. (2015). Promoting creativity and self-efficacy of elementary students through a collaborative research task in mathematics: A case study. *Journal of Curriculum and Teaching, 4*(1), 68–82.

Liljedahl, P., & Sriraman, B. (2006). Musings on mathematical creativity. *For the Learning of Mathematics, 26*(1), 17–19.

Mann, E. (2006). Creativity: The essence of mathematics. *Journal for the Education of the Gifted, 30*(2), 236–260.

Nadjafikhah, M., Yaftian, N., & Bakhshalizadeh, S. (2012). Mathematical creativity: Some definitions and characteristics. *Procedia—Social and Behavioral Sciences, 31,* 285–291.

National Governors Association Center for Best Practices, Council of Chief State School Officers. (2010). *Common core state standards (Mathematics).* Washington, D.C.: National Governors. Association Center for Best Practices, Council of Chief State School Officers. Retrieved from http://www.corestandards.org/Math.

Piaget, J. (1952). *The origins of intelligence in children.* New York, NY: International University Press. Retrieved from http://www.pitt.edu/~strauss/origins_r.pdf.

Piaget, J. (1973). Comments on mathematical education. In A. G. Howson (Ed.), *Developments in Mathematical Education: Proceedings of the Second International Congress on Mathematical Education* (pp. 79–87). Cambridge, UK: Cambridge University Press.

Piaget, J. (1981a). *Intelligence and affectivity. Their relationship during child development.* Palo Alto, CA: Annual Reviews.

Piaget, J. (1981b). Creativity: Moving force of society [Talk presented at 1972 Eisenhower Symposium, Johns Hopkins University, Baltimore, MD.] Appendix to J. M. Gallagher and D. K. Reid. *The learning theory of Piaget and Inhelder.* Monterrey, CA: Brooks/Cole.

Reynolds, B. E., & Fenton, W. E. (2011). *College geometry: Using the geometer's sketchpad* (1st ed.). New York, NY: Wiley.

Savic, M., Karakok, G., Tang, G., El Turkey, H., & Naccarato, E. (2016). Formative assessment of creativity in undergraduate mathematics: Using a creativity-in-progress rubric (CPR) on proving. In R. Leikin & B. Sriraman (Eds.), *Creativity and giftedness: Interdisciplinary perspectives from mathematics and beyond* (pp. 23–46). Germany: Springer.

Savic, M., El Turkey, H., Tang, G., Karakok, G., Cilli-Turner, E., Plaxco, D., et al. (2017). Pedagogical practices for fostering mathematical creativity in proof-based courses: Three case studies. In *Conference on Research in Undergraduate Mathematics Education (CRUME2017).* Retrieved from http://sigmaa.maa.org/rume/crume2017/Abstracts_Files/Papers/127.pdf.

Schumacher, C. S., & Siegel, M. J. (2015). *2015 CUPM curriculum guide to majors in the mathematical sciences*. Washington, DC: Mathematical Association of America.

Silver, E. (1997). Fostering creativity through instruction rich in mathematical problem solving and posing. *ZDM Mathematical Education, 3,* 75–80.

Tang, G., El Turkey, H., Savic, M., & Karakok, G. (2015). Exploration of undergraduate students' and mathematicians' perspectives on creativity. In T. Fukawa-Connelly, N. Engelke Infante, K. Keene, & M. Zandieh (Eds.), *Proceedings of the 18th Annual Conference on Research in Undergraduate Mathematics Education* (pp. 993–1000). Pittsburgh, PA: Mathematical Association of America.

Weller, K., Arnon, I., & Dubinsky, E. (2009). Preservice teachers' understanding of the relation between a fraction or integer and its decimal expansion. *Canadian Journal of Science, Mathematics and Technology Education, 9*(1), 5–28.

Weller, K., Arnon, I., & Dubinsky, E. (2011). Preservice teachers' understanding of the relation between a fraction or integer and its decimal expansion: Strength and stability of belief. *Canadian Journal of Science, Mathematics and Technology Education, 11*(2), 129–159.

Weller, K., Clark, J. M., Dubinsky, E., Loch, S., McDonald, M. A., & Merkovsky, R. (2003). Student performance and attitudes in courses based on APOS theory and the ACE teaching cycle. In *Research in collegiate mathematics education V. CBMS issues in mathematics education* (Vol. 12, pp. 97–131). Providence, RI: American Mathematical Society.

Zazkis, R., & Holton, D. (2009). Snapshots of creativity in undergraduate mathematics education. In R. Leikin, A. Berman, & B. Koichu (Eds.), *Creativity in mathematics and the education of gifted students* (pp. 345–365). Rotterdam, The Netherlands: Sense.

Chapter 19
The Nature of Knowledge and Creativity in a Technological Context in Music and Mathematics: Implications in Combining Vygotsky and Piaget's Models

Yves de Champlain, Lucie DeBlois, Xavier Robichaud and Viktor Freiman

Abstract Piaget and Vygotsky's prolific work continues to inspire many researchers in several areas of education. While these two authors are often referred to concurrently, sometimes as antagonists and sometimes as complementary theories from a developmental perspective, the debates regarding their epistemological stand and the interpretation of their research remain open. We propose a transdisciplinary approach to combining these two views of learning. Based on the results of two studies bringing technology and creativity together in music and mathematical education, we more specifically examine the transformation that occurs in knowledge when using technologies as a creative process for learners.

Keywords Creativity · Technology · Relationship with knowledge Transdisciplinary · Piaget · Vygotsky

---

Y. de Champlain (✉)
Université du Québec à Montréal, Québec, Canada
e-mail: de_champlain.yves@uqam.ca

L. DeBlois
Faculty of the Sciences of Education, Université Laval, Québec, Canada
e-mail: Lucie.DeBlois@fse.ulaval.ca

X. Robichaud
Faculty of the Sciences of Education, Université de Moncton, New Brunswick, Canada
e-mail: xavier.robichaud@umoncton.ca

V. Freiman
Faculty of the Sciences of Education, Université de Moncton, New Brunswick, Canada
e-mail: viktor.freiman@umoncton.ca

© Springer International Publishing AG, part of Springer Nature 2018
V. Freiman and J. L. Tassell (eds.), *Creativity and Technology in Mathematics Education*, Mathematics Education in the Digital Era 10,
https://doi.org/10.1007/978-3-319-72381-5_19

Réussir, c'est comprendre en acte
Comprendre, c'est réussir en pensée[1]

—Jean Piaget

19.1 Introduction

Piaget and Vygotsky's prolific work continues to inspire many researchers in several areas of education. These two innovators are often referred to side by side, sometimes as antagonists and sometimes as complementary theories from a developmental perspective. Piaget is part of a constructivist movement (Von Glaserfeld, 1982, 2013) and Vygotsky is part of a cultural-historical movement (Leontiev, 1981; Yasnitsky, 2011; Kozulin et al., 2003). The debates regarding their epistemological stands and the interpretation of their research are ongoing (Vergnaud, 2000; Rochex, 1997).

In this article, we propose a transdisciplinary approach to combine these two views of learning. Piaget introduces the transdisciplinary concept, which is a step further than interdisciplinary "that does not stop at attaining interactions or reciprocity between specialized studies, but that identifies these connections within a global system without stable boundaries between the disciplines" (Piaget, 1972; loose translation). Morin (1990, 1999) has worked a long time developing this idea of the importance of combining different theories of knowledge. Morin's work on complexity has in turn been used as the basis for developing a transdisciplinary epistemology (Nicolescu, 1996).

Today, we note that developing critical thinking, problem solving, communication, collaboration, and creativity requires developing transdisciplinary skills (Partnership for 21st Century Skills, 2011). The United States National Research Council highlights the pressing need to carry out studies that will better define these skills, also called "21st Century Skills," and support teachers and their pupils in the process of building efficient practices (ENRC, 2012). Recently, creativity has been attracting particular attention in various areas of education, including those considered as naturally creative such as the visual arts and music, as well as others generally viewed as less open to creativity such as science and mathematics (Kauffman & Bauer, 2004).

Our research team was created as part of the project entitled "Building the framework of creativity in digital learning spaces: A transdisciplinary perspective" (loose translation[2]). Our goal was to examine creativity in two disciplines: mathematics and music. This done, we identified inter and transdisciplinary elements,

[1]To succeed is to understand in action. To understand is to succeed at the level of thought (loose translation).

[2]Original title: *Construire les cadres de créativité dans les espaces numériques d'apprentissage: une perspective transdisciplinaire.*

notably four types of issues linked to the creative process in mathematics and music, that is, issues relative to precision and meaning in addition to heuristic and normative issues. In this chapter, we seek to understand the tensions inherent to a context that is both technological and school-related in order to define how these tensions contribute to the transformation of one's relationship with knowledge and how they foster creativity in pupils.

To shed light on the nature of these tensions and on their possible impact on the teaching-learning process and on research, we begin by explaining the roots of Piaget and Vygotsky's theories. This will enable us to discuss the nature of knowledge in general (*savoir*) and the development of specific knowledge (*connaissance*),[3] as well as their possible transformation in a technological context to describe an action mediated by tools and its effect on cognition. From a transdisciplinary perspective, we then question innovative approaches that integrate creativity in music and mathematics. This is a current research problem in education that has been seldom studied until now (Csikszentmihalyi & Wolfe, 2000; Robinson, 2005; Amadio et al., 2006). Finally, we wish to develop a crosscut model drawing from two studies in which we have identified, from our analyses: tensions "between appreciation of students' genuine creative process that may lead to a variety of solutions (not necessarily the correct ones) and assessment criteria that may value only culturally plausible products. These tensions and their implications to teaching call for deeper investigations" (Robichaud and Freiman 2016). Indeed, technology contains modes of interaction that encourage self-directed learning above all. And yet, these modes are generally structured to generate plausible cultural answers while also containing the resources to move beyond these beaten paths. The purpose of this chapter is to share the results that come from overlapping these models.

19.2 Piaget and Vygotsky on the Development of Specific Knowledge

The role of previous experience in the construction of general knowledge has not always had the same importance. Experience has often been relegated to a role after the fact of validating this general knowledge, in view of controlling subjectivity, which is characteristic of any human experience. However, both Piaget and Vygotsky have long ago recognized that action constitutes a knowledge by itself and that this knowledge can only be brought to the level of conscious thought through a specific process.

---

[3]Translator's note: Since there is, as of yet, no exact equivalent for the French distinction between "*connaissance(s)*" *and* "*savoir(s)*" in English (*knowledge*), we will refer here to *connaissance(s)* as the informal, personally devised, discrete elements of knowledge (*specific knowledge*), and to *savoir(s)* as the shared or institutional form of knowledge (*general knowledge*). See Warfield (2006) "Invitation to Didactique" for more on this linguistic gap (https://www.math.washington.edu/~warfield/Inv to Did667-22-06.pdf).

19.2.1 Piaget's Model of Conscious Realization

Piaget (1974a) notes two important elements in his model of conscious realization: "action constitutes a specific knowledge (know-how) that is autonomous. The conceptualization of said knowledge can only be reached through previous conscious realizations, which follow a rule of succession leading from the outskirts to the centre; in other words, from the zones of object accommodation to the internal coordination of actions" (Piaget, 1974b, p. 231–232; loose translation). To operationalize the passage from "exterior to interior," Piaget begins by using concepts of assimilation, accommodation, and "majoring equilibrium" to then develop the concept of "biological equilibrium," and finally the concept of "reflecting abstraction" (Piaget, 1974a, b, 1977a, b).

The first action, projection (*réfléchissement*), was identified by Piaget (1977a) and used again by Vermersch (2003) who emphasized the importance of sensorial experience from which representation can be created. Indeed, "the way that it [projection] exists at the level of representation is determined by the sensorial coding in which the subject represents his or her experiences" (Vermersch, 2003, p. 82; loose translation). We find here a point in common with Damasio (1999) and Varela, Thompson and Rosch (1993) concerning how awareness takes root in the senses. This convergence offers a new perspective on the sensorimotor stage described by Piaget, in that one's sensorial relationship with the world is not a developmental stage, but rather a lasting foundation of awareness. Thus, at the projection level, representation does not only exist in visual coding; it is also connected to all of the sensorial perceptions and to what is felt. The conceptualization process continues in the transposition of this into words—which starts with the product of projection. Reflected abstraction, which is to say reflection on the content of this representation, helps turn experience into an object of specific knowledge. Vermersch (2003) notes that the value of reflected abstraction "depends on the quality of the map initially drawn from the field (experience)" (p. 84; loose translation). We can therefore think that the quality of our presence in this experience (Legault, 2005, 2006) may play an important role in the mapping process.

Finally, for Piaget (1974a) the passage from the outskirts to the centre almost always comes with the neglect of negative aspects of the problem at hand in favour of the positive aspects, in a logical perspective, which leads to giving action a prominent place. Thus, on the one hand, an action of procedural nature is guided by perception, which is itself regulated according to sensations. On the other hand, a system of meaningful implications (ibid.) determines the reasons leading to meaning. The development of comprehension, viewed as a process of conceptualization, constitutes in this way an authentic process of creation or more specifically "a process of projecting the reality of a plan over another plan" (Piaget, 1977b; Vermersch, 2003, p. 81; loose translation).

19.2.2 Teleonomy as a Link from Conscious Realization to Creativity

Through ongoing interactions for each level of awareness, between more or less conscious purposefulness and equilibrium relationships open to the future (Piaget, 1974b), causal teleonomy (ibid.) is preferred over teleology that has predetermined purposefulness. In teleonomy, the meaning (*telos*) is based on an internal coherence, whereas in teleology, the meaning (*logos*) is expressed through logic. Teleonomy supposes therefore that our understanding of the world is influenced by how we see the future we hope to attain. This future is expressed through a representation based on balancing novel specific knowledge that is continually renewed until an end-purpose is achieved, a creation process taking part in reality overflows (*débordement du réel*) as mentioned by Piaget and that matches with the phenomenology of a creative act:

> Thus, the creative soaring, while pushing back the limits of the exterior world, is pushed by an inward drive towards not only the limits fixed by the world as a framework for our experience, but first of all the molds that we forged ourselves through our ways of feeling, of seeing, of evaluating what we passively perpetuate as the forms of our participation in the world, with others and with our own interiority […] By looking to reach beyond the world, the creative act aims for the world but, doing so, rebounds back to the self, to this interiority that was its starting point (Tymieniecka, 1972, p. 6, loose translation).

Although Piaget never emphasized the creative process, he nevertheless explained the phenomenon with the reflective process rooted in the experience that allows reality overflow as perception and conception in a teleonomic perspective of creating meaning.

19.2.3 Vygotsky's Cultural Model of Knowledge

For his part, inspired by the social context in which he lives, Vygotsky develops a conception of specific knowledge strongly rooted in culture. A number of constructs drawn from Vygotsky's work can also shed light on the nature of specific knowledge in relation to culture. The "active" learner's activity and development is accompanied by an equally "active" educator. During his relatively short career as a researcher, Vygotsky focuses on child development from a psychoeducational perspective, the nature of human cognition, and learning as a social and cultural rather than an individual phenomenon. He explores, among other things, the relationship between thought and language, teaching and development, and the formation of scientific and everyday life concepts. These themes bring him to identify teaching as "proactive" development that is not limited to the pupils' actual specific knowledge, but provokes situations leading pupils to reach beyond their specific knowledge, thanks to mediation by a parent, peer, or teacher enabling them to appropriate cultural tools (signs, symbols, structures, mental operations, etc.). It is then possible to identify the gap between every specific piece of knowledge as a zone of proximal development. This is why learning is viewed as the driving force behind pupils' development. The role of mediation in this process is therefore fundamental.

Vygotsky attributes an important place to the role of imagination and fantasy in the development of children, adolescents, and adults. Vygotsky highlights the need to provide pupils with multiple opportunities to have rich and relevant experiences that stimulate their imagination and creativity. He describes two types of imagination: a reproductive imagination that uses instruments from the dominant culture acquired through experience that grows with age, through interaction with others and the environment; and a productive imagination that makes it possible to establish more complex associations between the imagination's final product and a real complex phenomenon. For example, people who have never visited the desert in Africa may build their own idea, not as a result of lived experience, but as a combination of several "learned" elements, based on a number of concepts such as lack of water, presence of sand, vast spaces, or particular animals. Vygotsky considers this type of imagination as a result of a creative act (Vygotsky, 2004).

In essence, Piaget's work has culture at its core, particularly when he talks about the importance of sensorial experience in projection without exploiting it. However, Archambault and Venet (2007) remind us that Piaget explains the development of thought by taking someone distancing from guided subjectivity by developing logical reasoning. He has little interest in creativity. According to Piaget, children live in an imaginary world, but they will gradually leave it for the "real" world. Vygotsky gives a fundamental role to imagination in the development of thought. For him, imagination develops in parallel to rational thought and becomes a reflected activity. Moran (2010) adds that cultural resources make the repertoire of possibilities much more varied. Indeed, according to the meaning given through experience and subjectivity of the latter (that Vygotsky called *perezhivanie*), we perpetuate our culture by sharing our specific knowledge through teaching or the creation of cultural artifacts such as symbols in music and mathematics. This experience—an "externalization"—makes it possible to have a cultural agreement on its meaning, a change that reveals the creativity for each new externalization.

19.2.4 Reason and Inference as a Specific Knowledge

All of these cognitive turn-arounds mentioned up to now, whether the experience that comes before specific knowledge or the teleonomy that comes before teleology, bring us back to the fact that we are all acculturated to the precedence of formal reason, which functions through deductive inferences according to a tautology, based on a general rule that leads to what is particular. In fact, we are so used to this that we easily confuse formal reason with reason itself. But these changes highlight the need to refer to other types of inferences. In his study on the "art of doing," Denoyel (1999) develops the notion of *experiential reason*. This notion is linked to the Greek word *mètis* (cunning intelligence) that encompasses *kairos* (right or opportune moment), but Denoyel notes that we also find this notion in the Chinese *Art of Circumlocution* and in the Arabic *Book of Trickery*. The *Ojibway* tradition also refers to *mnopi*, the *real* time that corresponds to the *right moment* (Contré Migwans, 2013). In this way, experiential reason is a type of reason "without any reference point, which never justifies its process, but centres on the *situation's potential*. […] through a dialogic that is indissociable of […] sensitive reason (Maffesoli, 1996) and formal reason" (Denoyel, 1999, p. 116; loose translation).

Denoyel also notes that this *practical and cunning* intelligence operates through an inference similar to abduction, as Peirce theorized, in that it "discovers the relevant hypothesis by removing the multitude of possible hypotheses" (Op. cit., p. 118; loose translation). Experiential reason, which "seems to run across the different learning contexts" (Op. cit., p. 116; loose translation), therefore emanates from a dialogical logic with both inductive and abductive inferences. Experiential reason begins with the singular and works its way up to an existing rule (induction) or invents a new rule (abduction). Archimedes' *eureka* is doubtless the best-known example, but Denoyel notes the day-to-day aspect of abduction at the know-how level. For its part, sensitive reason leads to transduction, in that it operates from singular to singular and close to close, according to an analogical logic that is insensitive to contradictions because it is never in contact with a general rule. Experiential reason would thus be what is at work at a specific moment when one needs to say or do something right, which is in this sense similar to practical reason (Gadamer, 1995).

Each one of these types of reason and inferences not only constitutes a specific relationship with knowledge, they are also an open window to the creative process, where Piaget's *ability to create* and Vygotsky's *potential of the environment* meet, and yet are insufficient, so the question of *seizing an opportunity* must also be considered.

19.3 How Can the Nature of Knowledge Be Transformed?

The study of Piaget and Vygotsky's great theories leads to believe that the nature of knowledge can be transformed through actions viewed as regulations. In fact, on the basis of sensorial experiences, which is where Piaget and Vygotsky's work converge, procedures are developed according to the opportunities available to children and according to their sensitivity to some environmental[4] characteristics rather than others. Opportunities and sensitivity modify projection and the nature of the resulting procedures.

[4]"[the environment is] a model of a part of the world that refers to specific knowledge at stake and the interactions that it determines." The latter can be considered as "physical, social, cultural, or other" (Brousseau, 1988, p. 312; loose translation).

Sensitivity begs the question of how learners will be sensitive to their environment (René de Cotret, 1999). Where teachers often see pupils' whims, the biocognitive approach (Varela, Thompson and Rosch, 1993) invites teachers to see a continuous self-(re)organization. According to this approach, perception consists of actions, which are themselves guided by perception and based on biocognitive structures which contribute to build. In this sense, we can talk about an "enactive" approach to cognition (ibid.). This sensitivity will therefore be the starting point for learners to understand their environment and grasp the various opportunities that it potentially offers them. This objective and subjective mediation between the subject and the object in fact corresponds to the concept of affordances.

To begin with, these actions—which seek to transform the nature of knowledge—enhance projection. In light of trials and errors, like as many expressions of regulations, these actions lead to a conscious realization that the desired end-purpose has been reached. A new specific knowledge can then be structured or not according to children's sensitivity to the characteristics that should be remembered when becoming consciously aware of something. This sensitivity is part of the structures children build as well as the opportunities offered to them, and consequently of the culture in which they are growing. Thus new projections rely on culturally plausible or not specific knowledge in a given system (Bélanger et al., 2014a). The cultural artifacts become levers for creating new projections, particularly during the passage from sensorial to symbolic experiences. In addition, conscious realizations are transformed as they are adjusted through mediations between the adult and the child. In this way, the reasons of Piaget's system of meaningful implications not only reach beyond reality, but reach also beyond the individual in his sociohistorical constitution.

Brousseau (2010) distinguishes *milieu* from *environment*, by defining *milieu* as being "composed of objects (physical, cultural, social, and human) with which the subject interacts in a situation" (p. 2; loose translation). The *milieu* is therefore part of a situation or task and transforms according to the pupils' changing sensitivity taking into consideration the experimentation of their procedures and of the project in question. This is how the notion of *milieu* makes it possible to locate the pupils' mistakes as well as the teachers' decisions. "The *milieu* is an agent's (*actant*) antagonistic system" (Brousseau, 2010, p. 3; loose translation). From a transdisciplinary point of view (Nicolescu, 1996), the *milieu* is the element(s) of an environment that resist pupils, whose sensitivity in turn transforms the *milieu* by guiding a child toward zones with less resistance. We could therefore define creativity as what emerges from interactions between opportunities (the environment) and the pupils' sensitivity. This creativity is expressed in the procedures adopted by the pupils. But where does this sensitivity come from? They are the product of interactions on another scale, that is, those between culture and experimentation with the characteristics of the situation at stake. Sensitivity can therefore be viewed as part of a *milieu*'s local context or of a culture's global context.

And yet, the specificity of technology is to carry itself a culture and thus to let some conceptions related to itself emerge providing the *milieu* with a more or less restricted and controlled framework. It is in this sense that we ask how, in a technological context, the nature of knowledge (*savoirs*) is transformed? In other words, how does a *milieu* transform the procedures that result from the pupils' initial representations (projections or *réfléchissements*) and expectations? How does the cultural nature of information technology tools influence the pupils' affective and cognitive processes of learning music and a cognitive processes of learning mathematics?

19.4 Affordances and Creativity in Digital Music and Mathematics Problem-Solving Environments

"Affordance" is an object's ability to suggest how it should be used. Thus software offers pupils possibilities, but pupils transform how it is used. Software promotes the creation of a culturally plausible product, while also allowing one's own world to be created. Affordance therefore refers to the relationship between the environment and the observer (Gibson, 1977). By placing this ability in their theoretical model, researchers note that an instrument does not exist in isolation; it exists only when a person has been able to appropriate it. Affordance is not either an objective or a subjective property, it is both. Thus, an environment's or an object's potential to act can be objective, but it must always be placed in relation to the actor who will use it. "The term affordance refers to the perceived and actual properties of the thing, primarily those fundamental properties that determine just how the thing could possibly be used" (Norman, 1988, p. 9).

Information and communication technologies (ICT) refer to instruments that transmit and process information, mainly by using computers. Considering the technological shift that began in the 20th century and continues during our time, ICT are increasingly used in our society and more particularly in music (Burnard, 2007) and mathematics. The problem is that for most of us, the technical dimension of a digital book is only that, technical. In other words, technology appears intuitively neutral and completely subjected to our will. However, this is not the case. As sociologist and historian Melvin Kranzberg (1986) notes, "Technology is neither good nor bad; nor is it neutral" (p. 545). The technical dimension always holds keys to social, political, and economic transformations; the know-how that is conveyed always refers back to the social dynamics that shaped it in the first place. Digital books will result in particular transformations, just like the combustion engine led to its own unique environment (roads, highways, motels, gas stations, tourism, the power of oil companies, pollution, wars over oil, etc.). For the purpose of our research, the case studies were based in two distinct technological environments.

With respect to music, the researchers used the affordances specific to the field of music, but also those provided by technology that reinforces the perceived capacity to act (Faraj & Azad, 2012). In mathematics, pupils used a website for math problem solving called CAMI. We will discuss these two environments in the following paragraphs.

The CAMI website was created in 2000 as an interactive learning space for mathematics (Freiman & Lirette-Pitre, 2009) to support math problem solving among Francophone pupils in New Brunswick, Canada. Centered on the development of mathematical thinking and communication, this resource transformed through the years into an authentic virtual learning community that made it possible to keep a digital track of the creativity of the pupils who submitted their resolution process by means of an electronic interface. These tracks became a source of (1) formative feedback produced by students in pre-service teacher training, (2) discussions among pupils in the forum, and (3) analysis material regarding pupils' genuine mathematical thinking that becomes the object of a scientific study (Freiman & DeBlois, 2014; Freiman & Manuel, 2015; Freiman & Lirette-Pitre, 2009).

For the music case study, six 3rd grade pupils (age 8) from a primary school in Moncton, New Brunswick, gathered in the same classroom, individually created their own musical compositions with the Garage Band app on iPad. This digital audio workstation (DAW) includes sounds, visual signals (colours), writing, and pre-recorded musical samples. While using this sound bank, those composing can rely on their memory to link and organize the sounds. Watson (2011) highlights the creative benefits of using a DAW. This author claims that technology removes the obstacles caused by the absence of specific musical knowledge or practices that block creativity. Thus, people who neither know how to read nor write musical notation, as well as those who do not know how to play an instrument, can actually compose.

19.5 How the Nature of Knowledge—From the Point of View of Music and Math Learners—Is Transformed

19.5.1 Examples in Music

> [...] music is a coherent system invented by people, but that was built on the basis of their sensitive experience. This is why I don't teach the system as it is, but rather the discovery of a system based on one's sensations. A system is built on what one feels [...]
>
> I always say that before being able to play music, music must first and foremost exist. But the act of creating transcends music, which can also mean creating a working environment, creating living conditions that seem ideal to us, and giving oneself the right to be different and have dreams.
>
> —de Champlain (2010, p. 99; loose translation)

Gall and Breeze (2005, 2008) use the theory of affordances to shed light on the composition process in music. The simultaneous coexistence of the spoken word, writing, body movements, and the perception of sights and sounds is a phenomenon that connects musical creativity to the multimodality theory (Gall & Breeze, 2005, 2008): the more we can communicate in various ways, the better we become at composition. Since ICT are so flexible that each pupil can subjectively interpret the cultural elements that are present and that calls out to their imagination. A new "externalization" may take place that generates creativity. Indeed, to compose with a traditional instrument, one must know how to play or at least know music. However, anybody who can hear can compose with ICT. This statement reinforces what Rogers (1961) and Maslow (1954) wrote that all human beings possess a creative potential that is merely waiting for the opportunity to express itself. For these authors, the multimodal aspects of music software contribute, along with affordances, to making composition affordable.

For his part, Savage (2005) considers DAW an infinitely flexible "meta-instrument", a universal instrument with which one can shape and produce any kind of sound imaginable. It therefore promotes creativity as understood by Webster and Hickey (2001), Kratus (1990), and Giglio (2006). A DAW would have the property of being easily manipulated and making it possible to quickly create soundscapes. Savage (Ibid.) considers that composing with a DAW allows people to stick to the sounds rather than to the symbols that represent them (note that Savage calls this the "micro-phenomenon" of sound). In this sense, a DAW satisfies the requirements of multimodality.

Finally, it is important to mention that recording the work of pupils who are composing allows them to listen to their work with some with some detachment from the process of playing music, enabling them to enter a creative process. This corresponds to the distance needed with subjectivity discussed by Piaget. After conducting a study in a school, Gall and Breeze (2005) cite the example of a pupil who declared that the recording allowed him to recall what he had done and to present a more elaborate piece than the pupils who had composed with a traditional instrument. Giglio's (2006) findings corroborate these observations: listening to one's own work with some distance that gives room to objectivity makes it possible to improve creative musical production.

The DAW enabled pupils to talk about what they had done, making information available regarding their creative activity, and also kept observable traces of the work accomplished (Robichaud and Freiman 2016). At the beginning of the composition process, the pupils were not told how to use the software, but they nevertheless knew how to proceed. We concluded that this affordance was necessary in order to use correctly an instrument, whatever it may be, which is intended to promote creativity and imagination. We also observed that during the musical creative process, when pupils stuck to the limitations of the software's affordance, their creativity was "culturally plausible." This expression corresponds to what is accepted as probable in a given culture. On the contrary, when children stray from the limitations, their creations tend to become culturally implausible.

Indeed, by composing music, pupils may communicate their specific knowledge, imagination, culture, and the authenticity of their thinking process by using affordances. These results are consistent with the sociocultural model for an affordance theory of creativity of Glăveanu (2014) that considers that there are affordances of different natures issued from three factors. These factors overlap one another around the space of everyday actions, taking into account the physical, personal, and sociocultural limitations: intentionality (what a person would like to do), affordances (what a person could do), and normativity (what a person should do). These factors make it possible to see how creativity can be expressed according to affordances that are "not perceived" (because they are strangers to habits), "unexploited" (due to cultural conformity) or "not invented." This model of Glaveanu has contributed to the development of Fig. 19.1.

In this model, creations mostly result either from the pupils' system of specific knowledge, or from the ICT affordances. The system of specific knowledge may be linked to music as well as other fields, such as mathematical thinking or dance. ICT affordances correspond to the software's options as well as its multimodality. The pupils' system of specific knowledge may result in unperceived, unexploited, or un-invented software affordances. In this case, pupils will limit themselves to a particular schema without being able to either extricate themselves from it or enhance it.

The pupils' concern rests with specific musical knowledge (for example, playing with the notions of low and high pitch), corrections made when they do not like what they produced, the feeling of richness sought after, and the evaluation of what was previously created and the global satisfaction attained.

Our findings show that creativity can actually be developed by using ICT. Morgan and Cook (1998) built a software named *Metamuse* that makes it possible to teach the creative thinking process relying on composers' know-how and to help solve problems encountered by the learners.

In the conception of knowledge that can be taught, instruments like the guitar, drums, or keyboard require practice that involves a sustained effort in order to obtain an adequate technique to create and bring forth a culturally plausible creativity. The ICT representation of these instruments, through their easily perceived affordances and the possibility of working with sound without possessing a technical background, accelerates the pupils' possibility of acquiring new knowledge through creativity. In the next section, while presenting mathematics problem-solving environment, a very different one from that we used in music, we can find quite similar patterns of students' creativity related to affordances of the CAMI website, namely having the text of the problem presented on the computer screen along with an electronic form for communicating the solution which standard editing virtual tools (fonts, colours, tables, inserting image, etc.). When introducing their solutions in such environment, several of these affordances allow students to bring their personal touch to the solution (they need to choose, for example, to introduce a table or make a list of all possibilities). However, it is much more difficult to say which knowledge will emerge since there are no explicit instructions and since this situation generates a productive imagination that can

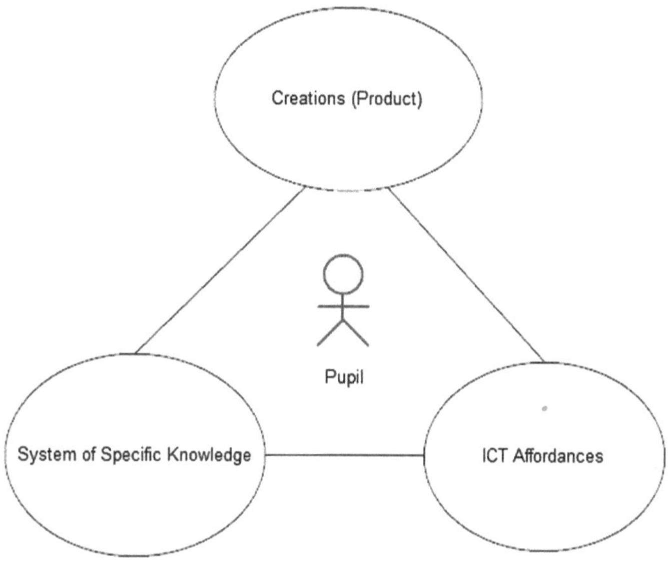

Fig. 19.1 Model of the creative process

move beyond culturally plausible mathematic knowledge, as we saw it in relation to the musical culture of reference.

19.5.2 Examples in Mathematics

19.5.2.1 DeBlois's Model for the Interpretation of Pupils' Cognitive Activities

Based on Piaget's (1977a) theory of reflecting abstraction, studies on the development of understanding number system and word problems with an additive structure for pupils with learning difficulties (DeBlois, 1996, 1997a, b) have led us to observe pupils' cognitive dynamic and the creativity they expressed. First, we describe a few cognitive activities encountered and then we specify the nature of mathematical knowledge expressed by pupils met while they were using the CAMI website.

In a study on the development of understanding how to count, the pupils' thinking alternated between the tokens in *an* opaque envelope and *many* envelopes as grouping method before becoming aware of equivalence relations, particularly between a dozen and 10 units, and 10 dozens and 1 hundred (DeBlois, 1995). These equivalence relations between the various representations of a single quantity have in turn contributed to using counting groups by 100, 10, or 1. Another study (DeBlois, 1997a) presenting a problem that deals with a complement of the set where the students develop an understanding of numbers as representing some quantity to then compare some dynamic of the story (before-after) (You have a box of 118 fruits. You know you have 37 apples in it. The remaining fruits are kiwis. How many kiwis do you have?). After that, they compare each of sub-sets to become aware of the missing quantity which corresponds to the addition with missing numbers.

The reflection that Piaget (1977a) uses to describe abstraction activities were called to mind by the pupils' representations of the situation. These are cultural artifacts (words, numbers, mathematical symbols, drawings, and graphics) to which pupils are sensitive when creating a set of relationships giving meaning to the problem's context. Mobilized for the situation, these initial representations will generate an action potential (things that they can or cannot do), that is, regulations of a pseudo-empirical abstraction (Piaget, 1977b). This action potential, revealed through procedures like illustrating and/or counting, transformed initial representations would enable generating new action potential (as result of back and forth mental work).

This coordination between procedures and initial representations reflects the creativity that stems from the pupils' culture. The actions carried out in relation to a system of personal knowledge, rather than in accordance with the rules of mathematical culture, are done with the material suggested by the pupils. This coordination enhance both the representations of the situation and the procedures that will eventually help pupils become consciously aware of something, what Piaget (1977a, b) called projection (*réflexion*[5]). For example, confronted with a problem of set comparison[6] (DeBlois, 1997b), a qualitative evaluation (few vs. many) first led pupils to establish a one-to-one correspondence procedure between the elements in both sets to become conscious of the "difference" among these sets. This conscious realization makes it possible to count what is left, or what is missing, to then build a relation by implication (if… then). This relationship leads to recognize that if they have 5 coloured pencils less than their teacher, then their teacher has 5 pencils more than them. This work has led to defining a model of interpretation of pupils' cognitive activities that focuses on the pupils' mental representations, procedures, conscious realizations, and expectations linked to one another by coordination that reflect the pupils' creativity. This model makes it possible for us to begin questioning based on the pupils' production rather than from the proposed situation (Fig. 19.2).

Placed in a situation in which a task is presented electronically, will the pupils' procedures be different from those traditionally produced within a paper-and-pencil format? We analyzed the productions of pupils having used the CAMI math problem solving website.

[5]In French.

[6]You have 8 colored pencils. You have 5 less than your teacher. How many pencils does your teacher have?

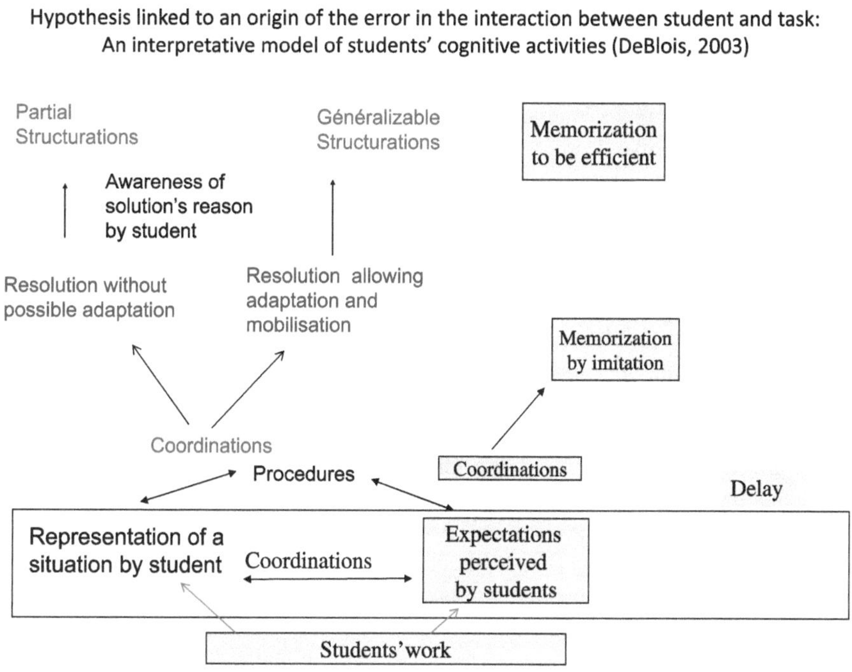

Fig. 19.2 Model for the interpretation of pupils' cognitive activities (DeBlois, 2003, 2014)

19.5.2.2 Analysis of Creativity in Students' Solutions

As mentioned in Sect. 19.3 of this chapter, the CAMI website offered mathematical problems to the pupils in New Brunswick, in close connection to that day's school curriculum. Since our approach is based on the didactics of mathematics school of thought, which seeks a genuine production resulting from the pupils' problem-solving process, we drew from the Problem of the Week model (Renninger & Shumar, 2002). Without targeting a specific grade, each problem attempted to offer pupils a sizeable challenge so they would call on various mathematical concepts and procedures relative to one or several fields of the curriculum: numbers and operations, relations and patterns, geometry and measurement, as well as statistics and probabilities. A dynamic structure of digital space presenting the problem enabled pupils to create their virtual profile, using it to access the problem's text (with a password and username), write a solution in an online form, save it, return to it at need, and submit it to the CAMI team or any other person (classmate or teacher) to whom pupils give access. Thus, contrary to "traditional" practices of problem solving with "paper and pencil," this virtual educational tool generated new didactical opportunities that can improve how we target and assess, among others, the pupils' creativity. As Freiman and Manuel (2015) note, an environment in which participants interact in order to collaboratively generate services and resources, build a foundation for new specific knowledge in mathematics, pedagogy, technology, and others, and establish a culture that emphasizes the skills necessary for asking and solving problems. These complex interactions generated by the virtual environment can lead to high-level cognitive learning that go beyond the "question-answer-feedback" triad. The researchers studying the NRICH (http://nrich.maths.org/frontpage) website claimed to firmly believe that problem solving is a creative process that is at the root of mathematics as a human activity (Piggott, 2007).

Contribution from ICT to support problem solving has been studied, among others, by Hersant (2003), who analyzed the use of a software program linked to a bank of structured problems related to proportionality. While in such an environment, problems are the space in which users apply their specific knowledge, messages, and explanations make it possible to enhance users' specific knowledge by giving them the opportunity to try to adapt themselves according to their errors, which are identified through an a priori analysis (Hersant, 2003). Kuntz (2007) notes that in such an environment, several aspects of pupils' work must be examined: for example, how to enhance and open exercises that are relatively closed; the complementarity of the virtual space relative to more traditional learning environments; the knowledge that pupils can acquire in such a space; and the means to evaluate what the pupils learn in these conditions.

In this way, by combining an open-ended problem that pupils try to solve and a virtual space that offers them tools for writing solutions on the CAMI website, we could anticipate a wider diversity in pupils' creative endeavors. The solutions analyzed in Manuel's (2018) study and the teachers' comments regarding pupils' work (Freiman & Manuel, 2015) seem to confirm that the problems posed on the website not only offered a range of strategies and types of mathematical reasoning, but also opened the way to developing other skills, such as critical thinking and even creative thinking, which has well-documented advantages (e.g., Leikin et al., 2006).

Our analyses led us to hypothesize that IT support creates a broader variety of registers (verbal, digital, symbolic, and geometric) in the pupils' production (Bélanger et al., 2014a). For example, with respect to the problem of cent distribution,[7] some pupils worked at the geometric register while others worked at the arithmetic register. Some pupils also worked with several microstructures (Kintsch & van Dijk, 1978) by using either trivial cases or different coins, while others respected the problem's macrostructure without necessarily taking into account the problem's limitations. Thus, Jeremy worked with four types of coins: pennies (1 cent), nickels (5 cents), dimes (10 cents), and quarters (25 cents). He counted the possibilities without following any particular order. Jeremy worked at a geometric

[7]Sasha has 75 cents in her piggy bank. She has pennies, nickels, dimes, and quarters. How many coins of each are in her piggy bank?

register (Douady, 1986) because the figures used were laid out without any reference to arithmetic operations.

Jeremy's solution (without corrections)
There can be: 25 25 5 5 10 1 1 1 1 1 , 5 5 5 5 5 5 5 10 1 1 1 1 1 25 , 5 5 5 1 1 1 1 1 10 10 10 25 , 25 5 5 5 10 10 1 1 1 1 1 1 1 1 1 1 1 1 1 1 1, 5 5 5 5 10 10 25 1 1 1 1 1 , 1 1 1 1 1 1 1 1 1 1 1 1 1 1 1 1 1 1 1 1 1 1 1 1 1 10 10 5 25 , 1 1 1 1 1 5 10 10 10 10 25 , 25 25 1 1 1 1 1 1 1 1 1 1 10 5 , 1 1 1 1 1 1 1 1 1 1 1 1 1 1 1 1 1 1 1 1 10 10 5 5 25 , 1 1 1 1 1 1 1 1 1 1 10 10 5 5 5 5 25 there are ***<u>10</u>*** possibilities

Jeremy may have interpreted the problem as a set that had to have each type of coin. This initial representation appears to have led him to adopt different procedures. He separated his answers with a comma instead of changing lines. Rather than going from the coins of higher value to the coins of lesser value, or the opposite, Jeremy built sets randomly. The absence of structure led to omitting possibilities.

Madelaine counted the possibilities at a basic level: N (75 cents| only pennies), N (75 cents | only nickels), and N (75 cents | only quarters). The pupil stayed at the arithmetic register (Douady, 1986) by using additions and multiplications. The purely arithmetic notation shows some kind of creativity.

Madelaine's solution (without corrections)
The possibilities are: • 25+25+25=75 • 1x75=75 • 5x15=75 • 10x7+5=75 • 10x6+5x3=75 • 25x2+10x2+5=75 • 25+10x5=75 • 5x10+25=75 • 1x50+25=75 • 1x25+25 Madelaine --

Although Madelaine used a structure, she did not use it optimally. First, she generated three trivial cases (using only quarters, only pennies, and only nickels) disregarding the increasing or decreasing value of the coins. Then, she moved from $10 \times 7 + 5 = 75$ to $10 \times 6 + 5 \times 3 = 75$ revealing the following structure: 10x(i) + 5x(j) à 10x(i−1) + 5x(j + 2). This structure made it possible to read $10 \times 5 + 5 \times 5 = 75$, but it was actually written $25 \times 2 + 10 \times 2 + 5 = 75$. Madelaine was able to free herself from first impressions (Smolucha, 1992) while keeping (self-)control over her action (Saboya, 2012) through the equality symbol ("=") except for the last answer.

Finally, Shawn worked with a macrostructure (Kintsch & van Dijk, 1978) by exploiting all of the possibilities. In so doing, he kept at least a coin of each type. When he changed one coin, he started with coins of the highest value. Creativity was expressed at a geometric register (Douady, 1986) because he did not write any arithmetic. Shawn treated coins like objects in a set. He moved the coins making sure, at each step, that the coin exchange between sets will keep equivalent value.

Shawn's solution (without corrections)

5	pennies	2	nickels	1	dime	2	quarters
5	pennies	3	nickels	4	dimes	1	quarter
5	pennies	5	nickels	2	dimes	1	quarter
5	pennies	7	nickels	1	dime	1	quarter
10	pennies	2	nickels	3	dimes	1	quarter
10	pennies	4	nickels	2	dimes	1	quarter
10	pennies	6	nickels	1	dime	1	quarter
15	pennies	1	nickel	3	dimes	1	quarter
15	pennies	3	nickels	2	dimes	1	quarter
15	pennies	5	nickels	1	dime	1	quarter
20	pennies	2	nickels	2	dimes	1	quarter
20	pennies	4	nickels	1	dime	1	quarter
25	pennies	1	nickel	2	dimes	1	quarter
25	pennies	3	nickels	1	dime	1	quarter
30	pennies	2	nickels	1	dime	1	quarter
35	pennies	1	nickel	1	dime	1	quarter

There are 16 possibilities

Shawn took a quarter and divided it from the first line to the second line. When he divided the quarter, the subjective fantasy took precedence over the objective fantasy because he wrote: "5 pennies 3 nickels 4 dimes 1 quarter." Then he wrote the following on the third and fourth lines: "5 pennies 5 nickels 2 dimes 1 quarter" and "5 pennies 7 nickels 1 dime 1 quarter." In short, when the quarter was divided, it was only a transfer between the dimes and the nickels. There was only an exchange into pennies when the transfer between these two sets was accomplished.

Examining these three solutions, we note differences in the way they were communicated by using the word processing tools suggested in the electronic form. While Jeremy seemed to use "plain text" without formatting, Madelaine and Shawn's solutions seemed to be more structured. Madelaine used bullet points to separate her solutions, which were complete mathematical sentences (including operation and equality signs). Shawn listed his solutions, one per line, each following the same structure. In this way, the number of coins of each type was followed by a space and the type of coin (e.g., nickel). We also noted that his list was built systematically (from the smallest to the most number of pennies).

The pupils' productions show different initial representations of the same problem. Moreover, the productions of the pupils who mainly used their subjective fantasy disregarded some data from the problem. Those who displayed a balance between their subjective and objective fantasies began solving the problem, but their balance was broken, which led them to stop taking account of certain hypotheses in their actions, as observed with Jeremy and Madelaine's productions. Pupils like Shawn maintained their balance between subjective and objective fantasy. This is why their procedures considered as many problems' constraints as possible and validated all of the hypotheses.

Pupils generated a large quantity of relations even if they did not have the rigor needed to carry out their ideas completely. Some pupils' sensitivity to certain artefacts created projections that were used as a springboard for procedures that contributed to regulations, which are part of a technological system and make new coordination that express creativity according to the regulation's characteristics. A greater sensitivity to daily experiences rather than to the problem's limitations would give rise to a precision issue whereas the opposite would give rise to a significance issue (Robichaud, 2016).

19.5.2.3 Complexity of Knowledge Construction in Digital Environments

While Piaget's work contributes to developing concepts pointing at the complexity of the phenomenon of specific knowledge construction, particularly scientific knowledge, it also focuses on the structure in question without delving deeper into the affective (emotional), cultural, and social factors involved. However, the sensitivity expressed by pupils toward cultural artifacts shows the importance of the concept of *milieu*. This is when what is conceivable lets the pupils' expectations intervene regarding a situation in a context like the classroom.

Technological tools also have a role in this learning process. According to Trouche (2005), an instrumental approach was developed based on Vygotsky (1934) and Rabardel's (1995) work. Vygotsky places all learning in the realm of culture in which instruments (material and psychological) play a crucial role.

Trouche (2005) recognizes that the artifact may or may not be developed by the subject, that this development is accomplished through a process (instrumental genesis) in which the nature of the user's activity and the context has a determining role. In addition, any instrument would include a "material" component (the part of the artifact that is called on during the activity) and a "psychological" component (incorporating the mobilization of schemas as described by Piaget and Vergnaud). By applying these ideas to our context of mathematical problem solving, we still noted—like Rabardel's work cited in Trouche (2005)—the dual nature of the instrumental genesis of the pupils' activity when confronted with a mathematical problem on the CAMI website. This dual nature included an instrumentalization process with respect to the subject's personalization of the artifact. In our case, the process is the choice of how to communicate a solution with the word processing tools in CAMI. Next, the instrumentation is relative to the emergence of schemas in the subjects. In this way, the artifact contributes to pre-structuring subjects' actions, in our case, the way subjects perceive the mathematical problem, in order to build their own problem-solving process. These two processes overlap each other and are simultaneous. With CAMI, we ignore in which context pupils are when problem solving. Pupils may be working in a classroom or at home, alone or with help, and the problem may have been discussed with the teacher, peers, or parents before pupils use a computer. According to Trouche, this context can contribute to a wide diversity of ("instrumentalized") processes among pupils. This diversity is also noted by Manuel (2018) and Bélanger et al. (2014b). Stemming from instrumentation and instrumentalization, this diversity can contribute to transforming the nature of knowledge.

19.6 Discussion

Could ICT go against the normativity and social status of a discipline centered on referential knowledge? By not taking into consideration the pupils' productive imagination, could this normativity produce a resistance to learning through the pupils' relationship with the knowledge thus generated?

As Piaget modeled it, conscious realization (*prise de conscience*) starts with the projection (*réfléchissement*) of lived experience. The product of this projection is then thematized in order to access the reflection (*réflexion*), and then at the moment of realization itself. And yet, projection is itself driven by the subject's particular sensitivity to his or her environment. ICT do indeed offer ways to diversify, or even enrich, one's relationship with experience according to the user's sensitivity. We have seen examples of this function with the CAMI site (in which math work is part of a knowledge development project) as well as with GarageBand (that enables pupils to record themselves). The process thematization will be greatly influenced by the procedures available to the subject since these procedures come with the words and meanings that make them usable in the both context, musical and mathematical.

In the case of CAMI, the environment suggests the importance of noting these procedures, and encourages to some extent a more elaborate transposition into words and mathematical symbols. GarageBand provides its own integrated procedures that the subjects use to find links between what they think and feel about what they are creating. This is followed by the application of regulations and self-regulations through which the concepts that emerged from the thematization process are tested in the real world until subjects are satisfied. This is a very important aspect of GarageBand: it provides immediate feedback and thus fosters self-regulation. However, this does not keep the feeling of satisfaction from remaining more linked to the conceptual level; for example, subjects who build their composition in a way to use all of the musical notes and tones available within a certain context. When conscious realization is attained, a structuring or restructuring of knowledge is transferred in the way a new projection will be processed. At that point, subjects turn again to GarageBand to reconsider their choices on one or several levels, either by changing an instrument, the tempo, or by starting a new composition. But what is important is the reversibility of each choice made with the software, which makes it easier for subjects to become consciously aware. In the case of mathematical problem-solving, the affordance of the CAMI environment allows students to save their solutions which become a part of their personal cyber-portfolio. Not only they can eventually return to their solutions and modify them, but when the problem is still active, their work remains in the archive and can be viewed, reflected and re-used in future work. For example, the teacher can eventually see the solutions of all his or her students and decide to bring some for the whole-class discussion, in a traditional form, or using an online discussion forum. Bélanger et al. (2014a) report benefits of such activities making students' creativity 'talk'. From this perspective, ICT can indeed influence the conscious realization process because they provide subjects with creative empowerment and when the system of specific knowledge of the subjects is the result of a creative activity with ICT we assist of the emergence of perceived, exploited, and invented affordances that will generate a new product.

Drawing from Vygotsky's work creatively transformed by successive generations of researchers (as understood by Kozulin, 2003), Martinovic et al. (2013) have combined the theories of affordance and activity in order to understand "visualized mathematics" and "elearning" phenomena. Thus, connections are established (and managed) between subjects (pupils) and objects (knowledge) through activity (with emphasis on context and environment) that mobilizes one's capacity to act and through affordances (wealth of interaction between subjects and their environments). The tools (in our case, technology) disrupt (and eventually transform) this learning process, which has the potential to become more collaborative, self-directed, democratic, communicative, and multimodal (Martinovic et al., 2013).

In this way, we were able to observe how the use of a technological tool transformed pupils' environment. The *milieu* to which pupils are sensitive may therefore reveal an accelerated process given the feedback provided by the technological tool. This accelerated process complicates teaching challenges in two ways. On the one hand, teachers cannot predict the knowledge that will emerge from the pupils' experimentations unless they have participated in them. On the other hand, pupils' conscious realizations require interpretations that place specific knowledge within a learning process rather than a teaching process. So, how can we prepare teachers for these new challenges?

We have already noted how experimenting with a co-built interpretation and intervention model (DeBlois, 2009) led some teachers to increase the complexity of their concerns regarding their conceptions of the object being evaluated, the intervention categories, and the possible flexibility. In order to enhance these concerns, it seems fundamental to make a distinction between a teaching process and a learning process. This may help awaken curiosity to know more about the meaning of the procedures chosen by their pupils (DeBlois, 2009) or about the fact that pupils' procedures were not a lack of conformity, but the product of an environment (DeBlois, 2006); a triggering element of their interpretation. In these circumstances, meaning may be linked to one's relationship with knowledge and learning (DeBlois, 2014), taking into account not only the characteristics of what is taught, but also its epistemic, social, and identity-related nature. In this context, the value and meaning of institutional knowledge provide a framework in which teachers practices can be situated. It is then possible to depart from the beaten tracks to discuss how adding constraints, rather than reducing them when faced with a mistake, may lead to the emancipation of pupils' epistemological position.

Finally, some educational concepts such as didactic time (Chopin, 2011), didactic memory (Centeno & Brousseau, 1991), and didactic contract (*contrat didactique*; Brousseau, 1986) may, among other things, contribute to establishing markers for interpreting the *milieu* in which pupils work. If pupils' expectations are an expression of their commitment and if their experimentations are expressions of their creativity, then it is crucial to train teachers to develop resources to help with interpretation, which can guide their resources for action.

19.7 Conclusion

In many ways, education consists of a complex juggling act: balancing pupils' general and specific knowledge, balancing approaches to provoke a transformation in the pupils' knowledge, balancing training and evaluation, etc. This equilibrium is increasingly fragile as the stakes become more complex, hence the need to develop cross-curricular approaches in education, that is, to create bridges and meeting places where elements appear to multiply. Our study suggested a new way of understanding creativity through two disciplines, based on the works of two leading thinkers, Piaget and Vygotsky. Our project can also be viewed as an attempt to reach a didactic-pedagogical balance between the internal and external processes involved when combining ICT and creativity in an educational setting.

On the one hand, we have shown that learners' relationship with knowledge can change because creativity involves rebuilding one's link of appropriation with what is being learned. This reconstruction, supported by using ICT, seems to be enhanced by the pupils' emancipation from the social status of knowledge. However, there is a new balance to be reached between the teacher's educational project or pedagogical device and the pupils' learning project. While Piaget's work requires teachers to be conscious of the various ICT frameworks in order to interpret their pupils' creativity, Vygotsky's work involves paying attention to the socio-historical construction of these relationships. Thus, rather than accounting for learning according to the learner-object relationship, we need to integrate a learner-object-colearner-tool relationship since each of these elements is a gateway to learners' creativity. The next step in our study will be to test our learning situation models in mathematics and music, as well as in any other discipline, but mainly in a cross-disciplinary setting.

Acknowledgements The research was supported by the *Faculté des études supérieures et de la recherche, Université de Moncton*. The preparation of the chapter was also supported by the Partnership Development Grant of the *Canadian Social Sciences and Humanities Research Council* (Grant #890-2013-0062).

References

Amadio, M., Truong, N., & Tschurenev, J. (2006). *Instructional time and the place of aesthetic education in school curricula at the beginning of the twenty-first century*. IBE Working Papers on Curriculum Issues, 1. Retrieved from http://unesdoc.unesco.org/images/0014/001462/146280e.pdf.

Archambault, A., & Venet, M. (2007). Le développement de l'imagination selon Piaget et Vygotsky: d'un acte spontané à une activité consciente. *Revue des sciences de l'éducation, 33*(1), 5–24.

Bélanger, J. P., DeBlois, L., & Freiman, V. (2014a). Interpréter la créativité du raisonnement dans les productions d'élèves en mathématiques d'une communauté d'apprentissages multidisciplinaires interactifs. *Éducation et Francophonie, XLII*(2), 44–63.

Bélanger, J.-P., DeBlois, L., & Freiman, V. (2014b). Interpréter la créativité manifestée dans les productions d'élèves en mathématiques. *Actes de la 36e rencontre du groupe canadien de didactique des mathématiques* (pp. 237–238). Toronto, Canada: CMESG/GCEDM.

Brousseau, G. (1986). Fondements et méthodes de la didactique des mathématiques. *Recherches en didactique des mathématiques, 7*(2), 33–115.

Brousseau, G. (1988). Le contrat didactique: Le milieu. *Recherches en didactiques des mathématiques, 9*(3), 309–336.

Brousseau, G. (2010). *Glossaire de quelques concepts de la théorie des situations didactiques*. Retrieved from http://guy-brousseau.com/wp-content/uploads/2010/09/Glossaire_V5.pdf.

Burnard, P. (2007). Reframing creativity and technology: Promoting pedagogic change in music education. *JMTE, 1*(1), 37–55.

Centeno, J., & Brousseau, G. (1991). Rôle de la mémoire didactique de l'enseignant. *Recherches en Didactique des Mathématiques, 11*(2, 3), 167–210.

Chopin, M.-P. (2011). *Le temps de l'enseignement. L'avancée du savoir et la gestion des hétérogénéités dans la classe*. Rennes, France: Presses universitaires de Rennes.

Contré Migwans, D. (2013). *Une pédagogie de la spiritualité amérindienne: Naa-ka-nah-gay-win*. Paris, France: L'Harmattan.

Csikszentmihalyi, M., & Wolfe, R. (2000). New conceptions and research approaches to creativity: Implications of a systems perspective for creativity in education. In K. Heller, F. J. Monks, R. J. Sternberg & R. F. Subotnik (Eds.), *International handbook of giftedness and talent* (pp. 81–93). Oxford, UK: Elsevier Science.

Damasio, A. R. (1999). *Le sentiment même de soi: Corps, émotions, conscience*. Paris, France: Odile Jacob.

DeBlois, L. (1995). Le développement de l'écriture des nombres chez Christine. *Revue des sciences de l'éducation, 21*(2), 331–351.

DeBlois, L. (1996). Une analyse conceptuelle de la numération de position au primaire. *Recherches en Didactique des Mathématiques, 16*(1), 71–128.

DeBlois, L. (1997a). Trois élèves en difficulté devant des situations de réunion et de complément d'ensembles. *Educational Studies in Mathematics, 3*(1), 67–96.

DeBlois, L. (1997b). Quand additionner ou soustraire implique comparer. *Éducation et Francophonie, 25*(1), 102–120 (http://ACELF.CA/revue).
DeBlois, L. (2003). Interpréter explicitement les productions des élèves: une piste. *Éducation et Francophonie, 31*, 176–192.
DeBlois, L. (2006). Influence des interprétations des productions des élèves sur les stratégies d'intervention en classe de mathématiques. *Educational Studies in Mathematics, 62*, 307–329.
DeBlois, L. (2009). Les contextes et les besoins à l'origine de la recherche collaborative. *Annales de didactique et de sciences cognitives, 14*, 213–229. Retrieved from http://turing.scedu.umontreal.ca/annales/documents/volume%2014/9DeBlois_mep.pdf.
DeBlois, L. (2014). Le rapport aux savoirs pour établir des relations entre troubles de comportements et difficultés d'apprentissage en mathématiques. In M.C. Bernard, A. Savard, & C. Beaucher (Eds.), *Le rapport aux savoirs: Une clé pour analyser les épistémologies enseignantes et les pratiques de la classe* (pp. 93–106). Retrieved from http://lel.crires.ulaval.ca/public/le_rapport_aux_savoirs.pdf.
de Champlain, Y. (2010). *Agir dans l'incrédulité: Investir des chemins de liberté pour inventer un lieu pédagogique*. Sarrebruc, Germany: Éditions universitaires européennes.
Denoyel, N. (1999). Alternance tripolaire et raison expérientielle à la lumière de la sémiotique de Peirce. *Revue française de pédagogie, 128*, 35–42.
Douady, R. (1986). Jeux de cadre et dialectique outil-objet. *Recherches en Didactique des Mathématiques, 7*(2), 5–31.
Education at the National Research Council (ENRC). (2012). *Education for life and work, developing transferable knowledge and skills in the 21st century*. Retrieved from http://www7.national-academies.org/bota/Education_for_Life_and_Work_report_brief.pdf.
Faraj, S., & Azad, B. (2012). The materiality of technology: An affordance perspective. In P. M. Leonardi, B. Nardi, & J. Kallinikos (Eds.), *Materiality and organizing: Social interaction in a technological world* (pp. 237–258). New York, NY: Oxford University Press.
Freiman, V., & DeBlois, L. (2014). Le site internet CAMI: Une ressource virtuelle pour soutenir la résolution de problèmes chez les élèves francophones du Nouveau-Brunswick. *Cahiers des Sciences de l'Éducation–Université de Liège (aSPe), 36*, 115–142.
Freiman, V., & Lirette-Pitre, N. (2009). Building a virtual learning community of problem solvers: Example of CASMI community. *ZDM—The International Journal in Mathematics Education, 41*(1–2), 245–256.
Freiman, V., & Manuel, D. (2014). Une ressource virtuelle de résolution de problèmes mathématiques: Les perceptions d'utilisateurs et les traces d'usage. *Sciences et Technologies de l'Information et de la Communication pour l'Éducation et la Formation, 20*. http://sticef.univ-lemans.fr/num/vol2013/18-freiman-reiah/sticef_2013_NS_freiman_18.htm.
Freiman, V., & Manuel, D. (2015). Relating students' perceptions of interest and difficulty to experts' views of richness of mathematical problems. *Quadrante*, 34(2), 61–84.
Gadamer, H.-G. (1995). *Langage et vérité*. Paris, France: Gallimard.
Gall, M., & Breeze, N. (2005). Music composition lessons: The multimodal affordances of technology. *Educational Review, 57*(4), 415–433.
Gall, M., & Breeze, N. (2008). Music and eJay: An opportunity for creative collaborations in the classroom. *International Journal of Education Research, 47*(1), 27–40.
Gibson, J. J. (1977). The theory of affordances. In R. Shaw & J. Bransford (Eds.), *Perceiving, acting and knowing: Toward an ecological psychology* (pp. 127–143). Hillsdale, N.J.: Lawrence Erlbaum Associates.
Giglio, M. (2006). Créativité musicale et réflexions partagées: Étude pilote chez des élèves d'école secondaire. *JFREM*, 195–207.
Glăveanu, V. P. (2014). *Thinking through creativity and culture, toward an integrated model*. New Brunswick, NJ: Transactions Publishers.
Hersant, M. (2003). L'utilisation de bases de problèmes en classe de mathématiques. Un exemple avec "La proportionnalité à travers des problèmes". *Colloque international Intégration des technologies dans l'enseignement des mathématiques*. Reims. Retrieved from http://www.reims.iufm.fr/Recherche/ereca/itemcom/.

Kauffman, J., & Bauer, J. (2004). Sure I'm creative—but not in mathematics; self-reported creativity in diverse domains. *Empirical Studies of the Arts, 22*(2), 143–155.
Kintsch, W., & Van Dijk, T. A. (1978). Toward a model of text comprehension and production. *Psychological Review, 85*(5), 363–394.
Kozulin, A. (2003). Psychological tools and mediated learning. In A. Kozulin, B. Gindis, V. Ageyev, & S. M. Miller (Eds.), *Vygotsky's educational theory in cultural context* (pp. 15–38). Cambridge, UK: Cambridge University Press.
Kozulin, A., Gindis, B., Ageyev, V., & Miller, S. (Eds.). (2003). *Vygotsky's educational theory in cultural context (Learning in doing: Social, cognitive and computational perspectives).* Cambridge, UK: Cambridge University Press.
Kranzber, M. (1986). Technology and history: Kranzberg's laws. *Technology and Culture, 27*(3), 544–560.
Kratus, J. (1990). Structuring the music curriculum for creative learning. *Music Educators Journal, 76*(9), 33–37.
Kuntz, G. (2007). Des mathématiques en ligne pour renouveler l'enseignement de mathématiques. *Repères—IREM, 66,* 104–113.
Legault, M. (2005). La symbolique en analyse de pratique. Pour une pratique de la présence au vécu de l'action et au vécu de la réflexion (1ere partie). *Expliciter, 62,* 35–44.
Legault, M. (2006). La symbolique en analyse de pratique. Pour une pratique de la présence au vécu de l'action et au vécu de la réflexion (2e partie). *Expliciter, 64,* 1–7.
Leikin, R., Levav-Waynberg, A., Gurevich, I., & Mednikov, L. (2006). Implementation of multiple solution connecting tasks: Do students' attitudes support teachers' reluctance? *Focus on Learning Problems in Mathematics, 28,* 1–22.
Leontiev, A. A. (1981). *Psychology and the language learning process.* Oxford, UK: Pergamon.
Maffesoli, M. (1996). *The times of the tribes: The decline of individualism in mass society.* London, UK: Sage.
Manuel, D. (2018). Virtual learning communities of problem solvers: A potential for developing creativity in mathematics? In V. Freiman & J. L. Tassell (Eds.), *Creativity and technology in mathematics education* (pp. 531–557).
Martinovic, D., Freiman, V., & Karadag, Z. (2013). Visual mathematics and cyberlearning in view of affordance and activity theories. In D. Martinovic, V. Freiman & Z. Karadag (Eds.), *Visual Mathematics and Cyberlearning. Mathematics Education in the Digital Era (MEDEra) Book Series* (vol. 1, pp. 209–238). The Netherlands: Springer.
Maslow, A. (1954). *Motivation and personality.* New York, NY: Harper & Row.
Moran, S. (2010). Commitment and creativity: Transforming experience into art. In C. Connery, V. John-Steiner, & A. Marjanovic-Shane (Eds.), *Vygotsky and creativity: A cultural-historical approach to meaning-making, play, and the arts* (pp. 141–160). New York, NY: Peter Lang.
Morgan, N. & Cook, J. (2014). *Teaching cycles in the creative community of inquiry.* Retrieved from https://www.researchgate.net/publication/253227964_Teaching_Cycles_in_the_Creative_Community_of_Inquiry.
Morin, E. (1990). *Introduction à la pensée complexe.* Paris, France: Éditions du Seuil.
Morin, E. (1999). *Les sept savoirs nécessaires à l'éducation du futur.* Paris, France: UNESCO.
Nicolescu, B. (1996). *La transdiciplinarité.* Manifeste. Monaco, Éditions du Rocher, 34–35. http://www.basarab-nicolescu.fr/BOOKS/TDRocher.pdf.
Norman, D. A. (1988). *The psychology of everyday things.* New York, NY: Basic books.
Partnership for 21st century skills. (2011). *Framework for 21st century learning.* Retrieved from http://www.p21.org/storage/documents/1.__p21_framework_2-pager.pdf.
Piaget, J. (1972). L'épistémologie des relations interdisciplinaires. *L'interdisciplinarité : problèmes de recherche et d'enseignement dans les universités* (pp. 154–171). Paris, France: OCDE.
Piaget, J. (1974a). *La prise de conscience.* Paris, France: Presses Universitaires de France.
Piaget, J. (1974b). *Réussir et comprendre.* Paris, France: Presses Universitaires de France.

Piaget, J. (1977a). *Recherches sur l'abstraction réfléchissante I*. Paris, France: Presses Universitaires de France.
Piaget, J. (1977b). *Recherches sur l'abstraction réfléchissante II*. Paris, France: Presses Universitaires de France.
Pigott, J. (2007). The nature of mathematical enrichment: A case study of implementation. *Educate, 7*(2), 30–45.
Rabardel, P. (1995). Qu'est-ce qu'un instrument? Appropriation, conceptualisation, mises en situation. *Les dossiers de l'ingénierie éducative, 19,* 61–65.
René de Cotret, S. (1999). Quelques questions soulevées par l'adoption d'une perspective "bio-cognitive" pour l'étude de relations du système didactique'. Dans *Séminaire DidaTech, Didactique et Technologies Cognitives en Mathématiques 1997* (pp. 161–178). Grenoble, France: Laboratoire Leibniz-IMAG.
Renninger, K. A., & Shumar, W. (2002). Community building with and for teacher at the math forum. In K. A. Renninger & W. Shumar (Eds.), *Building virtual communities: Learning and change in cyberspace* (pp. 60–95). Cambridge, UK: Cambridge University Press.
Robichaud, X. (2016). *Créativité musicale à l'ère numérique: Une étude d'explorations sonores conduites par les élèves du primaire à l'aide d'une station audionumérique*. Thèse de doctorat: Université de Moncton.
Robichaud, X., & Freiman, V. (2016). Exploring deeper connections between, mathematic, music and ICT: What avenues for research in education? In *Proceedings of the Symposium MACAS – Mathematics and its Connections to the Arts and Sciences*, 28 au 21 mai 2015, Schwäbisch Gmünd, Germany: University of Education.
Robinson, K. (2005). *Out of our minds: Learning to be creative*. Oxford, UK: Capstone.
Rochex, J. Y. (1997). Note de synthèse. L'oeuvre de Vygotski: Fondements pour une psychologie historico-culturelle. *Revue française de pédagogie, 120*(1), 105–147.
Rogers, C. (1961). *Le développement de la personne*. Paris, France: Dunod.
Saboya, M. (2012). Analyse d'une didactique d'intervention autour du développement d'une activité de contrôle: stratégies d'enseignement et indicateurs de contrôle chez les élèves du secondaire. In J. L. Dorier & S. Coutat (Eds.), *Enseignement des mathématiques et contrat social : enjeux et défis pour le 21e siècle – Actes du colloque EMF2012* (GT9, pp. 1246–1258). http://www.emf2012.unige.ch/index.php/actes-emf-2012.
Savage, J. (2005). Working towards a theory for music technologies in the classroom: How pupils engage with and organize sounds with new technologies. *British Journal of Education, 22*(2), 167–180.
Smolucha, F. (1992). A reconstruction of Vygotsky's theory of creativity. *Creativity Research Journal, 5*(1), 49–67.
Trouche, L. (2005). Des artefacts aux instruments, une approche pour guider et intégrer l'usage des outils de calcul dans l'enseignement des mathématiques. *Actes de l'Université d'été « Le calcul sous toutes ses formes »*, Ministère de l'Education Nationale, en ligne : http://www3.ac-clermont.fr/pedago/maths/pages/site_math_universite/CD-UE/Menu_pour_Internet.htm.
Tymieniecka, A. T. (1972). *Eros et Logos: Esquisse de phénoménologie de l'intériorité créatrice*. (illustrée par les textes de Paul Valéry). Louvain, Belgium: Nauwelaerts.
Varela, F. J., Thompson, E., & Rosch, E. (1993). *The embodied mind: Cognitive science and human experience*. Cambridge, MA: MIT Press.
Vermersch, P. (2003). *L'entretien d'explicitation* (Quatrième édition enrichie d'un glossaire) Issy-les-Moulineaux, France: ESF Éditeur.
Vergnaud, G. (2000). *Lev Vygotski pédagogue et penseur de notre temps*. Paris, France: Hachette Education.
Von Glaserfeld, E. (1982). An interpretation of piaget's constructivism. *Revue Internationale de Philosophie, 36*(4), 612–635.
Von Glasersfeld, E. (2013). *Radical constructivism*. Hoboken, NJ: Taylor and Francis.
Vygotsky, L. S. (1934). *Pensée et Langage*. Paris, France: La Dispute.

Vygotsky, L. S. (2004). Imagination and creativity in childhood. *Journal of Russian and East European Psychology, 42*(1), 7–97.
Warfield, V. (2006). *Introduction to didactique*. Seattle, WA: University of Washington. Retrieved from http://www.math.washington.edu/~warfield/Inv%20to%20Did66%207-22-06.pdf.
Watson, S. (2011). *Using technology to unlock musical creativity*. New York, NY: Oxford University Press.
Webster, P. R., & Hickey M. (2001). Creative thinking in music. *Music Educators Journal*, 19–23.
Yasnitsky, A. (2011). The Vygotsky that we (do not) know: Vygotsky's main works and the chronology of their composition. *PsyAnima, Dubna Psychological Journal, 4*(4), 53–61.

Chapter 20
Putting the Horses Before the Cart: Technology, Creativity, and Authorship Harnessed Three Abreast

Osnat Fellus and Yaniv Biton

Abstract Over the past few decades, educational systems have continually worked on integrating technology into mathematics education. Creativity, on the other hand, was—more often than not—less attended to. Building on Latour's (A sociology of monsters: Essays on power, technology, and domination. Routledge, London, pp. 103–131, 1991) perspective on technology, Vygotsky's (J Russ East Eur Psychol 42(1):7–97, 2004) treatment of creativity, and Bakhtin's (The dialogic imagination: four essays. University of Texas Press, Austin, Texas, 1981) perception of authorship, we contemplate a departure from traditional ways of viewing creativity in mathematics education—as arrogated to giftedness—and a shift of attention to a unified manifestation of apprenticeship and authorship to allow for expressions of creativity for all learners. The works of these three classic scholars, brought together, widen our conceptualization of technology, creativity, and learning; show how the very intertwining of these notions allows for a far wider perspective of creativity in mathematics education; and foster the learning and understanding of mathematics as a network of ideas. In this chapter, we first denote the terms technology, creativity, and authorship and then we showcase how these are manifested in an episode drawn from the work of the second author as a mathematics instructor in the virtual high school in Israel. The chapter concludes with a discussion of some of the implications this meeting point between Latour's, Vygotsky's, and Bakhtin's perceptions may have on the learning of mathematics.

Keywords TCA · Latour · Vygotsky · Bakhtin · Democratization of creativity

O. Fellus (✉)
University of Ottawa, Ottawa, ON, Canada
e-mail: osnat.fellus@uottawa.ca

Y. Biton
Shaanan College and Centre for Educational Technology (CET), Israel
e-mail: yanivb@cet.ac.il

© Springer International Publishing AG, part of Springer Nature 2018
V. Freiman and J. L. Tassell (eds.), *Creativity and Technology in Mathematics Education*, Mathematics Education in the Digital Era 10,
https://doi.org/10.1007/978-3-319-72381-5_20

20.1 Background

Research on the integration of digital technologies into schooling in general and mathematics education in particular has long been a subject of interest in the literature of mathematics education. In fact, it has been continually generating new —and renewed—explorations followed by instructive insights about the relationship between learning mathematics and using readily available technologies. To reflect the development and expansion of research on the use of technology in mathematics education, we start with Etlinger (1974), who pointed to the distinction between functional and pedagogical purposes for using digital technology. Functional purposes, he argued, emphasize the use of technology for computational processes while pedagogical purposes highlight its use for teaching and learning mathematical ideas. This rather simplistic, unidirectional view of the use of technology was expanded to represent a more complex relationship between technology and the teaching and learning of mathematics such as how thinking shapes—and is shaped by—our use of technology. Kieran and Drijvers (2006), for example, highlight the co-emergent dialectical relationship between thinking and technology. Similarly, Borba, Askar, Engelbrecht, Gadanidis, Llinares, and Aguilar (2016) point to additional, co-constitutive aspects such as affective-related notions as well as considerations of who teaches what, when, how, and why in technologically embedded contexts. Overall, despite continuous efforts to more systemically integrate digital technologies into teaching and learning, the practices, challenges, and frameworks associated with these efforts continue to be a concern for teacher educators (Hughes, 2013; Jimoyiannis, 2010; Lagrange, Artigue, Laborde, & Trouche, 2003; Polly, Mims, Shepherd, & Inan, 2010; Tassell, Stobaugh, & McDonald, 2013; Tondeur et al., 2012). This effort to integrate technology into mathematics classes is not surprising as there is growing research that associates the use of digital technology with learners' better understanding of mathematical concepts and their relationship (Hohenwarter, Hohenwarter, & Lavicza, 2008; Kieran & Damboise, 2007; Tan & Tan, 2014; Verzosa, Guzon, & Penãs, 2014; Yilmaza, 2015).

Research on creativity in mathematics education, on the other hand, has been scarce (Van Harpen, & Sriraman, 2013; Leikin & Pitta-Pantazi, 2013; Sriraman, Yaftian, & Lee, 2011) focusing mostly on the association between creativity and giftedness (Freiman & Sriraman, 2008; Leikin, Berman, & Koichu, 2009; Mann, 2006; Mhlolo, 2017; Sriraman, 2005). Having said that, a growing number of scholars argue that creativity in mathematics education is not necessarily coupled only with giftedness (Sriraman, 2005; Van Harpen & Sriraman, 2013), and that it is more about performance and passion than about an innate ability to mathematize (Sheffield, 2017) thus leaving open the exclusive association between creativity and giftedness in mathematics and generating new questions pertaining to how to habituate and nurture creativity in mathematics among all students (Prabhu, & Czarnocha, 2014; Sheffield, 2017).

This somewhat untraditional, democratic stance in regard to who can be mathematically creative opens promising new trajectories for research. For now, though, we can draw on the scant work done on the combination between mathematics and creativity. One example of such work can be seen in the mathematically creative work of students at Brock University, Canada, where the undergraduate core mathematics program follows a philosophy of learning mathematics in computer-based environments through a series of *Mathematics Integrated with Computers and Applications* (MICA) courses instituted there in 2001. The overarching two-tiered purpose of the MICA courses is to "encourage mathematical creativity" and to "develop mathematics concepts" (Marshall & Buteau, 2014, p. 50). Marshall and Buteau's report highlights the opportunities students get in the MICA courses to explore mathematical concepts, conjectures, and applications and to garner first-hand experience of the creative aspect of working in and through mathematics. This notion of creativity in mathematics through the use of available technologies seems to be the realization of the prediction articulated by Salomon, Perkins, and Globerson (1991), which envisioned that if technology becomes central in education, schooling will shift from "knowledge imparting to self-guided exploration and knowledge recreation" (p. 7), thus highlighting the benefits that digital technologies carry to support and enhance intellectual performance through free choice (Papert, 1992).

Against this backdrop, we frame our discussion within two metaphorical frameworks—one of a cart pulled by horses, harnessed three abreast; the other metaphor is that of a network that allows us to see the interconnectivity and intertwining relationship technology, creativity, and authorship in the context of mathematics education. We use these metaphors to suggest a broader framework for these constructs within mathematics education that allows us, on the one hand, to perceive technology, creativity, and authorship as paramount to the advancement of mathematics learning, and on the other hand, to see the relationships between the constructs as a networked, interconnected, and dynamic system where the whole is greater than the sum of its parts.

The Venn diagram (see Fig. 20.1) visually illustrates diverse areas of research that focus on different combinations of mathematics (M), technology (T), and creativity (C). While categorizing scholarly work into any of the intersection areas in the diagram below may yield a clearer picture as to which areas received adequate exploration and which still need to be further explored, we do not at all imply that such categorization is mutually exclusive. Work that largely focuses on the learning of mathematical ideas through the use of technology (see for e.g., Lawrenz, Gravely, & Ooms, 2006; Nguyen & Trinh, 2015; Novak & Tassell, 2015; Yerushalmy & Botzer, 2011) may be framed within the intersection area of $T \cap M$. There are copious publications that look into learning mathematics through technology. A review of cutting edge work on learning mathematics through a great variety of digital technologies including mobile devices, digital libraries, and personal spaces can be found in Borba et al. (2016). Other studies that focus more on creativity and mathematics such as Kwon, Levenson (2011), Haylock (1997), Livne and Milgram (2006), Bolden, Harries, and Newton (2010),

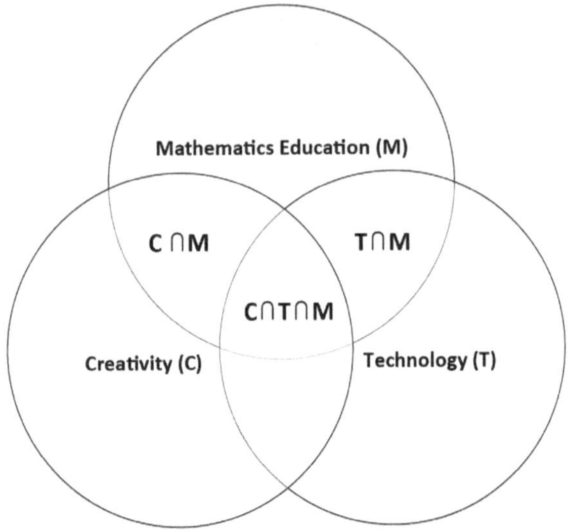

Fig. 20.1 A venn diagram showing the area of C ∩ T ∩ M as distinct from other areas

and Van Harpen and Sriraman (2013) reflect work in the intersection area of C ∩ M. However, very little research focuses on the combination of creativity, technology, and mathematics, C ∩ T ∩ M. In this chapter, we aim to shift attention to the underrepresented and not-yet-adequately explored C ∩ T ∩ M intersection area thus contributing to its emerging conversations and scholarly reflections.

In the following sections, we first define the concepts of technology, creativity and authorship. In order to exemplify a combination of the three, we then discuss an episode from the work of the second author in the virtual high school that is run by the Israeli Center for Educational Technology (CET). We conclude the chapter with possible implications of the unification of the three constructs.

20.2 Pulling the Cart of Mathematics Education with TCA (Technology, Creativity, and Authorship)

20.2.1 Harnessing Technology

In order to better understand the use of technology in mathematics education, we first define what we mean by the term *technology*. We use *technology* as the use of available tools to carry out actions. It is clear to us, though, that this is never a simple, unidirectional, or flat phenomenon. While the use of a tool is guided by the user's knowledge and experience, the tool itself carries affordances and constraints as to how it can be used. The former is dubbed *instrumentalization, which denotes agentive action*; the latter, *instrumentation, which references predesigned features to be used* (Artigue, 2002; Healy & Kynigos, 2010; Kieran & Drijvers, 2006). Working with this dyad of *instrumentalization* and *instrumentation* helps us perceive the co-emergent and co-constitutive nature of human-tool relationship. This relationship may explicate the rapid development and use of tools not only as an integral part of our sentient being but also as human-specific. In this respect, technology is perceived as a manifestation of the genealogical progression of the interaction between humans and tools.

If in the far past, understanding the role of tools in human life was left tacit and faded into the background, the works of Ellul (1964), Latour (1986, 1991), as well as those of Vygotsky (e.g., 1978, 1986) and Luria (e.g., 1960, 1994)—whose manuscripts were originally published in Russian as far back as the late 1920s—have made it possible for us to see the mutually constitutive nature of the relationship between humans and tools. What these scholars advocated for was a framework of reference that helps us perceive such tools as readily available extensions of being in the world. For example, Latour (1991) argues that while tools and humans are distinct, they are inseparable in everyday life. In a similar token, Ellul (1964) shows how simple tools, such as a pen, for example, function as human extensions to carry out tasks. Vygotsky's (1978) contribution to understanding the nature of human-tool relationship is equally relevant as he rigorously demonstrates how language, speech, and other semiotic artifacts are, in fact, used as tools and that they too are inseparable from human action.

The conceptualization of the mutually constitutive and co-emergent nature of human-tool relationship provides us a broader sense of what we mean by technology and functions as an important entry point in our work. We will thus, henceforth, use the term *technology* in this broader sense to reference any tool (digital, such as a computer, or non-digital, such as pencil-and-paper) that is used to carry out a task. Whereas we recognize the common denominator of such tools as essentially utilitarian, we argue that the way they are used may not be limited to their pre-determined purpose. Rather, instrumentalization and instrumentation are in constant process of co-emergent animation. As such, our attention to how technology becomes interwoven with our daily practices shifts from perceiving it as an end in and of itself to perceiving it as an ever changing human-tool interaction, which is exactly where creativity (defined later) comes into the picture. But before we discuss *creativity*, it will be helpful to understand what human-tool relationship means. We thus turn to research that sheds light on this relationship.

The co-constitutive nature of human-tool interaction is corroborated by research in the field of the anthropology of technology. Anthropologists of technology, Suchman, Blomberg, Orr, and Trigg (1999) and Suchman (2007), explored the relationship between everyday uses of technology and its design and development. By observing people's encounters with new technologies, they found that the successful integration of such technologies was less a matter of the users' "technological sophistication" and more a matter of a "familiarity with the particular features the technology offers" (Suchman, Blomberg, Orr, & Trigg, 1999, p. 394). Furthermore, they found that technology-infused practices involve complex manifestations of learning how to use these technologies—from more knowledgeable others (p. 399)—to address emerging needs in ways that were unpredicted and not previously designed thus expanding and improving the functionality of the tool itself. This finding of co-constitution is an important notion as it centralizes the human-tool relationship as interactive and puts into question the mere unidirectional reliance on tools and technology.

For the purpose of this chapter, we argue that technology, in its wider sense, is not only inseparable from our being in the world but also poses its own platform of interaction with humans, or in Latour's (1991) words, "When we consider social relations we need to weave them into a fabric that includes non-human actants, actants that offer the possibility of holding society together as a durable whole" (p. 103). It is this notion of technology as omnipresent and integral to human life that we wish to put forth here. We denote technology, then, as an all-encompassing view of human-tool interaction. As such, in our thinking about the use of technology in mathematics education, we propose the use of the verb *interweave* over *integrate* as the former carries a stronger reference to the notion of connections and relationship than the latter. Tools—digital and non-digital—are an extension of our being in the world or at least a partner in our task-oriented behaviour (Salomon, Perkins, & Globerson, 1991). Indeed in the context of mathematics education, the use of technology was found to carry more diverse anthropomorphic characteristics that translate into seeing technology as a master, a servant, or a partner (Geiger, 2009). Whichever the relationship perceived by the user, technology is something that cannot be ignored as it frames our actions, reactions, and interactions with it.

Such inter-relationship between humans and tools and its co-emergent and co-constitutive nature was highlighted by Borba et al. (2016) who have pointed out, "as humans develop and construct new media, these media seem to transform and 'construct' a new human" (p. 591). In a similar vein, Kieran and Drijvers (2006) note that what the tool allows the user to do affects the user's thinking, and in turn, the user's thinking shapes the way the tool is used. Within this context, our view of the use of technology in mathematics education shifts from perceiving it as a mere token of progress to seeing it as a venue to creatively interact with mathematical ideas. We suggest that technology can be used to germinate creativity, which we turn to in the next section.

20.2.2 Harnessing Creativity

Given Vygotsky's perspective of the work of education as an enterprise that is more oriented toward the future than the past, and toward what the student will be able to do tomorrow than what he or she already mastered yesterday (Davydov, 1986/2008, p. 40; Fellus & Biton, 2017), we draw on Vygotsky's (2004) work to understand the concept of creativity in terms of whether learners' activity is oriented toward producing new ways of representation of knowledge or reproducing replicas of previously formulated ways of knowing (Vygotsky, 2004). These terms of production and reproduction are important to our discussion because they frame acts of creativity as qualitatively distinct from and more collectively meaningful than acts of reproduction of representations that were already previously attained by others. The former is associated with creativity and progress, the latter with stagnation and regression, because, to quote Vygotsky, the latter "is very closely linked to memory; essentially it consists of a person's reproducing or repeating previously developed and mastered behavioral patterns or resurrecting traces of earlier impressions…[it refers to instances] where actions are based on more or less accurate repetition of something that already exists" (Vygotsky, 2004, p. 7).

Using Vygotsky's framework of creativity as actions that are more oriented toward the future (Fellus & Biton, 2017), we ask ourselves, how, then, do acts of creativity look like? Vygotsky's answer spans over the just-shy-of-a-hundred pages of his *Imagination and creativity in childhood,* where he brings in empirical support from different studies conducted with children to buttress his perception of *creativity*. Whereas Vygotsky mostly refers to creativity in writing, we believe that the principles and suggestions he pushes forth are equally relevant to mathematics education. In order to set up adequate conditions for creativity to happen, Vygotsky suggests that we "offer the greatest and widest choice of topics, without selecting those you think are particularly suited to children" (Vygotsky, 2004, p. 49). Such condition allows for free choice through *combinatorial* activity (Vygotsky, 2004, p. 9), which takes place "whenever a person imagines, combines, alters, and creates something new" (Vygotsky, 2004, p. 10). This can happen when students have opportunities "to combine elements to produce a structure, to combine the old in new ways" (Vygotsky, 2004, p. 10). According to Vygotsky, the ability for combinatorial thinking is human-specific and is an extension of one's memory and experience.

Vygotsky's framework to understanding *creativity* lends itself to using the term *bricoleur* (Levi-Strauss, 1962) to reference the role students play in promoting *combinatorial* activity by re-using attained material to create something new. According to Vygotsky (2004), the mechanism through which this combinatorial work can be carried out takes the form of acts of dissociation—i.e., breaking up a whole into discrete parts—and acts of association—i.e., knitting together discrete items to create new representations of networked systems of knowledge drawing on past experience, imagination, and emotions. Combinatorial thinking, according to Vygotsky (2004) can only be carried out through free choice because: "internal expression [is] associated with the choice of thoughts, images, and impressions" (Vygotsky, 2004, p. 18).

Framing creativity as combinatorial thinking, we ask ourselves: Who can practice it? What may happen if we do not allow it? And what are the conditions that are conducive to it? The question of whether creativity is an activity saved only for the abled few was negated by Vygotsky thus positioning him together with a growing number of scholars who call for the democratization of creativity and the appreciation of its manifestations among all learners. Vygotsky explains,

> There is a very widespread opinion that creativity is the province of the select few and that only those who are gifted with some special talent should develop it in themselves and have the right to consider themselves to have a vocation for creation. This is not true, as we have attempted to explain above. If we understand creativity in its true psychological sense as the

> creation of something new, then this implies that creation is the province of everyone to one degree or another; that it is a normal and constant companion in childhood (Vygotsky, 2004, p. 37-38).

As to the second question, the indelible effects of not scaffolding opportunities for creativity for all learners may be irreversible as Vygotsky explains,

> The entire future of humanity will be attained through the creative imagination. [O]rientation to the future, behavior based on the future and derived from this future, is the most important function of the imagination. To the extent that the main educational objective of teaching is guidance of school children's behavior so as to prepare them for the future, development and exercise of the imagination should be one of the main forces enlisted for the attainment of this goal (Vygotsky, 2004, pp. 87-88).

As to the third question, among potential conditions, the act of choice is underscored as necessary for combinatorial thinking. Choice can be possible if there are options to choose from, thus suggesting experience and available knowledge as pre-conditions for creativity. This idea of choice as a necessary condition for creativity was also recognized by MacKinnon (1966) who explains that for creativity to happen, one needs "a greater range of information and a greater fluency of combination" (p. 153). In a similar vein, Perkins (1988) suggests the term "*crossing boundaries*" as a condition for creative thinking thus buttressing the act of combinatorial thinking through choice as a manifestation of creativity.

Recent studies in neuroscience provide empirical evidence that corroborates Vygotsky's perception of creativity as acts of combinatorial thinking (see for example, Jung, et al., 2010; Thagard & Stewart, 2011; Vartanian, Bristol, & Kaufman, 2013; Wu, et al., 2015; Zhu et al., 2016). Bendetowicz, Urbanski, Aichelburg, Levy, and Volle, (2017), for example, used the Remote Associates Test (RAT) with 57 participants to record their ability to generate a word associated to three unassociated cue words. For example, participants were expected to generate the word "cheese" as an associative word for "*rat, cottage, blue.*" The researchers concluded that the very act of creativity blazes "long-range pathways" (p. 225), which is made possible through the brain's structural network thus contributing to the construct validity of combination in measuring creativity. In light of this research, even though Vygotsky's work references children, we can safely argue that his perception of creativity as a manifestation of combinatorial practices applies to topics across disciplines, is not object- or discipline-constrained, is human-specific, and, above all, learnable (Zhu et al., 2016).

Fast forward almost a century since Vygotsky wrote his seminal work on creativity in Russian before it was translated into English, however, creativity in mathematics cannot yet crown itself with a clear definition—not to mention operationalization (Czarnocha & Baker, 2015; Sriraman, 2008; Nadjafikhah & Yaftian, 2013). Nevertheless, mathematical activities that foster and nurture creativity have been continually explored. These include students posing mathematical problems (Shaffer, & Clinton, 2006; Silver, 1997; Van Harpen & Sriraman, 2013), students solving open-ended questions (Hashimoto, 1997; Kwon, Park, & Park, 2006; Li & Li, 2009; Stacey, 1995), students working on multiple-solution tasks (Leikin, 2009;

Levav-Waynberg & Leikin, 2012), and students writing mathematical riddles (Prusak, 2015). Of particular relevance is Csikszentmihalyi's (1988) distinction between problem finding—rather than problem solving—and creativity and his call to focus on a learner's "acute curiosity" (p. 166) as the very element that ignites and continuously fuels his or her motivation and perseverance in finding a problem and formulating a solution. Scrutinizing these studies, one cannot ignore the focus on the product-oriented perspective of creativity. The questions we pose are how might the process of creativity in a mathematics classroom look like? What particular actions do students who engage with creative mathematical thinking carry out? And how can creativity be nurtured in class?

Using these questions to guide our focus and building on the above-mentioned scholarship, we follow Johnson-Laird (1988) quip, "Freedom of choice occurs par excellence in acts of creation" to suggest that creativity is strongly associated with the act of choice and that, in fact, choice is a dominant and determining factor in the act of creativity (Koichu, 2015; Sriraman, 2008). We also argue that choice is dependent on knowledge of options and experience as more input facilitates better connections. This is supported by a recent comparative study working with three high school mathematics classes (two classes in China, one in the USA), where Van Harpen and Sriraman (2013) showed how those who had more content knowledge did better in creating more complex mathematics problems.

In light of this scholarship on combinatorial creativity, choice, and mathematics creativity, we ask what specific pedagogical practices can foster acts of choice in mathematics education? We argue that Bakhtin's act of authorship (Holquist, 1983) is an important condition for this process to which we turn next.

20.2.3 Harnessing Authorship

Authorship, according to Bakhtin (1981), refers to the appropriation of utterances that are heard spoken (or written) by others. When people apply their own intentions and meanings to utterances, they take ownership over them and appropriate them. Bakhtin (1981) writes:

> The word in language is half someone else's. It becomes 'one's own' only when the speaker populates it with his own intention, his own accent when he appropriates the word, adapting it to his own semantic and expressive intention. Prior to this moment of appropriation, the word does not exist in a neutral and impersonal language (it is not, after all, out of a dictionary that the speaker gets his words!), but rather it exists in other people's mouths, in other people's contexts, serving other people's intentions: it is from there that one must take the word, and make it one's own (pp. 293–294).

According to Bakhtin, "A speaker is to his utterance what an author is to his text" (Holquist, 1983, p. 315). Drawing on Bakhtin's notion of authorship, Wertsch (1998) distinguishes between mastery of content and the act of authorship explaining that one's authorship takes place only by the user's putting new meanings into the mastered content. In our context, such use goes beyond mastery of mathematical content by allowing students to develop ownership over the mathematical ideas through the active use of mastered content to address their own needs.

Within Bakhtin's notion of authorship, it would be hard to overstate Vygotsky's contribution to the understanding of creativity in educational contexts because it qualifies creativity as an innate characteristic shared by all and carried out by acts of authoring through combinatorial actions or *bricolaging* as we call it. For some, authorship and mathematics education might be read as an oxymoron. But what we suggest here is that mathematical authorship allows learners to take charge over how they mathematize, how they make sense of mathematical principles, and how they make connections between seemingly disconnected mathematical ideas. We thus use the term *authorship* to reference learners' ongoing engagement through acts of meaning making and bricolaging in mathematics. Pedagogically speaking, such acts of meaning making of mathematical ideas may take the form of engaging with multiple representations of mathematical concepts, for example. The concept of multiple representations is paramount in mathematical thinking (Ainsworth, 2006; Barmby, Harries, Higgins, & Suggate, 2009; Davydov, 1991; Dreher & Kuntze, 2015; Dreher, Kuntze, & Lerman 2012; Gagatsis & Shiakalli, 2004; Goldin & Shteingold, 2001) and one would expect to have multiple opportunities for authorship in this form of mathematical engagement.

However, simply being able to transform one mathematical representation to another does not necessarily generate mathematical understanding and authorship. Seufert (2003) emphasizes: "Learners must interconnect the external representations and actively construct a coherent mental representation in order to benefit from the complementing and constraining functions of multiple representations" (p. 228). In the context of teaching multiplication, for example, learners should fluidly author representations to show use of the commutative property, the distributive property, or the associative property (Barmby, Harries, Higgins & Suggate, 2009; Chi, De Leeuw, Chiu, & LaVancher, 1994; Davydov, 1991). Given that multiple representations are "linked through reasoning" (Barmby, Harries, Higgins, & Suggate, 2009, p. 4) and that reasoning is the building block of authoring mathematical ideas (Moseley, 2005; NCTM, 2000), we position both in the core of mathematical thinking:

> Being able to reason is essential to understanding mathematics. By developing ideas, exploring phenomena, justifying results, and using mathematical conjectures in all content areas and—with different expectations of sophistication—at all grade levels, students should see and expect that mathematics makes sense (NCTM, 2000, p. 56).

Pedagogical opportunities that allow learners of school mathematics to develop ideas, explore phenomena, and justify results, among other things, allow for opportunities to author mathematical ideas. In a recent research conducted in Ontario, Canada, Hillman (2014) describes how a student comes up with a not-pre-designed way to use TI-Nspire, a graphing calculator. Hillman reports how the student's suggestion changed the way the teacher approached the mathematics problem in other classes. He simply presented the student's suggestion in all the classes he taught thereafter thus solidifying the student's mathematical authorship. Hillman's (2014) study is relevant here as his work sheds light on an important intersection of mathematics, technology, creativity, and authorship. Hillman (2014) shows that students as well as teachers have a role in shaping the use of technology in the mathematics classroom and that this role needs to be acknowledged (Radford, 2003, 2012; Säljö, 2010, 2012). Even though Hillman himself did focus on the use of technology, he did not explicitly use the terms creativity and authorship in the sense we use them here. Nevertheless, his work exemplifies one such interaction in the context of mathematics education. Instead of merely reading values from a graph, the student in Hillman's research had an opportunity to rewrite the institutionally assigned roles. When the mathematics teacher asked the students to predict an unknown value in a line of best fit, the student already creatively formulated his own way of working with the graphing calculator, which was then adopted by the teacher and used in all his other classes thus reifying the unification of technology, creativity, and authoring in the context of school mathematics.

Compounded with the notion of authorship is its association with the phenomenon of making mistakes. MacKinnon (1966) declares, "The creative person, given to expression rather than suppression or repression thus has fuller access to his [sic] own experience both conscious and unconscious" (MacKinnon, 1966, p. 154). This line of thought is also used by Sheffield (2015, 2017), who explains that students who are allowed to make mistakes and "construct viable arguments and critique the reasoning of others…have greater enjoyment and a much deeper and long-lasting understanding of mathematical concepts as well as a willingness to attack difficult problems and persevere in their solutions" (Sheffield, 2015, p. 116). Indeed, making mistakes is part and parcel of learning. In fact, examining the biographies of great innovators, Weisberg (1988) provides a detailed account of what brought great innovators to generate their world-changing inventions. He shows how, contrary to the commonly shared narratives, it was actually a long process of trial and error, of perseverance, and commitment in the face of failure that drove those innovators rather than a fixed, pre-existing ability of creativity.

Perceiving the making of mistakes as an integral part of the process of authorship, we looked for a framework that can provide helpful concepts to understanding what it is exactly that students do when engaging in acts of authorship and creativity. That is, how does learning happen and how does authorship get a chance to surface? We thus bring in Rogoff's (1990) notion of *apprenticeship* that considers learners as active agents in socially generated processes of authorship of not-yet-mastered skills by means of observation as well as by guided or joint participation (Rogoff, Mistry, Göncü, & Mosier, 1993).

Etymologically speaking, the very meaning of the word apprenticeship references actions of learning and teaching, seizing, taking hold of, grasping, and apprehending (Klein, 1966, p. 95). Indeed, Rogoff (1990), who draws on Vygotsky's notion of social activity as the mechanism through which individuals develop, shows how humans teach and learn by collaboration, observation, involvement, development of realistic self-reliance, routine arrangements and interactions, transferring responsibility for roles and activities, and providing bridges from what is already known to what is not yet known. She provides numerous instructive examples that illustrate how the notion of apprenticeship cuts across cultures and geographical regions even though they carry distinctive historically, socially, and culturally shaped ways of performing tasks. It is through the act of apprenticeship that conditions for authorship can be cultivated and fostered as it makes knowledge more attuned, more accurate, more refined, thus allowing for creativity (read authorship) to take shape and direction.

20.3 TCA Harnessed Three Abreast—A Case Study

Within his work as a mathematics teacher educator, and as a teacher of mathematics in the Virtual High School that is operated through the Center for Educational Technology (CET) in Israel (see Biton, Fellus, Raviv, 2017a, b; Biton, Fellus, Raviv, Fellus, forthcoming; Fellus, Biton, Raviv, 2017), the second author has often noticed instances of learners' appropriation of mathematical ideas as it was manifested through the use of technology, the application of creativity, and the action of authorship. These instances have buttressed our perception of the three, equally prioritized, as conducive to the learning of mathematics. One such example is drawn from an after-the-lesson consultation between a tutor in the virtual high school and the second author.

Following a synchronous meeting between the class instructor (second author) and the class tutors, one of the tutors asked to continue the conversation at the end of the session to discuss an experience she had with her four Grade 10 tutees. She recounted that as she was reviewing and summarizing the unit on the Midsegment Theorem, her students provided the following proposition as a true Converse of the Midsegment Theorem: "A segment that connects the midpoint of one side with a point on another side and equals half of the third side, is a midsegment of that triangle." The tutor said that she knew this proposition was not true and that she simply told the students that it was not correct. Her students, however, wanted to know why the proposition was not a true Converse of the Midsegment Theorem.

Not waiting for her to prove them wrong, the students began to vehemently argue that this specific proposition was actually a true Converse of the Midsegment Theorem and embarked on constructing proofs on the shared computer screen to support their argument. The tutor told the course instructor (second author) that there were four students in the group and that three of those students formulated three different proofs to demonstrate that the proposition at hand was a true converse for the Midsegment Theorem. She then shared the three proofs with the course instructor (see below) noting that she was no longer sure whether these were incorrect. (See Table 20.1.)

Collectively, the students engaged with bricolaging through combinatorial thinking and brought in their knowledge of similar triangles, Thales' Theorem, and proof by contradiction to construct their mathematical proofs. The course instructor (second author) logged on GeoGebra (see Fig. 20.2) to show the tutor that the given proposition did not have the adequate conditions to support the Midsegment Theorem in a triangle.

Table 20.1 The students' proofs of the validity of a converse of Midsegment Theorem

Proof #1	**Proof #2**	**Proof #3**
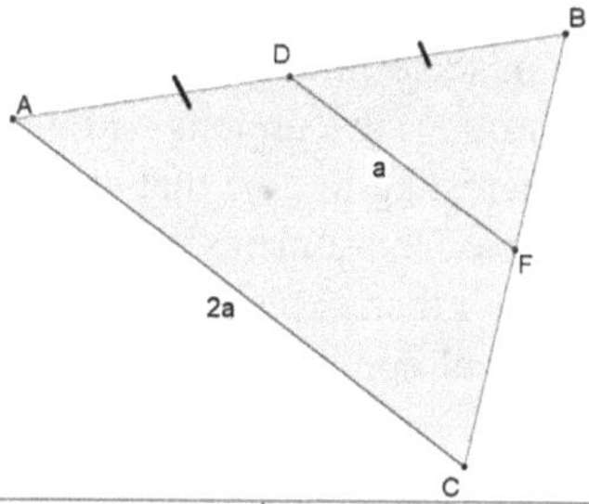		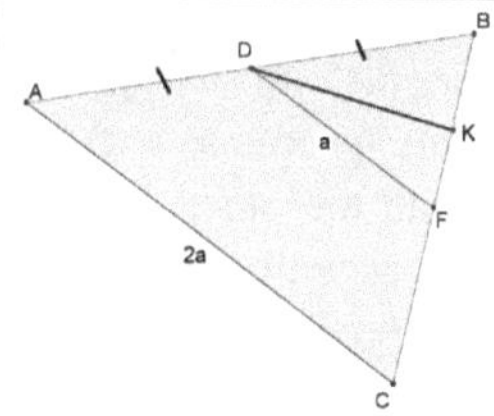
$\frac{BD}{BA}=\frac{DF}{AC}=\frac{1}{2}$ $\measuredangle B = \measuredangle B$ ΔDBF ΔABC by SSA $\frac{BF}{BC}=\frac{1}{2}$ Q.E.D.	$\frac{BD}{BA}=\frac{DF}{AC}=\frac{1}{2}$ DF ∥ AC by a Converse of the Thales Theorem $\frac{BF}{BC}=\frac{1}{2}$ by the Thales Theorem Q.E.D.	Let's assume that point F is not the midpoint This means that there is a point K, different than F, which is the midpoint of AC Let's connect points D and K. DK is a midsegment in the triangle and so: $DK=\frac{1}{2}AC$ But $DF=\frac{1}{2}AC$ And so: DK = DF This is true only if points K and F converge. This is a contradiction! This means the assumption is not true and so point F is the midpoint of BC Q.E.D.

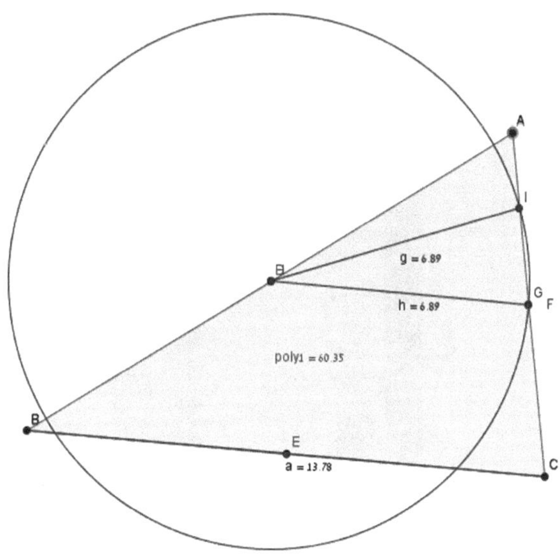

Fig. 20.2 The use of the dynamic tool GeoGebra to show that there are two segments that can have the given property

The representation generated by GeoGebra demonstrated that there are two line segments that fulfill the condition given in the proposition. Following this, the course instructor (second author) and the tutor continued the synchronous session to analyze each of the students' proofs and identify the mistake in each. They then decided to use the students' proofs as a trigger for synchronous discussions with the whole class as well as with the other five tutors. It is interesting to note that at first sight, the rest of the class, as well as the other tutors accepted the students' proofs as correct and thus the given proposition as true. The instructor then explained why the proposition was not true using the interactive mathematical application GeoGebra with these groups as well.

To attain experts' input on the proofs, the second author, who also teaches undergraduate and graduate courses in a mathematics education program, used the students' proofs in an assessment course of 49 mathematics teacher candidates, all of whom were asked to decide whether they accept the three high school students' proofs as clear and correct using a three-level Likert-type survey (Agree, Somewhat-agree, Disagree). They were also asked to provide the reason for their choice. Figures 20.3 and 20.4 show that while about 70% of the teacher candidates accepted proof #1 as clear, their acceptance of the proof as correct was almost equally divided across the three levels of acceptance. For proof #2, 40% of the teacher candidates rated it as clear, while 55% of them labeled it as incorrect. In regard to proof #3, 53% of the teacher candidates thought it was clear, and only 26% of them labeled it as incorrect.

Examining the reasons provided by the mathematics teacher candidates for their ratings, themes pertaining to creativity and originality were captured. The following are four quotations that demonstrate the acceptance of the proofs as original:

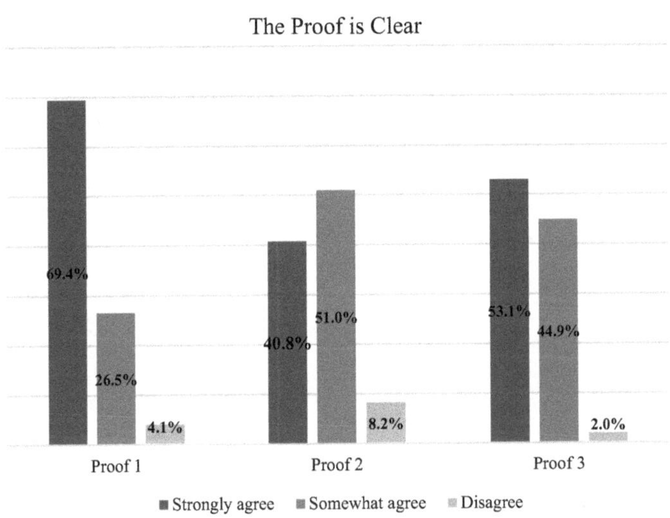

Fig. 20.3 Degree of clarity of the students' proofs assigned by the mathematics teacher candidates

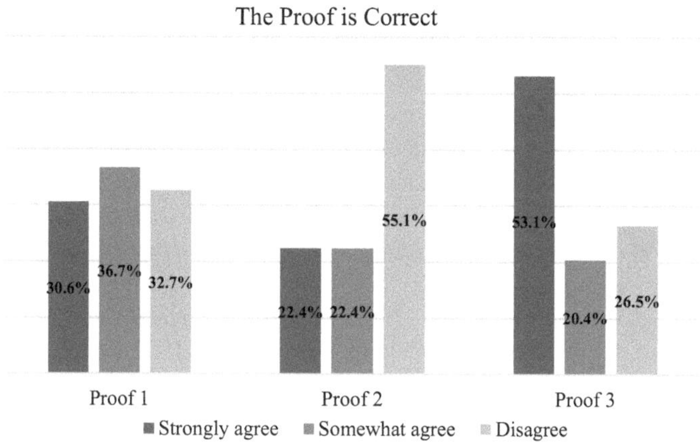

Fig. 20.4 Acceptance rate of the correctness of the three proofs by the mathematics teacher candidates

> Hadar: "It's a beautiful example to show students that you can get to the same result by using different tools. It is also a great opportunity to discuss Thales' Theorem."
>
> Gil-Ad: "The student solved the problem in a creative and sophisticated way. The thinking is generally correct and demonstrates advanced mathematical logic. However, the student was supposed to prove the given proposition. The proof that is provided still requires further explanations."
>
> Hadas: "One needs to give positive feedback for this elegant and original proof by contradiction. Having said that, this is an opportunity to explain to the students that proof by contradiction is always less intuitive to the human mind and more complex logically and thus more prone to logical mistakes."
>
> Hannah: "The proof is original and is constructed as a proof by contradiction, which is not used often. I would use this proof as a way to develop creative and spatial thinking. I would work with the students on identifying the mistake and on sketching cases where this proof would work and cases where this proof would not work."

Other threads that were identified in the teacher candidates' comments included the use of technology, expression of creativity, making mistakes as an integral part of learning, as well as the practice of apprenticeship in demonstrating the right way of using theorems of similar triangles, Thales' Theorem, and proof by contradiction. Descriptors such as "*proof that is considered prestigious*," "*elegant*," "*not trivial*," and "*beautiful*" demonstrate the teacher candidates' appreciation of the students' proofs as unique and novel. The teacher candidates highlighted the use of technology as a tool to effectively demonstrate these mathematical ideas. For example, one teacher candidate wrote: "I would use digital technology such as GeoGebra or artifacts such as magnets on the board to demonstrate that when segment AC is smaller than segment AB, we can find two points that show whether the proposition is true."

The context of the virtual high school (VHS), its technological design, pedagogical framework, and ongoing support to students (see Biton, Fellus, & Raviv,

2017a, b; Biton, Fellus, Raviv, Fellus, forthcoming; Fellus, Biton, & Raviv, 2017) create a unique environment where technology is used to carry out the task of learning and where opportunities for creativity, meaning making, bricolaging, and mathematical authorship are welcomed expressions of knowledge. By closely engaging with small groups of students (up to four), a tutor (a STEM-major university student) works with the students for two hours a week in the after-school hours. This is in addition to four-to-six hours of synchronous mathematics classes with up to 25 other students during school hours with a teacher of mathematics. The environment of the Israeli VHS provides learning opportunities not only to its students but also to its tutors and instructors thus constantly shifting the directionality of the apprenticeship model. In this episode, technology and its affordances made it possible for the expression of creativity and authorship to take place. When the students began to combine different mathematical ideas to construct their proofs using available technologies, they were driven by what Csikszentmihalyi (1988) calls "acute curiosity" (p. 166). By acts of bricolaging, of choice, and reasoning, they authored mathematical representations that were recognized as clear and correct by some mathematics teacher candidates—thus gaining what Stein (1953) and Sriraman (2008) perceive as the experts' approval of novelty.

The highly technological environment of the VHS where teachers, tutors, as well as students can equally use tools and applications, engage in acts of creativity, and learn through authorship creates an inseparable bundle of distinct elements the combination of which allows for ample opportunities to celebrate students' mathematical arguments and sense making.

20.4 Conclusion

The aim in this chapter was to offer a unification of technology, creativity, and authorship (TCA) within mathematics education. We draw on the works of Latour (1991), Vygotsky (2004), and Bakhtin (1981) to suggest a framework that has the potential to democratize creativity in mathematics education. Vygotsky's (2004) notion of creativity is essential to how we see creativity in mathematics education because it provides an operational framework of combinatorial work—corroborated by work in neuroscience; it underscores the role of the teacher or knowledgeable other to function as a mentor within the theory of apprenticeship in education; and it positions authoring as one of the building blocks of learning and development. We know how creativity looks like—it is a product that is perceived by others as unique, novel, and useful (Stein, 1953)—but we do not yet know, or have adequate practice of how to set up the conditions to allow for it to take place. The use of TCA as a framework, with the plethora of technologies, becomes more feasible and equitably accessible. In the episode that we described, we saw how students are providing their proofs during the synchronous meeting with their tutor. Their collective work demonstrates combinatorial thinking as they make connections between diverse mathematical concepts (similar triangles, Thales' Theorem, and proof by contradiction) to prove a proposition as a true converse of Midsegment Theorem. Whereas each of the students made a mistake in their respective proofs, their ability to draw on diverse mathematical ideas speaks to their collective ability to combine mathematical ideas in the process of making sense of the problem at hand.

Our purpose here is not to prescribe a solution. Rather, it is to invite the reader to reflect upon the concepts we brought forth so that we can consider the advantages and challenges of using these concepts as guidelines in the context of mathematics education. In writing this chapter, we asked ourselves the 'so what' question. We wanted to formulate a clear understanding of the reasoning behind putting the intersection area of $C \cap T \cap M$ in the limelight of research. We believe that allowing for creativity to have a voice, to take form and shape, and to make its mark may spark human agency in the doing of mathematics. In his discussion on mathematics from the perspective of positioning theory, David Wagner (2011) identifies five myths that need to be dispelled: that mathematics has no human subject—obscuring the fact that it is human beings who put together the theorems, the proofs, the graphs, the models, and the statistics; that mathematics is culture-free and values-free—obscuring the fact that there are other kinds of mathematics that are not privileged in schools; that mathematics is hard to do—obscuring the fact that doing math is not about doing it fast and right but about putting the time in exploring and understanding mathematical phenomena; that only a few can do math —obscuring the fact that notions of identity play an important part in building mathematical capabilities; and that mathematics is powerful—obscuring the fact that it can sometimes, intentionally or unintentionally, provide misleading information. We feel that using a unified framework of TCA in the sense that was presented in this chapter, may be effective in gradually dispelling these myths.

Within this TCA framework, we perceive mathematics education as being more about capacity building that is oriented toward the future rather than about ability—which is oriented toward the past (see Fellus & Biton, 2017). To summarize our message of the nature of TCA in mathematics education, we allude to two ideas. One is Seymour Papert (1992) use of the metaphor in the title of this book chapter; the second is Einstein's quip in a 1922 speech he gave in Kyoto. We will begin with Papert, whose work tying technology with creativity in education is instructive. He underscores the "volcanic explosion of creativity" (p. 33) with the use of technology and explains that focusing debates on whether the use of technology within educational contexts is justified is "more like attaching a jet engine to an old fashioned wagon to see whether it will help the horses. Most probably, it will frighten the animals and shake the wagon to pieces, 'proving' that jet technology is actually harmful to the enhancement of transportation" (p. 29). It is with this metaphor in mind that we chose the idiom of putting the horses before the cart, as it is our underlying assumption that the simultaneous harnessing of technology, creativity, and authorship as defined in this chapter will allow the horses to pull the cart of learning school mathematics. Without it, mathematics education may find it almost impossible to move toward what students will mathematically be able to do tomorrow. For us, then, to allude to Einstein's quip, describing the role of technology in mathematics education without reference to creativity and authorship "is similar to describing our thoughts without words" (Einstein, 1982, p. 47).

References

Ainsworth, S. (2006). DeFT: A conceptual framework for considering learning with multiple representations. *Learning and Instruction, 16*(3), 183–198.

Artigue, M. (2002). Learning mathematics in a CAS environment: The genesis of a reflection about instrumentation and the dialectics between technical and conceptual work. *International Journal of Computers for Mathematical Learning, 7*(3), 245–274.

Bakhtin, M. M., & Holquist, M. (1981). *The dialogic imagination: Four essays.* Austin, TX: University of Texas Press.

Barmby, P., Harries, T., Higgins, S., & Suggate, J. (2009). The array representation and primary children's understanding and reasoning in multiplication. *Educational Studies in Mathematics, 70*(3), 217–241.

Bendetowicz, D., Urbanski, M., Aichelburg, C., Levy, R., & Volle, E. (2017). Brain morphometry predicts individual creative potential and the ability to combine remote ideas. *Cortex, 86,* 216–229.

Biton, Y., Fellus, O., & Raviv, D. (2017a). Advanced mathematics and/or physics in a virtual high school: Why? Why not? Who? And so what? In Y. Eshet-Alkalai, I. Blau, A. Caspi, N. Geri, Y. Kalman, & V. Silber-Varod (Eds.), *Proceedings of the 12th Chais Conference for the Study of Innovation and Learning Technologies: Learning in the Technological Era* (pp. 9E–20E). Raanana, Israel: The Open University of Israel.

Biton, Y., Fellus, O., & Raviv, D. (2017b). *Extending scopes: Teachers' experience of teaching mathematics and physics in the Israeli virtual high school.* Paper presented at the 2017 annual meeting of the American Educational Research Association, AERA, San Antonio, TX.

Biton, Y., Fellus, S., Raviv, D., & Fellus, O. (Forthcoming). Yours virtually: Advanced mathematics and physics in the Israeli Virtual High School. *Interdisciplinary Journal of e-Skills and Lifelong Learning, 13,* 229–250.

Bolden, D. S., Harries, T. V., & Newton, D. P. (2010). Pre-service primary teachers' conceptions of creativity in mathematics. *Educational Studies in Mathematics, 73*(2), 143–157.

Borba, M. C., Askar, P., Engelbrecht, J., Gadanidis, G., Llinares, S., & Aguilar, M. S. (2016). Blended learning, e-learning and mobile learning in mathematics education. *ZDM Mathematics Education, 48*(5), 589–610.

Csikszentmihalyi, M. (1988). Motivation and creativity: Toward a synthesis of structural and energistic approaches to cognition. *New Ideas in Psychology, 6*(2), 159–176.

Chi, M. T. H., De Leeuw, N., Chiu, M.-H., & LaVancher, C. (1994). Eliciting self-explanations improves understanding. *Cognitive Science, 18*(3), 439–477.

Czarnocha, B., & Baker, W. (2015). Creativity and bisociation. In F. M. Singer, F. Toader, & C. Voica (Eds.), *The 9th Mathematical Creativity and Giftedness International Conference Proceedings*, 2015 June (pp. 80–85). Sinaia, Romania.

Davydov, V. V. (1986/2008). *Problems of developmental instruction: A theoretical and experimental psychological study.* New York, NY: Nova Science Publishers.

Davydov, V. V. (1991). A psychological analysis of multiplication. In L. P. Steffe (Ed.), *Psychological abilities of primary school children in learning mathematics. Soviet Studies in Mathematics Education Series* (Vol. 6, pp. 1–85). Reston, VA: NCTM.

Dreher, A., & Kuntze, S. (2015). Teachers' professional knowledge and noticing: The case of multiple representations in the mathematics classroom. *Educational Studies in Mathematics, 88*(1), 89–114.

Dreher, A., Kuntze, S., & Lerman, S. (2012). Pre-service teachers' views on using multiple representations in mathematics classroom-an intercultural study. In T. Y. Tso (Ed.), *Proceedings of the 36th Conference of the International Group of the Psychology of Mathematics Education: Opportunities to Learn in Mathematics Education* (Vol. 2, pp. 211–218). Taipei, Taiwan: PME.

Einstein, A. (1982). How I created the theory of relativity (Y. A. Ono, Trans.). *Physics Today, 35* (8), 45–47.

Ellul, J. (1964). *The technological society*. New York, NY: Knopf.

Etlinger, L. (1974). The electronic calculator: A new trend in school mathematics. *Educational Technology, 14*(12), 43–45.

Fellus, O., & Biton, Y. (2017). One is not born a mathematician: In conversation with Vasily Davydov. *International Journal for Mathematics Teaching and Learning, 18*(2), 136–160.

Fellus, O., Biton, Y., & Raviv, D. (2017). *Virtual terrains: Learners' experiences of mathematics and physics in the Israeli virtual high school*. Paper Presented at the 2017 annual meeting of the American Educational Research Association (AERA), April 2017, San Antonio, TX.

Freiman, V., & Sriraman, B. (2008). Does mathematics gifted education need a working philosophy of creativity? *Mediterranean Journal for Research in Mathematics Education, 6*(1–2), 23–46.

Gagatsis, A., & Shiakalli, M. (2004). Translation ability from one representation of the concept of function to another and mathematical problem solving. *Educational Psychology, 24*(5), 645–657.

Geiger, V. (2009). The master, servant, partner, extension-of-self framework in individual, small group and whole class contexts. In R. Hunter, B. Bicknell, & T. Burgess (Eds.), *Crossing Divides: Proceedings of the 32nd Conference of the Mathematics Education Research Group of Australasia* (pp. 201–208). Palmerston North, New Zealand: MERGA.

Goldin, G., & Shteingold, N. (2001). Systems of representation and the development of mathematical concepts. In A. A. Cuoco & F. R. Curcio (Eds.), *The role of representation in school mathematics* (pp. 1–23). Boston, VA: NCTM.

Hashimoto, Y. (1997). The methods of fostering creativity through mathematical problem solving. *ZDM Mathematics Education, 29*(3), 86–87.

Haylock, D. (1997). Recognizing mathematical creativity in schoolchildren. *ZDM, 29*(3), 68–74.

Healy, L., & Kynigos, C. (2010). Charting the microworld territory over time: Design and construction in mathematics education. *ZDM, 42*(1), 63–76.

Hillman, T. (2014). Finding space for student innovative practices with technology in the classroom. *Learning, Media and Technology, 39*(2), 169–183.

Hohenwarter, J., Hohenwarter, M., & Lavicza, Z. (2008). Introducing dynamic mathematics software to secondary school teachers: The case of Geogebra. *Journal of Computers in Mathematics and Science Teaching, 28*(2), 135–146.

Holquist, M. (1983). Answering as authoring: Mikhail Bakhtin's trans-linguistics. *Critical Inquiry, 10*(2), 307–319.

Hughes, J. (2013). Descriptive indicators of future teachers' technology integration in the PK–12 classroom: Trends from a laptop-infused teacher education program. *Journal of Educational Computing Research, 48*(4), 491–516.

Jimoyiannis, A. (2010). Designing and implementing an integrated technological pedagogical science knowledge framework for science teachers' professional development. *Computers and Education, 55*(3), 1259–1269.

Johnson-Laird, P. N. (1988). Freedom and constraint in creativity. In R. J. Sternberg (Ed.), *The nature of creativity: Contemporary psychological perspectives* (pp. 202–219). Cambridge, UK: Cambridge University Press.

Jung, R., Segall, J., Bockholt, J., Flores, R., Smith, S., Chavez, R., et al. (2010). Neuroanatomy of creativity. *Human Brain Mapping, 31*(3), 398–409.

Kieran, C., & Damboise, C. (2007). "How can we describe the relation between the factored form and the expanded form of these trinomials? We don't even know if our paper and pencil factorizations are right": The case for computer algebra systems (CAS) with weaker algebra students. In J. H. Woo, H. C. Lew, K. S. Park, & D. Y. Seo (Eds.), *Proceedings of the 31st*

Conference of the International Group for the Psychology of Mathematics Education (Vol. 3, pp. 105–112). Seoul, South Korea: PME.

Kieran, C., & Drijvers, P. (2006). The co-emergence of machine techniques, paper-and-pencil techniques, and theoretical reflections: A study of CAS use in secondary school algebra. *International Journal of Computers for Mathematical Learning, 11,* 205–263.

Klein, E. (1966). *A comprehensive etymological dictionary of the English language*. New York, NY: Elsevier.

Koichu, B. (2015). Problem solving and choice-based pedagogies. In F. M. Singer, F. Toader, & C. Voica (Eds.), *The 9th Mathematical Creativity and Giftedness International Conference Proceedings*, June 2015 (pp. 68–73). Sinaia, Romania.

Kwon, O. N., Park, J. S., & Park, J. H. (2006). Cultivating divergent thinking in mathematics through an open-ended approach. *Asia Pacific Education Review, 7*(1), 51–61.

Lagrange, J. B., Artigue, M., Laborde, C., & Trouche, L. (2003). Technology and mathematics education: A multidimensional study of the evolution of research and innovation. In A. J. Bishop, M. A. Clements, C. Keitel, J. Kilpatrick, & F. K. S. Leung (Eds.), *Second international handbook of mathematics education* (Vol. 1, pp. 237–269). Dordrecht, The Netherlands: Kluwer.

Latour, B. (1986). Visualization and cognition: Thinking with eyes and hands. *Knowledge and Society, 6*(1), 1–40.

Latour, B. (1991). Technology is society made durable. In J. Law (Ed.), *A sociology of monsters: Essays on power, technology, and domination* (pp. 103–131). London, UK: Routledge.

Lawrenz, F., Gravely, A., & Ooms, A. (2006). Perceived helpfulness and amount of use of technology in science and mathematics classes at different grade levels. *School Science and Mathematics, 106*(3), 133–139.

Leikin, R. (2009). Exploring mathematical creativity using multiple solution tasks. In R. Leikin, A. Berman, & B. Koichu (Eds.), *Creativity in mathematics and the education of gifted students* (pp. 129–145). Rotterdam, The Netherlands: Sense Publishers.

Leikin, R., & Pitta-Pantazi, D. (2013). Creativity and mathematics education: The state of the art. *ZDM, 45*(2), 159–166.

Leikin, R., Berman, A., & Kocihu, B. (2009). *Creativity in mathematics and the education of gifted students*. Rotterdam, The Netherlands: Sense Publishers.

Levav-Waynberg, A., & Leikin, R. (2012). The role of multiple solution tasks in developing knowledge and creativity in geometry. *The Journal of Mathematical Behavior, 31,* 73–90.

Levenson, E. (2011). Exploring collective mathematical creativity in elementary school. *The Journal of Creative Behavior, 45*(3), 215–234.

Levi-Strauss, C. (1962). *The savage mind* (George Weidenfeld & Nicolson Ltd., Trans.). Chicago, IL: The University of Chicago Press.

Li, Y., & Li, D. (2009). Open-ended questions and creativity education in mathematics. *Journal of the Korea Society of Mathematical Education Series D: Research in Mathematical Education, 13*(1), 23–30.

Livne, N. L., & Milgram, R. M. (2006). Academic versus creative abilities in mathematics: Two components of the same construct? *Creativity Research Journal, 18*(2), 199–212.

Luria, A. R. (1960). Experimental analysis of the development of voluntary action in children. In H. P. David & J. C. Brengelmann (Eds.), *Perceptions in personality research* (pp. 139–149). New York, NY: Springer.

Luria, A. R. (1994). The problem of the cultural development of the child. In R. V. Valsiner (Ed.), *The Vygotsky Reader* (pp. 47–56). Oxford, UK: Basil Blackwell.

MacKinnon, D. W. (1966). What makes a person creative? *Theory into Practice, 5*(4), 151–156.

Mann, E. L. (2006). Creativity: The essence of mathematics. *Journal for the Education of the Gifted, 30*(2), 236–260.

Marshall, N., & Buteau, C. (2014). Learning mathematics by designing, programming, and investigating with interactive, dynamic computer-based objects. *The International Journal for Technology in Mathematics Education, 21*(2), 49–65.

Mhlolo, M. K. (2017). Regular classroom teachers' recognition and support of the creative potential of mildly gifted mathematics learners. *ZDM, 49*(1), 81–94.
Moseley, B. (2005). Students' early mathematical representation knowledge: The effects of emphasising single or multiple perspectives of the rational number domain in problem solving. *Educational Studies in Mathematics, 60*(1), 37–69.
NCTM. (2000). *Principles and standards for school mathematics*. Reston, VA: National Council of Teachers of Mathematics.
Novak, E., & Tassell, J. L. (2015). Using video game play to improve education-majors' mathematical performance: An experimental study. *Computers in Human Behavior, 53*, 124–130.
Nguyen, N. D., & Trinh, T. T. P. (2015). An empirical research on the use of mobile phones to support students' mathematics learning. *International Journal of Learning, Teaching and Educational Research, 12*(1), 133–141.
Nadjafikhah, M., & Yaftian, N. (2013). The frontage of creativity and mathematical creativity. *Procedia-Social and Behavioral Sciences, 90*, 344–350.
Papert, S. (1992). *The children's machine: Rethinking school in the age of the computer*. New York, NY: Basic Books.
Perkins, D. N. (1988). The possibility of invention. In R. J. Sternberg (Ed.), *The nature of creativity: Contemporary psychological perspectives* (pp. 362–385). Cambridge, UK: Cambridge University Press.
Polly, D., Mims, C., Shepherd, C. E., & Inan, F. (2010). Evidence of impact: Transforming teacher education with preparing tomorrow's teachers to teach with technology. *Teaching and Teacher Education, 26*(4), 863–870.
Prabhu, V., & Czarnocha, B. (2014). Democratizing mathematical creativity through Koestler bisociation theory. In P. Liljedahl, S. Oesterly, C. Nicol, & D. Allan (Eds.), *Proceedings of the Joint Meeting of PME 38 and PME-NA 36* (Vol. 5, pp. 1–8). Vancouver, Canada: PME.
Prusak, A. (2015). Nurturing students' creativity through telling mathematical stories. In F. M. Singer, F. Toader, & C. Voica (Eds.), *The 9th Mathematical Creativity and Giftedness International Conference Proceedings*, 2015, June (pp. 16–21). Sinaia, Romania.
Radford, L. (2003). Gestures, speech, and the sprouting of signs: A semiotic-cultural approach to students' types of generalization. *Mathematical Thinking and Learning, 5*(1), 37–70.
Radford, L. (2012). On the cognitive, epistemic, and ontological roles of artifacts. In G. Gueudet, B. Pepin, & L. Trouche (Eds.), *From text to 'lived' resources* (pp. 283–288). New York, NY: Springer.
Rogoff, B. (1990). *Apprenticeship in thinking*. New York, NY: Oxford University Press.
Rogoff, B., Mistry, J., Göncü, A., & Mosier, C. (1993). Guided participation in cultural activity by toddlers and caregivers. *Monographs of the Society for Research in Child Development*, 58 (7, Serial No. 286).
Säljö, R. (2010). Digital tools and challenges to institutional traditions of learning: Technologies, social memory and the performative nature of Learning. *Journal of Computer Assisted learning, 26*(1), 53–64.
Säljö, R. (2012). Literacy, digital literacy and epistemic practices: The co-evolution of hybrid minds and external memory systems. *Nordic Journal of Digital Literacies, 7*(1), 5–19.
Salomon, G., Perkins, D. N., & Globerson, T. (1991). Partners in cognition: Extending human intelligence with intelligent technologies. *Educational Researcher, 20*(3), 2–9.
Seufert, T. (2003). Supporting coherence formation in learning from multiple representations. *Learning and Instruction, 13*(2), 227–237.
Shaffer, D. W., & Clinton, K. A. (2006). Tool for thoughts: Re-examining thinking in the digital age. *Mind, Culture, and Activity, 13*(4), 283–300.
Sheffield, L. J. (2015). Myths about "gifted" mathematics students: How widespread are they? In F. M. Singer, F. Toader, & C. Voica (Eds.), *The 9th Mathematical Creativity and Giftedness International Conference Proceedings*, 2015, June (pp. 114–119). Sinaia, Romania.
Sheffield, L. J. (2017). Dangerous myths about "gifted" mathematics students. *ZDM, 49*(1), 13–23.

Silver, E. A. (1997). Fostering creativity through instruction rich in mathematical problem solving and problem posing. *ZDM Mathematics Education, 29*(3), 75–80.

Sriraman, B. (2005). Are giftedness and creativity synonyms in mathematics? *The Journal of Secondary Gifted Education, 17*(1), 20–36.

Sriraman, B. (2008). The characteristics of mathematical creativity. *ZDM, 41*(1–2), 13–27.

Sriraman, B., Yaftian, N., & Lee, K. H. (2011). Mathematical creativity and mathematics education. In B. Sriraman & K. H. Lee (Eds.), *The elements of creativity and giftedness in mathematics* (pp. 119–130). Rotterdam, The Netherlands: Sense Publishers.

Stacey, K. (1995). The challenges of keeping open problem-solving open in school mathematics. *International Reviews on Mathematical Education, 27*(2), 62–67.

Stein, M. I. (1953). Creativity and culture. *The Journal of Psychology, 36*(2), 311–322.

Suchman, L. (2007). *Human–machine reconfigurations: Plans and situated actions.* Cambridge, UK: Cambridge University Press.

Suchman, L., Blomberg, J., Orr, J. E., & Trigg, R. (1999). Reconstructing technologies as social practice. *American Behavioral Scientist, 43*(3), 392–408.

Tan, C.-K., & Tan, C.-P. Graphic calculator instructional approach and students' probability performances. In A. N. M. Wahid & C. Reaiche Amaro (Eds.), *Australian Academy of Business and Social Sciences Conference*, 2014 August. Kuala Lumapur, Malaysia. Retrieved from https://www.aabss.org.au/research-papers/graphic-calculator-instructional-approach-and-students%E2%80%99-probability-performances.

Tassell, J.L., Stobaugh, R. R., & McDonald, M. (2013). Math and science teacher candidates' use of technology to facilitate teaching and learning during student teaching. *Educational Renaissance, 2*(1), 17–28.

Thagard, P., & Stewart, T. (2011). The AHA! experience: Creativity through emergent binding in neural networks. *Cognitive Science, 35*(1), 1–33.

Tondeur, J., van Braak, J., Sang, G., Voogt, J., Fisser, P., & Ottenbreit-Leftwich, A. (2012). Preparing pre-service teachers to integrate technology in education: A synthesis of qualitative evidence. *Computers and Education, 59*(1), 134–144.

Van Harpen, X. Y., & Sriraman, B. (2013). Creativity and mathematical problem posing: An analysis of high school students' mathematical problem posing in China and the USA. *Educational Studies in Mathematics, 82*(2), 201–221.

Vartanian, O., Bristol, A., & Kaufman, J. (Eds.). (2013). *Neuroscience of creativity.* Cambridge, MA: The MIT Press.

Verzosa, D., Guzon, A. F., & Penãs, M. (2014). Using dynamic tools to develop an understanding of the fundamental ideas of calculus. *International Journal of Mathematical Education in Science and Technology, 45*(2), 190–199.

Vygotsky, L. (1978). *Mind in society: The development of higher psychological processes.* Cambridge, MA: Harvard University Press.

Vygotsky, L. S. (1986). *Thought and language* (Rev. ed.). Cambridge, MA: The MIT Press.

Vygotsky, L. S. (2004). Imagination and creativity in childhood (M. E. Sharpe, Trans.). *Journal of Russian and East European Psychology, 42*(1), 7–97.

Weisberg, R. W. (1988). Problem solving and creativity. In R. J. Sternberg (Ed.), *The nature of creativity: Contemporary psychological perspectives* (pp. 148–176). New York, NY: Cambridge University Press.

Wagner, D. (2011). Intercultural positioning in mathematics. In B. Sriraman, & V. Freiman (Eds.), *Interdisciplinarity for the Twenty-first Century: Proceedings of the Third International Symposium on Mathematics and Its Connections to Arts and Sciences*, Moncton 2009 (Vol. 11, pp. 1–17). Charlotte, NC: Information Age Publishing.

Wertsch, J. (1998). *Mind as action.* New York, NY: Oxford University Press.

Wu, X., Yang, W., Tong, D., Sun, J., Chen, Q., Wei, D., et al. (2015). A meta-analysis of neuroimaging studies on divergent thinking using activation likelihood estimation. *Human Brain Mapping, 36*(7), 2703–2718.

Yerushalmy, M., & Botzer, G. (2011). Guiding mathematical inquiry in mobile settings. In O. Zaslavsky & P. Sullivan (Eds.), *Constructing knowledge for teaching secondary mathematics* (pp. 191–207). New York, NY: Springer.

Yilmaza, G. K. (2015). The effect of dynamic geometry software and physical manipulation on candidate teachers' transformational geometry success. *Educational Sciences: Theory and Practice, 15*(5), 1–20.

Zhu, W., Chen, Q., Tang, C., Cao, G., Hou, Y., & Qiu, J. (2016). Brain structure links everyday creativity to creative achievement. *Brain and Cognition, 103,* 70–76.

Chapter 21
Virtual Learning Communities of Problem Solvers: A Potential for Developing Creativity in Mathematics?

Dominic Manuel

Abstract Mathematics is viewed as a school subject that can develop creativity in students (Liljedahl and Sriraman in Learn Math 26:17–19, 2006; Sheffield in Creativity in mathematics and the education of gifted students. Sense Publishers, Rotterdam, The Netherlands, pp. 87–100, 2009; Sriraman in ZDM: Int J Math Educ 41:13–27, 2009). However, many authors have mentioned that mathematics learning is still based on applying routine procedures and already prescribed algorithms (Chan Chun Ming in The use of mathematical modeling tasks to develop creativity, 2008). Yet, studies have shown that open-ended problems can create opportunities for students to face more cognitive challenges and to develop different and original problem solving strategies, thus leading to more creative solutions (Leikin in Habits of mind associated with advanced mathematical thinking and solution spaces of mathematical tasks, pp. 2330–2339, 2007). Some researchers have also mentioned that virtual learning communities may support the development of creativity, but this does not seem to have been proved (Piggot in Math Teach 202:3–6, 2007). This study thus focuses on the richness of mathematical problems posted and the creativity of solutions submitted by members of the CAMI website, a virtual community of problem solving designed for Francophone students from New Brunswick, Canada, and elsewhere. I have developed a conceptual framework to: analyze the richness of the problems posted on the website; assess the mathematical creativity of the solutions submitted; and determine whether a link exists between these two variables. I created two grids to analyze the richness of the problems and the creativity of the solutions for 50 randomly selected problems. Then, using the Likelihood ratio, I determined whether there was a link between the two variables. Results show that, in general, richer problems seem to bring different correct answers and more original solutions.

Keywords Mathematical creativity · Richness of mathematical problems
Virtual communities of problem solvers · Problem solving

D. Manuel (✉)
McGill University, Montreal, Quebec, Canada
e-mail: dominic.manuel@mcgill.ca

© Springer International Publishing AG, part of Springer Nature 2018

V. Freiman and J. L. Tassell (eds.), *Creativity and Technology in Mathematics Education*, Mathematics Education in the Digital Era 10,
https://doi.org/10.1007/978-3-319-72381-5_21

21.1 Context of the Study

21.1.1 Creativity in Mathematics Education

For a long time, creativity was often cited along with artistic creations, musical compositions and scientific discovery. Thus, mathematics was not known as a topic associated with creativity (Chan Chun Ming, 2008). However, although it appears to be under-researched in mathematics and in mathematics education according to Sriraman (2009), mathematics educators view mathematics as a school subject that has the potential to develop creativity in students, and stress the importance of developing it inside classrooms (Leikin, 2011; Petrowski, 2000; Sheffield, 2009; Sriraman, 2009). In fact, they see creativity as an orientation or disposition toward mathematical activities that can be fostered broadly in the general school population (Sriraman, 2005). Creativity in school mathematics differs from that of professional mathematicians, but students can offer new insights or solutions to mathematical problems based on their previous experiences and to the performance of other students' contributions, the mathematics students previously learned, and the problems they solved (Leikin, 2009; Sriraman, 2005).

Mathematical creativity in school mathematics is usually connected with problem solving and problem posing (I will only address problem solving in this chapter) (Sheffield, 2009). Sheffield argued that, although learning basic mathematical facts remains important in classrooms, it is even more important to put emphasis on the development of higher cognitive abilities that enable students to be more creative, such as: recognizing and defining problems that emerge from the society; generate multiple solutions and strategies to different problems; and reasoning mathematically, creating and justifying conclusions, and communicating results. Students are not born with these abilities, and they don't develop automatically. It is therefore important to cultivate and nurture them in students (Mann, 2006; Sheffield, 2009).

The relationship between mathematical creativity and problem solving at the school level can be viewed as students engaging in a process that results in original solutions to a given problem or approaching the problems in new perspectives (Leikin, 2009). For Kwon, Park, and Park (2006), it involves open-ended tasks that focuses on: the creation of new knowledge, and flexible problem solving abilities. As opposed to problems only having one solution, open-ended tasks can bring students opportunities to solve these problems in their own ways and according to their specific abilities (Klavir & Hershkovitz, 2008). In addition, these authors added that these types of problems make a high cognitive demand for many reasons. First, they can have different interpretations. Second, they can have multiple strategies and correct answers. Third, they give students the chance to construct new knowledge in a variety of contexts. Fourth, students can confront the same problem in different perspectives, and represent the mathematical concepts and relations involved in it in their own and different ways until they discover an effective strategy that will permit them to rigorously solve it. Students could be more creative by bringing their own inventions and making their own discoveries (Mann, 2006). With open-ended problems, students can also take risks while finding different ways to solve a problem, test different possible answers, and possibly develop original strategies—all activities which characterize mathematical creativity (Mann, 2006). Chiu (2009) added that ill-structure problems can also support the development of mathematical creativity in students.

Educators have attempted to promote mathematical creativity in classrooms. Singer, Sheffield, Freiman, and Brandl (2016) cited joint publications of the National Council of Teachers of Mathematics, the National Association for Gifted Children, and the National Council of Supervisors of Mathematics, which suggested the inclusion of an additional standard to the Common Core Mathematics Curriculum focusing on mathematical creativity and innovation. This standard would encourage and support all students in "taking risks, embracing challenge, solving problems in a variety of ways, posing new mathematical questions of interest to investigate, and being passionate about mathematical investigations" (Johnsen and Sheffield, 2012, pp. 15–16). Despite the efforts, the change is still to come in classrooms (Chiu, 2009; Singer et al., 2016; Sriraman, 2009). The pedagogical approaches used in classrooms do not seem to foster the development of mathematical creativity (Chiu, 2009). In addition, open-ended problems, which are suggested for the development of mathematical creativity, do not seem to be proposed in classrooms (Freiman, 2006; Freiman & Sriraman, 2007). Yet, creative problem solving would be a step deeper than developing problem solving strategies (Liljedahl, Santos-Trigo, Malaspina, & Bruder, 2016). However, the word creativity does not seem to be part of the terminology in mathematics classrooms (Leikin, 2011).

Several aspects could explain this lack of opportunities in developing mathematical creativity in students. Chiu (2009) wondered whether the pedagogical approaches used in classrooms consider the specific needs of all students to express themselves, and to develop their talents and learning styles at their own pace. Hashimoto (1997) questioned students and teacher's beliefs on learning mathematics because some still think that each problem only has one correct answer and can only be solved one way, and consequently propose them to students as the "right and only way" to solve a particular problem. Mann (2006) added that teachers tend to focus mostly on questions and exercises that require quick and accurate answers. Yet, he argues that the development of creativity is a long process that demands considerable time and reflection. Some wondered if teachers tend to explicitly teach students specific strategies to solve particular types of problems (Chan Chun Ming, 2008; Chiu, 2009). By doing so, they prevent students from using divergent thinking, which means finding different solutions and strategies to solve problems. Teachers explicitly proposing problem-solving strategies to students limit the opportunities for them to create their own representations and strategies, and to possibly come up with different correct answers (Lithner, 2008). Meissner (2006) argued that explicitly teaching problem-solving strategies risk harming students' natural curiosity and enthusiasm toward mathematics as they move forward in their school years.

In conclusion, it appears that the pedagogies used in mathematics classrooms seem to put an emphasis on what students reproduce and transmitting pre-established knowledge instead of giving students opportunities to develop their own strategies and to find original solutions to problems. When students are placed in an environment that puts an emphasis on memorization of mathematical concepts and the application of strategies and algorithms presented by the teachers in order to find the one and only answer in exercises that demands practically no reflection on the part of students does not permit the development of mathematical creativity in students (Mann, 2006). If students had more opportunities to reflect on problems proposed, to question the problem, to develop different interpretations and strategies to the problem as well as original solutions to the problem, and to even task risks, what would the teaching and learning mathematics look like in classrooms? In such a case, mathematics would be, according to Mann (2006), a school subject that would not limit itself by a set of concepts and algorithms to be recognized and memorized. My research is aligned with this idea.

In this chapter, I explore the potential of problem solving in a different context: virtual communities (discussed in the following section), as an alternative option for developing mathematical creativity in students. The potential of information and communication technologies such as virtual communities in fostering mathematical creativity does not seem to have been explored empirically. Technologies consists of a different context since students interact with the resource and not the teacher like in classroom settings. Could this student technology relationship foster more creative thinking in students?

21.1.2 Virtual Communities: A Possible Alternative for Mathematical Creativity?

Problem solving has been studied for many decades (Liljedahl et al., 2016). Such studies include heuristic strategies in solving problems (Polya, 1945), how students solve mathematical problems (Schoenfeld, 1985, 1992), and how problem solving is taught in mathematics classrooms around the world (Tömer, Schoenfeld, & Reiss, 2007). However, these studies tend to focus more on paper and pencil tasks.

Information and communication technologies (ICT) have also been documented as a means to support the learning experiences of students in and beyond classrooms. In fact, according to Freiman, Kadijevich, Kuntz, Pozdnyakov, and Stedoy (2009),

> Technologies can: give access to resources that cannot be otherwise accessed; provide a free choice of resources based upon the level and particular needs, provide dynamic tools of mathematical investigation giving a chance to modify parameters of an activity in an interactive way, serve as a valuable tool of communication about mathematics with other people, and empower the people with the instruments by facilitating routine operations and more sophisticated mind tools (p. 129).

An example of this can be dynamic software like GeoGebra. According to (Liljedahl et al., 2016), these environments permit students to conceptualize and represent mathematical objects and tasks dynamically, and investigating the various parameters can identify properties and discover mathematical relations. Thus with dynamic tools, students can be engaged in mathematical activities in which they can formulate conjectures, and find arguments to support them.

Some researchers argue that ICTs like virtual learning communities could possibly develop mathematical creativity in students. Renninger and Shumar (2004) discussed the creative potential of the Math Forum website (www.mathforum.org), which is oriented toward solving authentic and contextualized problems, and requires students to communicate and reason in their own way. Researchers studying the NRICH website (http://nrich.maths.org/frontpage) envision the creative process as a human activity in problem solving (Piggott, 2007). However, the cited works do not define creativity nor how it can be measured in problem solving tasks. In addition, empirical studies on mathematical creativity in virtual communities appear to be absent.

This issue motivated me to study the potential of the CAMI (Communauté d'apprentissages multidisciplinaires interactifs, www.umoncton.ca/cami) website in developing mathematical creativity in students. A team of researchers from the University of Moncton created this virtual problem-solving community that is designed for Francophone schools in New Brunswick, Canada, and other counties. According to its creators, the CAMI website accentuates rich problem solving experiences in mathematics and other school subjects (Freiman & Lirette-Pitre, 2009; Freiman, Lirette-Pitre, & Manuel, 2008; Freiman, Manuel, & Lirette-Pitre, 2007). The main activity on this website is problem solving. Every two weeks, four mathematical problems are posted on the website, each of them having a different level of difficulty. The members of the website can choose the problems they want to solve and submit a solution using an electronic form. Once the two-week period is over, members of the CAMI team analyze each solution received and provide feedback to the author of the solution, commenting on the process used and the communication of the solution. This feedback is also written using an electronic form and the author of the solution can view his feedback. In addition, the members of the CAMI team also post a general feedback for each problem along with examples of exemplary solutions and the names of the members who solved it correctly. Other activities on the website include a discussion forum, and a space to create and submit problems to the CAMI team (Freiman & Lirette-Pitre, 2009). To become a member of the community, participants must register. This gives him access to his own e-portfolio in which all traces of his work (problems solved and feedback received, problems created, etc.) are placed (Freiman & Lirette-Pitre, 2009). Figure 21.1 presents the CAMI website homepage.

A few studies conducted on the CAMI website revealed signs of a potential in developing mathematical creativity. Freiman and Manuel (2007) conducted semi-structured interviews with a small sample of teachers and grade 7 and 8 students who used the CAMI website periodically. Results revealed that the students' attempt to solve certain problems resulted in using multiple strategies. In

Fig. 21.1 Homepage of the CAMI website

fact, teachers admitted being amazed and impressed by the fact that their students solved problems in ways that they did not think of. This result hinted at possible creative solutions. These results were perceptions from participants.

In another pilot study, I explored the mathematical creativity of the solutions for one problem and noted traces of creativity in the solutions (Manuel, 2009). The problem contained five different answers, and could be solved using different strategies. As results, 37.5% of the solutions submitted by the members of the community contained more than one correct answer. Moreover, multiple strategies were used, but these were mostly using trial and error and arithmetic properties to solve the problem compared to algebra. Although this study showed traces of creativity in the solutions to the problem, questions emerged from this study. Was the problem chosen rich enough? What makes a problem rich? The problem was open-ended since it contained multiple answers and strategies, which is what the literature seems to propose (Freiman, 2006; Klavir & Hershkovitz, 2008; Petrowski,

2000). Are those characteristics enough or do mathematical problems need to have more characteristics to bring creative solutions? Is there a link between the richness of mathematical problems and the creativity of the solutions submitted to those problems on the CAMI website? Although the creators of the website say that rich problems are posted on the CAMI virtual community (Freiman & Lirette-Pitre, 2009; Freiman et al., 2008; Freiman, Manuel, & Lirette-Pitre, 2007), this vision has never been studied empirically. The mathematical creativity of the solutions to the problems have not been studied as well. These questions and issues prompted me to study the mathematical creativity in the solutions to more problems posted on the CAMI virtual community. A more thorough empirical study would give more insight on the potential of the website in developing mathematical creativity in students. Some questions that guided my inquiry were: do students submit creative solutions to problems posted on the CAMI website, or do they submit similar ones?; are the problems posted on the CAMI website rich (as the creators of the website promote) enough to bring creative solutions?; and do more rich (open-ended, ill-structured, etc.) problems bring more creative solutions? In this chapter, I will present results of these questions.

21.1.3 Goals of the Study

In this book chapter, I present an exploratory study of the richness of problems posted and the mathematical creativity of the solutions submitted by members on the CAMI virtual community. I thus study mathematical creativity in a problem solving context within a virtual environment. As I mentioned, researchers argue that problems that are open-ended (Freiman, 2006; Klavir & Hershkovitz, 2008; Leikin, 2009; Mann, 2006; Petrowski, 2000) and ill-structured (Chiu, 2009) seem to support the development of mathematical creativity. Since no studies have been conducted on the problems of the CAMI website, I must address this issue. I argue that other characteristics that make problems a "good" one may also bring creative solutions. I conducted a review of the literature and found characteristics (or features) that make mathematical problems "good". I use the term *richness of mathematical problems* to describe those types of problems. I present the model of the richness of problems I created in the following section.

By having an indication as to how rich the problems posted on the CAMI website are, I can thus look at the mathematical creativity of the solutions to problems with different richness. This could permit me to determine whether there is a possible link between the richness of problems and the mathematical creativity of the solutions. Analyzing systematically the content of the mathematical problems posted in the CAMI virtual community and the creativity of the solutions submitted by its members would give initial insights into the CAMI's potential to promote mathematical creativity. The research goals are thus the following: (1) analyze the richness of the mathematical problems posted on the CAMI website; (2) assess the creativity of the solutions submitted to the problems on the website; and

(3) determine if there is a link between the richness of the problems and the mathematical creativity of the solutions submitted on the website. Results of this study cannot generalize whether ICTs support the development of mathematical creativity, but could give insights and possible future research to bring more answers to this aspect. Although mathematical creativity is often related to the education of gifted students (Leikin, 2011), I argue that creativity can be a personal characteristic that all students can develop. I will thus investigate the solutions from all students.

21.2 Conceptual Framework

21.2.1 Rich Mathematical Problems

Scholars view the concept of rich problems differently. For instance, researchers from the NRICH website (http://nrich.maths.org/frontpage) defined rich problems as "problems which have multiple entry points, force students to think outside the box, which may have more than one solution, and open the way to new territories for further exploration". Piggot (2007) saw a rich problem as one that possess many characteristics that altogether offer different opportunities to meet the needs of learners at different moments in an environment in which the problem is posed and is influenced by the questions asked by teachers and the expectations from students.

In order to investigate this variety of characteristics, I conducted a thorough study of the literature in order to identify features or characteristics in the text of problems that could consider them as rich (Manuel, 2010). I argue that a problem is rich when it respects as many of the following features: is open-ended (Diezmann & Watters, 2004; Takahashi, 2000); is complex (Diezmann & Watters, 2004); is ill-defined (Murphy, 2004); contextualized (Greenes, 1997); and has multiple possible interpretations (Hancock, 1995).

According to my definition, a problem is *open-ended* if it has multiple correct answers or can be solved using various strategies (Takahashi, 2000). Although some might argue that open-ended problems automatically bring both multiple answers and strategies, I saw those two criteria as rather disjoint ones suggesting that some problems could lead to multiple answers, but could be solved using the same strategy. For instance, if we consider the following problem:

Find all possible sums of 12 using whole numbers.

This problem has multiple answers. But it is possible that a student will use the same strategy, for example, a systematic approach: finding possible answers with two numbers, then three, and so on.

A *complex* problem is one which respects most of the following criteria: it requires more than one step to solve it (Diezmann & Watters, 2004); it implicitly or explicitly asks solvers to find patterns, generalize results or make mathematical proofs; it explicitly asks to make different choices and justify them; and it explicitly asks to create other problems or questions in order to explore further (Diezmann & Watters, 2004; Freiman, 2006).

A problem is *ill-defined* if it is missing certain data (information) which are necessary to be able to solve it. That data can either be found by searching other sources or can be explicitly defined by the problem solver (Murphy, 2004). It is also ill-defined if it contains unnecessary data or doesn't present enough information for solution. Although, it is plausible to suggest that an ill-defined problem can also have characteristics making it open-ended, I prefer to put this type of problems in a different category since some ill-defined problems can still have only one correct answer. For example, consider the following problem:

> When Bob got in the plain in Paris, the clock indicated 10 AM. When Bob got off the plain in Montreal, the clock indicated 11 AM. How long was Bob's flight?

This problem would be considered as ill-defined since the time difference between the two cities isn't indicated. The student must search this information and add the value to the difference between the times in the problem. However, it is not considered an open-ended problem since there is only one correct answer.

A problem with *multiple possible interpretations* is one that encourages different ways of thinking (can be seen in different ways) about the problem, leading to different possible answers (Hancock, 1995). These could automatically qualify them as open-ended. I argue that some problems can have multiple interpretations, but each interpretation ends up to one correct answer. For instance, consider the following problem about the game Snakes and Ladders. Students are given a Snakes and Ladders game board and asked the following questions:

1. Using one die, what is the minimum number of turns you would need to win the game if you always rolled the number you wanted on the die on your turn?
2. How many turns would you need to win the game?

Each question can have two different interpretations: if you begin on cell #1 on the board; or if you begin outside of the board. However, the first question would only have 1 correct answer for both interpretations. Thus, it would not be considered as open-ended.

A *contextualized* problem is one where the mathematics is presented in real life or fictive situations (Greenes, 1997). Exercises, such as ones asking to solve for x the equation $3x + 5 = 14$, or problems wrapping them in a kind of "artificial situations" like referring to a person that needs help to solve an equation from a mathematics textbook are not considered as contextualized problems.

Figure 21.2 presents a visual representation of my features of the rich mathematical problem. The rounded rectangles represent the five main features selected in the definition while the ovals represent the criteria used to assess different elements for every feature that can be found in the text of the problem. These criteria were used to assess the richness of each problem on the CAMI website.

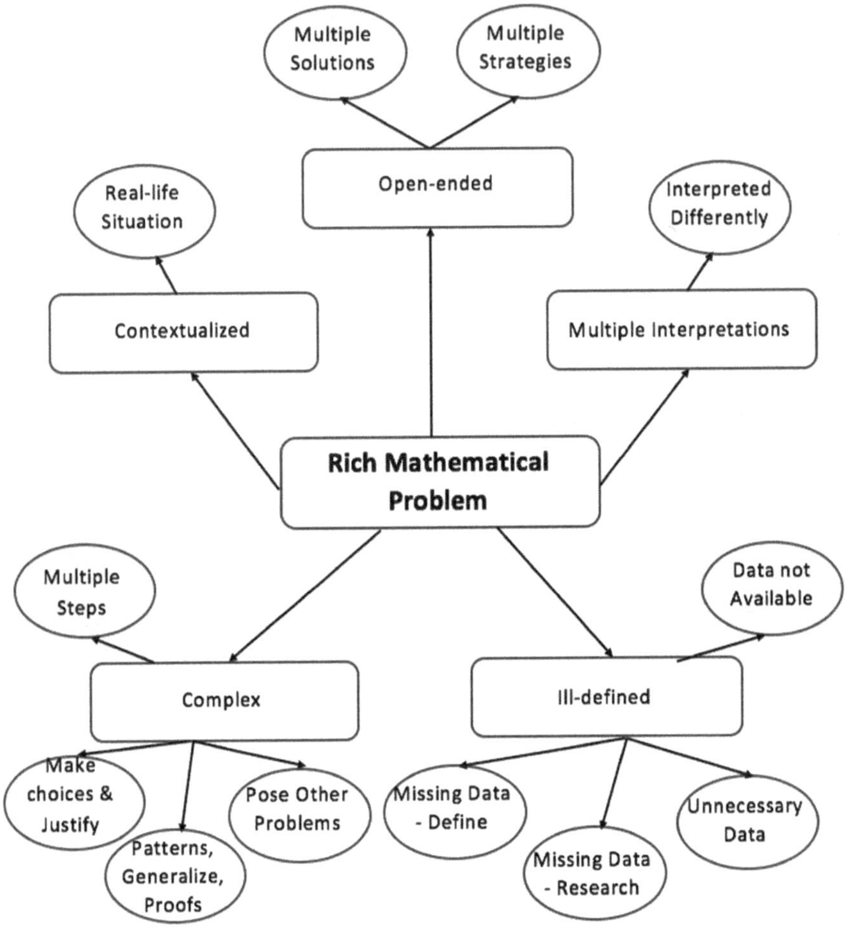

Fig. 21.2 Model of the richness of a mathematical problem (Manuel, 2010)

21.2.2 *Mathematical Creativity*

According to Sriraman (2009), the first known time in history that the term mathematical creativity appeared and was studied was in 1902 in a francophone periodical named *L'enseignement des mathématiques* (Teaching Mathematics) by a mathematician named Henri Poincaré. Researches on this concept were not popular from that time up to the first half of the century. In more recent times, researchers are getting more interested in the topic.

For a long time, creativity was often cited along with artistic creations, musical compositions and scientific discoveries. Thus, mathematics was not known to be a topic associated with the concept (Chan Chun Ming, 2008). In fact, this term was mostly used in relation to the education of gifted students. Some researchers like

Sternberg (1999a) viewed creativity as a type of giftedness while others like Renzulli (1986) viewed it as an essential component of giftedness. In fact, Renzulli (1986)'s three-ring model of giftedness (Fig. 21.3) considered giftedness as the intersection between above average abilities, task commitment and creativity where creativity is composed of fluency (finding different possible answers), flexibility (finding different possible strategies to a problem), insight and originality of ideas and strategies in problem solving and also the ability to create new problems.

However, other researchers like Sheffield (2009) argued that mathematical creativity can be developed by all students. This vision of creativity focuses on a production process which requires long periods of reflection and experimentations that any student can develop with considerable time and effort while solving non-familiar problems, a process that can be long, flexible, and deep (Holyoak & Thagard, 1997). This vision gives creativity different dimensions than being just a subset of giftedness according to Renzulli's model (Meissner, 2006). I lean towards this line of though in this study.

What are characteristics of mathematical creativity? How can creativity be detected and/or developed in mathematics? The next paragraphs will discuss existing theoretical views on these questions.

There are numerous ways to define mathematical creativity. Over 100 definitions can be found in the literature (Mann, 2006). The concept appears to be impossible to define (Liljedahl & Sriraman, 2006; Mann, 2006; Sriraman, 2009). For example,

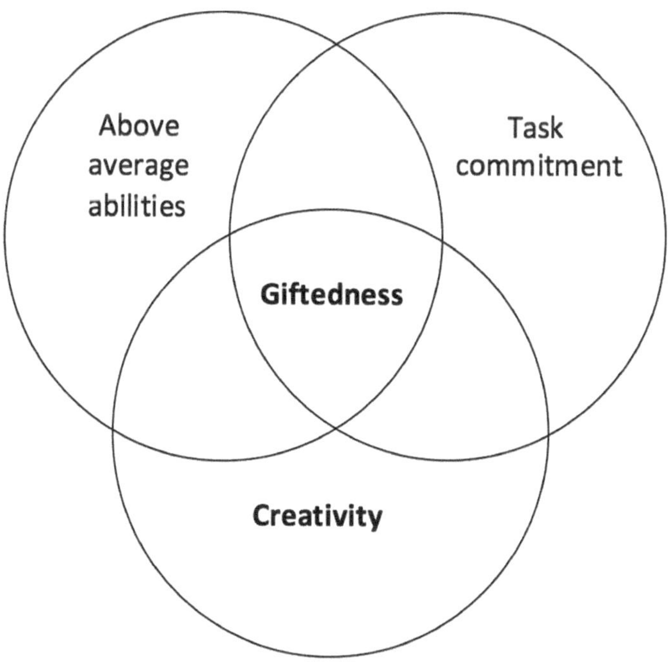

Fig. 21.3 Three-ring model of giftedness (Renzulli, 1986)

Runco (1993) described creativity as a multifaceted construct that involves convergent and divergent thinking, problem posing and problem solving, self-expression, intrinsic motivation, a questioning attitude and self-confidence. Haylock (1987) talked about the ability to find new relationships between techniques and areas of application and to make associations between possibly unrelated ideas. Krutetskii (1976) used the contexts of problem formation, invention, independence and originality to characterize mathematical creativity. Others have applied the concept of fluency, flexibility and originality to the concept of mathematical creativity (Haylock, 1997; Mann, 2006). Singh (1988) defined mathematical creativity as the process of formulating hypotheses concerning cause and effect in a mathematical situation, verifying them multiple times to make modifications and conclusions, and then communicating the results. Also, some authors focused on two main aspects: the originality and the utility of ideas (Anabile, 1989; Sternberg & Lubart, 1999).

An important part of studies on creativity have been conducted in the psychology and educational psychology communities. In a extensive review of the literature on creativity in those communities, Sternberg (1999b) claimed that creativity can be included under six categories: mystical, pragmatic, psychodynamic, psychometric, cognitive, and social-personality. Social and cultural dimensions of creativity can also be found in so-called "confluence" approaches, such as: *systems approach* (Csikszentmihalyi, 1999), which studies the interaction between the individual, the domain, and the field; the *case study as evolving systems approach* (Gruber & Wallace, 1999), which deals with a detailed analytical analysis of individual approaches to creativity; and the *investment theory approach* (Sternberg & Lubart, 1996), which in general described the investments of people attempting to convince others of ideas that might not be popular. Gruber and Wallace (1999) also argued in their theory that creative work is always the result of purposeful behavior. However, Sriraman (2009) argued that the discovery of penicillin as a counterexample to that statement. Can we consider accidently made illuminations as creative work? The resulting product might be classified as creative or innovative, but can we characterize the behavioral process as creativity? How can we also relate creativity to problem posing and solving which remain commonly accepted forms of productive (and potentially creative) mathematical activity? I present my working definitions of creativity in the following subsection.

21.2.3 Collective Solution Spaces: Linking Problem Solving with Mathematical Creativity

Arguing that solving problems in multiple ways can enhance the development of student's advanced mathematical thinking and creativity, Leikin (2007) defined a collection of solutions to a problem as a *solution space*. Those spaces are influenced by individual's experiences and memory, as well as the expectations of the problem. Leikin defined three types of solutions spaces: *expert solutions spaces*, *individual solution spaces*, and *collective solutions spaces*. Expert solution spaces are various solutions experts consider correct. Expert solution spaces can be divided into two categories: *conventional solution spaces*, which represents solutions proposed in the mathematics curricula; and *non-conventional*, which are solutions which are correct but may not be proposed as effective in the curricula. Individual solution spaces represent solutions proposed by an individual. Those solutions spaces can be either *personal solution spaces*, which are solutions an individual can submit without the help of experts, or *potential solution spaces*, which are solutions that solvers produce with the help of others. Collective solution spaces are solutions produced by a group of participants in a community of practice. Collective solution spaces are subsets of expert solution spaces, and are usually broader than individual solution spaces. Leikin argued that collective solution spaces are the main sources for the development of individual solution spaces within a community. Using collective solution spaces can permit the development of mathematical creativity because when students can bring various solutions to the personal solution spaces, they can connect between representations of mathematical concepts and tools (Leikin, 2007).

I argue that this notion of solution spaces can be extended in the context of virtual communities. Problems posted in a virtual community can create what I define as *virtual solution spaces*. Similar to Leikin (2007), these spaces can be expert *virtual solutions spaces*, which would represent the solutions accepted by mathematicians to a problem. However, with virtual communities, the expert virtual solution spaces are not local anymore. They become global, since members from all over the planet can contribute. In virtual communities, each member can have his or her *individual virtual solution space*. All member's contributions in the virtual community are conserved. For instance, in the CAMI virtual community, each member, once registered, has a personal e-portfolio in which all the problems solved, created, and the feedback received for the problems solved are saved (Freiman & Lirette-Pitre, 2009). Moreover, for each problem solved, a *collective virtual solution* space is created. A collective virtual solution space is a set of solutions found and submitted electronically by members of a virtual community. As I mentioned while describing the CAMI virtual community, each member can solve the problems he or she desires by using an electronic form (Freiman & Lirette-Pitre, 2009). Each member's individual virtual solution space is part of the collective virtual solution space. Similar to Leikin (2007), I argue that collective virtual solution spaces are the main sources for the development of individual virtual solution spaces within a virtual community. Using collective virtual solution spaces can permit the development of mathematical creativity because when members can bring various solutions to the virtual personal solution spaces, they can connect between representations of mathematical concepts and tools.

By linking creativity in a collective virtual solution space, I define it as the fluency, flexibility and the originality of solutions suggested to a mathematical problem (Haylock, 1997). As conceptual definitions, the fluency represents the number of correct answers or problems created in a collective virtual solution space, while the flexibility refers to the number of different appropriate strategies used to solve a problem in the same virtual collective solutions space. The difference between fluency and flexibility reflects on the fact that the same strategy can be used to get multiple different answers or that multiple distinct strategies can be used to get the same answer. The originality represents correct answers and strategies that are less frequent in the collective virtual sample space.

21.3 Method

This study is a quantitative exploratory one that follows Van der Maren (1996)'s typologies. It focuses on an analysis of the content (texts) of the problems posted in the CAMI website and the solutions submitted electronically by the members of this virtual community. In 2009 (when I began the study), 28,341 members were registered to the CAMI, and 88% of them were students. The members came from 40 different countries. More than half of the members were Canadians, and 83% of them are from New Brunswick. According to the rubric "Statistics" located in the administrator's interface of the website, members from New Brunswick are the most frequent visitors. These statistics were obtained from the administrative interface of the website.

The procedure of the study was divided into three phase, each of them focusing on one of its goals. First, I assessed the richness of the 180 problems posted in the CAMI website using a grid created and validated based on my working definition. The grid was constructed according to my working definition of the richness of a mathematical problem. For each feature in my working definition, I defined a series of criteria that could be assessed by analyzing the text of the problem. These criteria were based from my working definition for the feature (the ovals in the model in Fig. 21.2 or can be seen in Table 21.2). Each criterion was defined so that it could be answered by either yes or no. One point was given if it was respected in the problem. The richness of the problem was thus the total number of criteria the problem respected. At the beginning, the grid defined in an operational way the twelve criteria for each of the five features of a rich problem retained in the working definition. The original grid was validated using inter-rater's reliability as a measure before empirically assessing the problems. I, along with four members of the CAMI team, assessed the richness of ten problems selected randomly with the grid. After looking at the interpretive agreement percentages between myself and the other coders, two criteria were eliminated because the interpretive agreements were less than 80%. Those criteria were: problems having multiple interpretations, and ill-defined problems containing unnecessary data. In addition, during a debriefing session following the validation process, the assistants suggested that the two criteria in the category of ill-defined problems that addressed missing data be combined into one to avoid confusion. Also, one criterion was added in the grid in the category of open-ended problems: problems that explicitly ask for many solutions. Since problems with multiple answers or strategies were already criteria in the grid, this new one was just a question of interest: to compare between problems that did not explicitly ask to find multiple answers with those that did. This last one was not counted in the score for the richness of a mathematics problem. Table 21.1 presents the grid used to analyze the richness of a mathematical problem. The gray cells are criteria eliminated after validation. The grid was created using the Statistical Package for the Social Sciences (SPSS) software. I entered the result for each criterion for each problem in the software. The software calculated the total score (the richness of the problem).Rubric used to assess the richness of a mathematical problem

In the second phase, I examined the mathematical creativity of the collective virtual solution spaces of 50 problems selected randomly using SPSS. All the solutions to the problems selected were analyzed. The problems selected contained between 75 and 402 solutions with a mean of 187 solutions per problem. This second phase was conducted using a second grid that was created and validated. This grid presented the guidelines in order to give scores for the three variables of the mathematical creativity used in the working definition: fluency, flexibility, and originality. For the fluency, I gave one point for each different correct answer found

Table 21.1 Rubric used to assess the richness of a mathematical problem

Feature	Criterion	Respected (X)
Open-ended problem	Problem has multiple correct answers	
	**Problem explicitly asks to find multiple solutions	
	Problem has multiple appropriate strategies	
Complex problem	Problem requires using multiple steps to get answers	
	Problem asks to make and justify choices	
	Problem asks to create and explore other questions	
	Problem asks to find patterns, generalize or prove results	
Ill-defined problem	*Problem is missing necessary data	
	Problem contains unnecessary data	
	Problem contains insufficient or unrelated information	
Contextualized problem	Problem is centered around a real orfictive situation	
Problem with multiple interpretations	Problem can be interpreted in more than one way	
Richness of the problem (# of criteria the problem respected)		

*There were 2 criteria referring to the missing data in the model. After validation of the rubric, we combined them into one since it was too difficult to distinguish between the cases

**Criterion not counted in the score for richness

in the collective virtual solution space for the problem. For the flexibility, I gave 1 point for each strategy used to solve the problem in the collective virtual solution space. To determine the score of the originality of the collective virtual solution space, I used the following equation:

$$O = \frac{VO \times 2 + MO}{SOL}$$

O represented the score for originality, VO represented the number of correct answers and strategies that were very originals (used in 5% or less solutions), MO, the number of medially original answers and strategies (used in between 5 and 20% (included) in the solutions), and SOL, the total number of solutions in the collective virtual solution space. Table 21.2 presents the grid used for the analysis of the mathematical creativity for each variable.

Table 21.2 Analysis grid for the mathematical creativity of a collective virtual solution space to a problem posted in the CAMI virtual community

Variable	Descriptions
Fluency	Score attribution: 1 point for each correct answer or new correct problem posed in the collective virtual solution space.
	Exceptions: • Using neutral elements (0 for addition and 1 for multiplication) does not increase the number of solutions. • Equivalent fractions (ex: 1/2 and 3/6) are considered as 1 correct answer. • If the mathematical rule explicitly results to more than one answer (example, solving quadratic equations or absolute values), the answers are considered as 1 answer.
Flexibility	Score attribution: 1 point for each appropriate strategy used to solve a problem in the collective virtual solution space.
	Note: We inspired ourselves from the list from the following website; http://www.recreomath.qc.ca/lex_strategie.htm.
Originality	Score attribution: Use the following formula: $O = \frac{VO \times 2 + MO}{SOL}$, where O represents the score for originality, VO represents the number of solutions that are very original, MO, the number of medially original solutions, and SOL means the total number of solutions in the collective virtual solution space.
	Definitions: • VO = very original: the number or solutions that have an answer and/or a strategy used by 5% or less of members in the collective virtual solution space (highest level). • MO = medially original: the number or solutions that have an answer and/or a strategy used by 5.01–20% of members in the collective virtual solution space (second highest level). • If a solution has one element in each level (ex: correct answer VO and appropriate strategy MO), it is counted at the highest level.

Before using the grid, a member of the CAMI team and I applied the grid and assessed 100 solutions to one problem selected at random. Given the high interpretive agreement percentages, this grid was not modified. Each variable of creativity had its own score.

In the third phase, I determined if there existed a link between the richness of the problems posted in the CAMI website and the mathematical creativity of the solutions submitted. I used the Likelihood ratio test to do so (Field 2005). This test was selected after conducting preliminary tests on the data (Tabachnich & Fidell, 2005) that revealed that the distributions of the scores for the fluency and originality were not normal. The Z scores for the kurtosis and the skewness for those two variables were greater than the absolute value of 3.29. These results lead to transforming the scores by creating categories. The categories are presented in Table 21.3. For the richness of a mathematical problem, three categories were created: problems that are not rich, meaning those with a score of 3 or less; problems medially rich, meaning those with a score of 4; and problems that are rich, meaning those with a score of 5 and more. The categories were made according to the relative frequencies of the scores; the ones less frequent were combined into one category. For the mathematical creativity, the fluency and originality were both divided into two categories: one correct answer and more than one correct answer for the fluency; and original and non-original solutions for the originality. The flexibility was grouped into four categories according to the number of strategies used in the collective virtual solution space. It was the relative frequencies of the problems and solutions grouped in those categories that were submitted for the statistical analysis.

Table 21.3 Categories formed for the variables after preliminary tests

Richness of a mathematical problem		
Variable	Scores	Category name
Richness of a mathematical problem (N = 180)	1–3	Problems that are not rich
	4	Problems medially rich
	5–8	Problems that are rich
Mathematical creativity		
Variable	Scores	Category name
Fluency (N = 50)	1	One correct answer
	2 or more	Multiple correct answers
Fluency (N = 50)	1	1 appropriate strategy
	2	2 appropriate strategies
	3	3 appropriate strategies
	4 or more	More than 3 appropriate strategies
Originality (N = 50)	0	Non-original solution
	Other	Original solution

21.4 Results

I present the results in the following subsections that are based on my research goals, which I remind you were: to analyze the richness of the mathematical problems posted on the CAMI virtual community; to assess the mathematical creativity of the collective virtual solutions spaces; and determine if there is a link between the richness of the mathematical problems posted and the mathematical creativity of the solutions submitted.

21.4.1 Richness of the Mathematical Problems Posted on the CAMI Website

Normality tests revealed that the scores for the richness of the 180 problems posted on the CAMI website were almost perfectly distributed (Z score of 0.546 for skewness and 0.489 for the kurtosis). The mean of the scores was 4 out of a possibility of 8 with a standard deviation of 1.27. Those numbers indicated that the richness of the problems posted on the CAMI website was modest, meaning that the vast majority of the problems were medially rich.

Looking at the relative frequencies obtained for each of the criterion from the grid (see Table 21.4), it appeared that four criteria were respected in most of the problems. Those criteria were: problems with multiple correct answers, 61.1%; problems with multiple appropriate strategies, 94.4%; problems requiring multiple steps to get answers, 87.8%; and problems presented in real life or fictive contexts (contextualized problems), 90.6%. However, the four other criteria were neglected. Those criteria were: problems asking to make choices and justify them, 11.7%; problems asking to create and explore other questions, 2.8%; problems asking to

Table 21.4 Relative frequencies of each criterion of the richness of a mathematical problem, N = 180

Characteristics	Criteria	%
Open-ended problems	Problem has multiple correct answers	61.1
	Problem has multiple appropriate strategies	94.4
Complex problems	Problem requires multiple steps to get answers	87.8
	Problem asks to make and justify choices	11.7
	Problem asks to find and explore other questions	2.8
	Problem asks to find patterns and generalize results	28.9
Ill-defined problems	Some or all necessary data or information are missing in the text of the problem	22.8
Contextualized problems	Problem presented in a real or fictive context	90.8

find patterns and generalize results, 28.9%; and problems that are missing some data or information (ill-defined), 22.8%. The differences in the frequencies indicated that the vision of the CAMI team to create rich problems is restricted.

In summary, the problems posted on the CAMI virtual community were relatively modest in terms of their mathematical richness, and this richness was mostly based on four main criteria. The problems were mostly contextualized, open-ended (with multiple correct answers and/or multiple appropriate strategies), and required multiple steps to find one or more correct answers.

21.4.2 *Mathematical Creativity of the Collective Virtual Solutions Space of Problems*

The analysis of the collective virtual solution spaces of the 50 problems using the three variables in our definition of mathematical creativity revealed that most the members submitting solutions to problems limited themselves to one answer and one strategy, and those solutions were often not original. However, some traces of mathematical creativity were found in the collective virtual solution spaces. The relative frequencies for each of the three variables revealed these traces.

For the fluency, 48% of the problems contained two or more correct answers in their collective virtual solution space, while 52% of them only had one correct answer. Table 21.5 shows that 12% of problems had two correct answers, and 36% of them had more than two correct answers in their collective virtual solution spaces. In 30% of the ones containing more than two correct answers, the problem didn't explicitly ask to find multiple answers, while 6% of them explicitly asked to find multiple answers. Those data indicated that almost half of the problems posted on the CAMI website had more than one correct answer in the collective virtual solution spaces, and those answers were usually found without being asked in the text of the problem.

For the flexibility (see Table 21.5), there was a relatively important variation in the number of strategies used in the collective virtual solution spaces. In 24% of the problems, only one strategy was used to solve them. However, more than one strategy was used in 76% of the problems. Two, three and more different strategies were found in the collective virtual solution spaces in 36, 26, and 14% of the problems respectively. It was also noted that some strategies like trial and error were often used.

For the originality (see Table 21.5), 44% of the problems posted contained original solutions in the collective virtual solution spaces, while 56% of them had the same correct answers and were solve using similar strategies. However, many solutions were similar content wise. This made me hypothesize that some members might have solved the problem together, and they all submitted the same solutions. Therefore, those similar solutions probably influenced the results for this variable since they increased the frequency of the answers and the strategies.

Table 21.5 Relative frequencies of problems respecting the categories for each of the variables of mathematical creativity in the collective virtual solution spaces, N = 50

Variable	Categories		%
Fluency (N = 50)	1 correct answer		52
	Multiple correct answers	2 correct answers	12
		More than 2 correct answers without explicitly asking in the problem	30
		More than 2 correct answer and explicitly asking in the problem	6
		Total	48
Flexibility (N = 50)	1 appropriate strategy		24
	2 appropriate strategies		36
	3 appropriate strategies		26
	4 or more appropriate strategies		14
Originality (N = 50)	Problems with original solutions		44
	Problems without any original solutions		56

In terms of mathematical creativity, the results revealed some traces of it in the collective virtual solution spaces for less than half the problems posted on the CAMI virtual community. Between the three variables, the flexibility seemed to be the most important variable because more than one appropriate strategy were used in approximately three quarters of the problems posted, while the fluency and the originality were found in less than half of the problems.

21.4.3 Link Between the Richness of the Problems and the Creativity of Solutions

In responding to the goal concerning the existence of a link between the richness of the mathematical problems posted on the CAMI virtual community and the mathematical creativity in the collective virtual solution spaces of the problems, results of the Likelihood Ratio test (Chi square) (see Table 21.6) revealed a significant dependent link (L_2 [2] = 9.706, p = 0.008) between the fluency and the richness of problems. There was also a dependent link (L_2 [2] = 10.07, p = 0.007)

Table 21.6 Results of the Likelihood ratio tests between the richness of problems and the three variables of mathematical creativity, N = 50

Link between the richness of problems and	Likelihood ratio L_2	dl	Sign.	Cramer's V
Fluency	9.706	2	0.008	0.422
Flexibility	7.718	6	0.260	0.268
Originality	10.07	2	0.007	0.441

between the originality and the richness of the problems. The effect size between those variables, determined by Cramer's V, was medium in both cases: being respectively 0.422 and 0.441. However, there was no dependent link (L_2 [6] = 7.718, p = 0.260) between the flexibility and the richness of the problems. This last result could be explained by the fact that the criterion of problems that can be solved using multiple strategies was the one among the criteria of the richness of the problem that was the most frequent. This criterion was respected in approximately 95% of the problems.

21.5 Discussion

The results of this study showed that four among the eight criteria used to evaluate the mathematical richness of the problems posted on the CAMI virtual community are often used. They are: problems that has multiple correct answers; problems that can be solved using multiple appropriate strategies; problems that require more than one step to find answers; and problems that are contextualized. These results seem to confirm the initial intention for building the website, which was to create a problem-solving environment in which students could: solve complex, significant, and contextualized problems, permitting them to develop strategies; and to reason and communicate mathematically (Freiman & Lirette-Pitre, 2009). They are also aligned with the didactic principles promoted in the New Brunswick mathematics curricula. Those didactic principles explicitly suggest proposing complex, significant and contextualized situational problems to students so they can manage them by constructing a valid and justified (mathematical reasoning) mathematical approach and then communicate it clearly by using the proper vocabulary and various representations (New Brunswick Department of Education and Early Childhood Development, 2011). This process must also support students in developing the ability to make links between mathematics and real-life situations, between different mathematical concepts, and between mathematics and other disciplines. Moreover, these principles are promoted in order to create a mathematical culture in all students, and are similar to the ones defined by PISA (Aschleicher, 1999). They are also inscribed in socio-constructivist perspectives and are aligned with the principles and standards of the National Council of Teachers of Mathematics (National Council of Teachers of Mathematics, 2000).

However, the results also showed a weak percentage in problems that respect the other four criteria. Those criteria are: problems that demand to make and justify choices; problems that ask to find other questions to explore; problems that seek to find patterns and generalize results; and problems missing necessary data or information (ill-defined). These results showed a need to review the conceptual framework of the problems posted on the CAMI virtual community (Freiman & Lirette-Pitre, 2009; Freiman, Lirette-Pitre, & Manuel, 2007) by integrating criteria under-used, and to study some (or all) in more depth to determine their potential in developing student's mathematical creativity. Pallascio (2005) concluded that complex tasks that integrates those criteria support students in going beyond the simple application of rules and algorithms and engage them in a more creative process that involves discovering patterns, questioning phenomena, generalizing results, and making mathematical proofs. The criteria mentioned by those authors correspond to those under used on the virtual community.

Regarding the mathematical creativity, the results showed a presence of all the variables in approximately half of the problems posted on the CAMI website. This modest level of creativity brings up many questions. Do teachers train students by showing them the "right strategy" and the "correct way" to get to the only possible answer? Does the CAMI team need to enrich the problems posted by making them more open-ended and asking more questions that would lead to an investigation or different possibilities? Is it possible to create an environment in which students can create high-level questions that would stimulate their curiosity and critical thinking skills, and bring solutions to real life problems or lead to other advanced questions? This study does not give out clear ideas about ways to promote more creativity in solutions to problems posted on the CAMI website. However, authors like Sheffield (2009) mentioned that working on a more profound understanding of concepts and mathematical properties is necessary and this must be done by means of discussions between students. This form of collaboration is not only possible in classrooms but also in virtual communities like CAMI. Virtual communication forms like discussion forums can be examined in order to support creative problem solving by a group of students in a collaborative way (Stahl, 2009). One limit to the CAMI virtual is that the collective virtual solution space is not accessible to members of the community. Member submit their solutions. Those solutions are part of the member's individual collective solution space, but members do not get to see the other member's contribution. When the time frame of the problem posted is over, members have access to an analysis of the problem (Freiman & Lirette-Pitre, 2009). This analysis contains general comments about the collective virtual solution space, a few "exemplary solutions" from members, and the list of the members who solved the problem correctly. However, this analysis contains limited information about the entire collective virtual solution space. In addition, are the exemplary solutions posted as examples creative solutions? Or are they just solutions that would be considered as expert virtual solution spaces and well communicated solutions? As Leikin (2007) argued, collective solution spaces are the main sources for the development of individual solution spaces within a community, as they can permit the development of mathematical creativity because when students can bring various solutions to the personal solution spaces, they can connect between representations of mathematical concepts and tools. To support the development of mathematical creativity in a virtual community like CAMI, the collective virtual solution spaces to problems should be accessible to all members so they could enrich their individual collective solution spaces.

One possible way of making the collective virtual solution space accessible could be discussion forums The CAMI website also has a discussion forum, but unfortunately, it is hardly ever used (Freiman & Lirette-Pitre, 2009). Students only use this tool when a learning activity demands online exchanges. By posing problems that demand collaborative solving, the CAMI team could exploit more their discussion forum, which could give access to a rich environment of information as much for the teacher as for students. In a discussion forum, the written interactions that do not necessarily correspond to the anticipated results are still kept intact. Therefore, the discussion forum seems to be an interesting tool for conceptualizing since it presents information that not only permits teachers to better understand their students' reasoning, but also permits students to better understand their own reasoning. Studies on that subject have been conducted by the Math Forum website team (Stahl, 2009). In those studies, students grouped in *virtual mathematics teams* solve problems online and can make sense of the mathematical concepts included in them. More studies on this practice should be initiated.

Although the mathematical creativity found in the collective virtual solution spaces to the problems posted on the CAMI website is relatively restricted, the results did confirm a link between the richness of the problems posted and the fluency as well as the originality of solutions. These results seem to confirm the ideas of researchers in mathematics education who mention that rich problems have the potential of bringing more creative solutions (Chiu, 2009; Freiman, 2006; Freiman & Sriraman, 2007; Klavir & Hershkovitz, 2008; Leikin, 2009; Liljedahl & Sriraman, 2006; Mann, 2006; Sheffield, 2009). This result does invite the CAMI team to try to improve the richness of the problems posted on the CAMI website. I hypothesize that a growth in the richness of the problems posted on the CAMI virtual community could bring positive impacts on learning and on the development of mathematical creativity in students living in these settings.

It is also necessary to discuss and question the limits of this study, especially when it comes to the definitions and the criteria used. It is important to remember that the definitions and the criteria used to create the grids for the richness of a mathematical problem and for the mathematical creativity constitute choices in agreement with an epistemological quantitative stance that recalls having observable and measurable criteria already mentioned in the field's literature. Those two concepts constitute polysemous constructs and their study can be conducted in different angles and points of view. Other definitions and criteria are definitely possible. For instance, Piggot (2007) mentions the potential of enriching student's mathematical abilities and the discovery of new concepts as criteria of creativity. Such criteria could tend to be both a potential enrichment of a quantitative assessment grid or a qualitative investigation problematic. This study was also exploratory so no definitive conclusions could be made about the richness of the problems posted on the CAMI website nor on the mathematical creativity of the collective virtual solution spaces created by the members. However, this study looks at an emergent investigation field. The interest is less in provisionary answers.

21.6 Conclusion

The goal of the study was to determine a preliminary insight of the potential of the CAMI virtual community in supporting the development of mathematical creativity in students who solve the problems that are posted. I conduced this study by analyzing the richness of the problems posted on the website as well as the mathematical creativity of the solutions in order to determine whether there was a link between the two variables. This study provided evidence that the mathematical problems posted on the CAMI website are modest in general in terms of their richness (according to the criteria considered), and that there are traces of mathematical creativity in the collective virtual solution spaces. This study also revealed a link between the richness of the mathematical problems posted on the CAMI website and the mathematical creativity found in the collective virtual solution spaces in terms of fluency and originality. The results of this study point at the importance of improving the richness of the mathematics problems posted on the CAMI website. In addition, the results point at the importance of finding strategies so that the collective virtual solution space be accessible to all members. In a local community such as a classroom, creating multiple individual solution spaces can be done via interactions based on the flexibility of the teacher and students discussing the collective solutions spaces (Leikin, 2007; Stein, Engle, Smith, & Hughes, 2008). Strategies to create virtual discourses must be implemented.

Despite the limits, this present study shows an interest to conduct other studies on the adequacy of available resources and the missions of teaching and learning with regards to mathematical creativity. Two types of research can be considered. Firstly, more rigorous studies could be conducted on the criteria to consider when implementing rich problems in mathematics in link to the development of creativity in that subject. Secondly, the factors that help increase the efficacy of technological resources like virtual communities of problem solvers should be further explored and compared with the traditional ones like textbooks, with respect to the development of mathematical creativity. These types of studies could reinforce the theoretical frameworks and plan strategies to implement in mathematics.

References

Anabile, T. M. (1989). *Growing up creative: Nurturing a lifetime of creativity*. Buffalo, NY: C.E. F. Press.

Aschleicher, A. (1999). *Measuring student knowledge and skills: A new framework for assessment*. Paris, France: Organization for Economic Co-operation and Development.

Chan Chun Ming, E. (2008). *The use of mathematical modeling tasks to develop creativity*. Paper presented at the 11th International Congress on Mathematics Education (ICME), Monterrey, Mexico.

Chiu, M.-S. (2009). Approaches to the teaching of creative and non-creative mathematical problems. *International Journal of Science and Mathematics Education, 7*(1), 55–79.

Csikszentmihalyi, M. (1999). Implications of a systems perspective for the study of creativity. In R. J. Sternberg (Ed.), *Handbook of creativity* (pp. 313–338). Cambridge, UK: Cambridge University Press.

Diezmann, C., & Watters, J. J. (2004). *Challenges and connectedness in the mathematics classroom: Using lateral strategies with gifted elementary students*. Paper presented at the 10th International Congress on Mathematics Education (ICME), Copenhagen, Denmark.

Field, A. (2005). *Discovering statistics using SPSS* (2nd ed.). London, UK: Sage.

Freiman, V. (2006). Problems to discover and to boost mathematical talent in early grades: A challenging situations approach. *The Mathematics Enthusiast, 3*(1), 51–75.

Freiman, V., Kadijevich, D., Kuntz, G., Pozdnyakov, S., & Stedoy, I. (2009). Technological environments beyond the classroom. In E. J. Barbeau & P. J. Taylor (Eds.), *Challenging mathematics in and beyond the classroom: The 16th ICMI Study* (Vol. 12, pp. 97–131). Berlin, Germany: Springer.

Freiman, V., & Lirette-Pitre, N. (2009). Building a virtual learning community of problem solvers: Example of CASMI community. *ZDM: The International Journal on Mathematics Education, 41*(1 & 2), 245–256.

Freiman, V., Lirette-Pitre, N., & Manuel, D. (2008). Building virtual learning community of problem solvers: Example of CASMI community. In B. Sriraman, C. Michelsen, A. Beckmann, & V. Freiman (Eds.), *Interdisciplinary educational research in mathematics and its connections to the arts and sciences* (pp. 209–222). Charlotte, NC: Information Age Publishing Inc.

Freiman, V., & Manuel, D. (2007). Apprentissage des mathématiques. In S. Blain, J. Beauchamp, C. Esseimbre, V. Freiman, N. Lirette-Pitre, D. Villeneuve, H. Fournier, P. Clavet, M. Cormier, & D. Manuel (Eds.), *Les effets de l'utilisation des ordinateurs portatifs individuels sur l'apprentissage et les pratiques d'enseignement* (pp. 232–283). Moncton, Canada: Centre de recherche et de développement en éducation (CRDE), Université de Moncton.

Freiman, V., Manuel, D., & Lirette-Pitre, N. (2007). CASMI: A virtual learning collaborative environment for mathematical enrichment. *Understanding Our Gifted, 19*(4), 20–23.

Freiman, V., & Sriraman, B. (2007). Does mathematics gifted education need a working philosophy of creativity? *Mediterranean Journal for Research in Mathematics Education, 6*(1 & 2), 23–46.

Greenes, C. (1997). Honing the abilities of the mathematically promising. *The Mathematics Teacher, 90*(7), 582–586.

Gruber, H. E., & Wallace, D. B. (1999). The case study method and evolving systems approach for understanding unique creative people at work. In R. J. Sternberg (Ed.), *Handbook of creativity* (pp. 93–115). Cambridge, UK: Cambridge University Press.

Hancock, L. (1995). Enhancing mathematics learning with open-ended questions. *The Mathematics Teacher, 88*(6), 496–499.

Hashimoto, Y. (1997). The methods of fostering creativity through mathematical problem solving. *ZDM: The International Journal on Mathematics Education, 29*(3), 86–87.

Haylock, D. W. (1987). A framework for assessing mathematical creativity in schoolchildren. *Educational Studies in Mathematics, 18*(1), 59–74.

Haylock, D. W. (1997). Recognising mathematical creativity in schoolchildren. *ZDM: The International Journal on Mathematics Education, 29*(3), 68–74.

Holyoak, K. J., & Thagard, P. (1997). The analogical mind. *American Psychologist, 52*(1), 35–44.

Johnsen, S., & Sheffield, L. J. (Eds.). (2012). *Using the common core state standards for mathematics with gifted and advanced learners*. Washington, DC: National Association for Gifted Children.

Klavir, R., & Hershkovitz, S. (2008). Teaching and evaluating 'open-ended' problems. *International Journal for Mathematics Teaching and Learning*. http://www.cimt.org.uk/journal/klavir.pdf.

Krutetskii, V. A. (1976). *The psychology of mathematical abilities in schoolchildren*. Chicago, IL: University of Chicago Press.

Kwon, O. N., Park, J. H., & Park, J. S. (2006). Cultivating divergent thinking in mathematics through an open-ended approach. *Asia Pacific Education Review, 7*(1), 51–61.

Leikin, R. (2007). Habits of mind associated with advanced mathematical thinking and solution spaces of mathematical tasks. In D. Pitta-Pantazi & G. Philippou (Eds.), *Proceedings of the Fifth Congress of the European Society for Research in Mathematics Education* (pp. 2330–2339). Larnaca, Cyprus: The authors.

Leikin, R. (2009). Exploring mathematical creativity using multiple solution tasks. In R. Leikin, A. Berman, & B. Koichu (Eds.), *Creativity in mathematics and the education of gifted students* (pp. 129–145). Rotterdam, The Netherlands: Sense Publishers.

Leikin, R. (2011). The education of mathematically gifted students: Some complexities and questions. *The Mathematics Enthusiast, 8*(1 & 2), 167–188.

Liljedahl, P., Santos-Trigo, M., Malaspina, U., & Bruder, R. (2016). *Problem solving in mathematics education*. London, UK: Springer Open.

Liljedahl, P., & Sriraman, B. (2006). Musings on mathematical creativity. *For the Learning of Mathematics, 26*(1), 17–19.

Lithner, J. (2008). A research framework for creative and imitative reasoning. *Educational Studies in Mathematics, 67*(3), 255–276.

Mann, E. L. (2006). Creativity: The essence of mathematics. *Journal for the Education of the Gifted, 30*(2), 236–260.

Manuel, D. (2009). Does technology help building more creative mathematical environments? In B. Sriraman, V. Freiman, & N. Lirette-Pitre (Eds.), *Interdisciplinarity, creativity, and learning: Mathematics with literature, paradoxes, history, technology, and modeling* (pp. 233–247). Charlotte, NC: Information Age Publishing.

Manuel, D. (2010). *Étude de la créativité mathématique dans les solutions aux problèmes proposés dans la communauté virtuelle CASMI.* Maitrise-ès Arts en éducation (mention enseignement), Université de Moncton, Moncton, Canada.

Meissner, H. (2006). Creativity and mathematics education. *Elementary Education Online, 5*(1), 65–72.

Murphy, E. (2004). Identifying and measuring ill-structured problem formulation and resolution in online asynchronous discussions. *Canadian Journal of Learning and Technology, 30*(1). http://www.cjlt.ca/index.php/cjlt/article/view/26532/19714.

National Council of Teachers of Mathematics. (2000). *Principles and standards for school mathematics*. Reston, VA: National Council of Teachers of Mathematics.

New Brunswick Department of Education and Early Childhood Development. (2011). *Mathematics*. Fredericton, Canada: New Brunswick Government.

Pallascio, R. (2005). Les situations-problèmes: un concept central du nouveau programme de mathématiques. *Vie pédagogique, 134,* 32–35.

Petrowski, M. J. (2000). Creativity research: Implications for teaching, learning and thinking. *Reference Services Review, 28*(4), 304–312.

Piggott, J. (2007). Cultivating creativity. *Mathematics Teaching, 202*(May), 3–6.

Polya, G. (1945). *How to solve it*. Princeton, NJ: Princeton University.

Renninger, K. A., & Shumar, W. (2004). The centrality of culture and community to participant learning at and with the math forum. In S. Barab, R. Kling, & J. H. Gray (Eds.), *Designing for virtual communities in the service of learning* (pp. 181–209). Cambridge, UK: Cambridge University Press.

Renzulli, J. (1986). The three-ring conception of giftedness: A developmental model for creative productivity. In R. J. Stemberg & J. E. Davidson (Eds.), *Conceptions of giftedness* (pp. 332–357). New York, NY: Cambridge University Press.

Runco, M. A. (1993). Divergent thinking, creativity, and giftedness. *Gifted Child Quarterly, 37*(1), 16–22.

Schoenfeld, A. H. (1985). *Mathematical problem solving*. Orlando, FL: Academic Press.

Schoenfeld, A. H. (1992). Learning to think mathematically: Problem solving, metacognition, and sense-making in mathematics. In D. A. Grouws (Ed.), *Handbook of research on mathematics teaching and learning* (pp. 334–370). New York, NY: Simon and Schuster.

Sheffield, L. J. (2009). Developing mathematical creativity—Questions may be the answer. In R. Leikin, A. Berman, & B. Koichu (Eds.), *Creativity in mathematics and the education of gifted students* (pp. 87–100). Rotterdam, The Netherlands: Sense Publishers.
Singer, F. M., Sheffield, L. J., Freiman, V., & Brandl, M. (2016). *Research on and activities for the mathematically gifted students*. London, UK: Springer Open.
Singh, B. (1988). *Teaching-learning strategies and mathematical creativity*. Delhi, India: Mittal Publications.
Sriraman, B. (2005). Are giftedness and creativity synonyms in mathematics? *The Journal of Secondary Gifted Education, 17*(1), 20–36.
Sriraman, B. (2009). The characteristics of mathematical creativity. *ZDM: The International Journal on Mathematics Education, 41*(1–2), 13–27.
Stahl, G. (2009). *Studying virtual math teams*. New York, NY: Springer.
Stein, M. K., Engle, R. A., Smith, M. S., & Hughes, E. K. (2008). Orchestrating productive mathematical discussions: Five practices for helping teachers move beyond show and tell. *Mathematical Thinking and Learning, 10*(4), 313–340.
Sternberg, R. J. (1999a). The theory of successful intelligence. *Review of General Psychology, 3*(4), 292–316.
Sternberg, R. J. (Ed.). (1999b). *Handbook of creativity*. Cambridge, UK: Cambridge University Press.
Sternberg, R. J., & Lubart, T. L. (1996). Investing in creativity. *American Psychologist, 51*(7), 677–688.
Sternberg, R. J., & Lubart, T. L. (1999). The concept of creativity: Prospects and paradigmes. In R. J. Sternberg (Ed.), *Handbook of creativity* (pp. 3–15). Cambridge, UK: Cambridge University Press.
Tabachnich, B. G., & Fidell, L. S. (2005). *Using multivariate statistics* (5th ed.). Boston, MA: Pearson.
Takahashi, A. (2000). *Open-ended problem solving enriched by the internet*. Paper presented at the National Council of Teachers of Mathematics (NCTM) Annual Meeting, Chicago, IL. http://mste.illinois.edu/users/aki/open_ended/NCTM_Presentation/sld006.htm.
Tömer, G., Schoenfeld, A. H., & Reiss, K. M. (2007). Problem solving around the world: Summing up the state of the art. *ZDM: The International Journal on Mathematics Education, 39*(5 & 6), 363.
Van der Maren, J.-M. (1996). *Méthodes de recherche pour l'éducation* (2e ed.). Montréal, Canada: Presses de l'Université de Montréal.

Index

© Springer International Publishing AG, part of Springer Nature 2018

V. Freiman and J. L. Tassell (eds.), *Creativity and Technology in Mathematics Education*, Mathematics Education in the Digital Era 10,
https://doi.org/10.1007/978-3-319-72381-5

FSC
www.fsc.org
MIX
Papier | Fördert
gute Waldnutzung
FSC® C083411

Kolibri 360 GmbH
Ferdinand-Jühlke-Straße 7
99095 Erfurt, Deutschland
produktsicherheit@kolibri360.de